The Entolomataceae of Tasmania

Photograph was taken at Kermandie Falls, May 2010.
From *left to right*: David Ratkowsky, Genevieve Gates, Machiel Noordeloos, and Michael Pilkington.
Sitting: Fernanda Karstedt.

Fungal Diversity Research Series

Volume 22

Series Editor: Kevin D. Hyde

For further volumes:
http://www.springer.com/series/10408

Machiel E. Noordeloos • Genevieve M. Gates

The Entolomataceae of Tasmania

Springer

Machiel E. Noordeloos
Section National Herbarium of the Netherlands
Netherlands Centre for Biodiversity Naturalis
Leiden, The Netherlands

Genevieve M. Gates
School of Plant Science
University of Tasmania
Hobart, Australia

ISSN 1608-6791 ISSN 2212-6236 (electronic)
ISBN 978-94-007-4678-7 ISBN 978-94-007-4679-4 (eBook)
DOI 10.1007/978-94-007-4679-4
Springer Dordrecht Heidelberg New York London

Library of Congress Control Number: 2012943615

Printed on acid-free paper

Springer is part of Springer Science+Business Media (www.springer.com)

The Entolomataceae of Tasmania

by Machiel E. Noordeloos & Genevieve M. Gates

with photographs by the authors and with additional photographs by Michael Pilkington and with line drawings by Anita Walsmit-Sachs

Abstract

This book is the result of 14 years of collecting Entolomataceae in the native forests of Tasmania, Australia. Although initially involving only the Tasmanian residents Genevieve Gates and David Ratkowsky, who made twice- or thrice-weekly forays into the forests throughout the year, the project was subsequently joined by agaric specialist Machiel Noordeloos from the Netherlands and by fungi photographer Michael Pilkington from the United Kingdom. The international character of the project is further evidenced by the earlier contributions of American mycologist Tim Baroni to the Tasmanian *Rhodocybe* species which form the basis of the chapter on the now-expanded concept of *Clitopilus*, and a visit of several months in 2010 by Brazilian Ph.D. candidate Fernanda Karstedt, who helped to formulate the keys to the *Entoloma* species. Consequently, several thousand well-annotated collections were found during this inventory and form the basis of this monographic treatment of the *Entoloma* and *Clitopilus* of Tasmania. The resulting 90 *Entoloma* species and 10 *Clitopilus* species are well documented with standardized descriptions, line drawings of fruit bodies and diagnostic microscopic characters, and, when available, with colour photographs. Thanks to the intensive search, it was possible to illustrate most species in colour. Dichotomous keys facilitate identification of the species. The species concept used is morphologically based; in several cases, however, identification to species level is supported by molecular data.

The Entolomataceae mycota of Tasmania appears to be fairly unique, as 73 out of 90 species of *Entoloma* and 5 out of 10 *Clitopilus* species are new to science, with the majority of the remaining species shared with New Zealand. Only a few taxa have characteristics that match those of European species, and might have been introduced from Europe.

The large number of observations enabled the authors to use a statistical analysis of the phenological data, resulting in the recognition of five distinct fruiting patterns. Some species appear preferably in winter and spring, others in the summer and autumn months, where groups can be distinguished with a rather wide fruiting spectrum, encompassing eight months, whereas others have a typical autumnal appearance in the months of April–June.

The introductory part contains chapters focussed on the taxonomy, phylogeny, and biogeography of Entolomataceae in Tasmania, in which the current state of knowledge is discussed. There are chapters dedicated specifically to the study of Entolomataceae, giving instructions how to collect, document, and preserve specimens for identification, and a well-illustrated chapter on characters and character states that are used in Entolomataceae identification and taxonomy. The introductory part concludes with a chapter dedicated to the ecology, distribution, and phenology of the Tasmanian Entolomataceae, based on the very many observations during this study.

Full references to the cited literature are given, as well as an index of species names and synonyms.

Acknowledgments

This book is dedicated Dr. David Ratkowsky for his support, assistance during field trips and editing the final text, and to Johan van Dongen, for his support and patience.

Other persons played an important role in the production of this book. We are especially grateful to Michael Pilkington for his photographs, and also to Anita Walsmit-Sachs, who prepared the line drawings for publication.

We also wish to make special mention of the following individuals:

Alan Mills, formerly of the University of Tasmania, for his role in introducing G.M. Gates to many aspects of mycology and for his assistance in the laboratory;

Fernanda Karstedt, Instituto de Botânica de São Paulo, Brazil, for assisting with the development of the keys, useful comments and photographs;

Mayo Kajitani of Forestry Tasmania, for the preparation of Figure 1.11;

Kim and Peter Eastman of Forester Road, Scottsdale, for their collection and photographs of *E. albomagnum*;

Sarah Lloyd and Ron Nagorcka for the collection of *E. percrinitum* from their property at Black Sugarloaf, Birralee.

Many *Entoloma* collections were made from frequent visits to the Donnellys Road, Geeveston, property of Drs. Laurie Bishop and Fiona Lewis, whom we thank.

Microscopic photographs were made with a Nikon Coolpix 950, a gift from Dr. Tim Baroni of the State University of New York, Cortland, to G.M. Gates.

Type studies were made possible through the kindness of the curator of the New Zealand Fungal Herbarium (PDD), Auckland, New Zealand.

Dr. Brian Spooner of the Herbarium KEW, United Kingdom, kindly assisted M.E. Noordeloos during a visit to the herbarium in November 2010 to study the types of Greta Stevenson.

We thank the following institutions: the Schools of Plant Science and Agricultural Science at the University of Tasmania, Forestry Tasmania, the Cooperative Research Centre for Forestry, CSIRO Sustainable Ecosystems and the Tasmanian Herbarium (HO), for logistic support.

Author Biographies

Dr. Machiel E. Noordeloos was born in The Hague, The Netherlands, in 1949. He grew up in the outskirts of an expanding post-war town, with lots of opportunities to explore the plant life of the polders and nearby coastal dunes. Already as a small child he got interested in nature, particularly botany, and started collecting and making a herbarium. In 1967, he started his biology education at Leiden University and was trained as a mycologist by Dr. C. Bas working on a revision of *Marasmius* in the Netherlands. In 1981, he got his Ph.D. on a dissertation on the taxonomy and geographic distribution of *Entoloma* sensu lato in Europe. From 1987 to 1991 he was head of the Mycology Department of the Plant Protection Service in Wageningen where he studied plant pathogenic fungi and their ecology. From 1991 until his retirement in 2011, he was staff member and group leader of the Department for Plants and Cryptogams of the Netherlands and Europe at the National Herbarium of the Netherlands in Leiden. He is editor-in-chief of the series Flora Agaricina Neerlandica and has published many papers and books on various groups of Agaricales, including Entolomataceae, Marasmiaceae, and Strophariaceae, with a focus on Europe and Australia. In 2009, he was awarded the Clusius Medal of the Hungarian Mycological Society. He is an honorary member of the Dutch Mycological Society.

Dr. Genevieve Gates (née Piscioneri) was born in Pyramid Hill, Victoria, Australia, in 1952. Her father, an irrigation engineer, moved the family from the hot, dry, dusty Mallee of Victoria to the cooler climes of the island of Tasmania in 1959 where she continued her education, culminating in a B.Sc. degree majoring in botany and zoology at the University of Tasmania in 1974. She worked for several years as a laboratory technician at the Department of Agriculture at the University before becoming a full time mother. In 1998, in the middle of raising her three sons, she was drawn back to her botanical studies and became very interested in the taxonomy of Tasmanian fungi, particularly the family Entolomataceae. In 2009, she was awarded a Ph.D. in mycology and forest ecology for her study which investigated the macro-fungal assemblages associated with wood, soil, and litter in the wet eucalypt forests of southern Tasmania. Currently, she is an honorary research associate at UTAS.

Contents

Part I
Introduction

Chapter 1
Introduction

Abstract The introductory part gives a general introduction to the book, illustrating how the book came into being, based on 14 years of intensive collecting on a year-round basis by Genevieve Gates and David Ratkowsky, and how Machiel Noordeloos became involved in the project. The second chapter gives an account of the current state of knowledge of the taxonomy, phylogeny and biogeography of Entolomataceae, a species-rich agaric family with more than 1,500 described species worldwide, and their position in the Tasmanian mycota. In the book, special attention is paid to modern species concepts, and the infrageneric classification is considered in the light of current phylogenetic knowledge. A large chapter is devoted to methods used to study the family, from collecting to describing and conserving, as well as microscopic techniques, to facilitate identification. A fully illustrated guide to the characters used and their character states facilitate the use of the keys and help in the understanding of the descriptions. The last chapter of the introductory part, co-authored by David Ratkowsky, deals with the ecology, distribution and phenology of the Entolomataceae. The forest types in which this family has been studied are described. A map of Tasmania is given, showing the predominant forest types and the sites at which species of Entolomataceae were found. The size of the symbol reflects the number of species of the family found at each site. A dataset of about 4,000 collections forms the basis of an analysis of the fruiting pattern of *Entoloma* species in Tasmania. Statistical methods have been applied in an attempt to reveal differences between the monthly fruiting patterns. As a result, five distinct groups could be distinguished. Another database, containing information from ca. 1,000 forays, enabled a comparison to be made among Entolomataceae, ectomycorrhizal macrofungi, and soil-borne saprobic agarics, using the total number of records for each month as a percentage of the annual total. The results indicate that species of *Entoloma*, of which the majority are supposedly non-ectomycorrhizal, have a different emergence pattern from that of other saprobic agarics. Full references to the cited literature are given.

M.E. Noordeloos and G.M. Gates, *The Entolomataceae of Tasmania*,
Fungal Diversity Research Series 22, DOI 10.1007/978-94-007-4679-4_1,

General Introduction

This work started in 1998, when Genevieve Gates and David Ratkowsky began an inventory of the mycota in the forests of Tasmania on a regular basis, foraying throughout the year, and collecting as much data as possible about the taxonomy and distribution of macrofungi. This resulted in a number of reports and publications (Gates and Ratkowsky 2003, 2004a, b, 2005; Gates et al. 2005, 2009b; Ratkowsky and Gates 2002a, b, 2005, 2009).

Genevieve soon developed a particular interest in the family Entolomataceae and made throughout the years two and a half thousand well-annotated and documented collections, plus another 2,000 collections of now familiar species as voucher specimens from around Tasmania. When she realized how many of the taxa were unidentifiable, she took the initiative to contact known experts in this field, viz. Timothy J. Baroni, and Machiel E. Noordeloos. Her contact with Baroni resulted in a study of the genus *Rhodocybe* in Tasmania (Baroni and Gates 2006), and with Machiel Noordeloos the current project was undertaken and focused on a monographic treatment of the Entolomataceae of Tasmania. Since 2002, Machiel has jointly collected with Genevieve in Tasmania during visits on an almost yearly basis, each visit lasting several weeks. On some of these visits, they were accompanied by Michael Pilkington, a photographer friend from the United Kingdom who specializes in photographing fungi and who has provided photos for this monograph. This collaboration has led to a series of preliminary accounts of the new taxa encountered (Gates and Noordeloos 2007; Gates et al. 2009a; Noordeloos and Gates 2009).

The current publication is an account of all work spent on the study of the Tasmanian Entolomataceae to date, including also the results of Baroni and Gates (2006). We present keys and full descriptions of 90 *Entoloma* species, some with more than one variety, and 10 taxa belonging to the emended genus *Clitopilus*. We are well aware of the fact, however, that this surely is not the last word on this agaric family in Tasmania. More species will be discovered and described, and further molecular-phylogenetic studies will help also to understand more of the variability of species, and will make it possible, eventually, to make a sounder phylogeny-based classification. See also the remarks on species concept, below.

Taxonomy, Phylogeny and Biogeography of the Entolomataceae

Definition of the Entolomatoid Fungi and the Current State of Knowledge

The euagaric family Entolomataceae Kotl. & Pouzar is very species-rich. It is composed of more than 1,500 species and occurs worldwide from arctic to tropical habitats (Baroni 1981; Gates and Noordeloos 2007; Gates et al. 2009a; Horak 1980,

2008; Largent 1977, 1994; Noordeloos 1988, 1992, 2004; Noordeloos and Gates 2009; Noordeloos and Hausknecht 2007; Romagnesi and Gilles 1979). Recent molecularly based phylogenetic studies have revealed that the family is monophyletic, and sister to the Lyophyllaceae (Matheny et al. 2006; Moncalvo et al. 2002; Co-David et al. 2009).

The Entolomataceae is highly variable in terms of sporocarp morphology (tiny to large pleurotoid, omphalioid, collybioid, mycenoid, and tricholomatoid, as well as sequestrate), and micromorphology (spore shape, pileipellis structures, pigmentation types, presence and shape of cystidia, etc.; see Noordeloos 2004). Lifestyles are equally varied. Most species are saprotrophic on soil, wood or moss, but some are parasitic on other mushrooms (Noordeloos 2004) or plants or are ectomycorrhizal (Antibus et al. 1981; Agerer and Waller 1993; Agerer 1997; Kobayashi and Hatano 2001; Montecchio et al. 2006). The family traditionally is divided into three main agaricoid genera: *Rhodocybe* Maire, *Clitopilus* (Fr. ex Rabenh.) P. Kumm. and *Entoloma* (Fr.) P. Kumm. sensu lato. The latter genus is sometimes split into more genera (e.g. 13 genera, including *Entoloma* sensu stricto, *Nolanea*, *Leptonia*, *Inocephalus*, *Trichopilus*, *Claudopus*, etc.; see Largent 1994). Additionally, three smaller non-agaricoid genera have been distinguished on the basis of habit, namely, the monotypic *Rhodocybella* T.J. Baroni & R.H. Petersen (with a cyphelloid habit), *Rhodogaster* E. Horak (secotioid) and *Richoniella* Costantin & L.M. Dufour (gasteroid).

It is no surprise that Entolomataceae, being such a large and highly variable family, raised questions that analysis of morphological characters alone cannot answer, either due to scarcity of characters and/or difficulty in interpreting the significance of the characters. Co-David et al. (2009) applied molecular phylogenetic methods in an attempt to clarify the inter-generic relationships within the Entolomataceae, as well as the evolution of the characteristic spore morphology within the family.

The entolomatoid fungi form morphologically a rather heterogeneous group of agarics and secotioids that share a unique character in the spore wall. The genus *Entoloma* has spores with an internal skeleton of connected ribs. When viewed under the light microscope they have an angular appearance. In addition, the spores are pink to pinkish brown in mass which, usually, is easily noticed by the pinkish colour on the lamellae. The two other genera, *Clitopilus* and *Rhodocybe*, have a similar structure in the spore wall. In *Clitopilus* this structure is present in the form of longitudinal ribs, which are usually clearly visible under the microscope. When seen from above, the spore appears angular. In *Rhodocybe*, the internal structure consists of isolated warts or bumps, which cause the irregular warted outline of the spores under the microscope. Co-David et al. (2009) found that in the three-gene phylogeny they reconstructed, *Clitopilus* is nested within *Rhodocybe*. This had also been suggested earlier in a paper by Moncalvo et al. (2004). As a result it was decided to merge the two genera into one. For nomenclatural reasons, *Clitopilus*, being the eldest name, has priority for the merged genus. However, a recent study by Baroni and Matheny (2011) suggests that the emended genus *Clitopilus* may be broken up again into *Clitopilus* sensu stricto, the genera *Rhodophana* and *Clitopilopsis*, and an emended concept of *Rhodocybe* sensu stricto.

The study of Co-David et al. (2009) also demonstrated that the gasteroid genera *Richoniella* and *Rhodogaster* should be combined with genus *Entoloma*. Baroni and Matheny (2011) confirmed the polyphyly of gasteroid *Richoniella*, and also transferred the cyphelloid genus *Rhodocybella* to the genus *Entoloma* where it was placed in the basal Entoloma clade close to *E. pluteisimilis* and *E. zuccherellii*. Thus the emended concept of the genus *Entoloma* now encompasses agaricoid, gasteroid, and cyphelloid basidiocarp types.

Current Taxonomy

Species Concept

This monograph essentially uses a morphological species concept, and not a phylogenetic one, because current knowledge of the phylogenetic position of most species does not allow us to use a phylogenetic species concept. We therefore follow Kuyper (in Bas et al. 1988) in defining a species on two supposedly independent characters in which it differs from similar (related) taxa. In the long taxonomic practice of the Leiden agaricologists, this has proved to be a practical and feasible approach.

Similarity in morphology may easily lead to the use of European or North American names, as has been done frequently in the past. However, our increasing knowledge of other groups makes it clear that many agaric species do not have a world-wide distribution, but are often geographically restricted (Geml et al. 2004, 2006, 2008; Nuytinck et al. 2006; Matheny et al. 2009). Ongoing phylogenetic studies in *Entoloma* also show that the Tasmanian species often form a clade of their own, sister to a clade with another geographic origin.

Infrageneric Classification

When the project started, we initially classified the Tasmanian species as best we could in the taxonomic framework of Noordeloos (2004), which essentially goes back to the earlier classifications by Romagnesi (1937, 1974). Increasing insight into the phylogeny of *Entoloma*, however, made it necessary to adjust the current classification considerably, although, since the studies currently are not finished, a final, phylogenetically-based infrageneric classification cannot be presented. The recently published phylogenies of Entolomataceae (Baroni and Matheny 2011; Co-David et al. 2009) led to the following new taxonomic insights:

Subgenus *Entoloma* as conceived by Noordeloos (2004) is polyphyletic. The /prunuloides clade, which contains the type species *E. prunuloides*, is phylogenetically distant from the /rhodopolioid clade, which includes *Entoloma sinuatum* and *E. rhodopolium*. As a result, subgenus *Entoloma* must be emended, excluding the *rhodopolium*-group, which can better be considered a subgenus in its own right, characterized among other things by the mainly tricholomatoid habit, and possibly also an ectomycorrhizal life mode. The emended subgenus *Entoloma* now not only includes species from the old subgenus, like *E. prunuloides*, *E. bloxamii*, *E. nitidum* and related species, but also section *Turfosa* (= section *Trachyospora*), and surprisingly also species with a differentiated pileipellis (trichoderm, hymeniderm and so-called calliderm), including the type species of the genus *Calliderma* (Romagn.) Largent. Good supporting morphological characters for the emended subgenus *Entoloma* are the often relatively small, weakly angled, relatively thin-walled, isodiametric to subisodiametric spores, presence and abundance of clamp-connections, intracellular pigments, and a usually well-developed bi-layered pileipellis, with inflated elements in the subpellis. Recently, Baroni et al. (2011) created the new genus *Entocybe* for part of the /prunuloides clade, mainly based on the occurrence of pustulate-angular spores, encompassing former section *Turfosa* (= *Trachyospora*) and *Entoloma nitidum*. Whether this generic concept will stand in future phylogenetic analyses remains to be seen.

Subgenus *Leptonia*, traditionally split up in three sections, viz. *Leptonia*, *Cyanula*, and *Griseorubida* (Noordeloos 2004), is polyphyletic. Section *Leptonia* belongs to the /nolanea-claudopus clade, *Cyanula* and *Griseorubida* to the /inocephalus-cyanula clade.

Claudopus, considered by many mycologists as a genus in its own right (e.g. Largent 1974; Horak 1980, 2008), based on the pleurotoid habit, is within the /nolanea-claudopus clade, and pleurotoid species turn out to be mixed with omphalioid and collybioid species in the same clade. This proves that the pleurotoid habit cannot be considered a justification for generic rank. Furthermore, species with a pleurotoid habit are also found in the /inocephalus-cyanula clade.

Pouzarella is strongly supported as a monophyletic clade within the large genus *Entoloma* with a highly specialized morphology (Noordeloos 1979; Baroni and Matheny 2011), and can best be considered at a subgeneric rank.

Trichopilus also falls out as a good monophyletic group within the bigger /inocephalus-cyanula clade, uniting species with capitate cheilocystidia and a fibrillose or subsquamulose pileus.

Calliderma, an assemblage of species with a velvety-cracked pileal surface, which consists of a hymenidermal or pallisade-like layer, is also polyphyletic. The type species *E. callidermum* is in the /prunuloides clade, close to our *E. indigoticoumbrinum.*

Species assigned to *Alboleptonia* appear in different clades in the phylogeny. *Entoloma sericellum* is sister to "*Trichopilus porphyrescens*", but the morphologically very similar *Entoloma albidosimulans* from Tasmania is placed in the /inocephalus-cyanula clade. *Entoloma cephalotrichum* (P.D. Orton) Noordel., however, is in the /nolanea-claudopus clade. This proves that the genus *Alboleptonia* is polyphyletic.

Richoniella pumila, a secotioid species with cuboid spores, from Australia and New Zealand, is within the /inocephalus-cyanula clade, and has therefore been incorporated in the genus *Entoloma*. Since the epithet *pumila* already was given to an *Entoloma* species, it has been renamed *Entoloma gasteromycetoides* Noordel. & Co-David.

Some Tasmanian taxa, such as *Entoloma camarophyllus* in our analyses are in a big unresolved group within the /inocephalus-cyanula clade, and cannot be classified satisfactorily at the moment.

In the synopsis in the taxonomic part of this book the species are arranged according to these new insights. We want to stress, however, that it is still a preliminary classification that surely will be emended in the future due to ongoing morphological and phylogenetic research.

The apparent monophyletic status of some of the clades (e.g. /pouzarella, /inocephalus-cyanula, /entocybe) may tempt one to distinguish them at a generic level. However, the existing phylogenetic reconstructions are far from resolved, and more work has to be done to get a better view of the status of many clades. Therefore, we stick to the broad concept of *Entoloma* s.l. bearing in mind what Romagnesi in Romagnesi and Gilles (1979) wrote on this subject: "one should not be surprised by the difficulty of delimiting sections, let alone subgenera. It happened often that a new collection made our classification questionable, and it is extremely difficult to define natural groups, because all characters seem to mix. Therefore, at this stage, and without any doubt for a long time to come, it is perfectly vain to elevate one of these taxa to the rank of genus". Despite the introduction of molecular characters and phylogenetic methods, this still seems to be true.

Biogeographical Considerations

Although the family Entolomataceae has received considerable attention in the past decades, not many publications deal with the Australian Entolomataceae. Grgurinovic (1997), in her account of the macrofungi of South Australia, devotes a chapter to the Entolomatales, including 2 species of *Rhodocybe* and 10 of *Entoloma*. However, only one of these species, viz. *Entoloma viridomarginatum* (Cleland) E. Horak, has been recorded from Tasmania with certainty. Recently, the North American *Entoloma* expert David L. Largent started an investigation of the Entolomatoid fungi of Queensland and New South Wales, together with a number of Australian mycologists (Largent and Abell-Davis 2011; Largent et al. 2011a, b). While exploring the Tasmanian Entolomataceae, it became clear that many species appeared to be undescribed. Some of them could be identified with the works of Horak (1973, 1980, 2008), but others had to be published as new species (Gates and Noordeloos 2007; Gates et al. 2009a; Noordeloos and Gates 2009, and this book).

The current state of knowledge, despite intensive collecting by the authors in Tasmania, and by E. Horak in New Zealand, is still insufficient to make reliable statements about endemism. Of the more than 100 taxa in the current treatise, at least 15 are also found in New Zealand, including several species that Horak (2008) cited as endemic to New Zealand. With further work more species in common can be expected. Some of our species may well occur on mainland Australia in similar habitats (Largent and Abell-Davis 2011; Largent et al. 2011a, b), and in addition we have unpublished information that this is the case with some of our species. Eventually, detailed biogeographical studies in a larger area, including the *Nothofagus* zone of South America (i.e. Patagonia), will make it possible to decide whether or not certain species are endemic.

Materials and Methods

The species descriptions in this work are all based on extensive collecting by the authors. Colours are matched with Methuen's Handbook of Colour (Kornerup and Wanscher 1978) (e.g. 6A2 or 19E5–19F5), and Munsell Soil Colour Charts (Munsell 1975) (e.g. 2.5Y 8 or 10YR 7/6). Photos were taken in the field by Genevieve Gates with a Nikon Coolpix 5000 and by Machiel Noordeloos with a Nikon D90. Michael Pilkington used a Canon 5D Mark II, in conjunction with CombineZ stacking software. Microscopic observations were made with Olympus and Leica microscopes, using standard techniques. Fresh material was observed in water to determine the true colour and characteristics of the pigmentation. Fresh and dried material were mounted in 5 % KOH or 10 % ammonia or a Congo Red solution to observe, draw and measure microscopic details (for details see section "Microscopic Techniques").

How to Study the Entolomataceae

Although the study of the Entolomataceae requires the same techniques as for other genera of Agaricales, some experience and skill are needed with the microscope, since microscopic characters are essential for identification in many cases. A good microscope equipped with an oil-immersion objective is therefore necessary.

How to Collect

Good material facilitates identification. The following recommendations are made:

- Collect fresh specimens only, preferably both young and mature fruit bodies. Be sure to collect the entire mushroom, including the stipe base and any adherent rhizomorphs.
- Be careful not to mix up different species. Different *Entoloma* species can grow together on quite a small spot. Make a note of the habitat: type of vegetation, dominant trees, shrubs, herbs and mosses. Characterise, if possible, the soil type.

- Be aware of colour changes during transport and storage. In many taxa the colour can change rapidly, not only because of the sometimes hygrophanous nature of the pileus, but also subtle colours (especially blue, violaceous, pink etc.) may change or disappear. The best way is to make notes in the field, compare the fresh material with a colour code, and take a colour photograph of the fruit bodies (including their undersides) in the field.
- Note the smell and taste in the field and compare it with the smell and taste after transport in a small container and after dissecting the fruit body. Temperature may affect these characters. Also, odour changes with the age of the fruit body.
- Transport the fruit bodies of each collection separately in small boxes, carefully packed with some moist mosses or other soft material, so that they arrive home relatively hydrated and undamaged. Do not collect small species in a basket.

What to Do at Home

If you want to name the species at a later stage after they have been dried, or if you want to save your material to deposit in your herbarium, make a full description of all macroscopic characters that may disappear with drying. Make a sketch of the fruit body using colour pencils. This is very useful even if you took a photograph. No matter how good your colour photo may be, diagnostic characters often cannot be verified with certainty from it. Making a standard description for all your collections makes critical comparison easier and facilitates identification.

How to Make a Macroscopic Description

There are several ways of doing this, but it is important to do it always in the same way. Some people use pre-printed forms, other people write the description down for each collection. One of the advantages of a pre-printed form is that you do not easily forget to note a character. When making a description without a form, always use the same order of characters. This also helps you to remember the relevant and important morphological features. The authors generally use the following order in their descriptions in this book:

Pileus [size in mm, shape, margin, centre, hygrophanity, translucency, colour when moist, colour when dry, surface structure]. Lamellae [insertion (attachment) on stipe, shape, spacing, width and colour of sides, shape and colour of edge, other characteristics, e.g. L = number of entire lamellae, l = number of short lamellae (lamellulae), veins on sides (transvenose) or veins between lamellae (intervenose)]. Stipe [dimensions in mm, length x width, shape: cylindrical, compressed etc., shape of base e.g. swollen, tapering etc., colour and surface structure e.g. glabrous, fibrillose-striate, squamulose etc., colour of basal tomentum]. Fruit body context [colour, consistency, stuffed, fistulose]. Odour, taste and spore print colour.

Drying and Conservation of the Material

To make sure that your freshly collected specimens can be studied in a dried state, it is important to dry them as quickly as possible after collecting and describing. Use a drying method with good ventilation and a moderate temperature, preferably not above and not too much below 40 °C. Store the material after drying in envelopes with a label. On this label write with permanent ink or type the following data: Species name, collector, locality, date, and collecting number. In case you keep your notes and descriptions separately, give all items the same collecting number. Store your herbarium in a cool, dry place. Beware of insects. Freezing for 3 days at –25 °C has been proved to be a good conservation method.

Microscopic Techniques

For the study of the various microscopic characters the following methods are recommended:

Spores, Hymenial Structures (Basidia, Cystidia) and Clamp-Connections

Stain with Congo Red (1 % Congo Red in concentrated ammonia):

- Fresh material: put a fragment of the lamella including the edge in Congo Red, let it stain for a few minutes then observe.
- Dried material: similarly put a fragment of the lamella in Congo Red, let it stain for a few minutes. It may be necessary, particularly with old or badly dried material, to warm it a bit with a small flame. Remove the excess stain with filter paper, and replace it with a 3–10 % KOH solution. Observe in this medium.

For both fresh and dried material, be careful not to press the cover slip too much, as you want to leave the fragment intact. Observe first with low magnification to locate the structures you want to make visible. If you cannot see them properly, push gently on the cover slip until the hymenial elements lie free in the medium. The KOH makes this process easier.

Pileipellis Structure and Pigmentation

- Fresh material: make a radial section of the pileipellis, and observe in water to see the true colour of the pigments, and also make another mount in a saturated sugar or salt solution to make the intracellular pigment more visible.
- Dried material: make a radial section of the pileipellis, and observe in 10 % ammonia. Do not use KOH, except on very old material that does not respond well to the ammonia treatment.

Be careful not to press the cover slip too much, so as not to change the orientation of the hyphae. When you have determined the structure of the pileipellis, you may press a bit to have a close look at the size and shape of the hyphal elements, pigmentation, and for the presence of clamps.

The Characters Used for the Delimitation of Taxa in *Entoloma*

Macroscopic Characters

Habit types are illustrated in Figs. 1.1 and 1.2.

Habit

Habit types are often used to characterise higher taxa (subgenera and sections), and are frequently used in the keys in this monograph. Five main types are distinguished (see Figs. 1.1 and 1.2).

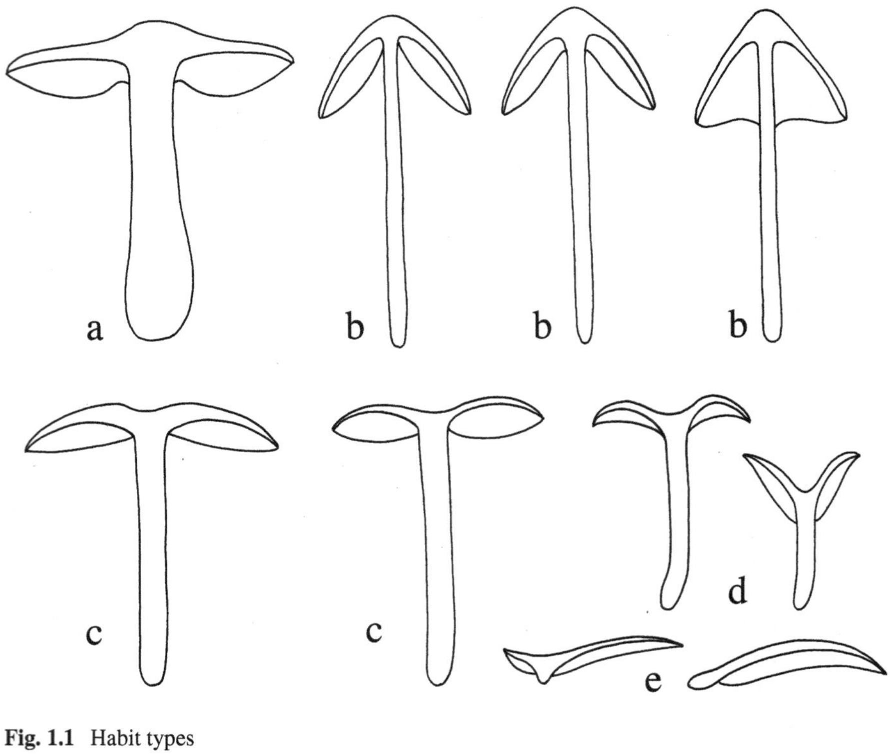

Fig. 1.1 Habit types

Fig. 1.2 Habit types

a – tricholomatoid: resembling a *Tricholoma* with convex, umbonate pileus, free to adnate-emarginate lamellae, a relatively thick stipe, and thick context.
b – mycenoid: resembling a *Mycena* with conical or campanulate pileus, free or adnate lamellae and a relatively long and slender stipe and thin context.
c – collybioid: resembling a *Collybia* with convex or plano-convex pileus with blunt or slightly depressed centre; lamellae adnate-emarginate or almost free.
d – omphalioid: resembling an *Omphalia* (or *Clitocybe*) with depressed to infundibuliform pileus and decurrent lamellae.
e – pleurotoid: resembling a (small) *Pleurotus* or *Crepidotus* with a lateral stipe, often reduced or absent.

Colour

Colour is one of the important diagnostics in describing fruit bodies. It is essential to know how the colour changes from young to mature to overmature, and from moist to dry. Many species are hygrophanous, i.e. the pileus changes colour from moist to dry. Usually this goes in a pattern of radial streaks starting from the centre of the pileus towards the margin. Many other species are not hygrophanous in this way, but become more or less equally paler on drying. They are called weakly hygrophanous. Colour changes may also occur when the specimen is dried for conservation. This is different to being hygrophanous. For changes involved with age, young, mature and overmature fruit bodies must be compared. In particular, the species of subgenus *Leptonia* show remarkable differences. In this subgenus one must be careful to look for blue tinges in the pileus and lamellae of young specimens. The colour of the stipe may change quickly, and, for example, lose the blue or violaceous tinge even immediately after being removed from the substrate. Many *Leptonia* species have a grey-blue or violaceous stipe when young, fading to grey or even grey-brown with age. Good collecting and colour notation in the field may be of considerable help.

Surface Structures

The surfaces of the pileus and stipe must be studied and noted for the use of the keys. The following types are found (see Fig. 1.3):

a. glabrous – bald.
b. fibrillose – with fibrils that usually are arranged radially. They can be innate, as is seen in most glabrous species, or superficial, where the fibrils are lying on the surface, often giving a silvery impression. In descriptions of the stipe you may find the description 'stipe silvery striate', which means that the surface is covered with silvery, shining, superficial fibrils.
c. velutinous (velvety, felted, tomentose) – covered with fine short hairs.
d. squamulose – covered with minute, appressed or somewhat erect scales.
e. micaceous – covered with glistening particles.
f. radially veined – covered with low vein-like elevations (as in *Pluteus* species).
g. squamules – scales.
h. squarrose – with relatively coarse, erect scales.

In addition (not depicted):
pruinose – dusted with a fine bloom.
rugulose – minutely wrinkled, rough.

Lamella Edge and Face

Quite a few species have a lamella edge that has a colour different from the faces, usually brown or blue. In addition, many taxa have an irregular lamella edge, varying from finely fimbriate to coarsely toothed. Staining or mottling may also occur on the edges and faces of the lamellae. Intervenose ribbing may also occur on the lamella faces.

Stipe Surface

The structure of the stipe surface is often characteristic and of diagnostic value, particularly in *Nolanea* and *Cyanula*. The following main types are distinguished (see Fig. 1.4):

Fig. 1.3 Surface structures of pileus. (**a**) Glabrous. (**b**) Fibrillose. (**c**) Velutinous. (**d**) Minutely squamulose. (**e**) Micaceous. (**f**) Radially veined. (**g**) With raised squamules in central part. (**h**) Squarrose

Microscopic Characters

Spores

The spores of all *Entoloma* species are angular, as seen from all views under the microscope (see Figs. 1.5a, b, c, d, e, f, g). It is very easy to make a preparation to study the spores. A small part of the lamella, gently pressed under the cover slip and

Fig. 1.4 Stipe surface. (**a**) Glabrous and polished. (**b–d**) Longitudinally fibrillose-striate (covered with paler, glistening fibrils, sometimes twisted). (**e**) Pruinose. (**f–g**) Squamulose. (**h**) Hairy-squarrose with erect hairs

observed in water (with fresh material) usually will reveal a large number of spores. This is in contrast to genera such as *Marasmius* and *Collybia*, where spores can be very sparse. Since the spores of most *Entoloma* species are relatively thick-walled and slightly straw-coloured using light microscopy, they can be fairly easily observed. Because of the relatively thick wall, it is also easy to study dried material. Soak a very small piece of lamellae (0.5 × 0.5 mm) in a 10 % ammonia solution for approximately 10 min and observe under the microscope. Normally you will find as many spores as you want, or even more!

Spore Size and Shape

For measurements only choose spores in your preparation that lie in profile (i.e. side) view. Measure at least 10 spores to see the variation in length and width. It is useful to calculate the length/width ratio of individual spores (Q), and its average, as this

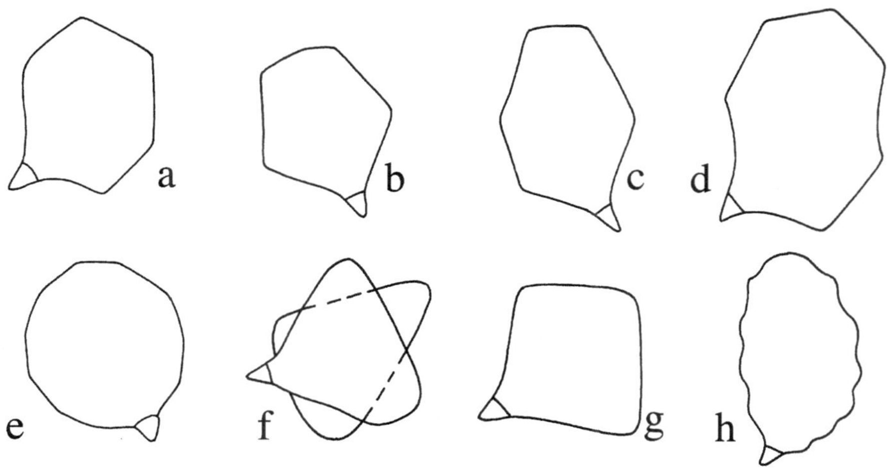

Fig. 1.5 Spore shape. (**a–g**) Examples of *Entoloma* spore shapes

is often used in the keys. Do not include the apiculus, which is the triangular-shaped point of attachment of the spore to the sterigma of the basidium.

The overall shape of the spores, in combination with the number of angles in side view, and the thickness of the spore wall are important diagnostic characters. When the Q value is 1.0–1.1, the overall shape is called **isodiametric** (Figs 1.5b, e), and when the Q value is larger, **heterodiametric** (Figs 1.5a, c, d). A special shape is the so-called **cuboid** spore (Fig. 1.5g), which has a rectangular shape in side view, and in three-dimensional views resembles a cube. **Cruciform** spores (Fig. 1.5f) have a very complex structure, more or less derived from the cuboid type, where the upper part of the spore is twisted along the side axis. Some species have spores with many angles which are blunt and these are called **nodulose** spores (Fig. 1.5h). The number of angles is counted in side view, i.e. in the same position as when the size is measured. The apiculus is included as one of the angles. Furthermore, it is important to note whether the angles are sharp (pronounced) or weak, and the relationship to the wall thickness.

Clamp-Connections

Basidiomycetes typically possess clamp-connections, which are often formed, and are visible, on the septae of the hyphae, and also at the base of the hymenial elements, such as the basidia and cystidia. In *Entoloma* they are usually easy to find and are fairly large. It is far more difficult, in some cases, to establish the absence of clamps. Many species in subgenus *Entoloma* have abundant clamps, almost on every septum in the hymenium, but also in the trama and cutis layers. However, in subgenus *Nolanea*, for example, clamp-connections are often only present in the hymenium, and rare or absent in the pileipellis and subpellis. Many species in subgenus *Cyanula* do not have clamp-connections at all.

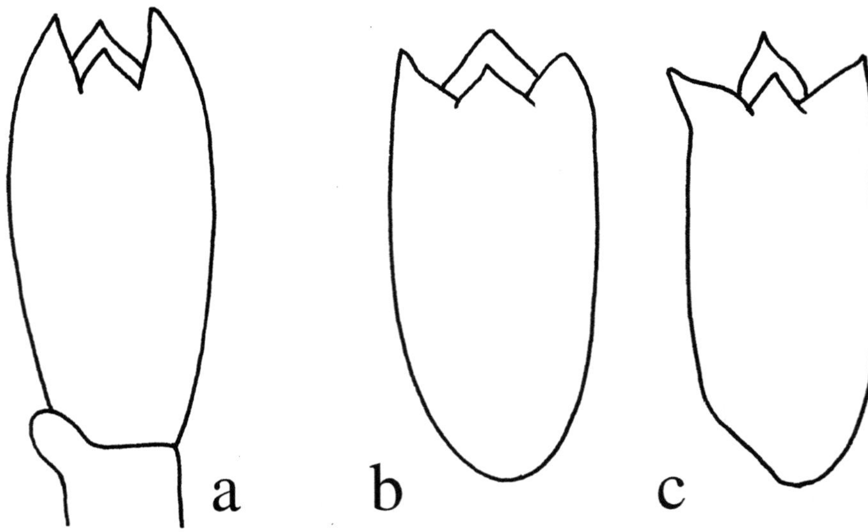

Fig. 1.6 Basidial shape

In our keys the statement 'clamps present' always means: clamp-connections present in the hymenium, at the base of basidia. Therefore, for this character it is sufficient to look for clamp-connections in a preparation of a small piece of the lamella. To facilitate the finding of clamp-connections, it is recommended to stain the hyphal wall with Congo Red as described above. Having done so, there are two possibilities (see Fig. 1.6):

1. Clamp-connection can clearly be observed at the base of the basidium (a).
2. There are no clear clamp-connections at the base of the basidium (b, c). In this case there are two possibilities: there are no clamp-connections or they are not visible as complete clamps. In many species of *Entoloma* mature basidia have lost the basal clamp because it has become part of a new basidium or has degenerated. Therefore, it is important to look for young basidia to see the clamp-connections. The base of a mature basidium is also important. Clampless species have basidia with an equally rounded base (b), whereas basidia that did have a basal clamp that degenerated usually show a slight flattening at the base (c).

Basidia

In *Entoloma* most species have 4-spored basidia, with only a few species having exclusively 2-spored basidia. An important diagnostic feature is the presence or absence of a clamp at the base (see Fig. 1.6a). The size of the basidia usually ranges from 20–40 × 8–12 μm. Members of section *Dysthales* have basidia that range from 40–60 × 10–20 μm. Basidia with brown, granulose, intracellular pigment may occur

Fig. 1.7 Lamella edge of serrulatum-type

in subgenera *Allocybe* and *Pouzarella*. This has also been called necropigment (Mazzer 1976), but that is not considered a good characteristic of this phenomenon, since this type of pigment can also occur in living basidia that produce normal spores. In many species of subgenus *Entoloma*, skeletobasidia can be found mixed with normal basidia. In these basidia the wall is between 1 and 3 μm thick, colourless and refringent. In Congo Red these walls stain very intensively. No taxonomic value has been attributed to these structures so far.

Structure of the Lamella Edge

The lamella edge has various characteristics:

Entirely fertile – formed as a hymeniform layer of young and mature basidia.

Entirely sterile – the lamella edge is composed of a hymeniform layer of variously shaped sterile cells called cheilocystidia (see Fig. 1.7). These may be filled with brown, green, grey or blue intracellular pigment and the edge appears coloured under a hand lens.

Serrulatum-type – a special type of lamella edge is encountered in subgenus *Cyanula*, the so-called serrulatum-type, after *Entoloma serrulatum* (Fr.) Helser. The lamella edge is completely sterile, and consists of a strand of hyphae that runs along the lamella edge from the insertion of the stipe to the margin of the pileus. At regular or irregular intervals, dense clusters of cylindrical, clavate or vesiculose terminal elements, often filled with coloured intracellular pigment, arise and

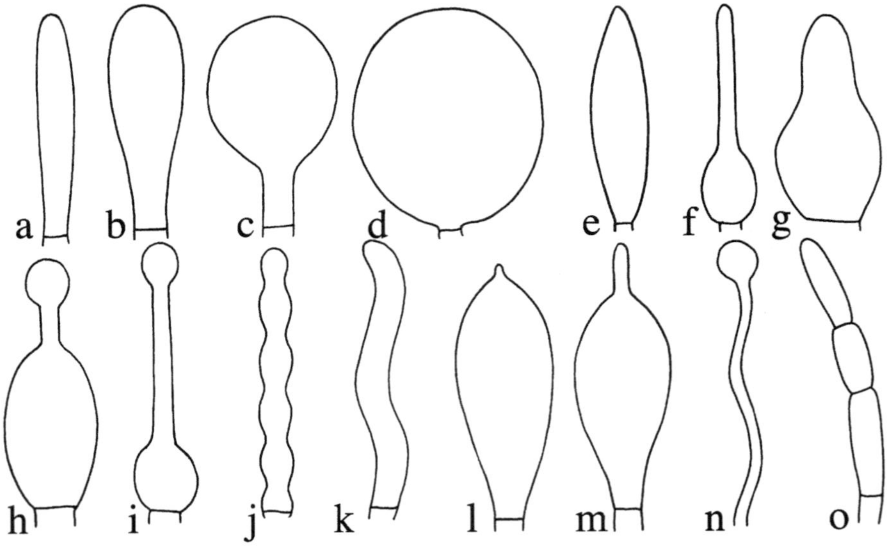

Fig. 1.8 Cystidial shapes. (**a**) Cylindrical. (**b**) Clavate. (**c**) Sphaeropedunculate. (**d**) Globose. (**e**) Fusiform (fusoid). (**f**) Lageniform. (**g**) Utriform. (**h**) Lecythiform. (**i**) Tibiiform. (**j**) Moniliform. (**k**) Flexuose. (**l**) Clavate with mucronate apex. (**m**) Clavate with rostrate apex. (**n**) Capitulate. (**o**) Septate

form the typical fringed edge that can be observed with a hand lens in fresh basidiocarps. These terminal elements frequently are called cheilocystidia, but it is not clear whether they are homologous with 'normal' cheilocystidia.

Heterogeneous – in this type of lamella edge, cystidia occur mixed with basidia. They can be scattered or numerous, and inconspicuous, difficult to distinguish from immature basidia, or very prominent, often much longer than the basidia, and protruding from the edge.

Cystidia

Cystidia can be present in the lamellae (hymenial cystidia), sometimes also on the stipe (caulocystidia) or on the pileus (pileocystidia). In *Entoloma* hymenial cystidia are usually present as cheilocystidia along the edge of the lamellae. Pleurocystidia, situated on the sides, are far less frequent. The technique for studying these elements is similar to that for the clamp-connections, and can be done in the same preparation. Congo Red helps to make the cystidia visible, as it stains the hyphal wall. Thin-walled cystidia, in particular, are easier to find in this way. Unfortunately, in some species of *Entoloma* hymenial cystidia are very thin-walled and are best observed in fresh material. In well-dried specimens they can usually be re-inflated by careful staining with Congo Red. In specimens that are dried when wet or overmature or dried under too hot conditions, cystidia are often difficult to find. See Fig. 1.8 for illustrations of the various types of cystidia that may be encountered.

Caulocystidia and Hairs on Stipe

Caulocystidia occur frequently, especially at the apex of the stipe. They are usually simple, cylindrical-flexuose, sometimes slightly to distinctly capitate. Caulocystidia are easily observed on a small lengthways-cut of the cuticle of the stipe (also known as the stipitipellis). Always look for caulocystidia when the stipe appears pruinose when observed with a hand lens. In subgenus *Pouzarella* the stipe is often covered with long, septate, encrusted hairs.

The Structure of the Pileipellis

The structure of the pileipellis and its pigmentation are best observed on a radial section of the pileus. This can be done on fresh as well as dried specimens. It is very important to have specimens in a good state, because age and weather conditions may result in a damaged pileipellis structure and may accordingly lead to false identification.

Field observations with a hand lens on the nature of the surface of the pileus, whether it is, for example, glabrous, fibrillose or squamulose may give an indication of the structure you may expect under the microscope. This also helps you to decide from which part of the pileus you must make a section to study the microscopic nature of the pileipellis. As a rule it is recommended to make a radial section about halfway between the margin and the centre of the pileus, and in addition also one from the centre, particularly when the field observation reveals a glabrous pileus with a slight villosity of small squamules at the centre.

It needs some experience to make a good radial section. Try to use a razorblade that you have broken or cut (with a pair of scissors) into a small pointed knife. It helps a great deal if you make your section under a binocular stereo-microscope at a magnification of about 25x. With fresh material you can observe the fragment in water, as this gives the true colour of the pigment, and also in a saturated sugar or salt solution in order to determine the nature of the pigmentation. The sugar or salt causes plasmolysis of the contents of the hyphae, which may help you to observe an intracellular pigment. It is recommended to observe dried material in a weak ammonia solution. Do not stain the fragments in Congo Red or another dyeing agent, because then it is more difficult to see the pigments.

Describe the pileipellis structure and compare it with the standard types. Often you will note a difference between the structures at halfway between the pileus margin and the centre. Frequently transitions between pellis types can be observed. Note the size and shape of the hyphae and terminal elements.

Basic Types of Pileipellis

The basic types of pileipellis that are found in the genus *Entoloma* are as follows (also, see Fig. 1.9):

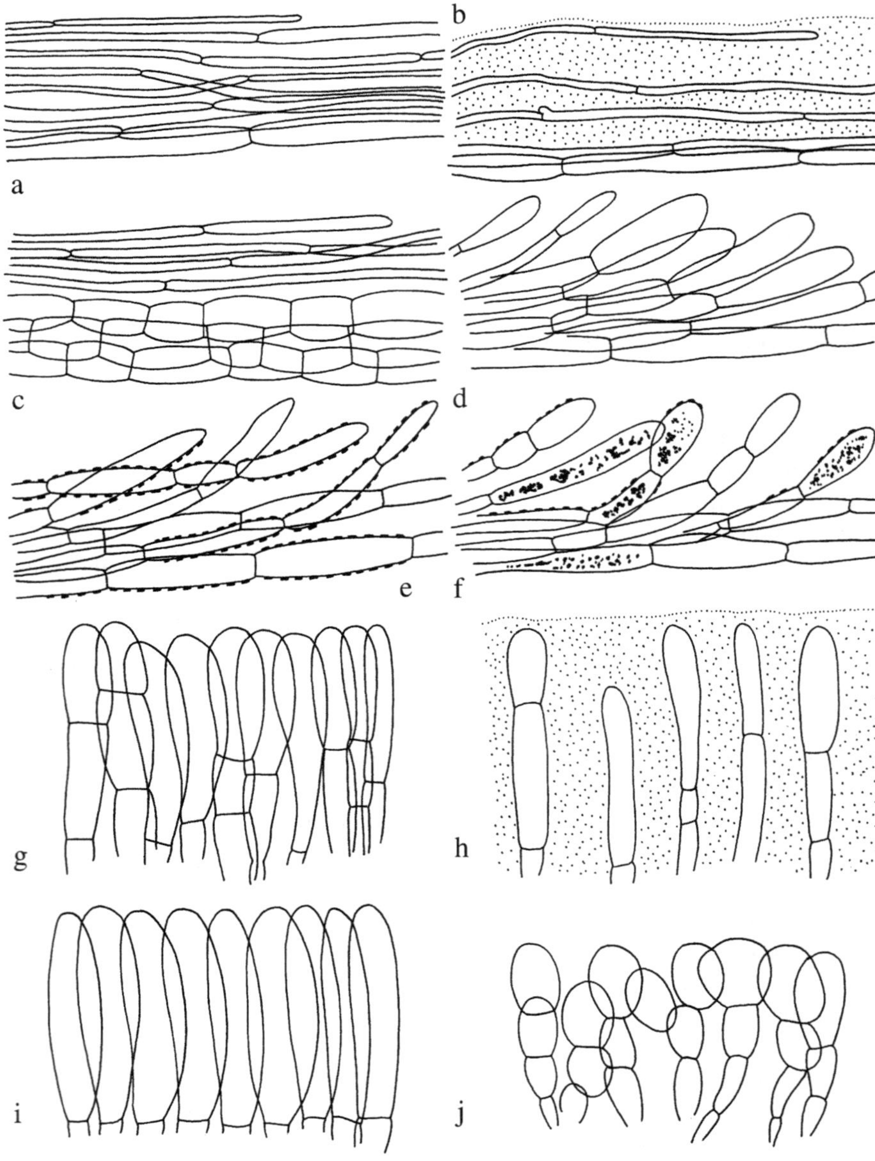

Fig. 1.9 Pileipellis types. (**a**) Cutis. (**b**) Ixocutis. (**c**) Cutis with subpellis. (**d**) Transition between cutis and trichoderm. (**e**) as (**d**), with encrustations. (**f**) Showing intracellular pigmentation. (**g**) Trichoderm. (**h**) Ixotrichoderm. (**i**) Hymeniderm. (**j**) Epithelium

cutis – a pileipellis consisting of radially arranged, repent hyphae (Fig. 1.9a). When these hyphae are embedded in a gelatinous layer, the pileipellis is called an ixocutis (Fig. 1.9b). In large groups (especially subg. *Entoloma* and *Nolanea*) a rather simple type of cutis occurs, made up of narrow, cylindrical hyphae, 2–10 μm wide. In other subgenera a cutis can be made up of inflated, much wider hyphae. There can be a gradual transition to the underlying trama, or a well-developed subpellis may be present, usually composed of short, inflated elements (Fig. 1.9c).

trichoderm – pileipellis consisting of erect septate elements, not originating at the same level (Fig. 1.9g). When the elements are embedded in a gelatinous layer, it is called an ixotrichoderm (Fig. 1.9h).

hymeniderm – pileipellis consisting of a layer of cylindrical, clavate or globose elements that originate in the same level, and are arranged in the same way as basidia in a hymenium (Fig. 1.9i). A palisadoderm is like a hymeniderm, except that the hyphae are not so even in height and are more loosely arranged. Occasionally in literature the term *calliderm* is used, named after *Entoloma callidermum* (Romagn.) Noordel. & Co-David, which has a perfect hymeniderm of clavate elements. Using this term, however, may lead to confusion, as it suggests that it is different from a true hymeniderm.

epithelium – pileipellis composed of erect chains of globose elements (Fig. 1.9j).

Transitions are frequently found, especially between a cutis and a trichoderm (Figs 1.9d, e, f). Many species can have a cutis at the margin, and a trichoderm or even a hymeniderm at the centre of the pileus. Other differences (Figs 1.9e, f) may involve encrustation and pigmentation types.

Pigmentation Types

The pigmentation of the pileipellis and upper trama is one of the most used diagnostic characters, and it is essential that one learns how to locate and recognize the different types. If possible, always observe this character on fresh material, since pigments, especially the blues, may change colour in dried material, and are often far more difficult to observe in reconstituted material. When observing dried (herbarium) material, always start by observing your sections in a weak ammonia solution. Only when the material is too difficult to remoisten, then the more strongly alkaline 5 or 10 % KOH should be used. Be aware that pigment may change or dissolve in that medium. One should also be aware of the possibility that more than one type of pigment can be found in any one section.

The following pigmentation types may be present (see Fig. 1.10):

encrusting pigment – the pigment is present on the hyphal wall, often in the form of brownish crusts, patches or rings, but sometimes only visible as very fine particles on the outer hyphal wall that break the light. In many cases you must look carefully at the very narrow hyphae of the upper layer of the pileipellis to find this type of pigment which sometimes is only present close to the septum.

intracellular pigment – this type of pigment, also known as plasmatic pigment, is soluble and present in the contents of the hyphae. Sometimes it is very diffuse and spread all over the content of the hyphae; in other cases it is present in the form of small or larger granules or clots.

parietal or membranal pigment – the pigment is located in the wall of the hyphae, not forming encrusting patterns. It is often pale, and not easy to observe.

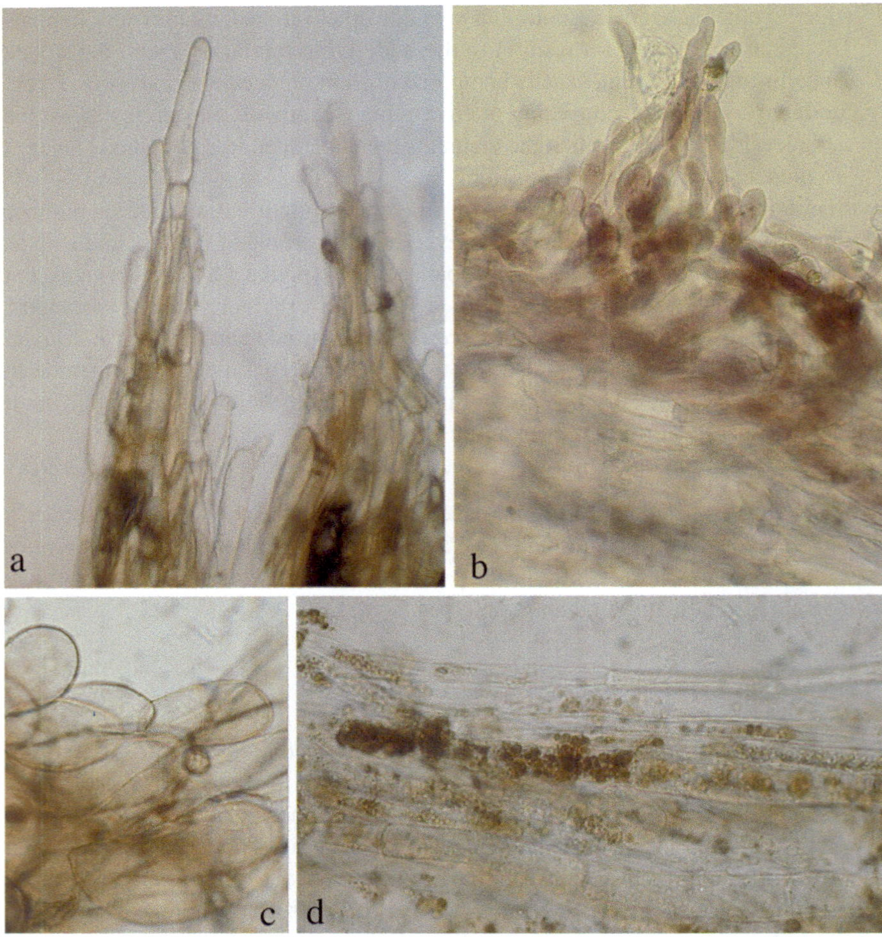

Fig. 1.10 Pigments. (**a**) Trichoderm with encrusting pigment. (**b**) Trichoderm with intracellular pigment. (**c**) Trichoderm with globose, encrusted elements. (**d**) Cutis with granular intracellular pigment

diffuse pigment – the pigment does not appear to be confined within the hyphae, i.e. it is intercellular rather than intracellular.

Structure of the Context (Trama)

The context of the pileus (pileitrama) is usually regular, and made up of parallel hyphae. Towards the centre of the pileus and in the layer just under the pileipellis, irregular structures may be observed. The size and shape of the fundamental hyphae are of diagnostic value. In the subgenera *Entoloma* and *Alboleptonia*, the fundamental hyphae are composed of relatively short, inflated elements, ca. 40–100 × 10–25 μm.

In the subgenera *Nolanea*, *Allocybe*, *Trichopilus* and *Clitopiloides*, very long, more or less fusiform elements form the base of the trama, ranging from 150–500 (or more) × 10–40 μm. In *Leptonia* the elements of the pileitrama are more or less cylindrical, and in size more or less intermediate. The same applies roughly for the context of the lamellae (hymenophoral trama). A special feature occurs in subgenus *Pouzarella*, where the hymenophoral trama is differentiated into a layer of narrow, cylindrical hyphae just below the subhymenium, and a central hymenopodium of distinctly wider, inflated hyphae. The structure of the trama is therefore of important diagnostic value for the distinction of subgenera.

Brilliant Granules

In certain cases, numerous granules or refractive oily droplets may be observed in the trama of the lamellae and pileus. They are called brilliant granules. Romagnesi (1937, 1974) was the first to discover the taxonomic value of this character. This character is mainly or exclusively present in two phylogenetically close clades, viz. */cyanula* and */inocephalus*. In */cyanula* the granules are present, usually distinctly so, but sometimes weakly so. In */inocephalus* in the strict sense (i.e. the species with cuboid spores), the granules can be so abundant that they hamper proper observation of the tissues.

Ecology, Distribution and Phenology (Co-author David Ratkowsky)

Tasmania, the island state of Australia, lies between 40° and 43° 40′ south of the equator and is separated from the mainland of Australia by Bass Strait. The island, with an area of 68,200 km^2, is approximately the same size as Sri Lanka and a little smaller than Ireland. Its climate can be described as modified marine Mediterranean. Broadly speaking, the vegetation can be classified either as austral montane, temperate rainforest or sclerophyll forest. However, within these three broad formations there is a complex mosaic of vegetation types across the island due to the geology, edaphic factors, precipitation and fire history of the island. Furthermore, the vegetation has plant species in common with New Zealand and South America that are relicts from an ancient Gondwanan flora or resulting from a more recent dispersal from these regions.

Forest types in Tasmania are determined by the long-term rainfall that an area receives (see the Forest Botany Manuals of the Forest Practices Authority, 2005, for a full description of Tasmanian forest typing). In general, the island of Tasmania is divided into a wet, western half and a dry, eastern half, although there are gradations within this broad division. For example, although extensive and diverse areas of rainforest occur in river valleys, alpine and subalpine environments of the west and southwest, the north-eastern highlands also contain patches of rainforest. Rainforests vary greatly in their composition, including extremes where stands contain pure *Nothofagus cunninghamii* (Hook.) Oerst. or pure *Atherosperma moschatum* Labill. Other characteristic rainforest species of lowland forests include *Eucryphia lucida* (Labill.) Baill., *Anodopetalum biglandulosum* (Hook.) Hook. and *Anopterus gladulosus* Labill. Their continued existence depends wholly upon the absence of fire, on a time scale of hundreds of years. In a mixed forest, a term used to describe a forest dominated by *Eucalyptus*, i.e. having a ground cover of at least 5 % (Forest Botany Manual, Module 6) but with an understorey of rainforest species, more frequent extreme fires (once or twice a century) allow the regeneration of *Eucalyptus* species and the continuation of the eucalypt dominated forest types.

Mixed forests are extensive in the Florentine and Styx valleys of the west and in the Arve and Esperance valleys of the south. Wet sclerophyll denotes forests with a high cover of eucalypts, often of a single species, such as *Eucalyptus obliqua* L'Hér., *E. regnans* F. Muell. or, at higher altitudes, *E. delegatensis* R.T. Baker, but which lack rainforest understorey species. Wet sclerophyll dominates as humidity increases in high rainfall areas and in more shaded areas and less fire-prone landforms. Dry sclerophyll forests and woodlands occur in less fertile, drier lowland areas of lesser rainfall, or at exposed sites at higher altitudes. Other localised broad forest types occur in Tasmania, such as swamp forests, which may be dominated by *Acacia melanoxylon* R.Br., up to three species of *Leptospermum* J.R. Forst. & G. Forst., or two species of *Melaleuca* L.

Because the majority of *Entoloma* species are decomposer species, they are often associated with broad-leaved (soft-leaved) species such as *Pomaderris apetala* Labill., *Olearia argophylla* (Labill.) F. Muell., *Bedfordia salicina* DC and *Zieria arborescens* Sims. In contrast, the narrow-leaved (hard-leaved) understorey species of the genera *Pultenaea* Sm., *Epacris* J.R. Forst. and *Hakea* Schrad. & J.C. Wendl. appear to be unfavourable for species of *Entoloma.*

Collecting was not restricted to native forest but included disturbed areas either naturally disturbed as by wildfire or anthropogenic as in domestic gardens and silvicultural treatments. The following vegetation types best describe the sites from which the species in this book were collected:

Native forest including:

- Wet sclerophyll and mixed forest – the tall wet eucalypt species *E. obliqua* and *E. regnans* dominate, on soils of good to moderate fertility, with an understorey of *Acacia*, *Olearia*, *Bedfordia*, *Pomaderris* and *Phebalium.* When the understorey consists of rainforest species rather than just broad leaved shrubs then the forest is termed a mixed forest rather than wet sclerophyll.
- Lowland temperate rainforest – dominated by *Nothofagus cunninghamii*, with *Atherosperma moschatum*, *Eucryphia lucida*, *Phyllocladus aspleniifolius* (Labill.) Hook.f. and *Anodopetalum biglandulosum.* This forest type is widespread in areas with summer rainfall in excess of 50 mm a month and is the climax of a wet sclerophyll or mixed forest that has not been affected by wildfire for 400–500 years.

- Dry sclerophyll forests and woodlands – these forests are found in parts of Tasmania (predominantly the midlands, the East and North-East) which, as a result of topographic attributes or edaphic factors, are prone to drought. Eucalypts or *Allocasuarina* spp. are dominant above an understorey of xeromorphic species such as *Acacia* spp., *Banksia*, *Exocarpos*, members of the Fabaceae Lindl., and small ground covers (Epacridaceae R.Br.) and grasses, e.g. *Poa* spp.
- Austral-montane – e.g. Mt Barrow (an *Entoloma* species was found in the shelter of a shrub on an exposed mountain top) and during a walk along the Overland Track, a 65 km bush walk through Cradle Mountain and Lake St Clair National Parks in the highlands of Tasmania, which includes buttongrass plains (see below) interspersed with forested areas. The majority of species of Entolomataceae that were collected from the forested areas along the walk were the same as those found in forests at lower altitudes.
- Buttongrass plains – moorlands, being areas of treeless vegetation with the sedge *Gymnoschoenus sphaerocephalus* (R.Br.) Hook.f. as the dominant species. *Entoloma camarophyllus* was collected along the edge of the track in the buttongrass plains of Cradle Mountain National Park, although that species is not exclusive to that environment, as it has also been collected in wet sclerophyll and mixed forest.
- Heath – dense to mid-dense vegetation dominated by sclerophyll scrubs less than two metres tall with a ground cover of, for example, species of epacrids, native grasses, sedges and rushes, orchids and lilies.
- Coastal – wet or dry sclerophyll forests and even rainforest (where there is both shelter and a long fire-free period) can grow to the high water mark. Although the soil may be quite sandy and depauperate in such areas, a good mycota is present including many of the same species of Entolomataceae that are found in forests on more fertile soils.

Disturbed sites:

- Gardens – the Northern Hemisphere species *Entoloma saundersii* and *E. clypeatum* were found associated with introduced members of the family Rosaceae Juss. (i.e. hawthorn and roses) and roadside elms (family Ulmaceae Mirb., which, like Rosaceae, is another family of the order Rosales), indicating their mycorrhizal association with these plants even if it is not considered to be a sustainable symbiosis (see Agerer and Waller 1993). Species of Entolomataceae were also collected from domestic lawns, football ovals and marsupial lawns (i.e. grassed areas of high moisture that are maintained by the grazing of wallabies and other marsupials).
- Roadside verges amid weeds and introduced grass species.
- Wildfire sites were surveyed in native forests that had been recently burnt.
- CBS coupes. Native forests that had been subjected to the silviculture treatment of 'clearfell, burn and sow' were surveyed in the early stages (10–26 months) of regeneration.

The most intensive collecting took place in the south and south-west of Tasmania within a radius of a 2 hour drive from Hobart. In addition, foraying trips to other parts of the island, although less frequent, have contributed to extending the range and habitats of known species of Entolomataceae in Tasmania as well as the discovery of some new species. Large areas of the mainland of Tasmania are inaccessible by road and were not surveyed, nor were the majority of the off-shore islands. Figure 1.11 shows the location of all sites that were surveyed, indicated by symbols superimposed upon a map of Tasmania that shows the major forest types, with the size of the symbol reflecting the number of species of Entolomataceae that were found at each site. However, the apparent predominance of species in the south of Tasmania in Fig. 1.11 is probably a reflection of the more intensive survey effort in that area than in other parts of Tasmania.

Collections were made from the following substrata:

- Soil
- Litter
- Wood

The majority of native species of Entolomataceae are decomposers and were found growing in soil, litter or, more rarely, on wood, especially wood that was in an advanced stage of decay and approaching a humus-like state. The introduced Northern Hemisphere species have some mycorrhizal association with the introduced hosts (see Agerer and Waller 1993). There appears to be some correlation between the presence of Entolomataceae and the presence of the litter of *Pomaderris apetala*, as many species of *Entoloma* and *Clitopilus* were found in association with this understorey plant species. *P. apetala* is an early coloniser after disturbance and grows on soils derived from dolerite. The leaves have a high calcium content but whether this affects the mycota is as yet unproven.

Fruiting Patterns of Species of Entolomataceae

A dataset of ca. 4,000 collections from 13 years of foraying in Tasmania was available for statistical analysis. Bar graphs depicting the fruiting patterns on a month by month basis are presented for each individual species. A non-metric multidimensional scaling ordination (nMDS) and a cluster analysis were applied to the dataset in an attempt to reveal differences between the monthly fruiting patterns. These analyses were confined to the 33 most frequently collected species, i.e. ones that had been collected a minimum of 40 times, as these species are more likely to reveal a stable phenology than infrequently collected ones. The results are depicted in Fig. 1.12, which indicates that the species of this dataset fall into five distinct groups. Group 1 consists of a single species, *E. fuligineopallescens*, this species appearing to have a unique pattern of maximum fruiting in late winter and early spring. The other four groups all have their maximum fruiting in the spring months but display subtle differences in their fruiting patterns. These differences are best revealed by comparing the monthly fruiting patterns using bar graphs (Fig. 1.13).

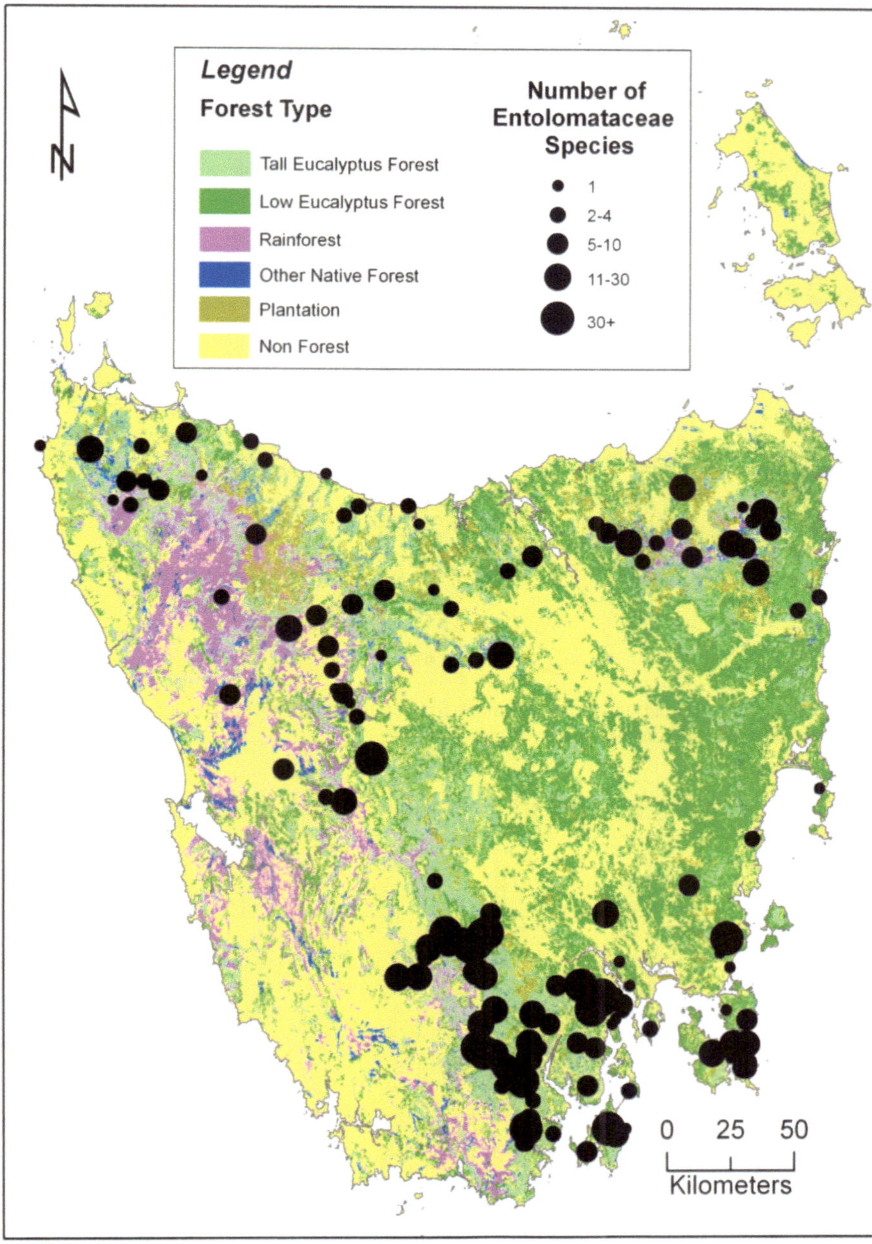

Fig. 1.11 Map of Tasmania showing the predominant forest types and the sites at which species of Entolomataceae were found. The size of the symbol reflects the number of species of the family found at each site

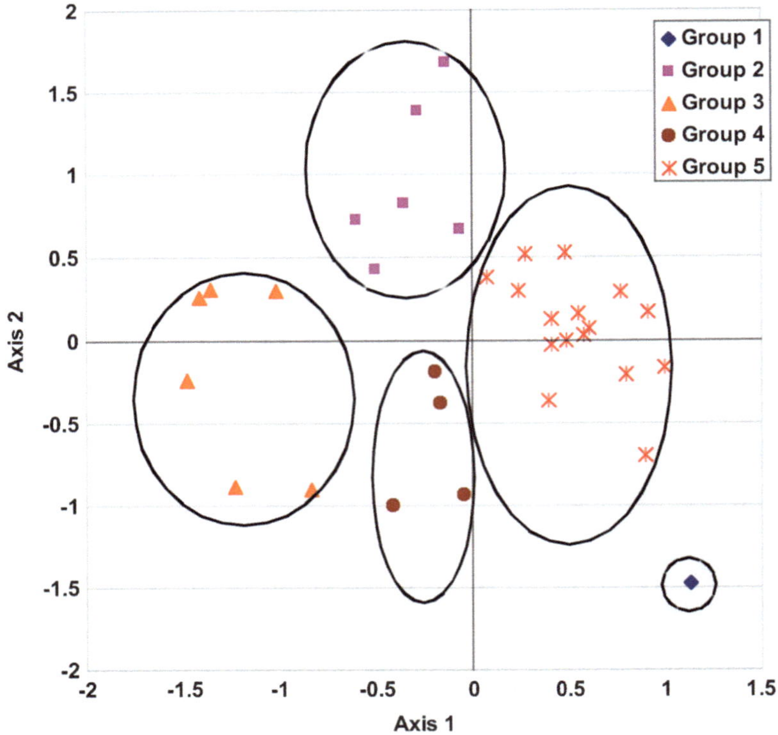

Fig. 1.12 nMDS ordination of the 33 most frequently occurring species of Entolomataceae, grouped according to their month of fruiting

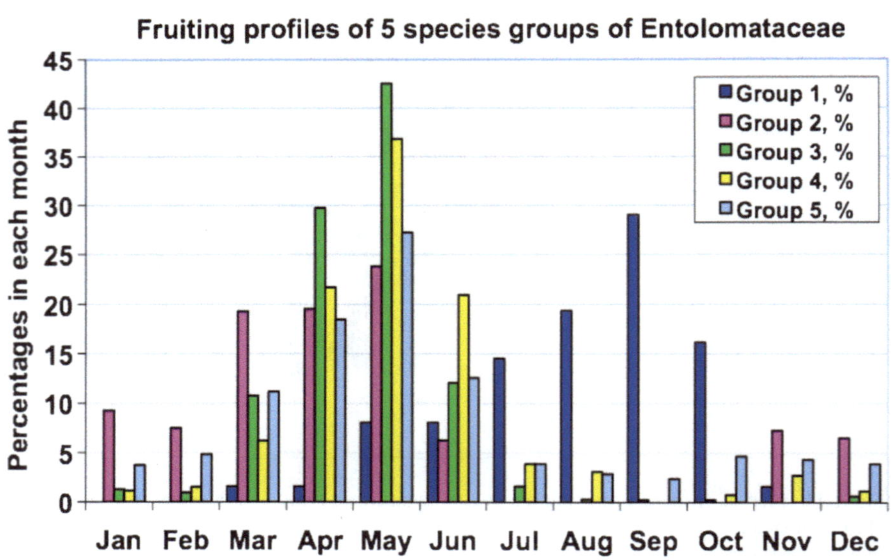

Fig. 1.13 Bar graphs showing the percentages of records of fruiting for each of five groups of species of Entolomataceae identified by their differences in fruiting pattern using cluster analysis and ordination. For each group separately, the percentages sum to 100 %

The most obvious difference in fruiting pattern revealed by Fig. 1.13 is the spring fruiting peak for Group 1 (*E. fuligineopallescens*) compared with all other groups. Amongst the other groups, Group 2 (*E. austroprunicolor*, *E. melanophthalmum*, *E. porphyrescens*, *E. saponicum*, *E. stramineopallescens*, *E. tasmanicum*) has its fruiting spread out over a period of almost 8 months, encompassing late spring, summer, autumn and early winter. Group 3 (*E. albidocoeruleum*, *E. camarophyllus*, *E. haastii*, *E. indigoticoumbrinum*, *E. panniculus*, *E. readiae* var. *sulphureum*) has over 70 % of its fruiting records occurring in the 2 months April and May, with almost all of the remaining records occurring in the surrounding months of March and June. Group 4 (*E. austrorhodocalyx*, *E. cystidiosum*, *E. fumosopruinosum*, *E. obscureotenax*) differs from Group 3 less overtly, as both groups have their peak fruiting in April and May, but fruiting in the winter months June–August is relatively more frequent for Group 4 than for Group 3. Group 5 (*E. albidosimulans*, *E. amarum*, *E. aromaticum* var. *aromaticum*, *E. asprellopsis*, *E. aurantiolabes*, *E. austronitens*, *E. brevispermum*, *E. chrysopus*, *E. conferendum*, *E. discrepans*, *E. fibrosopileatum*, *E. phaeomarginatum*, *E. readiae* var. *readiae*, *E. rodwayi*, *E. transmutans*, *E. viridomarginatum*) differs from Group 4 in having its fruiting somewhat more spread out over the 12-month period, with a fair number of records occurring in the spring and summer months. This may reflect a response by litter-inhabiting genera, e.g. *Mycena* and *Marasmius* (Gates et al. 2011), to rainfall events that occur in warmer conditions. Although there do appear to be distinct groups, as Fig. 1.12 indicates, physiological reasons for their presence are yet to be discovered.

Comparison of Fruiting Patterns of Entolomataceae with Other Groups of Fungi

A comparison was made between the fruiting phenology of Entolomataceae and those of other fungal groups from the same database of records collected throughout Tasmania over a 13-year period. This database was compiled from lists made of all macrofungi (i.e. species of fungi visible to the naked eye) encountered during ca. 1,000 forays made by G. Gates (with or without the company of others). Two groups of fungi were chosen with which to compare the Entolomataceae records. One of these was the ectomycorrhizal (EcM) fungi (incl. *Amanita*, *Boletus*, *Cortinarius*, *Descolea*, *Hydnum*, *Hydnellum*, *Inocybe*, *Laccaria*, *Lactarius*, *Ramaria*, *Russula*, *Tricholoma* and many other genera), for which there were 7,439 records. The objective here was to compare the largely saprotrophic Entolomataceae species, living on decaying organic matter, with a large group of species whose relationship with the plant host species is a symbiotic one. To ensure that the Entolomataceae records are those for saprobic species only, the combined seven records of *E. saundersii* and *E. clypeatum* were excluded, as those species may be ectomycorrhizal. This comparison between 3,947 Entolomataceae records and the records from EcM species of unrelated genera is displayed in Fig. 1.14a. The other group of fungi for comparison with Entolomataceae was the soil-borne saprobic "agarics", with agarics being defined in the traditional sense of being gilled fungi irrespective of the modern taxonomic

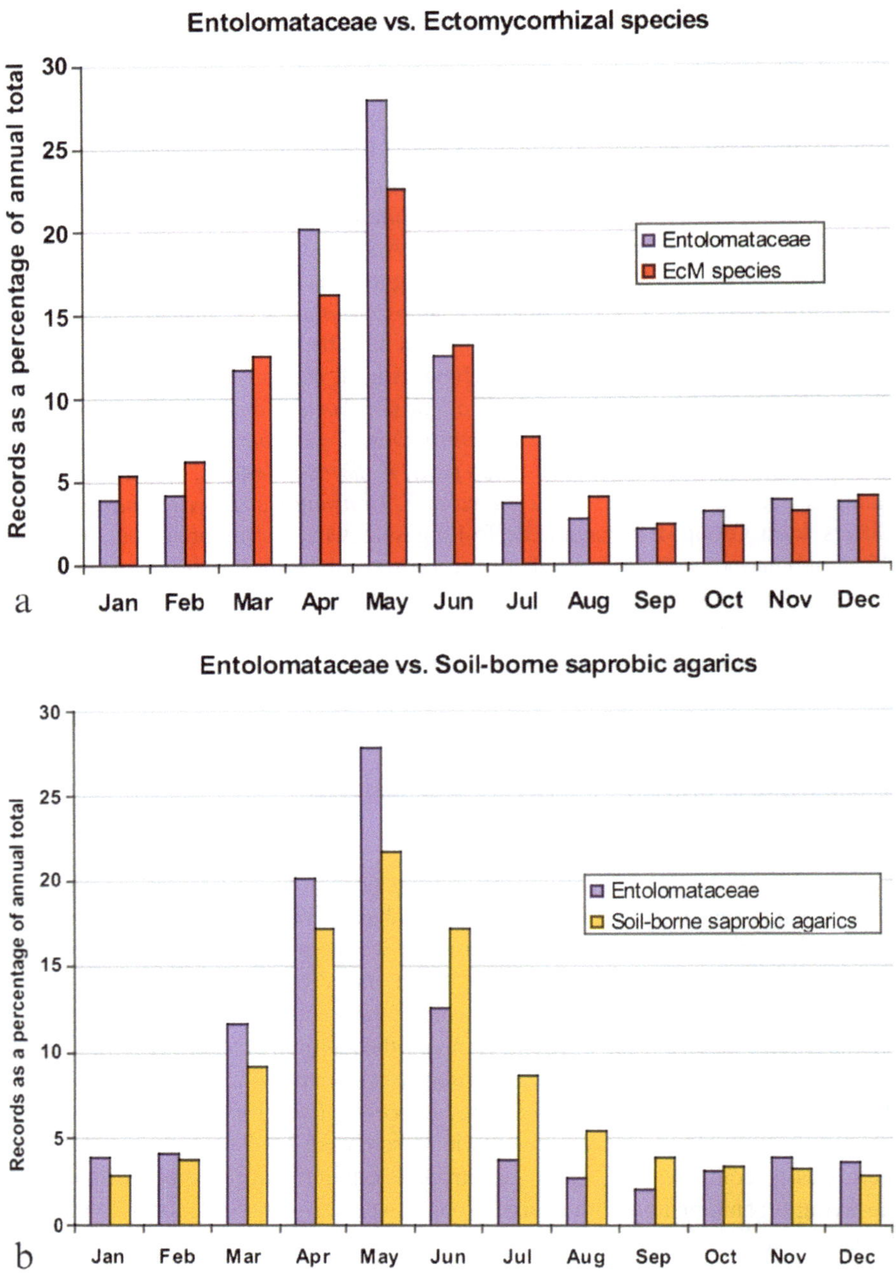

Fig. 1.14 Comparison between Entolomataceae and (**a**) ectomycorrhizal macrofungi, (**b**) soil-borne saprobic agarics, based on the total number of records for each month as a percentage of the annual total

groups into which they may now be classified. The EcM species, including the suspected species of the Entolomataceae, were excluded from this group of saprobes. Hence, the comparison here was between the 3,947 records of soil-borne saprobic Entolomataceae species and 6,395 records of other soil-borne saprobic agarics. Monthly percentages of emergence are shown in Fig. 1.14b.

The paired bar graphs in Fig. 1.14 show that the Entolomataceae has large fruiting spikes in April and May, with March and June also contributing substantially to the total number of records, so that those 4 months account for more than 72 % of the total number of Entolomataceae records. In contrast, the records for EcM fungi (Fig. 1.14a) are somewhat more spread out, with higher percentages both in the summer months (December–February) and in the winter months (June–August) than the Entolomataceae. One reason for the higher percentages of EcM fungi in summer is that many species of genera such as *Amanita*, *Boletus*, *Lactarius*, *Russula* and *Tricholoma* are early season fungi, provided that there is sufficient rainfall during these months. The need for substantial rainfall results in great annual variation, so that in years that experience a dry summer, the number of records of both EcM fungi and of Entolomataceae will be correspondingly low. From Fig. 1.14b, which compares the Entolomataceae with the saprobic "agarics", we see some similarities of the latter with the EcM fungi, with both groups having a higher percentage of emergence in the winter months (June–August) than the Entolomataceae, but the summer months are quite different, with less saprobic agarics being recorded than the EcM species. Nevertheless, the differences apparent in Fig. 1.14b lead to the conclusion that the non-EcM species of Entolomataceae have a different emergence pattern from that of the other saprobic agarics.

References

Agerer, R. (1997). *Entoloma sinuatum* (Bull.: Fr.) Kummer+Salix spec. *Descriptions of Ectomycorrhizae, 2*, 13–18.

Agerer, R., & Waller, K. (1993). Mycorrhizae of *Entoloma saepium*: Parasitism or symbiosis? *Mycorrhiza, 3*, 145–154.

Antibus, R. K., Croxdale, J. G., Miller, O. K., & Linkins, A. E. (1981). Ectomycorrhizal fungi of *Salix rotundifolia* III. Resynthesized mycorrhizal complexes and their surface phosphatase activities. *Canadian Journal of Botany, 59*, 2456–2463.

Baroni, T. J. (1981). *A revision of the genus Rhodocybe Maire (Agaricales)* (Beihefte zur Nova Hedwigia, Vol. 67). Vaduz: Cramer.

Baroni, T. J., & Gates, G. M. (2006). New species and records of *Rhodocybe* (Entolomataceae, Agaricales) from Tasmania. *Australian Systematic Botany, 19*, 343–358.

Baroni, T. J., & Matheny, P. B. (2011). A re-evaluation of gasteroid and cyphelloid species of Entolomataceae from eastern North America. *Harvard Papers in Botany, 16*(2), 293–310.

Baroni, T. J., Hofstetter, V., Largent, D. L., & Vilgalys, R. (2011). *Entocybe* is proposed as a new genus in the Entolomataceae (Agaricomycetes, Basidiomycota) based on morphological and molecular evidence. *North American Fungi, 6*, 1–19.

Co-David, D., Langeveld, D., & Noordeloos, M. E. (2009). Molecular phylogeny and spore evolution of *Entolomataceae*. *Persoonia, 23*, 147–176.

Gates, G. M., & Noordeloos, M. E. (2007). Preliminary studies in the genus Entoloma in Tasmania – I. *Persoonia, 19*, 157–226.

Gates, G. M., & Ratkowsky, D. A. (2003). The fungal genus *Lentinellus* in Tasmania. *Tasmanian Naturalist, 125*, 9–13.
Gates, G. M., & Ratkowsky, D. A. (2004a). Some interesting fungal records for Tasmania. *Tasmanian Naturalist, 126*, 2–5.
Gates, G. M., & Ratkowsky, D. A. (2004b). A preliminary census of the macrofungi of Mt Wellington, Tasmania – the non-gilled Basidiomycota. *Papers & Proceedings of the Royal Society of Tasmania, 138*, 53–59.
Gates, G. M., & Ratkowsky, D. A. (2005). A preliminary census of the macrofungi of Mt Wellington, Tasmania – the Ascomycota. *Papers & Proceedings of the Royal Society of Tasmania, 139*, 49–52.
Gates, G. M., Ratkowsky, D. A., & Grove, S. J. (2005). A comparison of macrofungi in young silvicultural regeneration and mature forest at the Warra LTER site in the southern forests of Tasmania. *Tasforests, 16*, 127–152.
Gates, G. M., Horton, B. M., & Noordeloos, M. E. (2009a). A new *Entoloma* (Basidiomycetes, Agaricales) from Tasmania. *Mycotaxon, 107*, 175–179.
Gates, G. M., Ratkowsky, D. A., & Grove, S. J. (2009b). Aggregated retention and macrofungi, a case study from the Warra LTER site, Tasmania. *Tasforests, 18*, 33–54.
Gates, G. M., Carpenter, D. M., Ratkowsky, D. A., & Dalton, P. J. (2011). Wood-inhabiting macrofungal assemblages in 43-year-old regenerating wet *Eucalyptus obliqua* L'Hér. forest. *Papers & Proceedings of the Royal Society of Tasmania, 145*, 65–76.
Geml, J., Geiser, D. M., & Royse, D. J. (2004). Molecular evolution of *Agaricus* species based on ITS and LSU rDNA sequences. *Mycological Progress, 3*, 157–176.
Geml, J., Laursen, G. A., O'Neill, K., Nusbaum, H. C., & Taylor, D. L. (2006). Beringian origins and cryptic speciation events in the fly agaric (*Amanita muscaria*). *Molecular Ecology, 15*, 225–239.
Geml, J., Tulloss, R. E., Laursen, G. A., Sazanova, N. A., & Taylor, D. L. (2008). Evidence for strong inter- and intracontinental phylogeographic structure in *Amanita muscaria*, a wind-dispersed ectomycorrhizal basidiomycete. *Molecular Phylogenetics and Evolution, 48*, 694–701.
Grgurinovic, C. (1997). *Larger fungi of South Australia* (p. 725). Adelaide: Botanic Gardens of Adelaide and State Herbarium/Flora and Fauna of South Australia Handbooks Committee.
Horak, E. (1973). *Fungi Agaricini Novaezelandiae. I. Entoloma (Fr.) and related genera* (Beihefte Nova Hedwigia, Vol. 43, pp. 1–86). Vaduz: Cramer.
Horak, E. (1980). *Entoloma (Agaricales) in Indomalaya and Australasia* (Beihefte Nova Hedwigia, Vol. 65, pp. 1–352). Vaduz: Cramer.
Horak, E. (2008). Agaricales of New Zealand 1: *Pluteaceae – Entolomataceae*. In *The Fungi of New Zealand* (Fungal Diversity Research Series, no. 19, Vol. 5). Hong Kong: Fungal Diversity Press.
Kobayashi, H., & Hatano, K. (2001). A morphological study of the mycorrhiza of *Entoloma clypeatum* f. *hybridum* on *Rosa multiflora*. *Mycoscience, 42*, 83–90.
Kornerup, A., & Wanscher, J.H. (1978). *Methuen handbook of colour* (3rd ed.). London: E. Methuen (Reprinted in 1989, pp 252 by E. Methuen, London)
Kuyper, T. W. (1988). Specific and infraspecific delimitation. In C. Bas, T. W. Kuyper, M. E. Noordeloos, & E. C. Vellinga (Eds.), *Flora agaricina neerlandica* (Vol. 1, pp. 30–37). Rotterdam: Balkema.
Largent, D. L. (1974). New or interesting species of *Claudopus* and *Entoloma* from the Pacific Coast. *Madrono, 22*, 363–373.
Largent, D. L. (1977). *The genus Leptonia on the Pacific Coast of the United States: Including a study of the North American types* (Bibliotheca mycologica, Vol. 55, p. 286). Vaduz: J. Cramer.
Largent, D. L. (1994). *Entolomatoid fungi of the Western United States and Alaska* (p. 516). Eureka: Mad River Press.
Largent, D. L., & Abell-Davis, S. E. (2011). Observations on *Inocephalus virescens* comb. nov. and *Alboleptonia stylophora* from northeastern Queensland. *Mycotaxon, 116*, 231–245.
Largent, D. L., Abell-Davis, S. E., Cummings, G. A., Ryan, K. L., & Bergemann, S. E. (2011a). Saxicolous species of *Claudopus* (Agaricales, Entolomataceae) from Australia. *Mycotaxon, 116*, 253–264.

Largent, D. L., Bergemann, S. E., Cummings, G. A., Ryan, K. L., Abell-Davis, S. E., & Moore, S. (2011b). *Pouzarella* (Agaricales, Entolomataceae) species from New South wales (Barrington Top National Park) and Northeastern Queensland (Australia). *Mycotaxon, 117*, 435–483.
Matheny, P. B., Curtis, J. M., Hofstetter, V., Aime, M. C., Moncalvo, J.-M., Ge, Z.-W., Yang, Z.-L., Slot, J. C., Ammirati, J. F., Baroni, T. J., Bougher, N. L., Hughes, K. W., Lodge, D. J., Kerrigan, R. W., Seidl, M. T., Aanen, D. K., Denitis, M., Daniele, G., Desjardin, D. E., Kropp, B. R., Norvell, L. L., Parker, A., Vellinga, E. C., Vilgalys, R., & Hibbett, D. S. (2006). Major clades of Agaricales: A multilocus phylogenetic overview. *Mycologia, 98*, 982–995.
Matheny, P. B., Aime, M. C., Bougher, N. L., Buyck, B., Desjardin, D. E., Horak, E., Kropp, B. R., Lodge, D. J., Soytong, K., Trappe, J. M., & Hibbett, D. S. (2009). Out of the Palaeotropics? Historical biogeography and diversification of the cosmopolitan ectomycorrhizal mushroom family Inocybaceae. *Journal of Biogeography, 36*, 577–592.
Mazzer, S. J. (1976). *A monographic study of the genus Pouzarella: a new genus in the Rhodophyllaceae, Agaricales, Basidiomycetes* (p. 191). Vaduz: Cramer.
Moncalvo, J.-M., Vilgalys, R., Redhead, S. A., Johnson, J. E., James, T. Y., Aime, M. C., Hofstetter, V., Verduin, S. J. W., Larsson, E., Baroni, T. J., Thorn, R. G., Jacobsson, S., Clémençon, H., & Miller, O. K., Jr. (2002). One hundred and seventeen clades of euagarics. *Molecular Phylogenetics and Evolution, 23*, 357–400.
Moncalvo, J.-M., Baroni, T. J., Bhatt, R. P., & Stephenson, S. L. (2004). *Rhodocybe paurii*, a new species from the Indian Himalaya. *Mycologia, 96*, 859–865.
Montecchio, L., Rossi, S., Causin, R., & Grendene, A. (2006). *Leccinum lepidum* (H. Bouchet ex Essette) Bon and Contu+*Quercus ilex* L. *Descriptions of Ectomycorrhizae, 9*(10), 55–60.
Munsell Soil Color Charts. (1975). *MacBeth Division of Kollmorgen Instruments Corporation*. Baltimore: Munsell Soil Color Charts.
Noordeloos, M.E. (1979). *Entoloma* subgenus *Pouzaromyces* emend. in Europe. *Persoonia, 10*, 207–243.
Noordeloos, M. E. (1988). Entolomataceae. In C. Bas, T. W. Kuyper, M. E. Noordeloos, & E. C. Vellinga (Eds.), *Flora agaricina neerlandica* (Vol. 1, pp. 77–182). Rotterdam: Balkema.
Noordeloos, M. E. (1992). *Entoloma* s.l. In *Fungi Europaei* (Vol. 5). Alassio: Edizioni Candusso.
Noordeloos, M. E. (2004). *Entoloma* s.l. In *Fungi Europaei* (Vol. 5a). Alassio: Ed. Candusso. 618 pp.
Noordeloos, M. E., & Gates, G. M. (2009). Preliminary studies in the genus *Entoloma* of Tasmania II. *Cryptogamie Mycologie, 30*, 107–140.
Noordeloos, M. E., & Hausknecht, A. (2007). The genus *Entoloma* (Basidiomcyetes, Agaricales) of the Mascarenes and Seychelles. *Fungal Diversity, 27*, 111–144.
Nuytinck, J., Miller, S. L., & Verbeken, A. (2006). A taxonomical treatment of the North and Central American species in *Lactarius* sect. Deliciosi. *Mycotaxon, 96*, 261–307.
Ratkowsky, D. A., & Gates, G. M. (2002a). Keys to the Tasmanian families and genera of gilled fungi. *Tasmanian Naturalist, 124*, 2–24.
Ratkowsky, D. A., & Gates, G. M. (2002b). A preliminary census of the macrofungi of Mount Wellington, Tasmania – the Agaricales. *Papers & Proceedings of the Royal Society of Tasmania, 136*, 89–100.
Ratkowsky, D. A., & Gates, G. M. (2005). An inventory of macrofungi observed in Tasmanian forests over a six year period. *Tasforests, 16*, 153–168.
Ratkowsky, D. A., & Gates, G. M. (2009). Macrofungi in early stages of forest regeneration in Tasmania's southern forests. *Tasforests, 18*, 55–66.
Romagnesi, H. (1937). Essai d'une sectionnement du genre Rhodophyllus. *Bulletin de la Société mycologique de France, 53*, 319–338.
Romagnesi, H. (1974). Essai d'une classification des Rhodophylles. *Bulletin mensuel de la Société Linnéenne de Lyon, 43*, 325–332.
Romagnesi, H., & Gilles, G. (1979). *Les Rhodophylles des forêts côtières du Gabon et de Côte d' Ivoire* (Beihefte zur Nova Hedwigia, Vol. 59, pp. 1–649). Vaduz: Cramer.

Part II
Taxonomic Part

Chapter 2
Family Entolomataceae

Abstract A definition of the family *Entolomataceae* is given with full synonymy followed by a key to the genera, and a taxonomic synopsis of the species encountered in Tasmania. Full references to the cited literature are given.

Entolomataceae Kotl. & Pouzar, Česká Mykol. 26: 218. 1972 emend Noordel. & Co-David 2009. — *Rhodogoniosporaceae* Heim, Treb. Mus. Cienc. Nat. Barcelona 15: 86. 1934 (nom. nud.). — *Jugasporaceae* Singer, Ann. Mycol. 34: 327. 1936 (inval.). — *Clitopilaceae* PD Orton, Trans. Brit. mycol. Soc., suppl.: 6. 1960 (nom. nud.). — *Rhodophyllaceace* Singer, Lilloa 22: 218. 1972.

Basidiomata agaricoid, cyphelloid, secotioid or gasteroid; spore print pink; spores provided with either bumps or undulate ridges, or longitudinal ridges, and then angular in polar view, or faceted and angular in all views; spore wall in TEM (transmission electron microscopy) with a distinct epicorium forming the bumps and ridges. Saprotrophic, mycorrhizal or parasitic. Widespread all over the world.

Type-genus: *Entoloma* (Fr.) P. Kumm. emend. Noordel. & Co-David

Genera: *Entoloma* (Fr.) P. Kumm., *Clitopilus* (Fr. ex Rabenh.) P. Kumm.

The family *Entolomataceae* includes the gasteroid genus *Richoniella* Costantin & L.M. Dufour, the secotioid *Rhodogaster* E. Horak, and the cyphelloid genus *Rhodocybella* T.J. Baroni & R.H. Petersen, all now incorporated into the genus *Entoloma*.

Key to the Genera

1. Spores angular when seen in all views	1. *Entoloma*
1. Spores pustulate or with longitudinal ribs, appearing angular in polar view only	2. *Clitopilus*

M.E. Noordeloos and G.M. Gates, *The Entolomataceae of Tasmania*,
Fungal Diversity Research Series 22, DOI 10.1007/978-94-007-4679-4_2,

Synopsis of the Tasmanian Species

Subgenus *Entoloma*
Section *Entoloma*
Subsection *Entoloma*
1. *Entoloma baronii* G.M. Gates & Noordel.

Section *Albida* Noordel.
2. *E. cretaceum* G.M. Gates & Noordel.
3. *E. albomagnum* G.M. Gates & Noordel.

Subsection *Gelatinosa* Noordel.
4. *E. gelatinosum* E. Horak

Section *Nitida* (Romagn.) Noordel.
5. *E. coeruleogracile* G.M. Gates & Noordel.
6. *E. contrastans* G.M. Gates & Noordel.
7. *E. gracilior* Noordel. & G.M. Gates
8. *E. haastii* G. Stev.

Section *Fibrillosa* Noordel.
9. *E. kermandii* G.M. Gates & Noordel.
10. *E. perbloxamii* Noordel., D.L.V. Co-David, G.M. Gates & Morgado
11. *E. fuligineoviolaceum* G.M. Gates & Noordel.

Section *Calliderma* (Romagn.) Noordel.
12. *E. indigoticoumbrinum* G.M. Gates & Noordel.
13. *E. coeruleomagnum* G.M. Gates & Noordel.

Section *Luteifolia* Noordel.
14. *E. manganaense* G.M. Gates & Noordel.
15. *E. mathinnae* G.M. Gates, B.M. Horton & Noordel.

Section *Nolanidea* (Fr.) Quél.
16. *E. saundersii* (Fr.) Sacc.
17. *E. clypeatum* (L.) P. Kumm.

Subgenus *Inocephalus* Noordel.
Section *Erophila* (Romagn.) Noordel.
18. *E. plebejum* (Kalchbr.) Noordel.

Subgenus *Nolanea* (Fr.: Fr.) Noordel.
Section *Austrofernandae* Noordel.
19. *E. chrysopus* G.M. Gates & Noordel.
20. *E. maldea* G.M. Gates & Noordel.

Section *Papillata* (Romagn.) Noordel.
21. *E. obscureotenax* G.M. Gates & Noordel.
22. *E. psilocyboides* G.M. Gates & Noordel.
23. *E. phaeophthalmum* Noordel. & G.M. Gates

24. *E. fumosopruinosum* G.M. Gates & Noordel.
25. *E. fibrosopileatum* G.M. Gates & Noordel.

Section *Cosmeoexonema* (Largent & Thiers) Noordel.
26. *E. fuligineopallescens* G.M. Gates & Noordel.
27. *E. austronitens* Noordel. & G.M. Gates
28. *E. sericeum* (Bull.: Fr.) Quél.
29. *E. readiae* G. Stev.
29a. *E. readiae* var. *readiae*
29b. *E. readiae* var. *sulphureum* (E. Horak) Noordel. & G.M. Gates
30. *E. convexum* G. Stev.
31. *E. aromaticum* E. Horak
31a. *E. aromaticum* var. *aromaticum*
31b. *E. aromaticum* var. *aromaticellum* (E. Horak) G.M. Gates & Noordel.
32. *E. blandiodorum* E. Horak

Section *Staurospora* (Largent & Thiers) Noordel.
33. *E. conferendum* (Britzelm.) Noordel.
34. *E. brevispermum* G.M. Gates & Noordel.
35. *E. stellatum* G.M. Gates & Noordel.
36. *E. procerum* G. Stev.

Section *Lepiotoidei* Noordel. & G.M. Gates
37. *E. lepiotoides* G.M. Gates & Noordel.
38. *E. sepiaceovelutinum* G.M. Gates & Noordel.
39. *E. strigosum* G.M. Gates & Noordel.

Subgenus *Leptonia* (Fr.: Fr.) Noordel.
Section *Leptonia* (Fr.: Fr.) Noordel.
40. *E. panniculus* (Berk.) Sacc.
41. *E. tomentosolilacinum* G.M. Gates & Noordel.
42. *E. violascens* G.M. Gates & Noordel.
43. *E. endotum* Noordel. & G.M. Gates

Subgenus *Trichopilus* (Romagn.) Noordel.
44. *E. porphyrescens* E. Horak

Subgenus *Cyanula* (Romagn.) Noordel.
Stirps *Serrulatum*
45. *E. tasmanicum* G.M. Gates & Noordel.
46. *E. splendidum* Noordel. & G.M. Gates
47. *E. violaceocoeruleum* Noordel. & G.M. Gates
48. *E. natalis-domini* G.M. Gates & Noordel.
49. *E. sassafras* G.M. Gates & Noordel.
50. *E. viridomarginatum* (Cleland) E. Horak

Stirps *Transmutans*
51. *E. transmutans* G.M. Gates & Noordel.

Stirps *Austroprunicolor*
52. *E. austroprunicolor* G.M. Gates & Noordel.

Stirps *Roseum*
53. *E. austroroseum* G.M. Gates & Noordel.
54. *E. carminicolor* G.M. Gates & Noordel.

Stirps *Obscureovirens*
55. *E. obscureovirens* G.M. Gates & Noordel.

Stirps *Albidae*
56. *E. albidosimulans* G.M. Gates & Noordel.
57. *E. totialbum* G.M. Gates & Noordel.

Stirps *Roseoluteum*
58. *E. roseoluteolum* G.M. Gates & Noordel.

Stirps *Sarcitulum*
59. *E. aurantiolabes* G.M. Gates & Noordel.
60. *E. rufobasis* G.M. Gates & Noordel.
61. *E. stramineopallescens* G.M. Gates & Noordel.

Stirps *Phaeomarginatum*
62. *E. phaeomarginatum* E. Horak
63. *E. persimile* E. Horak
63a. *E. persimile* var. *persimile*
63b. *E. persimile* var. *macrosporum* Noordel. & G.M. Gates

Stirps *Amarum*
64. *E. amarum* G.M. Gates & Noordel.
65. *E. moabus* Noordel. & G.M. Gates
66. *E. obscureogracile* Noordel. & G.M. Gates
67. *E. saponicum* G.M. Gates & Noordel.

Stirps *Formosum*
68. *E. pyropus* Noordel. & G.M. Gates

Stirps *Indistinctum*
69. *E. indistinctum* Noordel. & G.M. Gates
69a. *E. indistinctum* var. *indistinctum*
69b. *E. indistinctum* var. *macrosporum* Noordel. & G.M. Gates
70. *E. austrosarcitulum* Noordel. & G.M. Gates

Stirps *Melanocephalum*
71. *E. melanocephalum* G. Stev.
72. *E. discrepans* Noordel. & G.M. Gates
73. *E. purpureofuscum* Noordel. & G.M. Gates
74. *E. duplocoloratum* E. Horak

Stirps *Asprellum*
75. *E. albidocoeruleum* G.M. Gates & Noordel.

76. *E. asprellopsis* G.M. Gates & Noordel.
77. *E. melanophthalmum* G.M. Gates & Noordel.
78. *E. tenuicystidiatum* G.M. Gates & Noordel.
79. *E. griseosquamulosum* G.M. Gates & Noordel.

Stirps *Incanum*
80. *E. rodwayi* (Massee) E. Horak

Section *Griseorubida* (Romagn.) Noordel.
81. *E. cystidiosum* G.M. Gates & Noordel.

Subgenus *Claudopus* (Gill.) Noordel.
Section *Claudopus* (Gill.) Noordel.
82. *E. pitereka* Noordel. & G.M. Gates

Section *Undati* (Romagn.) Noordel.
83. *E. austrorhodocalyx* G.M. Gates & Noordel.
84. *E. lillirium* Noordel. & G.M. Gates
85. *E. percrinitum* G.M. Gates & Noordel.

Subgenus *Pouzarella* (Mazzer) Noordel.
Section *Dysthales* (Mazzer) Noordel.
86. *E. farinosum* (Largent & Skye Moore) Noordel. & Gates

Species of uncertain position
87. *E. camarophyllus* G.M. Gates & Noordel.
88. *E. choanomorphum* G.M. Gates & Noordel.
89. *E. uliginicola* E. Horak

Subgenus *Richoniella* (Costantin & L.M. Dufour) Noordel.
90. *E. gasteromycetoides* Noordel. & Co-David

Genus *Clitopilus* (Fr. ex Rabenh.) P. Kumm.
Section *Clitopilus*
1. *C.* cf. *prunulus* (Scop.) P. Kumm.
2. *C. hobsonii* (Berk.) P.D. Orton

Section *Rhodocybe* (R. Maire) Contu
3. *C. tasmanicus* (T.J. Baroni & G.M. Gates) Noordel. & Co-David
4. *C. fuligineus* (E. Horak) Noordel. & Co-David
5. *C. acerbus* Noordel. & Co-David

Section *Rufobrunnei* (T.J. Baroni) Contu
6. *C. lateritius* (T.J. Baroni & G.M. Gates) Noordel. & Co-David
7. *C. pseudopiperitus* (T.J. Baroni & G.M. Gates) Noordel. & Co-David

Section *Decurrentes* (Konrad & Maubl.) Contu
8. *C. pallidogriseus* (T.J. Baroni & G.M. Gates) Noordel. & Co-David
9. *C. reticulatus* (Cleland) Noordel. & Co-David

Section *Crepidotoides* (Singer) Contu
10. *C. conchatus* (E. Horak) Noordel. & Co-David

Keys to the Species of Entoloma

1. Basidiocarps gasteromycetoid, like a small white truffle (subgenus *Richoniella*)	**90. Entoloma gasteromycetoides**
2. Basidiocarps agaricoid	**Main Entoloma Key**

Reference

Horak, E. (2008). Agaricales of New Zealand 1: Pluteaceae – Entolomataceae. In *The fungi of New Zealand* (Fungal Diversity Research Series, no. 19, Vol. 5). Hong Kong: Fungal Diversity Press.

Chapter 3
Genus *Entoloma*

Abstract This chapter deals with the genus *Entoloma* in Tasmania. Dichotomous keys to the species precede a taxonomic treatment of all known Tasmanian *Entoloma* taxa. Each species is presented with full synonymy, reference to existing literature, followed by detailed macroscopic and microscopic descriptions, data on habitat, distribution, and phenology, usually followed by notes on taxonomic position, similar species, and other relevant facts. Line drawings are made for each species, consisting of a habit drawing, as well as diagnostic microscopic features, such as spores, basidia, cystidia, and pileipellis structures. For almost every species, one or more coloured photographs are included. In total, 90 species of Entoloma have been accepted, of which 73 are known only from Tasmania. Other species appear to be similar to taxa described from New Zealand. Only a few taxa match those known from Europe, and these are probably alien, introduced with plant material. Full references to the cited literature are given.

Entoloma (Fr.) P. Kumm. emend Noordel. & Co-David 2009.

Agaricus tribus *Entoloma* Fr., Epicrisis: 143. 1838; *Agaricus* subgenus *Entoloma* (Fr.) Rabenh., Deutschl. KryptogFl. 1: 508. 1844; *Entoloma* (Fr.) P. Kumm., Führ. Pilzk.: 23. 1871. — *Nolanea* (Fr.) P. Kumm., Führ. Pilzk.: 23. 1871. — *Leptonia* (Fr.) P. Kumm., Führ. Pilzk.: 23. 1871. — *Eccilia* (Fr.) P. Kumm., Führ. Pilzk.: 23. 1871. — *Claudopus* Gill., Hymenom. Fr.: 426. 1876. — *Rhodophyllus* Quél., Enchiridion: 56. 1886. — *Latzinia* O. Kuntze, Rev. Gen. Pl. 2: 857. 1891. — *Leptoniella* Earle, Bull. N.Y. Bot. Gdn 5: 424. 1909. — *Lanolea* Nieuwl., Amer. Midl. Nat. 4: 381. 1916. — *Richoniella* Costantin et Dufour, Nouv. Fl. Champ. ed. 5. 203. 1916. — *Pouzaromyces* Pilát, Acta Mus. Nat. Prag. (B) 9 (2): 60. 1953. — *Rhodogaster* E. Horak, Sydowia 17: 190. 18 Jun 1964 '1963'). — *Alboleptonia* Largent & Benedict, Mycologia 63: 439. 1970. — *Pouzarella* Mazzer, Bibltca Mycol. 46: 69. 1978. — *Rhodocybella* T.J. Baroni & R.H. Petersen, Mycologia 79(3): 358. 1987. — *Trichopilus* (Romagnesi) P. D. Orton, Mycologist 5(4): 175. Oct 1991. — *Paraleptonia* (Romagnesi ex M. E. Noordeloos) P. D. Orton, Mycologist

M.E. Noordeloos and G.M. Gates, *The Entolomataceae of Tasmania*,
Fungal Diversity Research Series 22, DOI 10.1007/978-94-007-4679-4_3,

5(4): 174. Oct 1991. — *Clitopiloidea* (Romagn.) Largent, Entolomatoid fungi of the Western United States and Alaska (Eureka): 31. 1994. — *Calliderma* (Romagn.) Largent, Entolomatoid fungi of the Western United States and Alaska (Eureka): 31. 1994. — *Fibropilus* (Noordel.) Largent, Entolomatoid fungi of the Western United States and Alaska (Eureka): 32. 1994. — *Paraeccilia* Largent, Entolomatoid fungi of the Western United States and Alaska (Eureka): 368. 1994. — *Entocybe* Baroni et al., North American Fungi 6(12): 8. 2011.

Lectotype (Donk, Beih. Nova Hedwigia 5: 95. 1962): *Entoloma prunuloides* (Fr.) Quél.

Main characters: basidiomata agaricoid, secotioid or gasteroid; agaricoid habit variable: tricholomatoid, mycenoid, collybioid, clitocyboid, omphalioid or pleurotoid; lamellae almost free, adnexed to adnate, adnate-emarginate or adnate-decurrent; veil none or rarely present in the shape of traces of cortina at margin of pileus; spores thin- to thick-walled, angular in all views; facets formed by completely or incompletely interconnecting ridges; spore print pink or pinkish brown; basidia usually 4-, rarely 2-spored; cheilocystidia present or absent; pleurocystidia rare and only occurring in combination with cheilocystidia; hymenophoral trama regular; subhymenium usually thin, filamentous, in some cases (subgen. *Pouzaromyces*) well differentiated and subcellular, rarely somewhat gelatinized (subgenus *Entoloma*); pileipellis varying from a simple cutis and then sometimes with a well-differentiated, subcellular subpellis, to a trichoderm of cylindrical or inflated hyphal tips or a trichoderm of well differentiated, attenuate, encrusted hyphae hairs, or a hymeniderm or epithelium of clavate to globose elements; pigment intracellular-diffuse or intracellular-granular, encrusting or parietal; brilliant granules sometimes very abundant in the trama; vascular hyphae present or absent; clamp-connections absent or present.

Habitat & distribution: saprotrophic or mycorrhizal, rarely parasitic, in soil and raw humus or on plant remains and woody debris, rarely on living plants or on other fungi. Widespread, cosmopolitan, occupying almost the whole range of climate conditions, from the arctic through the boreal zone into the Mediterranean and tropics, in wet and dry climates.

Entoloma, in this phylogenetic concept, includes the secotioid and gasteroid genera *Rhodogaster* and *Richoniella*, and the cyphelloid genus *Rhodocybella*.

Main Entoloma Key

(For the convenience of the users, some very similar New Zealand species, as described by Horak (2008) are included. Characteristics can be found in the appendix.)

1. Fruit bodies crepidotoid with lateral stipe often strongly reduced or lacking, then pileus sessile on substrate	**82. E. pitereka**
1. Fruit bodies with well-developed, central stipe	2.

2.	Pileus and stipe densely hairy with ochre-grey hairs on grey-brown background, pileipellis and stipitipellis a trichoderm with long, heavily incrusted, septate hyphae, spores up to 20 μm long, nodulose-angular (subgenus Pouzarella)	**86. E. farinosum**
2.	Pileus and stipe different; spores smaller, not or rarely nodulose-angular	3.
3.	Pileus white, beige or yellow	**Key one**
3.	Pileus differently coloured	4.
4.	Pileus pink, pinkish red, carmine or with green or olivaceous tinges	**Key two**
4.	Pileus differently coloured	5.
5.	Pileus and/or stipe with blue, blue-grey or pink, or violaceous with brown tinges or uniformly pale brown but not as pallid as beige	6.
5.	Pileus and stipe pale to dark yellow- or red-brown, purplish red, brown-grey, dark brown, grey or blackish	7.
6.	Habit tricholomatoid or mycenoid; pileus conical to convex with umbo; clamp-connections present; spores often isodiametric or subisodiametric, with weak angles	**Key three**
6.	Habit collybioid rarely mycenoid; pileus conico-convex to plano-convex, usually slightly to distinctly depressed at centre, rarely with small acute papilla; clamp-connections absent; spores usually heterodiametric with pronounced angles	**Key four**
7.	Habit tricholomatoid or mycenoid; pileus conical to convex, usually umbonate, never distinctly umbilicate	**Key five**
7.	Habit collybioid or omphalioid	**Key six**

Key one: Pileus white, beige or yellow

1.	Pileus pure white when fresh	2.
1.	Pileus beige, greyish blonde yellow when fresh	6.
2.	Habit tricholomatoid with relatively thick flesh; spores isodiametric to subisodiametric, $Qav = 1.0–1.1$	3.
2.	Habit collybioid, mycenoid, or omphalioid, relatively thin-fleshed; spores heterodiametric, $Qav > 1.2$	5.
3.	Appearing in spring under cultivated fruit trees, probably introduced; pileus with micaceous patches; spores 8–12 μm in diameter	**16. E. saundersii**

3. Appearing in summer and autumn in native forest; pileus without micaceous patches, spores 6–8 μm in diameter	4.
4. Fruit bodies small; pileus 15–45 mm, chalky white, opaque; stipe hyaline-watery white, indistinctly fibrous; smell rancid-farinaceous	**2. E. cretaceum**
4. Fruit bodies large; pileus up to 100 mm diam.; stipe densely silvery white fibrillose; smell soapy at first	**3. E. albomagnum**
5. Clamp-connections abundant; lamella edge heterogeneous	**57. E. totialbum**
5. Clamp-connections absent or very rare; lamella edge sterile	**56. E. albidosimulans**
6. Pileus and stipe strong yellow; pigment encrusting	**29b. E. readiae var. sulphureum**
6. Pileus beige or greyish yellow	7.
7. Fruit bodies omphalioid with deeply decurrent gills	**83. E. austrorhodocalyx**
7. Fruit bodies collybioid	8.
8. Pileus and stipe concolorous greyish blonde, strong spermatic or bleach odour	**61. E. stramineopallescens**
8 Pileus beige or whitish, stipe blue contrasting with pileus; without distinct odour	9.
9. Spores isodiametric	**6. E. contrastans**
9. Spores heterodiametric	**75. E. albidocoeruleum**

Key two: Pileus pink, pinkish red, carmine or with green or olivaceous tinges

1. Pileus green or with olivaceous tinges	2.
1. Pileus pink, pinkish red or carmine	4.
2. Pileus and stipe very dark brown-olivaceous	**55. E. obscureovirens**
2. Pileus green to dark green; stipe green or yellow	3.
3. Lamella edge green or blue-green	**50. E. viridomarginatum**
3. Lamella edge concolorous with sides	**80. E. rodwayi**
4. Pileus pink to pinkish red	5.
4. Pileus and stipe deep carmine-red	**54. E. carminicolor**
5. Stipe pink	**53. E. austroroseum**
5. Stipe yellowish	**58. E. roseoluteolum**

Key three: Pileus and/or stipe with blue, blue-grey or violaceous tinges

1.	Pileus without distinct blue or violaceous colour, rather some shade of yellow-, red- or purple-brown, pinkish brown, brown-grey, grey or black; stipe blue, blue-grey or violaceous	2.
1.	Pileus with pronounced blue or violaceous colour, sometimes mixed with brown, stipe with or without blue or violaceous tinges	7.
2.	Lamellae deep yellow	3.
2.	Lamellae without yellow tinges	4.
3.	Pileus dark red-brown; stipe blue	**14. E. manganaense**
3.	Pileus pale brown; stipe white or with slight violet tinges	**15. E. mathinnae**
4.	Pileus strongly viscid	**4. E. gelatinosum**
4.	Pileus dry, innately fibrillose to fibrous	5.
5.	Lamellae white then flesh-coloured; pileus greyish ruby to brown with purple tinges (plum-coloured) or pinkish grey	**9. E. kermandii**
5.	Lamellae dark with lilac or violaceous tinges	6.
6.	Slender species; pileus 5–20 mm diam., translucently striate, smooth, fibrillose; lamellae grey-lilac; spores 7–10 × 6–8 µm	**7. E. gracilior**
6.	Robust species; pileus 20–60 mm diam., tomentose, not translucently striate; lamellae dark brown-violet to violet; spores 5.5–7.5(–8) × 5.5–6.5(–7) µm	**11. E. fuligineoviolaceum**
7.	Pileus strongly viscid	**4. E. gelatinosum**
7.	Pileus dry or slightly lubricous	8.
8.	Pileus velutinous, tomentose or minutely squamulose	9.
8.	Pileus glabrous, fibrillose or fibrous	15.
9.	Robust species with tricholomatoid or mycenoid habit; pileus velutinous or minutely squamulose	10.
9.	Slender, mycenoid species with tomentose to minutely squamulose pileus	13.
10.	Pileus deep blue or dark violet-blue	11.

10.	Pileus dark brown mixed with indigo or violaceous tinges	12.
11.	Pileus deep blue; lamellae pink with violaceous tinge; stipe blue, innately fibrous; pileipellis a palisadoderm of erect hyphae with cystidioid terminal elements	**13. E. coeruleomagnum**
11.	Pileus deep blue to violet-blue, tomentose then minutely squamulose; lamellae golden brown or orange-brown; stipe fibrillose-squamulose; pileipellis a trichoderm	**43. E. endotum**
12.	Pileipellis a hymeniderm (calliderm); pileus brown mixed with indigo-blue	**12. E. indigoticoumbrinum**
12.	Pileipellis a trichoderm; pileus with violaceous tinges	**11. E. fuligineoviolaceum**
13.	Pileus and stipe deep blue; stipe fibrillose-squamulose all over	**40. E. panniculus**
13.	Pileus and stipe with violaceous tinges	14.
14.	Stipe polished; spores 8–12 × 7–9.5 μm	**42. E. violascens**
14.	Stipe fibrillose; spores very small (5.5–)6–7 × 5.5–7 μm	**41. E. tomentosolilacinum**
15.	Habit mycenoid, very slender, with a thin, relatively long stipe	16.
15.	Habit tricholomatoid or mycenoid, but then relatively robust	18.
16.	Pileus very pale, almost white, with slightly darker brown centre	**6. E. contrastans**
16.	Pileus dark brown or blue	17.
17.	Pileus very dark brown, occasionally with violaceous tinge	**7. E. gracilior**
17.	Pileus and stipe deep blue	**5. E. coeruleogracile**
18.	Pileus and stipe deep blue, or covered in blue fibrils	**10. E. perbloxami**
18.	Pileus brown mixed with violaceous blue or indigo	19.
19.	Lamellae dark brown-violet	**11. E. fuligineoviolaceum**
19.	Lamellae pale, pink, sordid pink, or with faint blue or violaceous tinge	20.
20.	Pileus strongly radially fibrous, centre sometimes somewhat tomentose; stipe equally fibrous	**9. E. kermandii**
20.	Pileus glabrous, greasy to touch, later on somewhat innately fibrous or rugulose; stipe subfibrillose, often shiny and appearing smooth	**8. E. haastii**

Key four: Species with collybioid, rarely mycenoid habit and blue or violaceous tinges in any part of the basidiocarp.

1.	Pileus blue, blue-pink, blue-black or violaceous	2.
1.	Pileus blackish brown, grey-brown, dark brown, red-brown, purple-brown, purple-black, yellow-brown, pale brown almost white, or grey-dark grey	12.
2.	Spores many-angled, nodulose; stipe with strigose hairs at base	**89. E. uliginicola**
2.	Spores not nodulose; stipe without strigose hairs	3.
3.	Spores (5.5–)6.0–7.0 × 5.5–7.0 μm, Q = 1.0–1.2, isodiametric to subisodiametric, with very thin walls, slightly but distinctly 7–many-angled in side-view	**41. E. tomentosolilacinum**
3.	Spores larger, heterodiametric and more pronouncedly angled; basidiocarps differently coloured	4.
4.	Spores 7.5–8 × 5.5–6 μm	**72. E. discrepans (form with small spores)**
4.	Spores larger	5.
5.	Spores from 8–10.5 μm	6.
5.	Spores larger, 10–12 μm long or more	8.
6.	Fruit bodies blue, remaining so, lamellae blue with concolorous or brown edge	**72. E. discrepans**
6.	Pileus initially blue, but soon with either pinkish or violaceous tinges; lamellae never deep blue	7.
7.	Pileus dark grey-blue, when mature showing delicate pink-purple tinges between the squamules and fibrils, not or indistinctly translucently striate; stipe bluish pink with age; lamellae edge pinkish brown	**51. E. transmutans**
7.	Pileus blue-violet at first, soon violaceous pink with blackish blue centre, deeply translucently striate; lamella edge violet-red	**47. E. violaceocoeruleum**
8.	All parts of fruit body staining orange when bruised; pileus and stipe with violet-blue tinges that disappear with age	**59. E. aurantiolabes**
8.	No orange staining; pileus persistently blue or violaceous	9.
9.	Lamella edge concolorous with sides; pileus typically bi-coloured with dark central spot	**77. E. melanophthalmum**
9.	Lamella edge coloured	10.
10.	Pileus blue, distinctly translucently striate	**46. E. splendidum**

10. Pileus dark blue-black, not translucently striate — 11.
11. Cheilocystidia lageniform with short or long, occasionally moniliform neck (**E. atrellum** E. Horak, extralimital, described from New Zealand, see Horak, 2008)
11. Cheilocystidia cylindrical to clavate — **71. E. melanocephalum**
12. Spores in majority >10 μm long — 13.
12. Spores 8–10(–11) μm long — 19.
13. Pileus translucently striate, at least at margin — 14.
13. Pileus not translucently striate — 17.
14. Stipe grey or grey-brown with slight blue hue (steel blue) — 15.
14. Stipe distinctly blue, blue-grey, dark grey or grey-violet — 16.
15. Lamella edge brown — **70. E. austrosarcitulum**
15. Lamella edge concolorous with sides — **69b. E. indistinctum var. macrosporum**
16. Pileus brown; stipe blue to blue-grey; lamella edge brown — **76. E. asprellopsis**
16. Pileus dark grey to grey-black; stipe grey-violet; edge concolorous with sides — **79. E. griseosquamulosum**
17. Stipe white, often with delicate violet-grey tinge; pileus plum-coloured — **52. E. austroprunicolor**
17. Stipe violet-grey to blackish blue; pileus very dark brown-black, occasionally with violet or carmine tinges — 18.
18. Stipe delicate violet-grey; pileus entirely tomentose, not translucently striate; taste mild — **66. E. obscureogracile**
18. Stipe grey-blue to purplish grey; pileus minutely squamulose at centre, fibrillose towards margin, translucently striate at least at margin; taste often distinctly bitter — **64. E. amarum**
19. Pileus distinctly translucently striate — 20.
19. Pileus not or faintly translucently striate — 22.
20. Cheilocystidia absent; pileus pale isabel to pale yellow-brown; stipe very pale blue (**E. consanguineum** E. Horak**,** extralimital, described from New Zealand, see Horak, 2008)
20. Cheilocystidia present; pileus and stipe darker — 21.
21. Pileus pale red- or yellow-brown, sometimes with greyish tinge — **74. E. duplocoloratum**
21. Pileus purple-brown — **73. E. purpureofuscum**

22.	Lamella edge concolorous with sides	**65. E. moabus**
22.	Lamella edge coloured	23.
23.	Stipe dark grey-violet to bluish black with distinct yellow-green coloured base, fibrous; lamella edge brown-black, serrulate	**45. E. tasmanicum**
23.	Stipe blue to blue-grey, without yellow-green tinges, polished; lamella edge with blue or violaceous tinges	24.
24.	Pileus conical with acute papilla, dark brown velutinous; lamellae edge deep blue; stipe deep blue	**48. E. natalis-domini**
24.	Pileus umbilicate, entirely black at first then yellow-brown with small but very conspicuous blackish brown, erect squamules; lamella edge deep carmine to violet-black; stipe dark grey, polished	**49. E. sassafras**

Key five: Habit mycenoid or tricholomatoid

1.	Habit tricholomatoid, relatively thick-fleshed	2.
1.	Habit mycenoid, relatively thin-fleshed or robust	6.
2.	Pileus and stipe with red-brown (porphyraceous) colour; cheilocystidia present, capitate	**44. E. porphyrescens**
2.	Pileus and stipe differently coloured; cheilocystidia absent	3.
3.	Lamellae deep yellow	**15. E. mathinnae**
3.	Lamellae white then sordid pink	4.
4.	Pileus rather light brown; spores small, 6.5–8 µm in diam., very thin-walled, many-angled in side-view; in native forest	**1. E. baronii**
4.	Pileus either pale brown or strong brown, spores larger and thick-walled; in cultivated land (parks, gardens), under fruit trees, etc.	5.
5.	Pileus usually rather pale brown, with micaceous patches, especially when young; spores 10–13 × 9–11 µm, many-angled	**16. E. saundersii**
5.	Pileus brown, without micaceous patches; spores 9–11 × 7–9.5 µm, 5–7-angled in side-view	**17. E. clypeatum**
6.	Spores at least in part, cuboid or cruciform	7.
6.	Spores differently shaped, either isodiametric or heterodiametric	11.

7.	Spores cuboid	**36. E. procerum**
7.	Spores cruciform	8.
8.	Spores small, 6.5–9 × 5.5–8.0 μm	**34. E. brevispermum**
8.	Spores larger	9.
9.	Cheilocystidia present, lageniform to fusiform, 30–90 μm long	**35. E. stellatum**
9.	Cheilocystidia absent	10.
10.	Spores very irregularly shaped, some 5–6-angled, some more or less cruciform; pigment in pileipellis intracellular, both diffuse and in form of agglutinated golden brown granules and minutely encrusting	**19. E. chrysopus**
10.	Spores all distinctly cruciform, pigment diffuse, intracellular, golden brown granules absent	**33. E. conferendum**
11.	Pileus glabrous, radially fibrillose, wrinkled or shiny-micaceous, frosty	12.
11.	Pileus tomentose, velutinous, strigose, or squamulose	27.
12.	Smell strong, aromatic, like fruit drops or bubble-gum	13.
12.	Smell different (farinaceous, cucumber, spermatic, rancid or indistinct)	16.
13.	Smell rather like bubble-gum; spores 9.5–14 × 6.5–9 μm, spores irregularly shaped, often twisted and almost cruciform	**19. E. chrysopus**
13.	Smell like fruit drops, rather strong; pileus reddish brown or yellowish brown, markedly translucently striate; spores smaller than above and rather regularly 5–7-angled, heterodiametric to subisodiametric in side view	14.
14.	Rather robust, dark brown species, margin of pileus hardly paler than centre	**32. E. blandiodorum**
14.	Thin-fleshed, light to moderately dark brown species; pileus distinctly translucently striate, much paler at margin than at centre	15.
15.	Spores 9–11 × 7–8 μm, Q = 1.25–1.5, heterodiametric	**31a. E. aromaticum var. aromaticum**
15.	Spores 7.5–8.5 × (5.5–)6–6.5 μm, Q = 1.15–1.3, isodiametric to subheterodiametric	**31b. E. aromaticum var. aromaticellum**
16.	Spores isodiametric, Qav = 1.0–1.1	17.
16.	Spores heterodiametric, Qav = 1.2 or more	20.

17.	Spores large, average 10–13 μm in length	**30. E. convexum**
17.	Spores smaller, up to 10 μm in length	18.
18.	Spores 6–7 × 5.5–7 μm	**26. E. fuligineopallescens**
18.	Spores 7–10 × 7–10 μm	19.
19.	Pileus very dark brown; stipe dark brown, fibrous; lamellae dark brown; pigment in pileipellis only, coarsely brown-encrusting	**28. E. sericeum**
19.	Pileus moderately dark brown, pallescent on drying, becoming strongly silky-shiny, lustrous; stipe moderately dark brown, subfibrillose to polished; lamellae moderately dark brown-pink; pigment minutely encrusting in suprapellis; abundant brown intracellular pigment in subpellis and upper pileitrama	**27. E. austronitens**
20.	Basidiocarps tough; pileus dark brown, very hygrophanous, becoming a bright yellow; lamellae dark brown, hardly pink when mature	**21. E. obscureotenax**
20.	Basidiocarps not particularly tough; pileus not drying yellow; lamellae paler, distinctly pinkish when mature	21.
21.	Cheilocystidia present	22.
21.	Cheilocystidia absent	25.
22.	Pileus fibrillose to tomentose, often with frosty or micaceous patches	23.
22.	Pileus smooth	24.
23.	Habit robust mycenoid; pileus opaque, fibrillose with micaceous patches; stipe pallid, almost white, fibrillose-striate	**18. E. plebejum**
23.	Habit slender, mycenoid; pileus fibrillose at margin, centre tomentose or distinctly frosted; stipe greyish, about same colour as pileus, subpolished or finely striate	**24. E. fumosopruinosum**
24.	Cheilocystidia large, slender, lageniform, 50–160 × 4–9 μm; lamellae white then flesh pink	**23. E. phaeophthalmum**
24.	Cheilocystidia smaller, clavate, rarely lageniform, 30–65 × 5–15 μm; lamellae chocolate brown	**22. E. psilocyboides**
25.	Pileus distinctly radially fibrillose, very minutely and darkly velutinous to squamulose at centre, sometimes radially ribbed at margin; pigment intracellular; clamp-connections present	**25. E. fibrosopileatum**

25.	Pileus more or less smooth or innately radially fibrillose; pigment intracellular, diffusely and also as golden brown, agglutinated clots in pileipellis; and minutely encrusting in lower parts of pileipellis and upper pileitrama; clamp-connections absent	26.
26.	Spores 7–10×6–8 μm, regularly 5–7-angled; smell slightly to distinctly farinaceous	**20. E. maldea**
26.	Spores 9–13(−14)×6.5–9 μm, 5–7-angled, often twisted, irregularly shaped and almost cruciform; smell slightly to distinctly aromatic, like bubblegum	**19. E. chrysopus**
27.	Spores 7–9×7–8.5 μm	**37. E. lepiotoides**
27.	Spores larger	28.
28.	Pileus with strigose tufts of erect hairs; cheilocystidia absent; spores 10–12×7–8 μm	**39. E. strigosum**
28.	Pileus velutinous, breaking up into small squamules, not hygrophanous; cheilocystidia present, spores 10–14×7–9 μm	**38. E. sepiaceovelutinum**

Key six: Pileus and stipe pale to dark brown, yellow-brown, red-brown, brown-grey or blackish

1.	Habit omphalioid with distinctly decurrent lamellae	2.
1.	Habit collybioid	6.
2.	Lamella edge with conspicuous cheilocystidia	3.
2.	Lamella edge fertile without cystidia	4.
3.	Fruit bodies resembling a *Camarophyllus* with distant, thickish, waxy lamellae; cheilocystidia clavate,	**87. E. camarophyllus**
3.	Fruit bodies typically omphalioid with thinner lamellae; cheilocystidia lageniform to fusiform, strong sulphurous odour	**88. E. choanomorphum**
4.	Pileus pale to moderately dark brown, smooth, distinctly translucently striate; lamellae distant	**83. E. austrorhodocalyx**
4.	Pileus dark grey-brown, fibrillose to tomentose, not translucently striate; lamellae not distant	5.
5.	Pileus densely tomentose then squamulose with minute, erect squamules	**84. E. lillirium**
5.	Pileus hoary with loose fibrils, somewhat concentrically zoned	**85. E. percrinitum**

6.	Pileipellis a cutis of narrow hyphae with fine incrustations	**29a. E. readiae var. readiae**
6.	Pileipellis a trichoderm, at least at centre, of inflated elements with intracellular pigment	7.
7.	Lamella edge distinctly coloured	8.
7.	Lamella edge concolorous with sides or very faintly coloured brown	15.
8.	Pileus yellow-brown, tan-brown, or bronze-brown	9.
8.	Pileus dark brown-black; sepia-brown, grey-brown or grey	10.
9.	Spores 10–11 × 7.0–8.0 μm, average 10.4 × 7.4 μm; stipe very pale brown, almost white, contrasting with the warm tan-brown pileus, frequently with a conspicuous pinkish orange blotch	**60. E. rufobasis**
9.	Spores 8.5–11 × 5.5–7.5 μm, average 9.1 × 6.5 μm; stipe yellow to grey-yellow, pileus vivid orange-yellow, bronze, stipe base without orange blotch	**68. E. pyropus**
10.	Pileus distinctly translucently striate up to half the radius, squamulose at centre only	11.
10.	Pileus entirely tomentose or squamulose, not translucently striate at least when fresh; stipe base without orange blotch	14.
11.	Spores 8.5–10(–10.5) × 6–7.5 μm	12.
11.	Spores 10–12.5 × 7–9 μm	13.
12.	Lamellae blue-grey with dark blue serrulate edge	(**E. asprelloides,** extralimital, described from New Zealand, see Horak, 2008)
12.	Lamellae grey-pink with brown edge	**63a. E. persimile var. persimile**
13.	Stipe distinctly paler than pileus, buff to pale brown, without orange blotch	(**E. persimile,** as described from New Zealand, notes)
13.	Stipe almost the same colour as pileus, with orange blotch at base	**63b. E. persimile var. macrosporum**
14.	Spores 6–10 × 5–7 μm	**67. E. saponicum**
14.	Spores 10–13 × 7–8 μm	**62. E. phaeomarginatum**
15.	Pileus rather pale brown with darker centre (“eye”)	16.
15.	Pileus brown, reddish brown or ochre-brown/violet-brown	19.

16. Spores small, 8.5–10(–10.5) μm in length	**69a. E. indistinctum var. indistinctum**
16. Spores larger, 10–12(–15) μm in length	17.
17. Strong odour of bleach, cheilocystidia 20–40 × 5–12 μm, clavate	**61. E. stramineopallescens**
17. Odour none	18.
18. Cheilocystidia 40–110 × 5–9 μm, slender, cylindrical; pileus pale brown; stipe white	**78. E. tenuicystidiatum**
18. Cheilocystidia 30–60 × 4–10 μm, cylindrical to clavate; stipe pale grey	**69b. E. indistinctum var. macrosporum**
19. Cheilocystidia absent	**28. E. sericeum**
19. Cheilocystidia present	20.
20. Basidiocarps not staining orange when bruised; cheilocystidia up to 90 μm long, variably shaped from clavate to fusiform, utriform or sphaeropedunculate	**81. E. cystidiosum**
20. Basidiocarps staining orange when bruised; cheilocystidia up to 50 μm long, cylindrical, fusiform or clavate	**59. E. aurantiolabes**

Species Descriptions

Subgenus ***Entoloma*** emend Co & Noordeloos

Agaricus Tribus *Entoloma* Fr., Epicr.: 143. 1838: *Agaricus* subgenus *Entoloma* (Fr.) Rabenh., Deutschl. KryptogFl. 1: 508. 1844: *Entoloma* (Fr.) Kumm., Führ. Pilzk.: 23. 1871; *Rhodophyllus* subgenus *Entoloma* (Fr.) Quél., Enchir.: 57. 1886: *Hyporrhodius* subgenus *Entoloma* (Fr.) Schroet. in Cohn, KryptogFl. Schles. 3(1): 616. 1889. — *Rhodophyllus* subgenus *Romagnesia* Singer in Annls mycol. 41: 3. 1943.

Habit tricholomatoid, mycenoid or collybioid; stipe often fusiform with tapering base; hymenophoral and pileitrama regular, made up of short elements with abundant clamp-connections; spores usually isodiametric to subisodiametric, often small, 5.5–8(–9) μm in diameter, rarely up to 11 μm, thin- or thick-walled, often with weak angles. Pileipellis often distinctly bilayered: suprapellis structures vary from a simple (ixo) cutis to (ixo) trichoderm or hymeniderm; subpellis usually compact, well-differentiated from the underlying trama, made up of very inflated to globose elements.

Lectotype (Donk 1962): *Entoloma prunuloides* (Fr.) Quél.

Section ***Entoloma***

Agaricus sect. *Genuini* Fr., Epicr.: 143. 1838; *Entoloma* sect. *Genuini* (Fr.) Quél. in Mém. Soc. Émul. Montbéliard, sér. II.5: 116. 1872: *Rhodophyllus* sect. *Genuini* Quél., Enchir.: 57. 1886. Lectotype (Donk 1962): *Entoloma prunuloides* (Fr.) Quél.

Pileipellis bilayered, suprapellis an ixocutis of narrow hyphae, subpellis of inflated hyphal elements, well differentiated from pileitrama.

Plate 3.1 *Entoloma baronii* (Photo Michael Pilkington)

1. **Entoloma baronii** G.M. Gates & Noordel., spec. nov.

MycoBank #564492.

Diagnosis: Habit tricholomatoid, robust; pileus up to 100 mm diam, pale brown to greyish brown, subviscid, glabrous; lamellae whitish pink then tinged with brown; stipe 60–67 × 9–17 mm, broadened towards base, pallid, fibrillose; smell and taste farinaceous. Spores 6.0–8.5 × 6.0–8.0 μm, isodiametric, poorly angled to subpustulose; pileipellis bilayered cutis with well-developed subpellis; pigment brown, intracellular; clamp-connections abundant. On the ground in mixed forest.

Holotype: Australia, Tasmania, Bruny Island, Mt Mangana, 43° 22′ S, 147° 17′ E, 17 April 2011, G. Gates E 2292 (HO 561472, isotype in L).

Etymology: named in honour of Dr. Timothy J. Baroni, to acknowledge his contributions to the study of Entolomataceae.

Main characters: habit tricholomatoid; fruit bodies large, pale brown to greyish brown; spores small, isodiametric; pileipellis bilayered (Plate 3.1).

Description:
Pileus 60–100 mm diam, plano-convex with low broad umbo and straight, then uplifted and rimose margin, not hygrophanous, slightly translucently striate at outer margin, light brown to light greyish brown (dark blonde), paler at margin, glabrous, lubricous when wet, very smooth. Lamellae narrowly adnexed, deeply emarginate to free, moderately crowded to crowded, up to 15 mm broad, flesh pink (6A2) when young then tinged with pale brown, with entire, concolorous edge, lamellulae in two tiers. Stipe 60–67 × 9–17 mm, cylindrical, broadened towards base (21–22 mm), pale brown, darkening towards whitish pruinose base, longitudinally fibrillose-striate. Context white, rather firm. Smell and taste farinaceous.

Spores 6.5–8(–8.5) × 6.0–8.0 µm, Q = 1.0–1.2, isodiametric, 6-angled in side-view with rather thin walls, sometimes slightly pustulate. Basidia 32–45 × 8–12 µm, clavate, 4-spored, clamped. Lamella edge fertile, without cystidia. Hymenophoral trama regular, made up of short, barrel-shaped elements, 40–90 × 6–18 µm. Pileipellis bilayered, suprapellis a narrow ixocutis of 3–5 µm wide, cylindrical hyphae; subpellis made up of very inflated cylindrical to barrel-shaped elements, 24–50 × 18–25 µm. Pigment pale brown, intracellular, mainly located in suprapellis; some hyphae with internal encrustations. Pileitrama regular, made up of cylindrical to barrel-shaped elements, 30–140 × 12–30 µm. Stipitipellis a cutis of narrow, cylindrical hyphae, 3–8 µm wide, with pale brown intracellular pigment. Clamp-connections abundant in all tissues (Fig. 3.1).

Habitat & distribution: in wet sclerophyll forest litter, on a small mountain on an island in the D'Entrecasteaux Channel; known only from a single collection.

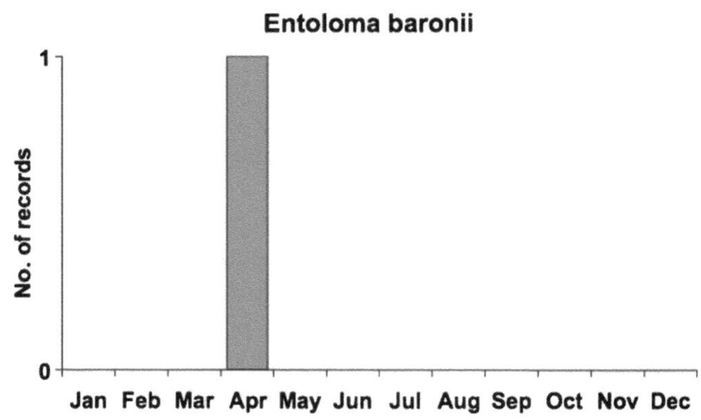

Collection examined. Australia, Tasmania, Bruny Island, Mt Mangana, 43° 22′ S, 147° 17′ E, 17 April 2011, G. Gates E 2292 (holotype, HO 561472).

Entoloma baronii belongs to subgenus *Entoloma* on account of the large, tricholomatoid habit, small, isodiametric spores, and bilayered pileipellis. As such it is close to *E. cretaceum* and *E. albomagnum*, differing from both by the distinct pigmentation of pileus and stipe. The light medium brown to grey-brown

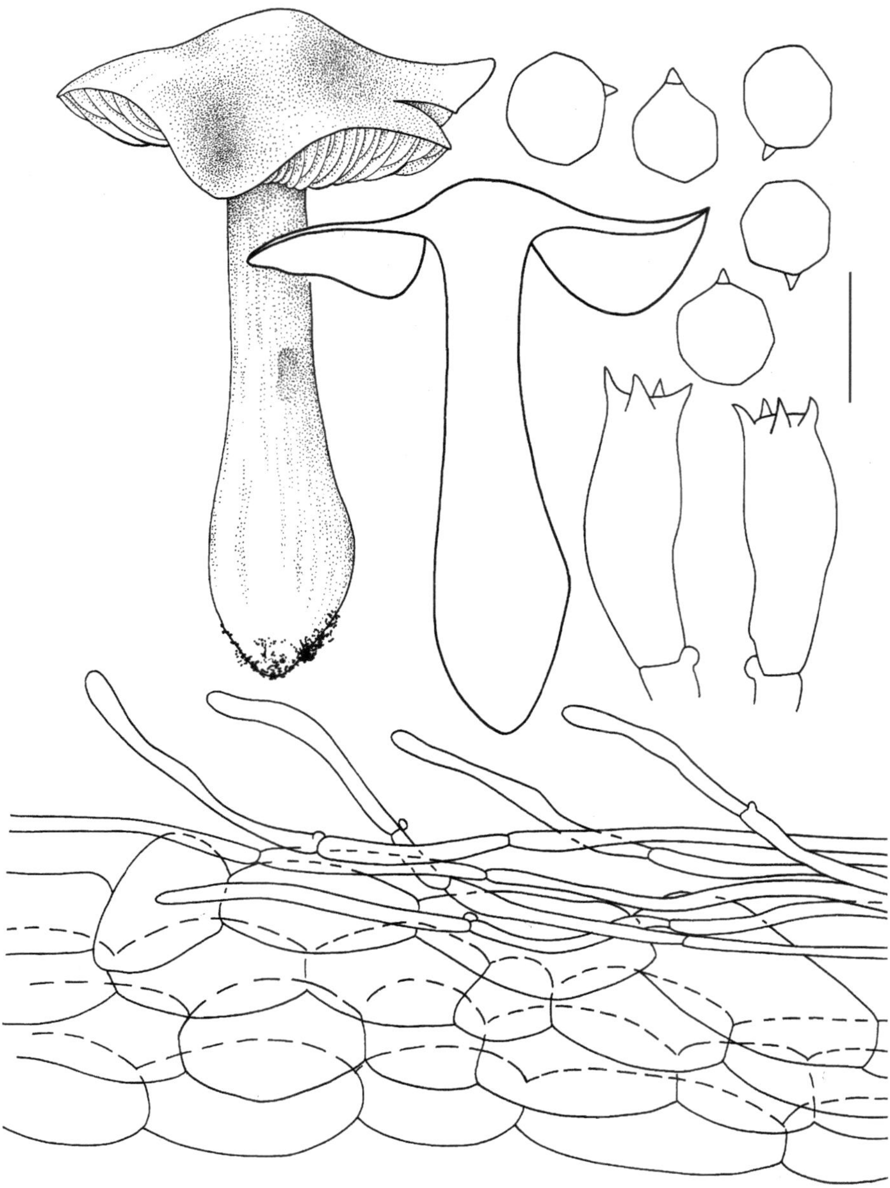

Fig. 3.1 *Entoloma baronii.* Habit, spores, basidia, and pileipellis. Bar = 10 mm or 10 μm

pileus is reminiscent of *Entoloma prunuloides* and related taxa, which are widespread in the Northern Hemisphere. *Entoloma corneri* E. Horak from Singapore and Papua New Guinea is similar, but differs by a darker, somewhat rimose pileus.

Plate 3.2 *Entoloma cretaceum* (Photo Machiel Noordeloos)

Section ***Albida*** Noordel., sect. nov.
Diagnosis: Habit tricholomatoid, small to large, totally white.
Type species: *Entoloma cretaceum* G.M. Gates & Noordel.

2. **Entoloma cretaceum** G.M. Gates & Noordel., Persoonia 19: 158. 2007.

Original diagnosis: Habitus tricholomatoideus sed parvus. Pileus 15–45 mm latus cretaceus glaber. Stipes 15–40 × 2–10 mm hyalinus flavitinctus. Sporae 6.0–8.5 × 6.0–8.0 μm isodiametrales. Cystidia desunt. Pileipellis ixocutis hyphis epigmentatis 2–7 μm latis. Fibulae adsunt.

Holotype: Australia, Tasmania, Bruny Island, Mt Mangana, 43° 22′ S, 147° 17′ E, 26 May 2001, G. Gates E 1181 (HO 543528; isotype in L).

Etymology: cretaceus (Lat.) = chalky, chalky white, referring to the colour of the basidiocarp.

Main characters: pileus chalky white; stipe hyaline to fibrous; spores small, very thin-walled, isodiametric (Plate 3.2).

Description:
Pileus 15–50 (–70) mm diam, conical to conico-convex, expanding with age to convex, with or without low umbo, with deflexed then straight margin, not hygrophanous, not translucently striate, pure white when fresh, developing yellowish or ochraceous tinges or spots, slightly viscid when moist, subfelted, appearing almost

glabrous. Lamellae adnate to adnexed, sometimes deeply emarginate, narrowly ventricose, close or crowded, up to 5 mm broad, white to creamy white, becoming pink with age, slightly irregular, sometimes with a thickish, concolorous edge, L = 40–50, l = 3–5. Stipe 15–40 × 2–10 mm, cylindrical, tapering at base, hyaline white with a yellow-ochre flush at base, dry, innately fibrillose to somewhat scurfy with age, glabrous. Context compact, firm, white. Odour like cucumber becoming farinaceous. Taste like cucumber, becoming farinaceous.

Spores 6.0–8.5 × 6.0–8.0 μm, on average 7.0–7.6 × 6.6–7.2 μm, Q = 1.0–1.1(–1.2), isodiametric, many-angled, rather thin-walled. Basidia 17–35 × 7–12 μm, 4-spored, clavate, clamped. Lamella edge fertile. Cystidia absent. Pileipellis an ixocutis to ixotrichoderm of narrow, cylindrical hyphae 2–7 μm in diameter, in a thin layer overlaying a subpellis composed of inflated elements, 22–70 × 5–19 μm, gradually passing into a pileitrama of inflated elements 30–120 × 5–30 μm. Pigment not present. Stipitipellis a compact cutis of narrow hyphae 2–9 μm in diameter. Caulocystidia absent, except for a few subcylindric loose terminal elements of surface hyphae. Clamp-connections abundant in all tissues (Fig. 3.2).

Habitat & distribution: widespread, but infrequently encountered, in Tasmanian wet eucalypt forests.

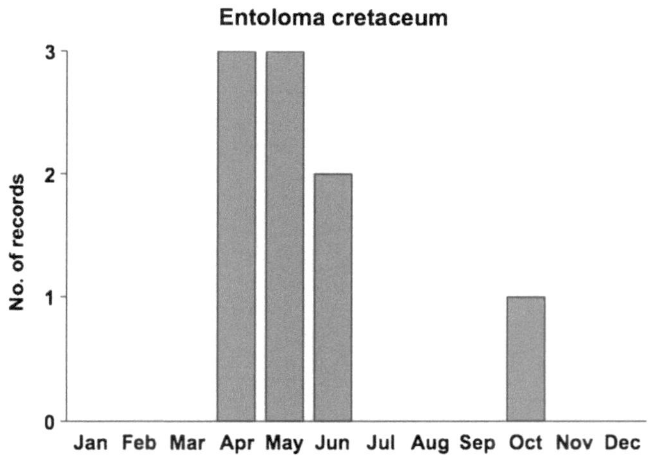

Collections examined. Australia, Tasmania, Kermandie Falls, Lower Track, 43° 12′ S, 146° 52′ E, 8 Oct. 1998, G. Gates E 69; Bruny Island, Mt Mangana, 43° 22′ S, 147° 17′ E, 1 June 1999, G. Gates E 555; —, 26 May 2001, G. Gates E 1181 (holotype, HO 543528; isotype, L); —, 17 May 2010, M.E. Noordeloos 2010039; —, 17 April 2011, G. Gates E 2293; Wielangta, 42° 42′ S, 147° 51′ E, 13 April 2004, M.E. Noordeloos 2004015a (L).

Entoloma cretaceum is a striking chalky white member of subgenus *Entoloma,* usually rather small, with a pileus rarely exceeding 50 mm (although recently a large specimen with pileus of 70 mm diameter, and stipe 73 mm long was found). The hyaline stipe makes a good distinguishing character from *E. albomagnum*, which has a fibrillose-scurfy stipe surface. Both taxa are supported by molecular data: together

Fig. 3.2 *Entoloma cretaceum*. Habit, spores, basidia, and pileipellis. Bar = 10 mm or 10 μm

with *E. albomagnum*, *E. cretaceum* forms a small monophyletic clade within subgenus *Entoloma*, viz. section *Albida*. The small, isodiametric spores that are weakly angled, and sometimes also pustulate, are reminiscent of those of the closely related section *Turfosa* (*Entocybe*). So far this species has been found only a few times in Tasmania. However, we have examined material from Queensland that appears very similar macroscopically and microscopically and which may prove to be the same species. Horak (2008) does not record any similar species from New Zealand.

3. **Entoloma albomagnum** G.M. Gates & Noordel., Persoonia 19: 160. 2007.

Original diagnosis: Habitus robustus. Pileus usque ad 100 mm latus albus glaber. Stipes 95 × 26 mm albus. Sporae 6.0–8.5 × 6.0–8.0 μm isodiametriceae paulisper

Plate 3.3 *Entoloma albomagnum* (Photo Peter Eastman)

angulatae. Pileipellis ixocutis hyphis parvis 6 µm latis haud pigmentatis. Fibulae abundantes.

Holotype: Australia, Tasmania, Scottsdale, Forester Road, 41° 06′ S, 147° 37′ E, 9 June 2004, G. Gates E 2030 (HO 543532; isotype in L).

Etymology: albus (Lat.) = white; magnus (Lat.) = big, referring to the robust, white basidiocarp.

Main characters: large, white, tricholomatoid species; odour strong, soapy; spores small, isodiametric (Plate 3.3).
Description:
Pileus 40–100 mm diam, conico-convex to convex, expanding to plano-convex with a low, broad umbo and undulating entire margin, not distinctly hygrophanous, not translucently striate, pure white with faint yellow tinge at centre with age, glabrous, somewhat lubricous when moist. Lamellae adnate to emarginate, segmentiform, ventricose, crowded, up to 9 mm broad, very thin, pale pink with entire, concolorous edge, lamellulae in two tiers. Stipe 60–95 × 26 mm at apex, up to 14 mm at base, tapering towards base, stuffed, white, silky fibrillose with scattered loose fibrils or fibrillose flocks, pruinose at base. Context white. Odour soapy at time of collecting, then like cucumber. Taste like cucumber.

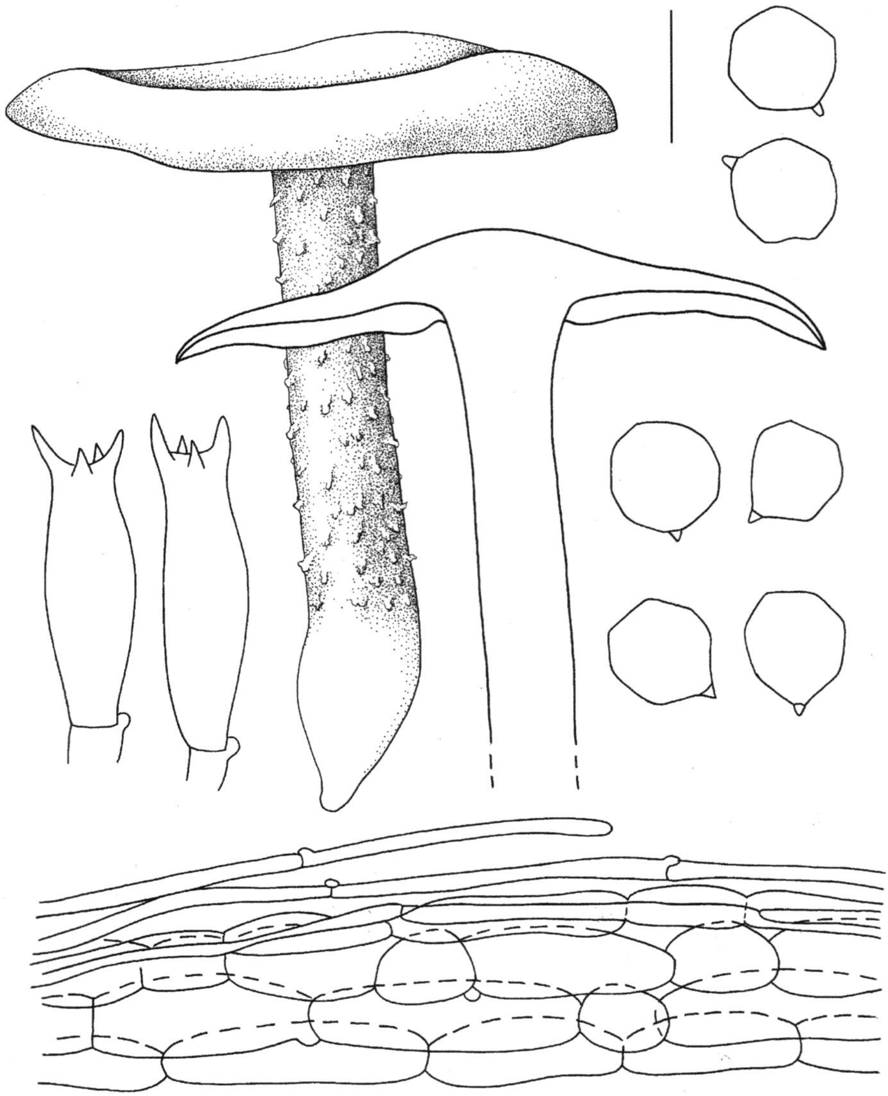

Fig. 3.3 *Entoloma albomagnum*. Habit, spores, basidia, and pileipellis. Bar = 10 mm or 10 μm

Spores 6.0–8.5 × 6.0–8.0 μm, on average 7.0–7.6 × 6.5–7.2 μm, Q = 1.0–1.2, isodiametric with 6 or 7 rather weak angles, relatively thin-walled. Basidia 4-spored, clamped. Lamella edge fertile. Cystidia absent. Pileipellis an ixocutis of narrow, cylindrical hyphae 6 μm in diameter. Pigment not observed. Hymenophoral and pileitrama regular, made up of relatively short, barrel-shaped elements 20–65(–90) × 4–15 μm. Clamp-connections present in all tissues (Fig. 3.3).

Habitat & distribution: known only from one location along a river in Tasmania's northeast.

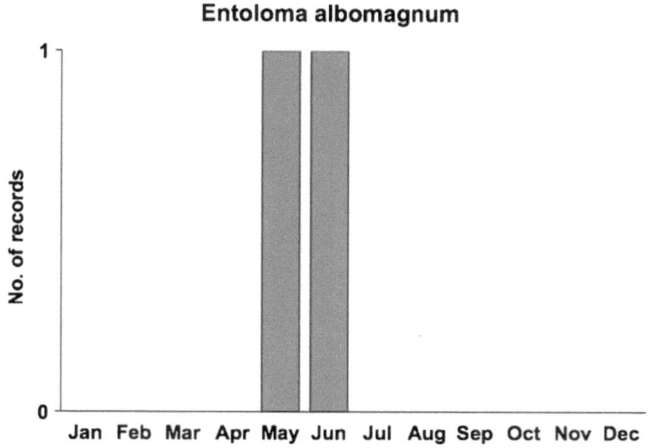

Collections examined. Australia, Tasmania, Scottsdale, Forester Road, 41° 06′ S, 147° 37′ E, 9 June 2004, G. Gates E 2030 (holotype, HO 543532; isotype, L); —, 31 May 2005, G. Gates s.n.

Entoloma albomagnum resembles *E. cretaceum* in the white colour of the basidiocarps, but differs in the very large size of the fruit body, the non-hyaline, more or less fibrillose to floccose stipe, and the soapy odour. In our phylogenetic studies they group together as a distinct clade within /entoloma. For that reason they were placed in a separate section *Albida*. Similar white, tricholomatoid species in the literature differ in having larger, more conspicuously angular spores, typical of subgenus *Rhodopolia* (Noordeloos 2004; Co-David et al. 2009).

Subsection ***Gelatinosa*** Noordel., subsect. nov.
MycoBank #564514.
Diagnosis: Differs from other species of section *Entoloma* by its very viscid-gelatinous pileus.
Type species: *Entoloma gelatinosum* E. Horak

4. **Entoloma gelatinosum** E. Horak, Beih. Nova Hedwigia 43: 40. 1973.

Original diagnosis: Pileo 30–50 mm lato, e convexo late umbonato, brunneo, reflexo lilacino tinctu, gelatinoso, cuticula cartilaginea instructo, glabro, vix striato. Lamellis adnatis, ventricosis, primo albidis dein roseis, acie concolori instructis. Stipite 35–50×8–10 mm, cylindraceo, basim versus attenuato, albidulo fibrillis lilacinis instructo, sicco, cavo. Odore saproque nullo. Sporis 8–9×6.5–7.5 μm, 6-angulatis. Basidiis 4-sporigeris. Epicute ex hyphis erectis, gelatinosis, fasciculatis palisadam formantibus, pigmento brunneo et plasmatico instructis. Hyphis fibuligeris. Ad terram in silvis. Novazelandia.

Plate 3.4 *Entoloma gelatinosum* (Photo Genevieve Gates)

Holotype. New Zealand, Buller, Taylorville, Sewell Park Track, 1 April 1968, E. Horak (PDD 27029).

Etymology: gelatinosus (Lat.) = gelatinous, referring to the pileipellis.

Selected literature: Horak, Beih. Nova Hedwigia 43: 40. 1973; Horak, Beih. Nova Hedwigia 65: 262–263. 1980; Horak, Agaricales of New Zealand 1: 238–239, Figs 90, 114.3. 2008.
Main characters: stout, tricholomatoid species; pileus very glutinous, mixed brown and blue; spores small, isodiametric, thin-walled (Plate 3.4).

Description:
Pileus 20–60 mm diam, conico-convex, expanding with age to convex or plano-convex with low umbo, not hygrophanous, not translucently striate, reddish brown with lilac-violet tinge, viscid, frosty at centre, radially rugulose towards margin. Lamellae deeply emarginate to adnexed, ventricose, crowded, up to 8 mm broad, pale grey, sometimes with faint lilac tinge, then sordid greyish or bluish pink, with concolorous eroded or entire edge, L = 30–50, l = 3–7. Stipe 30–50 × 4–12 mm, cylindrical or fusiform, tapering towards base, blue-violet longitudinal fibrils on whitish-yellowish background, yellow at pointed base, firm, stuffed. Context brown just below pileal surface in pileus, elsewhere white to grey. Odour and taste like cucumber.

Spores 6–8 × 6–7 µm, Q = 1.0–1.2, isodiametric with 5–6 or more rather weak angles, relatively thin-walled. Basidia 20–40 × 6–11 µm, 4-spored, clamped. Lamella edge fertile. Cystidia absent. Hymenophoral trama regular, made up of rather short, sausage-shaped elements, 40–90 × 5–14 µm. Pileipellis an ixocutis up to 150 µm thick with transitions to an ixotrichoderm, made up of 2–6 µm wide, cylindrical hyphae, overlaying a well differentiated subpellis of short, up to 20 µm wide, inflated elements. Pigment intracellular, pale blue in suprapellis, brown in subpellis. Pileitrama similar to hymenophoral trama. Stipitipellis a cutis of narrow, cylindrical hyphae, 2–7 µm wide, with bluish intracellular pigment. Clamp-connections abundant (Fig. 3.4).

Habitat & distribution: widespread in Tasmanian wet eucalypt forests.

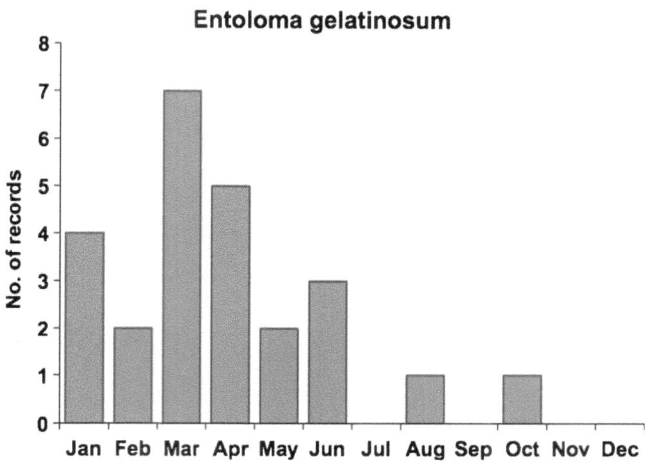

Collections examined. Australia, Tasmania, Collinsvale, Myrtle Forest, 42° 52′ S, 147° 09′ E, 18 March 1999, G. Gates E 240; —, 14 March 2002, G. Gates E 1455; —, 6 April 2004, G. Gates E 1965; —, 3 Jan. 2006, G. Gates E 2233; Douglas Apsley National Park, 41° 52′ S, 148° 11′ E, 2 May 2011, G. Gates (M.E. Noordeloos 2011027); Growling Swallet, 42° 41′ S, 146° 30′ E, 21 April 2007, G. Gates s.n.; —, 17 Feb. 2001, G. Gates E 1025; —, 1 May 2001, G. Gates E 1151; —, 11 April 2002, G. Gates E 1468; Mt Wellington, Jacksons Bend, 42° 55′ S, 147° 15′ E, 25 Jan. 1999, G. Gates E 119; Kermandie Falls, Upper Track, 43° 12′ S, 146° 52′ E, 16 March 1999, G. Gates E 228; —, 26 April 2001, G. Gates E 1133; —, 10 Jan. 2002, E 1393; North West Bay River, 42° 57′ S, 147° 12′ E, 7 June 2001, G. Gates E 1223; —, 16 June 2001, G. Gates E 1251; —, 1 Jan. 2002, G. Gates E 1385; —, 24 June 2003, G. Gates E 1885; —, 12 Aug. 2003, G. Gates E 1093; —, 25 Oct. 2005, G. Gates E 2208; Reuben Falls, 43° 00′ S, 146° 40′ E, 15 May 1999, G. Gates E 509; —, 4 March 2000, G. Gates E 792; Styx Valley, Christmas Tree Grove, 42° 50′ S, 146° 41′ E, 19 March 2009, M.E. Noordeloos 2009018 & 2009019; Wielangta, 42° 42′ S, 147° 51′ E, 20 March 1999, G. Gates E 263. — New Zealand, Buller, Taylorville, Sewell Park Track, 1 April 1968, E. Horak (holotype, PDD 27029).

The Tasmanian collections fit well with *Entoloma gelatinosum* E. Horak from New Zealand, which was confirmed by the study of the holotype. Horak (2008)

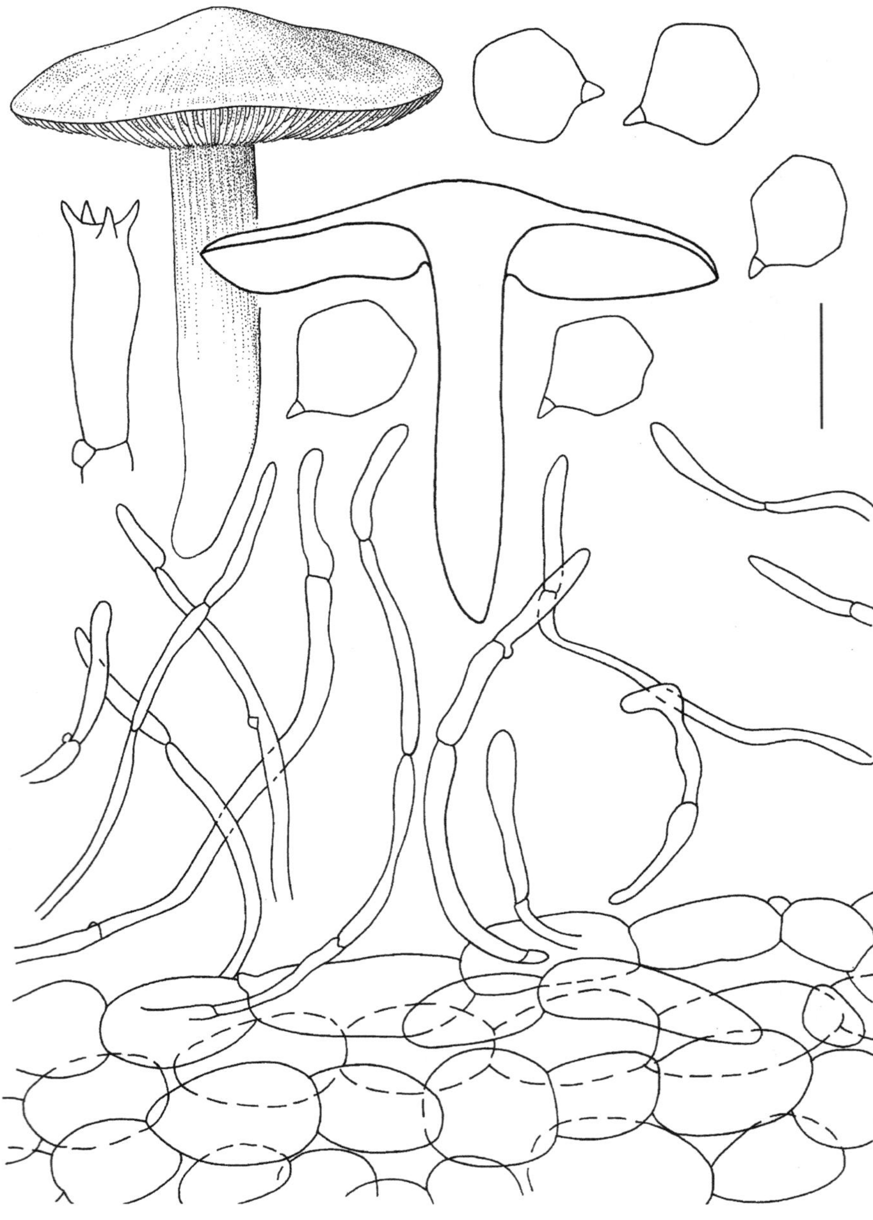

Fig. 3.4 *Entoloma gelatinosum*. Habit, spores, basidia, and pileipellis. Bar = 10 mm or 10 μm

suggests that *E. gelatinosum* in New Zealand is only known from the type locality, whereas it is not uncommon in Tasmania. *Entoloma gelatinosum* has a strong resemblance to *Entoloma bloxamii* and *E. prunuloides* from Europe, from which it differs particularly by the very viscid, rugulose pileus with, at least at centre, a true ixotrichoderm. Molecular studies have confirmed its status as a good species in that group (Co-David et al. 2009; Morgado-Neves et al. 2012, in prep.). *Rhodophyllus glutiniceps* Hongo, described from the Bonin Islands, Japan, also has a bluish grey, glutinous pileus and similar microscopy, but differs by the white stipe.

Section ***Nitida*** (Romagn.) Noordel. comb. nov.

Basionym: *Rhodophyllus* sect. *Nitidi* Romagn. in Bull. trimest. Soc. mycol. Fr. 53: 326. 1937.
MycoBank #562976.
Diagnosis: Habit slenderly tricholomatoid, mycenoid or collybioid; frequently with blue or violaceous colours; pileipellis bilayered: (1) suprapellis a cutis or a trichoderm of cylindrical hyphae, rarely an ixocutis, (2) subpellis well-developed, made up of short, inflated elements.
Type species: *Entoloma nitidum* Quél.

Section *Nitida* comprises a relatively large group of related species, characterised by a slenderly tricholomatoid to mycenoid habit, often with a fusiform stipe, and blue, violaceous or brown colours. The spores are small, isodiametric to subisodiametric, usually within the range of 5.5–9 (10) μm in diameter, with 6 or more weak angles; clamp-connections are abundant. There is a remarkable likeness of the species in section *Nitida* to those of section *Turfosa* (syn. section *Trachyospora*) occurring in the Northern Hemisphere, both in general habit, i.e. slenderly tricholomatoid, with fusiform stipe, but in particular with respect to microscopic characters such as size and shape of the spores, and structure of the pileipellis as a bilayered cutis or ixocutis. For these reasons Baroni et al. (2011) created the genus *Entocybe* for this group. Indeed, molecular studies confirm that section *Nitida* is close to section *Turfosa* ("*Entocybe*"), but both represent separate and well-supported lineages in the phylogeny (Morgado-Neves et al. 2012, in prep.). Within section *Nitida*, the four known Tasmanian species are well supported as distinct monophyletic entities. They also have counterparts in the mycoflora of New Zealand and South America (Horak 1978, 1980, 1982, 2008).

5. **Entoloma coeruleogracile** G.M. Gates & Noordel., Persoonia 19: 167. 2007.

Original diagnosis: Habitus mycenoideus. Pileus 8–22 mm latus profunde coeruleus innate fibrilloso-virgatus. Stipes 30–60 × 1–3 mm concolorus vel profunce indigotius innate fibrillosus. Sporae 6.0–8.0 × 5.5–7.5 μm isodiametrales paulisper multiangulatae tenuitenucatae. Pileipellis cutis hyphis tenuis 3.0–9.0 μm latis constituis pigmento intracellulari. Fibulae numerosae.

Holotype: Australia, Tasmania, Timbs Track, 42° 44′ S, 146° 25′ E, 8 May 2003, G. Gates E 1777 (HO 543550; isotype in L).

Etymology: coeruleus (Lat.) = sky blue, gracilis (Lat.) = slender, referring to the slender, mycenoid, blue basidiocarp.

Main characters: habit mycenoid; pileus and stipe deep blue with violet hues (Plate 3.5).

Plate 3.5 *Entoloma coeruleogracile* (Photo Michael Pilkington)

Description:
Pileus 8–22 mm diam, conico-convex with or without small umbo, with straight, entire margin, not hygrophanous, sometimes faintly translucently striate at margin, dark blue (19E5–19F5), with hint of dark violet at margin or a greyish violet (18E5) in young specimens, dry, innately radially fibrillose. Lamellae adnexed, ventricose, moderately crowded, up to 5 mm broad, thin, flesh-coloured pink when young, then buff-pink, with entire, concolorous edge. Stipe 30–60 × 1–3 mm, cylindrical, broadened at base to 2–3 mm, slender, brittle, indigo-blue or deep violet, cream to pale yellow at base, longitudinally innately fibrillose, mostly glabrous with a sheen, but with some sparse superficial violet fibrils, dry, stuffed with dark blue-violet inner part, yellowish at base, fistulose with age, with some white basal tomentum. Odour not distinctive. Taste like green grass.

Spores 6.0–8.0 × 5.5–7.5 µm, on average 6.5–7.3 × 5.8–6.8 µm, Q = 1.0–1.25, 6–9-angled in side-view, very thin-walled. Basidia 25–35 × 7–9 µm, 4-spored, clamped. Lamella edge fertile. Cystidia absent. Hymenophoral trama regular, made up of short, inflated elements. Pileipellis a thin cutis of cylindrical hyphae, 3.0–9.0 µm in diameter; subpellis of inflated elements, 16–45 × 7–22 µm. Pigment blue, intracellular in supra- and subpellis. Clamp-connections present in all tissues (Fig. 3.5).

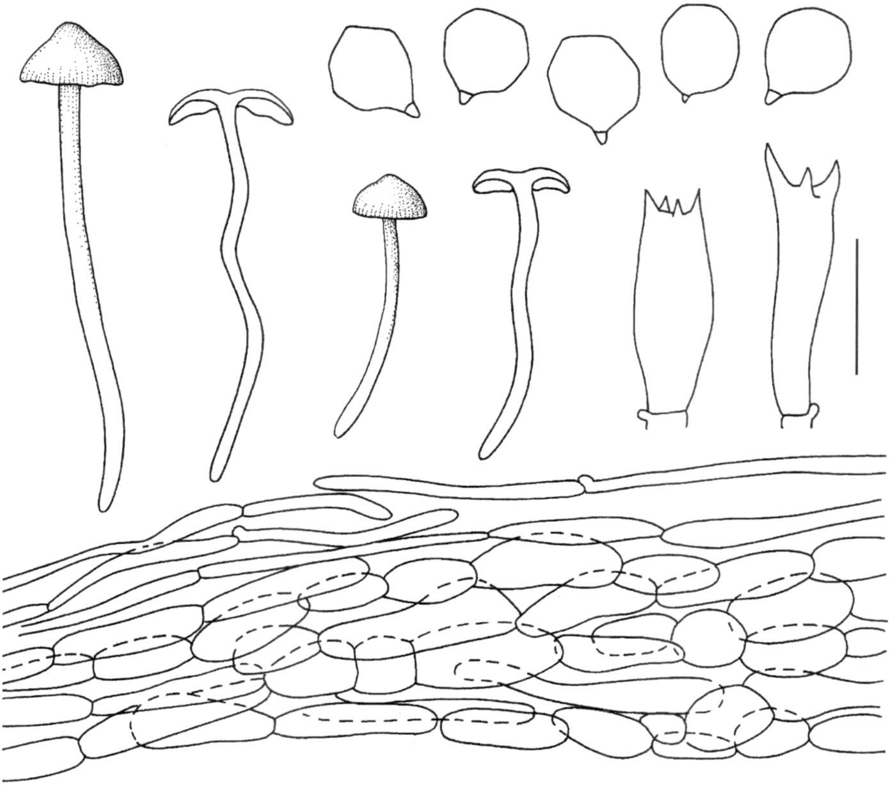

Fig. 3.5 *Entoloma coeruleogracile*. Habit, spores, basidia, and pileipellis. Bar = 10 mm or 10 μm

Habitat & distribution: widespread in wet sclerophyll forests.

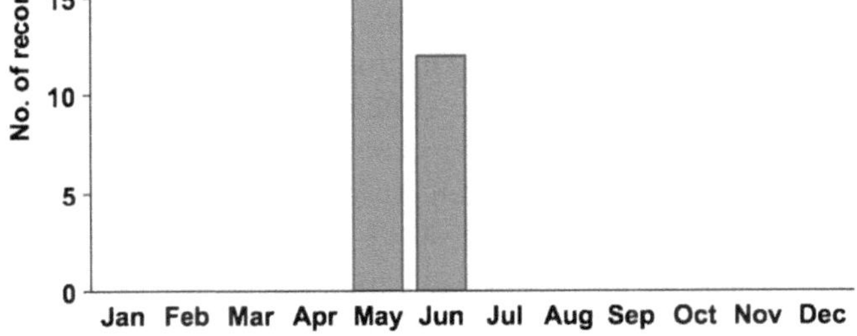

Collections examined. Australia, Tasmania, Cape Pillar Track, 43° 09′ S, 147° 56′ E, 29 May 1999, G. Gates E 550; Koonya, Clarks Cliffs, 43° 06′ S, 147° 47′ E, 5 June 2001, G. Gates E 1220; Timbs Track, 42° 44′ S, 146° 25′ E, 8 May 2003, G. Gates E 1777 (holotype, HO 543550; isotype, L); Warra LTER site, 43° 06′ S, 146° 39′ E, numerous observations between May 2006 and June 2007.

Entoloma coeruleogracile is a relative of *E. bloxamii* and *E. nitidum* with its bluish colours and small, isodiametric spores, but differs in its small, mycenoid basidiocarps and relatively small spores. *Entoloma gracile* G. Stev. differs by the deep fuliginous or fuscous brown pileus, larger spores and farinaceous to rancid odour (Horak 2008). *Entoloma inops* E. Horak differs by having a whitish, isabel or buff coloured pileus, which has, however, a distinct but slight blue tinge at margin when fresh (Horak 2008). *Entoloma alcedicolor* Arnolds & Noordeloos, described from Europe (Noordeloos 2004) is similar, but differs in having a distinctly fibrillose-squamulose pileus with a trichodermal layer of rather wide hyphae. The North American species *E. trachyosporum* Largent has very similar spores and habit, but differs in the colour of the basidiocarp, which is either grey-brown or has violaceous tinges (Largent 1974) and belongs to a different clade in the phylogeny (Morgado-Neves et al. 2012, in prep.).

6. **Entoloma contrastans** G.M. Gates & Noordel., Persoonia 19: 168. 2007.

Original diagnosis: Habitus mycenoideus. Pileus 8–20 mm latus pallide brunneus vel albidulus centro obscurior haud vel leviter hygrophanus versus marginem translucidus striatus glabrus. Stipes 30–50×2 mm intense violaceus vel coeruleus, politus. Sporae 6.0–8.0×5.5–7.5 μm isodiametrales paulisper multiangulatae tenuitenucatae. Pileipellis cutis hyphis tenuis 3–9 μm latis constituis pigmento intracellulari. Fibulae numerosae.

Holotype: Australia, Tasmania, Creepy Crawly, 42° 50′ S, 146° 22′ E, 6 May 2003, G. Gates E 1774 (HO 543538; isotype in L).

Etymology: contrastans refers to the strong contrast in colour between pileus and stipe.

Main characters: habit mycenoid; basidiocarps thin-fleshed; pileus pallid brown, often appearing almost white, but then often with darker centre, without blue or violaceous tinges, glabrous; stipe deeply violaceous or blue, contrasting with the pale pileus, polished; spores small, isodiametric (Plate 3.6).

Description:
Pileus 8–20 mm diam, conical, then expanding to plano-convex with acute centre, with straight margin, not or only slightly hygrophanous, slightly translucently striate in marginal zone, very pale beige-brown, often appearing almost white in the field, then often with somewhat darker brown centre, dry, glabrous. Lamellae adnexed, ventricose, crowded or close, up to 3 mm broad, thin, pink with concolorous, entire edge, lamellulae in two tiers. Stipe 30–50×1–2 mm, cylindrical, flexuous, deeply violaceous or blue (16–18A4), markedly contrasting with colour of pileus, shining, polished, lower part paler with appressed fibrils, with yellowish tinges at base. Context thin. Odour and taste mild.

Plate 3.6 *Entoloma contrastans* (Photos Genevieve Gates)

Spores 6.0–8.0×5.5–7.5 µm, Q=1.0–1.2, Qav=1.05, isodiametric with 5 to many weak angles, thin-walled. Basidia 14–20×4–10 µm, 4-spored, clamped. Lamella edge fertile. Cystidia absent. Hymenophoral trama regular, made up of short, cylindrical elements, 40–90×8–20 µm. Pileipellis a thin (ixo)cutis of cylindrical hyphae, 3–9 µm in diameter; subpellis well-developed, made up of inflated elements, 20–50(–90)×6.0–20 µm. Pigment pale brown, intracellular in pileipellis. Clamp-connections frequent (Fig. 3.6).

Habitat & distribution: uncommon, known only from three collections in wet eucalypt forest and rainforest.

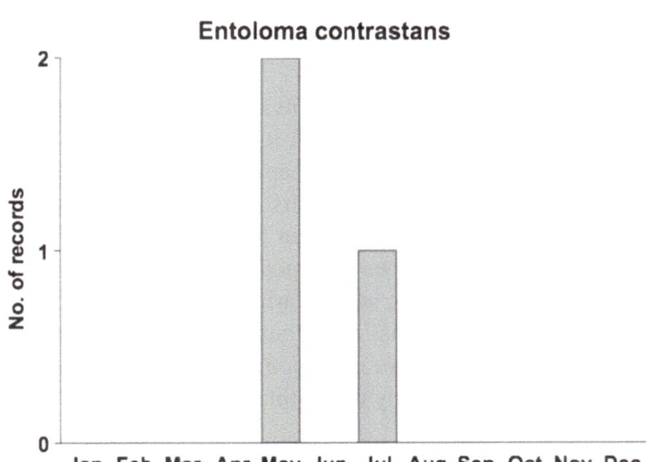

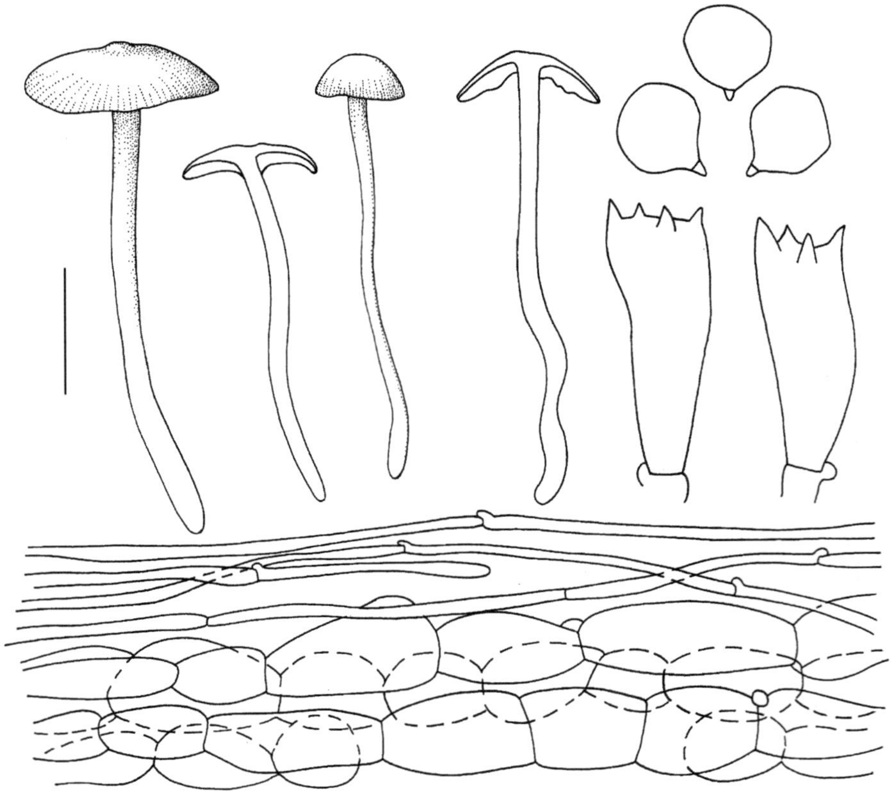

Fig. 3.6 *Entoloma contrastans*. Habit, spores, basidia, and pileipellis. Bar = 10 mm or 10 μm

Collections examined. Australia, Tasmania, Creepy Crawly, 42° 50′ S, 146° 22′ E, 6 May 2003, G. Gates E 1774 (holotype, HO 543538; isotype, L); —, 23 May 2002, G. Gates E 1528; Mt Wellington, Radfords Track, 42° 55′ S, 147° 15′ E, 18 July 2000, G. Gates E 987.

Entoloma contrastans is characterized by its slender, mycenoid habit, very similar to that of *E. coeruleogracile*, from which it differs in its pallid beige-brown, translucently striate, glabrous pileus and deep violet, polished stipe. It resembles *Entoloma gracile* G. Stev. very closely in its original description and the colour plate of Stevenson fits especially well with our fungus. However, Horak (2008) who studied additional material, describes *E. gracile* as a species with a fuliginous or fuscous, very hygrophanous pileus, strong farinaceous-rancid smell and taste, and larger spores. A single collection has been found in Tasmania that fits the description of Horak (see photos of *E. gracile*). *Entoloma inops* E. Horak from New Zealand is also very similar, differing only by the distinctly fibrillose stipe, and may well be a synonym of *E. contrastans*.

Plate 3.7 *Entoloma gracilior* (Photo Genevieve Gates)

7. **Entoloma gracilior** Noordel. & G.M. Gates, spec. nov.

MycoBank #564493.

Diagnosis: Habit mycenoid; pileus 9–23 mm diam, blackish brown, slightly hygrophanous, translucently striate at margin; lamellae greyish lilac; stipe 35–50 × 2–4 mm, deep blue with yellow tinge at base, fibrillose-striate; spores 7–10 × 6–8 μm, isodiametric, weakly angled, sometimes also pustulate; cheilocystidia usually absent; pileipellis a bilayered cutis with inflated hyphae in subpellis; pigment intracellular.

Holotype: Australia, Tasmania, Mt Field, 42° 41′ S, 146° 42′ E, 20 Sept. 2002, G. Gates E 1639 (HO 564348; isotype in L).

Etymology: gracilior (Lat.) = slender.

Main characters: habit mycenoid; pileus brown-black with violaceous tinge, translucent-striate; lamellae grey-lilac; stipe indigo-blue; spores subisodiametric to heterodiametric, with 5–8 weak angles (Plate 3.7).

Description:
Pileus 9–23 mm diam, conico-convex to hemispherical, expanding to plano-convex, weakly hygrophanous, faintly translucent-striate at margin when moist, very dark brown, almost black, then very dark brown, with a slight violaceous tinge at margin,

smooth, lubricous when moist. Lamellae adnate with decurrent tooth, ventricose, crowded, up to 6 mm broad, thickish, sometimes forked near edge, lilac-grey when young then grey-buff with violaceous pink tinge, with concolorous, entire edge. Stipe 35–50×2–4 mm, cylindrical, equal, sometimes attenuated downwards, very dark indigo-blue (20F8) and very silky when young, becoming greyer with age, yellowish at base, with white basal tomentum. Stuffed context dark indigo-blue. Odour farinaceous-rancid. Taste oily.

Spores 7–10×6–8 μm, Q=1.1–1.3, Qav=1.25, subisodiametric to short heterodiametric, with 5–8 angles, variably shaped with weak angles. Basidia 42–50×11–12 μm, 4-spored, clamped. Lamella edge fertile or heterogeneous. Cheilocystidia absent or present, scattered among basidia, 40–55×5–9 μm, cylindrical to narrowly clavate. Hymenophoral trama regular, made up of cylindrical elements, 7–10 μm in diameter. Pileipellis a cutis of cylindrical, 5–7 μm wide hyphae; subpellis made up of inflated, short elements, 40–70×10–19 μm, gradually passing into a pileitrama of longer, cylindrical, 5–20 μm wide elements. Pigment dark brown, intracellular in pileipellis; blue, granular and intracellular in subpellis. Clamp-connections abundant in all tissues (Fig. 3.7).

Habitat & distribution: widespread, but infrequently encountered, in Tasmanian wet eucalypt and mixed forests.

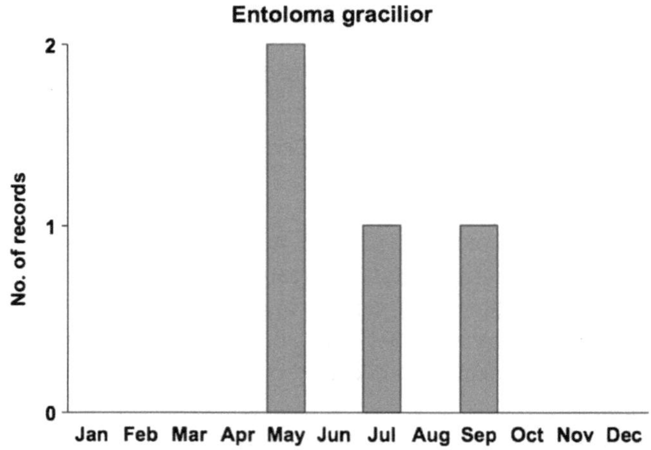

Collections examined. Australia, Tasmania, Black River Picnic Area, 40° 75′ S, 144° 51′ E, 6 May 2011, M.E. Noordeloos 2011042 (L); Mt Field, 42° 41′ S, 146° 42′ E, 20 Sept. 2002, G. Gates E 1639 (HO 564348, holotype); Mt Wellington, Myrtle Gully, 42° 54′ S, 147° 15′ E, 31 July 2001, G. Gates E 1266.

Entoloma gracilior is similar to *E. gracile* G. Stev. in the concept of Horak (2008), particularly in habit, size and shape of spores. The basidiocarps of *E. gracile* are smaller, and the pileus turns rather strong yellow-brown on drying. *Entoloma haastii* and *E. fuligineoviolaceum* are more robust species with a non-striate pileus, and smaller, more isodiametric spores. Phylogenetically, *E. gracilior* is well-supported as a species, forming a monophyletic clade with *E. coeruleogracile* (Morgado-Neves et al. 2012, in prep.).

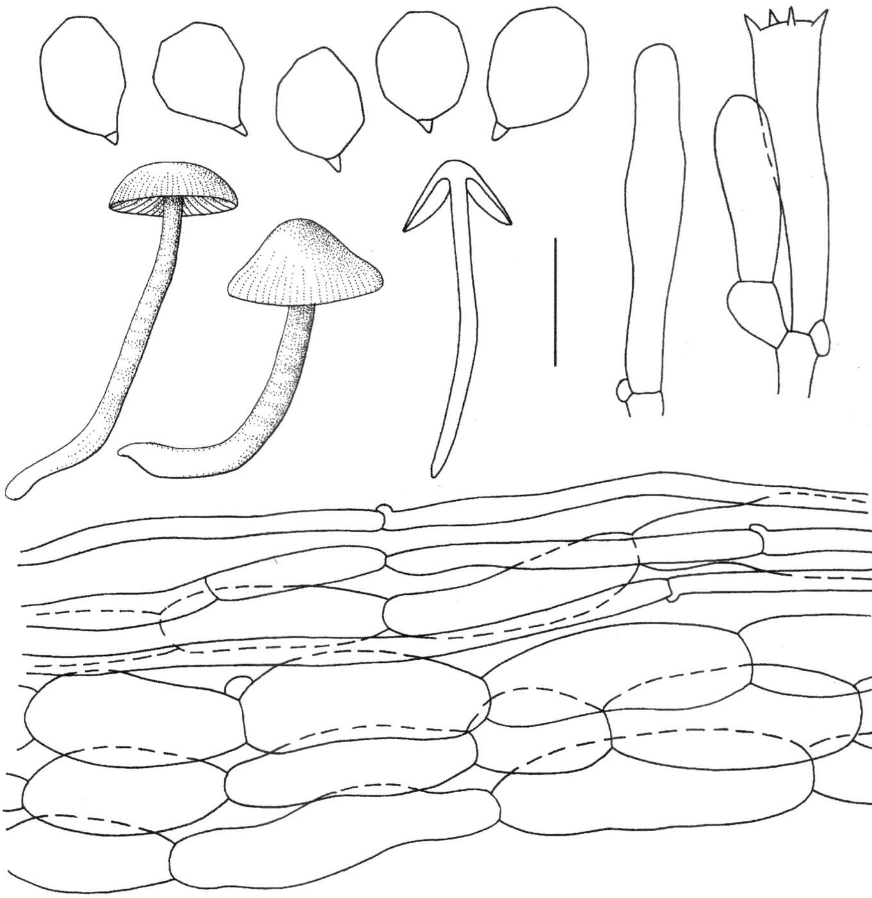

Fig. 3.7 *Entoloma gracilior.* Habit, spores, basidia, cheilocystidium, and pileipellis. Bar = 10 mm or 10 μm

8. **Entoloma haastii** G. Stev., Kew Bull. 16(2): 229 (1962).
Rhodophyllus haastii (G. Stev.) E. Horak, Flora Criptogámica de Tierra del Fuego 11(6): 96 (1980) [1979]

Selected literature: Horak, Beih. Nova Hedwigia 43:26. 1973; Horak, Beih. Nova Hedwigia 65: 276–277. 1980 (as *E. nitidum*); Horak, Agaricales of New Zealand 1: 158–159, Figs. 56, 114.4, Plate 4G. 2008.

Original diagnosis: Pileus 3.5 cm diam, atrolazulinus, late conicus, humidus, subtiliter rugulosus; caro tenuis, grisea. Lamellae adnexae, subdistantes, griseo- caeruleae, dein roseo-tinctae. Stipes 4.5–6×0.3–1 cm, sursum lazulinus, deorsum albidus, sericeo-striatus, basin bulbosam versus luteo-tinctus, cavus. Sporae 8×10 μm, angulatae, subglobosae, in cumulo roseae. Cystidia nulla.

Holotype: New Zealand, Nelson, Dun Mountain Track, 25 April 1949, Stevenson 509 (K).

Plate 3.8 *Entoloma haastii* (Photos Michael Pilkington)

Etymology: named after Johann F.J. von Haast, a German explorer.

Main characters: habit slenderly tricholomatoid; pileus brown-blue; stipe blue; spores small, isodiametric (Plate 3.8).

Description:
Pileus 20–50 mm diam, conical to conico-convex to plano-convex with distinct umbo, with deflexed margin, not hygrophanous, not or faintly translucent-striate at margin, dark purplish brown always mixed with dark indigo-blue, particularly in marginal zone, with darker centre, glabrous, shining, greasy to touch, eventually innately radially fibrillose or slightly rugulose. Lamellae free or narrowly adnexed, ventricose, moderately crowded, up to 7 mm broad, white then sordid pink, with suberoded, concolorous edge, L = 26–50, l = 3–7. Stipe 50–80 × 3–7 mm (middle), gradually broadening towards base, then tapering again in a fusoid base, blue-grey all over, tinged yellow at very base, innately fibrillose but often shining, as if polished. Odour and taste indistinct.

Spores 6.0–8.5 × 5.5–7.0 µm, on average 6.5–7.5 × 6.0–6.8 µm, Q = 1.0–1.25, 6–9-angled in side-view, very thin-walled. Basidia 22–40 × 7–11 µm, 4-spored, clamped. Lamella edge fertile. Cystidia absent. Hymenophoral trama regular, made up of short, inflated elements, 40–110 × 6–15 µm. Pileipellis a thin cutis of cylindrical hyphae, 2.0–8.0 µm in diameter; subpellis of inflated elements, 20–55 × 8–20 µm.

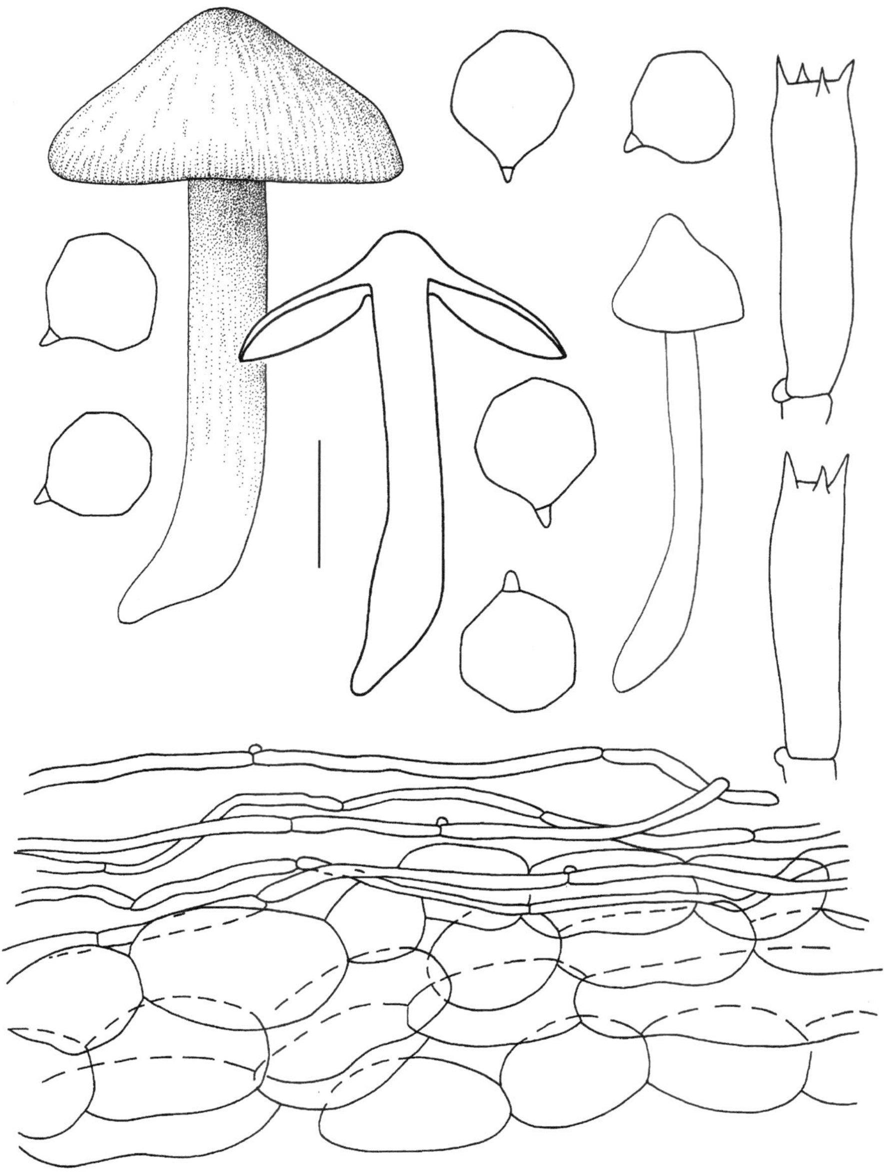

Fig. 3.8 *Entoloma haastii.* Habit, spores, basidia, and pileipellis. Bar = 10 mm or 10 μm

Pigment blue and brown, intracellular in suprapellis and subpellis. Clamp-connections present in all tissues (Fig. 3.8).

Habitat & distribution: widespread in wet sclerophyll forest, fruiting in autumn and early winter (Group 3, in Chap. 1, see section "Fruiting Patterns of Species of Entolomataceae").

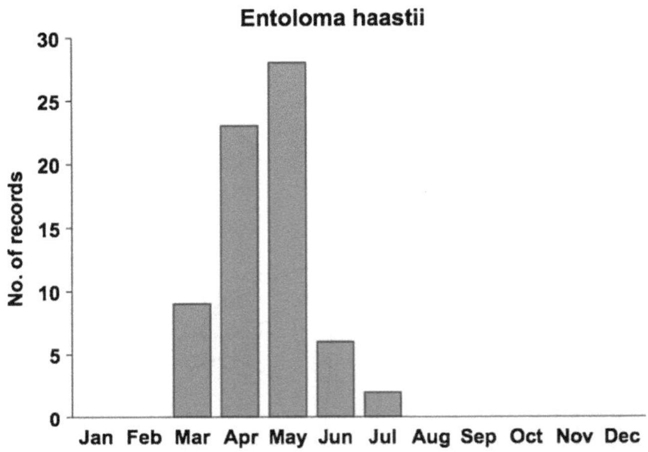

Collections examined. Australia, Tasmania, Bermuda Road, 43° 04′ S, 146° 54′ E, 13 July 2000, G. Gates E 986; Bluff River State Forest, 42° 31′ S, 147° 40′ E, 9 June 2002, G. Gates E 1569; Bruny Island, Mt. Mangana, 43° 22′ S, 147° 17′ E, 27 May 2004, G. Gates s.n.; Creepy Crawly, 42° 50′ S, 146° 22′ E, 20 March 2001, G. Gates E 1030; Duckhole Lake Track, 43° 22′ S, 146° 53′ E, 8 April 2003, G. Gates E 1676; —, 16 April 2005, G. Gates E 2116; Franklin River Nature Trail, 42° 13′ S, 146° 01′ E, 15 April 2008, M.E. Noordeloos 2008023 (L); Garden Island Creek, 43° 13′ S, 147° 11′ E, 16 June 2005, G. Gates E 2184; Growling Swallet, 42° 41′ S, 146° 30′ E, 20 April 1999, G. Gates E 435; —,19 May 1999, G. Gates E 518; —, 6 April 2000, G. Gates E 821; Junee Caves State Reserve, 42° 44′ S, 146° 36′ E, 12 March 2005, G. Gates E 2089; Kermandie Falls, Lower Track, 43° 12′ S, 146° 52′ E, 20 April 2005, G. Gates s.n.; Lake Skinner Track, 42° 57′ S, 146° 44′ E, 11 April 1999, G. Gates E 395; Lake St Clair National Park, Echo Point to Cynthia Bay, 42° 04′ S, 146° 09′ E, 19 April 2003, G. Gates E 1730; Mt Field National Park, Lyrebird Track, 42° 41′ S, 146° 40′ E, 31 March 2001, G. Gates E 157b; —, 16 March 2002, G. Gates E 1457; —, 3 May 2003, G. Gates E 1765; Mt Wellington, Silver Falls, 42° 55′ S, 147° 15′ E, 27 March 2003, G. Gates E 1658; Oldina Forest Reserve, Noel Jago Track, 41° 00′ S, 145° 40′ E, 1 May 2005, G. Gates E 2154; Timbs Track, 42° 44′ S, 146° 25′ E, 8 May 2003, G. Gates E 1778; Warra LTER site, 43° 06′ S, 146° 41′ E, G. Gates s.n., numerous observations between April 2006 and June 2007; Wielangta, 42° 42′ S, 147° 51′ E, 14 July 1998, G. Gates E 17. — New Zealand, Nelson, Dun Mountain Track, 25 April 1949, Stevenson 509 (holotype, K).

Entoloma haastii is distinguished by having a pileus with a mixture of dark reddish or purplish brown and indigo-blue tinges. Particularly, the centre of the pileus has predominantly purple-brown colour, and the marginal zone usually is deep indigo-blue. The stipe has blue-grey to blue tinges, is conspicuously fibrillose-striate and very frequently has yellow tinges at the pointed base. Stevenson (1962), in her original description, described it as a sooty blue species, close to the Northern Temperate *Entoloma nitidum* Quél. Horak (1980) initially considered both taxa synonymous; however, Horak (2008) later acknowledged *E. haastii* as a species in its own right, but without any explanation. Molecular studies confirm that *E. haastii* is a distinct species, sister to the clade of *E. coeruleogracile* and *E. gracilior* (Morgado-Neves et al. 2012, in prep.).

Section ***Fibrillosa*** Noordel., sect. nov.
MycoBank #562977.
Diagnosis: Habit tricholomatoid; pileus surface dry, fibrillose; pileipellis a cutis with transitions to a trichoderm or a true trichoderm.
Type species: *Entoloma kermandii* G.M. Gates & Noordel.

9. **Entoloma kermandii** G.M. Gates & Noordel., Persoonia 19: 164. 2007.

Original diagnosis: Pileus 20–70 mm latus stipiteque purpureo-ianthinus innato-fibrillosus. Stipes 46–105×9–10 mm. Sapore amaro. Sporae 6.0–8.5×(5.5–)6.0–7.0 μm paulisper angulatae tenuitunicatae. Pileipellis cutis pigmento intracellulari spadiceus. Fibulae adsunt.

Holotype: Australia, Tasmania, Kermandie Falls Lower Track, 43° 12′ S, 146° 52′ E, 16 March 1999, G. Gates E 227 (HO 543530; isotype in L).

Etymology: referring to the type locality.

Main characters: relatively stout tricholomatoid species; pileus and stipe violet-blue; taste bitter (Plate 3.9).

Description:
Pileus 20–70 mm diam, hemispheric at first, becoming convex to plano-convex with age, weakly to distinctly umbonate or blunt, with straight, entire margin becoming undulate with age, slightly hygrophanous, not translucently striate, deep blue-purple when young, then ranging from greyish ruby (12D3) with brownish centre to brown with violet or blue hues at margin or entirely violaceous brown (15E8), when old fading to a rather pale grey-pink in the largest basidiocarps, innately radially fibrillose, centre sometimes fibrillose-subsquamulose, dry. Lamellae adnexed, ventricose, up to 13 mm broad, whitish in youth becoming flesh-coloured pink with concolorous edge, sometimes with violet or grey-violet hues, L=30–45, l=3–5. Stipe 40–105×5–10 mm, cylindrical, gradually broadening towards fusiform base (up to 12 mm in diameter), firm, blue, violet to purple-red (18E8, 15C8, 17A8), whitish at base or in lower part, dry, innately fibrillose throughout. Odour faint cucumber. Taste like cucumber and bitter.

Spores 6.0–8.5×(5.5–) 6.0–7.0 μm, on average 7.1×6.4 μm, Q=1.0–1.2, 5–8-angled in side-view with rather weak angles. Basidia 38–51×4–8 μm, 4-spored, relatively slender clavate, clamped. Lamella edge fertile. Cheilocystidia absent. Hymenophoral trama consisting of short elements. Pileipellis a differentiated cutis, cylindrical to narrowly clavate terminal elements, 25–55(–70)×4–7.5(–9) μm; subpellis of short elements, gradually passing into pileitrama of short elements. Pigment very pale brownish, intracellular in pileipellis. Stipitipellis a cutis with some ascending loose terminal hyphal elements, 4.0–15 μm in diameter with pale greyish violet intracellular pigment. Caulocystidia absent. Clamp-connections present (Fig. 3.9).

Plate 3.9 *Entoloma kermandii* (Photos Michael Pilkington)

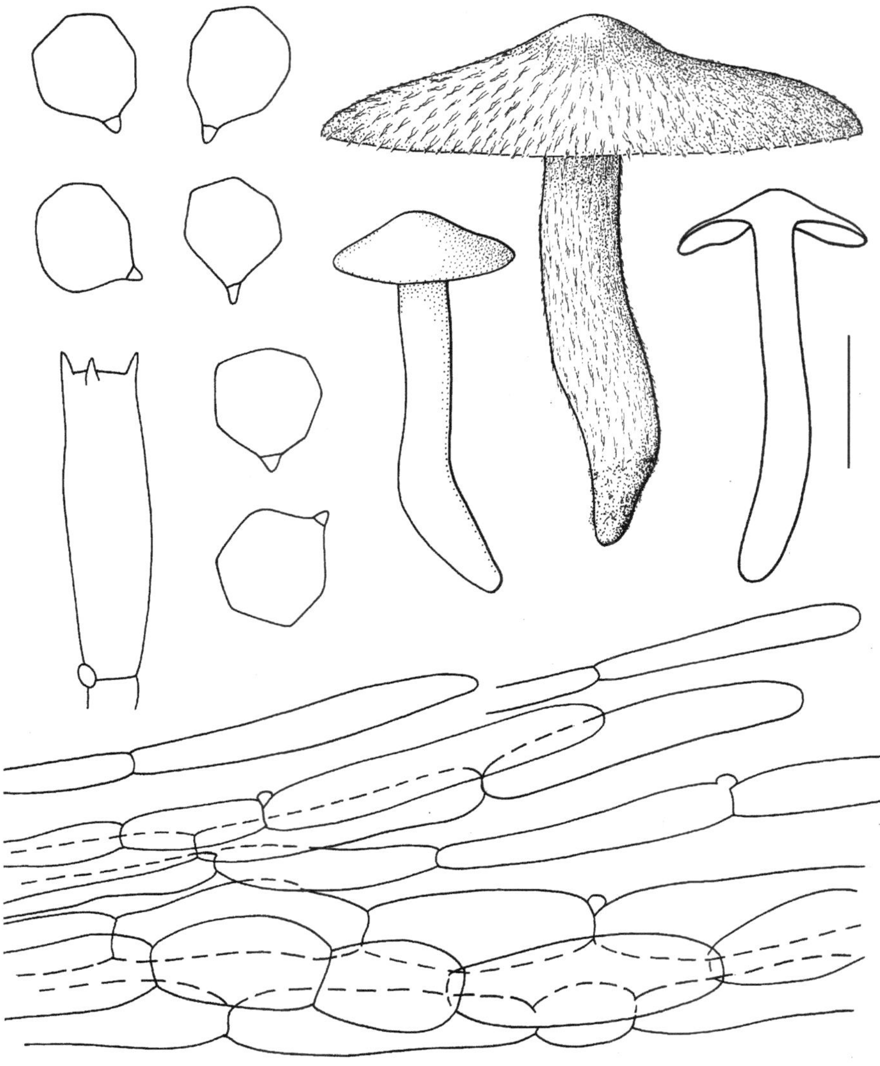

Fig. 3.9 *Entoloma kermandii.* Habit, spores, basidia, and pileipellis. Bar = 10 mm or 10 μm

Habitat & distribution: widespread in wet sclerophyll forest.

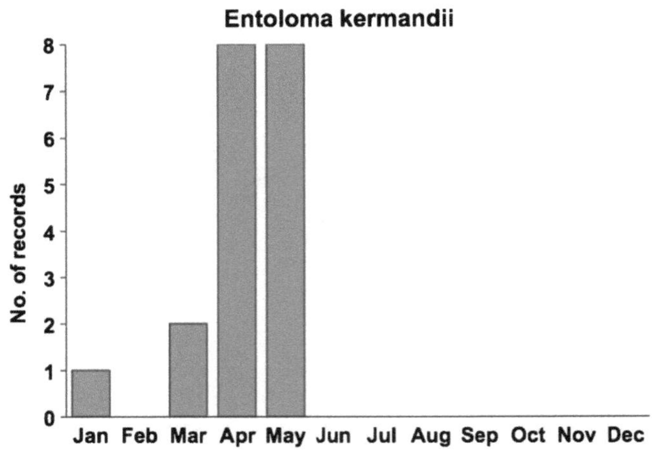

Collections examined. Australia, Tasmania, Bruny Island, Mt. Mangana, 43° 22′ S, 147° 17′ E, 8 April 1999, G. Gates E 370; —, 17 May 2010, M.E. Noordeloos 2010037; Cradle Mtn Lake St Clair NP, Frog Flats, 41° 50′ S, 146° 01′ E, 1 May 2004, G. Gates E 1985; Duckhole Lake Track, 43° 22′ S, 146° 53′ E, 15 May 2003, G. Gates E 1805; —, Jan. 2004, G. Gates E 1921; Geeveston, Donnellys Rd, 43° 07′ S, 146° 54′ E, 19 May 2005, G. Gates E 2171; Kermandie Falls Lower Track, 43° 12′ S, 146° 52′ E, 16 March 1999, G. Gates E 227 (holotype, HO 543530; isotype, L); —, 5 April 2001, G. Gates E 1067; —, 16 May 2002, G. Gates E 1510; —, 20 April 2005, G. Gates E 2119; —, 19 May 2010, M.E. Noordeloos (not conserved); Mt Wellington, Myrtle Gully, 42° 52′ S, 147° 09′ E, 18 April 2011, G. Gates E 2295; North West Bay River, 42° 57′ S, 147° 12′ E, 2 April 2011, G. Gates (not conserved); Tahune, Stringybark Loop, 43° 06′ S, 146° 44′ E, 27 May 2000, G. Gates E 945.

Entoloma kermandii is a distinctive and attractive species with its fairly robust basidiocarps, fibrillose purple-blue pileus, and small spores. The rather small, many-angled spores are distinctive for the *E. nitidum* group, but in *E. kermandii* the spores are more pronouncedly angled than in *E. nitidum* and *E. indigoticoumbrinum.* The stout basidiocarps are reminiscent of *E. bloxamii*, which usually lacks distinct purple-violet tinges, and has a smooth to rugulose pileal surface. *Entoloma kermandii* also has some similarity to *E. assimulatum* Corner & E. Horak, described from tropical rainforest in Singapore, Sabah and Peninsular Malaysia, but that species differs in the absence of clamp-connections and a more pronouncedly farinaceous odour.

10. **Entoloma perbloxamii** Noordel., D.L.V. Co-David, G.M. Gates & Morgado, Cryptogamie Mycologie 30: 109. 2009.

Original diagnosis: Pileus 30–45 mm latus, conico-convexus, umbonatus, haud hygrophanus, haud translucido-striatus, obscure lazulinus, glaber vel fibrillosus centro minute squamulosus. Lamellae adnexe-sinuatae, confertae, albae demum

Plate 3.10 *Entoloma perbloxamii* (Photos Genevieve Gates)

roseae. Stipes 30–70 × 3–8 mm, cylindraceus, lazulinus, innate fibrillosus. Sporae 6.5–9 × 5.5–6.5 µm, isodiametricae. Acies lamellarum fertilis. Cheilocystidia desunt. Pileipellis cutis vel trichoderma, ex elementis cylindraceis, 6–13 µm latis, constituta pigmentis intracellularibus. Granula lucentia absentia. Fibulae abundantes.

Holotype: Australia, Tasmania, Macgregor Peak, 42° 59′ S, 147° 57′ E, 10 April 2004, M.E. Noordeloos 2004037 (HO 558246; isotype L).

Etymology: named for its resemblance to *Entoloma bloxamii* (Berk. & Broome) Sacc.

Main characters: habit tricholomatoid; totally dark blue species with dry, fibrillose pileus, minutely squamulose at centre; spores very small, many-angled, isodiametric to subisodiametric (Plate 3.10).

Description:
Pileus 30–45 mm diam, conico-convex with umbo, with deflexed margin, not hygrophanous, not translucent-striate, deep ultramarine blue, not fading with age, dry, innately radially fibrillose at margin, subtomentose at centre. Lamellae adnate-emarginate, rather crowded, white then pink with irregularly serrulate, concolorous edge, L = 40–60, l = 3–5. Stipe 30–70 × 3–8 mm, cylindrical with tapering base, ultramarine blue, concolorous with pileus, innately fibrillose all over, with white base. Context thin, blue in cortex, white in inner part. Odour not distinct. Taste not distinct.

Spores 6.5–8(–9) × 5.5–6.5 µm, on average 7.5 × 6.5, Q = 1.02–1.40, Qav = 1.15, subisodiametric, 6–8-angled, with thick walls. Basidia 34–40 × 8–10 µm, 2–4-spored, clamped. Lamella edge fertile. Cystidia absent. Hymenophoral trama regular, made up of cylindrical elements, 64–152 × 11–23 µm. Pileipellis bilayered, suprapellis a cutis with transition to a trichoderm, made up of cylindrical elements 18–47 × 6–13 µm, with blue to brown, intracellular pigment; subpellis well-differentiated from the trama, made up of inflated elements, 40–74 × 11–26 µm, with pale yellow, intracellular pigment. Pileitrama regular, made up of cylindrical hyphae, 36–96 × 5–18 µm, with diffuse pale yellow, intracellular pigment. Stipitipellis a cutis with transitions to a trichoderm, made up of 3–5 µm wide, cylindrical hyphae with distinct terminal elements, 22–70 × 3–5 µm with pale blue, intracellular pigment. Stipititrama regular, made up of cylindrical elements, 64–152 × 11–23 µm with pale yellow, intracellular pigment. Vascular hyphae present in pilei- and stipititrama. Clamp-connections abundant in all tissues (Fig. 3.10).

Habitat & distribution: solitary or in small groups; terrestrial in mixed forest and wet sclerophyll forest, including remnant rainforest; uncommon.

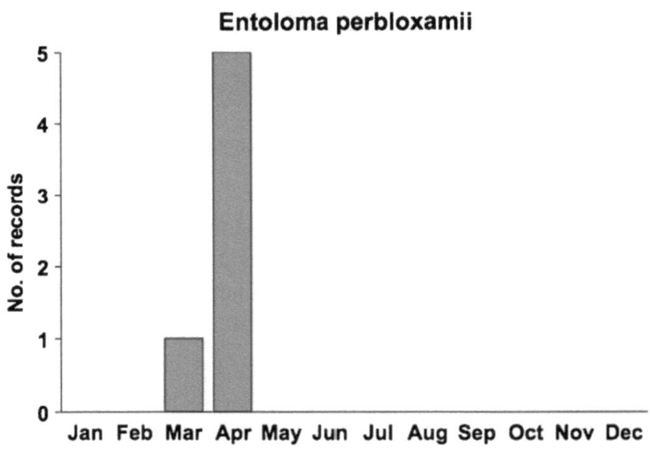

Collections examined. Australia, Tasmania, Bruny Island, Labillardiere Estate, 43° 27′ S, 147° 11′ E, 15 May 2010, M.E. Noordeloos 2010034. Lake St Clair National Park, 42° 04′ S, 146° 09′ E, 19 April 2003, G. Gates E 1728 (HO 548301); Macgregor Peak, 42° 59′ S, 147° 57′ E, 10 April 2004, M.E. Noordeloos 2004037 (holotype, HO 558246).

The striking characters of *Entoloma perbloxamii* are the dry, fibrillose surface of the pileus and stipe, persistent ultramarine blue colour, and the relatively thick-walled spores, as compared with *E. bloxamii* (Berk.) Sacc. *Entoloma assimulatum* Corner & E. Horak from Papua New Guinea has a somewhat similar pileipellis structure, but a brown pileus, and clampless hyphae. *Entoloma chalybs* E. Horak from Malaysia has similar dark blue basidiocarps, but differs in the structure of the pileipellis, which is an ixocutis of 2–5 µm wide, cylindrical hyphae. It has a strong spermatic or farinaceous smell, very similar to *Entoloma bloxamii*.

Fig. 3.10 *Entoloma perbloxamii*. Habit, spores, basidia, and pileipellis. Bar = 10 mm or 10 μm

11. **Entoloma fuligineoviolaceum** G.M. Gates & Noordel., Cryptogamie Mycologie 30: 110–111. 2009.

Original diagnosis: Pileus 23–30 mm latus, plano-convexus, umbonatus, ad marginem rimosus, haud hygrophanus, haud translucido-striatus, sordide violaceo brunneus, toto velutinus centro minute squamulosus. Lamellae adnexe sinuatae, moderate confertae, brunneo-violaceae roseotinctae. Stipes 30–35 × 4–6 mm, cylindraceus, coeruleo-violaceus, paulisper fibrillosus vel politus. Sporae 5.5–7.5(–8) × 5.5–6.5(–7) μm, isodiametricae. Acies lamellarum fertilis. Cheilocystidia desunt. Pileipellis cutis vel trichoderma ex elementis cylindraceis, 30–60 × 2–9 μm latis, constituta pigmentis intracellularibus. Granula lucentia absentia. Fibulae abundantes.

Plate 3.11 *Entoloma fuligineoviolaceum* (Photo Machiel Noordeloos)

Holotype: Australia, Tasmania, Bruny Island, Mt Mangana, 43°22′ S, 147°17′ E, 19 May 2007, G. Gates E 2253 (HO 548302; isotype L).

Etymology: fuligineus (Lat.) = very dark brown; violaceus = violet, referring to the colour of the pileus.

Main characters: habit slenderly tricholomatoid; pileus dark brown with distinct violaceous tinge; lamellae violet-brown; stipe blue-violet; spores small, isodiametric; clamp-connections abundant (Plate 3.11).

Description:
Pileus 23–60 mm diam, conico-convex to plano-convex with small umbo, with deflexed then straight, finally rimose margin, not hygrophanous, not translucent-striate or at margin only when young, dark brown with distinct dark violaceous hues, velutinous to finely squamulose all over, dry. Lamellae adnexed-sinuate, ventricose, moderately crowded, up to 5 mm broad, brown-violet to violet with pink tinge, with entire, concolorous edge. Stipe 30–65 × 4–8 mm, cylindrical, up to 10 mm at base, blue-violet, fibrillose-striate to almost polished, paler almost white at base, with white basal tomentum. Context concolorous in cortex, whitish in inner parts. Odour faint, sometimes reminiscent of grass or iodine. Taste mild to bitter.

Spores 5.5–7.5(–8) × 5.5–6.5(–7) μm, Q = 1.0–1.2, Qav = 1.05, isodiametric to subisodiametric, thin-walled with many blunt angles. Basidia 23–32 × 5–9 μm, 4-spored, clamped. Lamella edge fertile. Cystidia absent. Hymenophoral trama regular, made up of cylindrical to slightly inflated elements, 60–120 × 6–20 μm with pale greyish, intracellular pigment. Pileipellis a cutis with transitions to a trichoderm, made up of

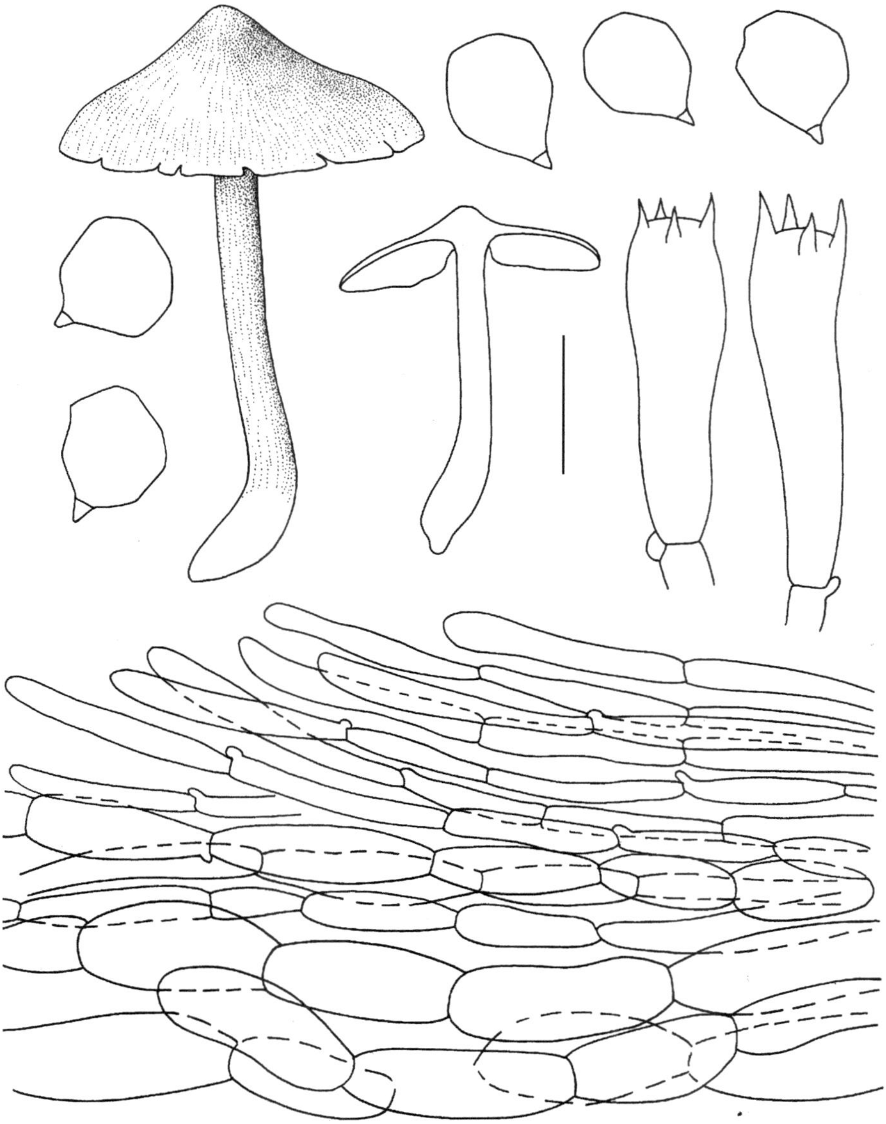

Fig. 3.11 *Entoloma fuligineoviolaceum*. Habit, spores, basidia, and pileipellis. Bar = 10 mm or 10 μm

cylindrical to narrowly clavate or fusiform terminal elements, 30–60 × 2–14 μm; subpellis not very differentiated, made up of cylindrical hyphae, 4–10 μm wide; pigment strong brown, intracellular, diffuse and granular in pileipellis. Stipitipellis a cutis of cylindrical, 4–7 μm wide hyphae with pale bluish, intracellular pigment. Brilliant granules absent. Clamp-connections abundant in all tissues (Fig. 3.11).

Habitat & distribution: in wet sclerophyll and mixed forest. Known from only three localities.

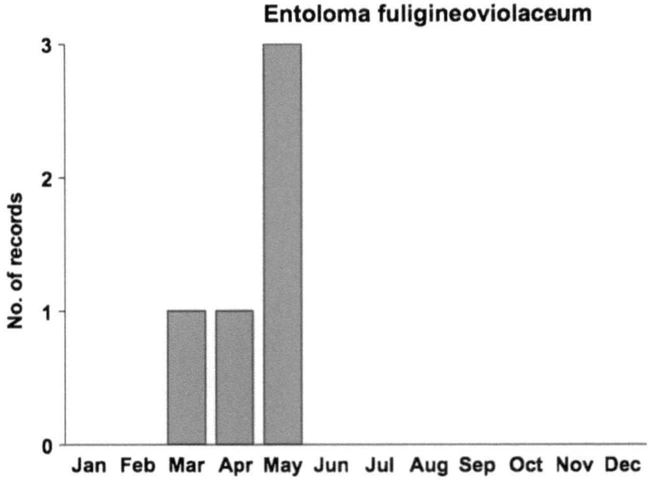

Collections examined. Australia, Tasmania, Bruny Island, Mt Mangana, 43°22′ S, 147°17′ E, 19 May 2007, G. Gates E 2253 (holotype, HO 548302; isotype, L); Growling Swallet, 42° 41′ S, 146° 30′ E, 24 May 2008, G. Gates E 2270; Junee Caves State Reserve, 42°44′ S, 146°36′ E, 24 May 2003, G. Gates E 1816; —, 21 April 2007, G. Gates E 2251; —, 28 March 2009, M.E. Noordeloos 2009071.

Entoloma fuligineoviolaceum belongs to the rather species-rich group of *Entoloma nitidum* and *E. bloxamii*. The very dark, brownish-black, tomentose pileus distinguishes it from *E. kermandii* G.M. Gates & Noordel. and *E. indigoticoumbrinum* G.M. Gates & Noordel. *Entoloma haastii* G. Stev. has a smoother, bicoloured pileus with a smooth, non-velutinous, almost viscid surface, and slightly larger, 5–7-angled spores. *Entoloma fabulosum* E. Horak from New Zealand is similar, but has a palisade-like pileipellis, vesiculose cheilocystidia, and a brown to almost black pileus without a distinct blue or violaceous tinge (Horak 2008).

Section ***Calliderma*** (Romagn.) Noordel.

Rhodophyllus sect. *Calliderma* Romagn. in Bull. mens. Soc. linn. Lyon 43: 329. 1974; *Entoloma* sect. *Calliderma* (Romagn.) Noordel. in Persoonia 12: 76. 1983.

Pileus pruinose to velutinous; pileipellis a hymeniderm or palisadoderm, made up of clavate or fusiform elements.

Type species: *Entoloma callidermum* (Romagn.) Noordel.

One of the surprising results of the phylogenetic research of Co-David et al. (2009); Morgado-Neves et al. 2012, in prep.) appears to be that some *Entoloma* species, incl. *Entoloma callidermum*, with a hymenidermal or palisade-like pileipellis ("calliderm") fall into one big monophyletic clade together with species with a relatively simple cutis or ixocutis. Apart from this incongruency, most other morphological characters show much resemblance: the structure of the tramal layers with abundant clamp-connections, and in particular the relatively small, often thin-walled and weakly angled spores are very similar.

12. **Entoloma indigoticoumbrinum** G.M. Gates & Noordel., Persoonia 19: 162. 2007.

Original diagnosis: Pileus 30–80 mm latus conico-convexus expansus leviter umbonatus brunneus indigoticus tinctus toto tomentosulus demum rimosus vel rugulosus. Stipes 30–70×5–7 mm (apex)×12 mm (basis) coeruleo-fibrillosus. Sporae 6.0–7.5(–8.0)×5.5–7.5 μm paulisper 5–7-angulatae. Cystidia desunt. Pileipellis trichoderma vel hymenoderma elementis terminalis clavatis 10–30×20 μm. Pigmentis intracellulosis. Fibulae adsunt.

Holotype: Australia, Tasmania, Growling Swallet, 42° 41′ S, 146° 30′ E, 22 April 2004, M.E. Noordeloos 2004063 (HO 543559; isotype in L).

Etymology: indigoticus (Lat. & Gr.) = indigo-blue; umbrinum = umber-brown, referring to the colour of the pileus.

Main characters: habit tricholomatoid; pileus and stipe with dark indigo-blue tinges, often mixed with dark brown, pileal surface finely tomentose or velutinous, cracking with age (Plate 3.12).

Description:
Pileus 20–80 mm diam, conico-convex when young, with involute margin, expanding to convex, then plano-convex with deflexed margin, with low umbo, not hygrophanous, not translucently striate, uniformly dark brown with indigo-blue tinges particularly when fresh and in older specimens often very distinct near margin (7F1–2, 11F2–3, 19F2–3), entirely finely pruinose to velutinous, when old sometimes cracked or faintly radially wrinkled. Lamellae adnate-emarginate, narrowly ventricose, moderately crowded, flesh-coloured pink, often tinged blue especially near the margin of the pileus, then grey-pink, with irregular, concolorous edge, L=28–42, l=3–7. Stipe 30–70×5–7 mm (apex)×12 mm (at base), cylindrical, distinctly broadened at base to subbulbous, often with tapering point attached to mycelium, sometimes flexuous, covered with grey-blue, brown, or dark blue to purplish blue fibrils (in colour different from pileus) on a paler background (19E2–4, 17–18E3), shining, base glabrous and pallid, often tinged yellow. Context concolorous in cortex, grey-blue in context of pileus and most of the stipe, pale yellow at stipe base. Odour none or like raw beans. Taste none or like raw beans.

Spores 6.0–7.5(–8.0)×5.5–7.5 μm, on average 6.5–7.5×6.1–6.7 μm, Q=1.0–1.2, isodiametric to subisodiametric, with 5–7 weak angles and relatively thick walls. Basidia 20–34×8–15 μm, (broadly) clavate to sphaeropedunculate, clamped. Lamella edge fertile. Cystidia absent. Pileipellis a trichoderm with transitions to a hymeniderm of short, clavate terminal elements, 10–30×20 μm. Pigment brown, intracellular in most elements of the suprapellis and subpellis, some elements with dark blue pigment. Pileitrama and hymenophoral trama made up of relatively short elements, 40–120×7–20 μm. Brilliant granules absent. Stipitipellis a cutis of narrow, cylindrical hyphae, 4–10 μm wide, with brown intracellular pigment and blue crystalline pigment clots. Caulocystidia absent. Clamp-connections abundant in all tissues (Fig. 3.12).

Plate 3.12 *Entoloma indigoticoumbrinum* (Photos Machiel Noordeloos)

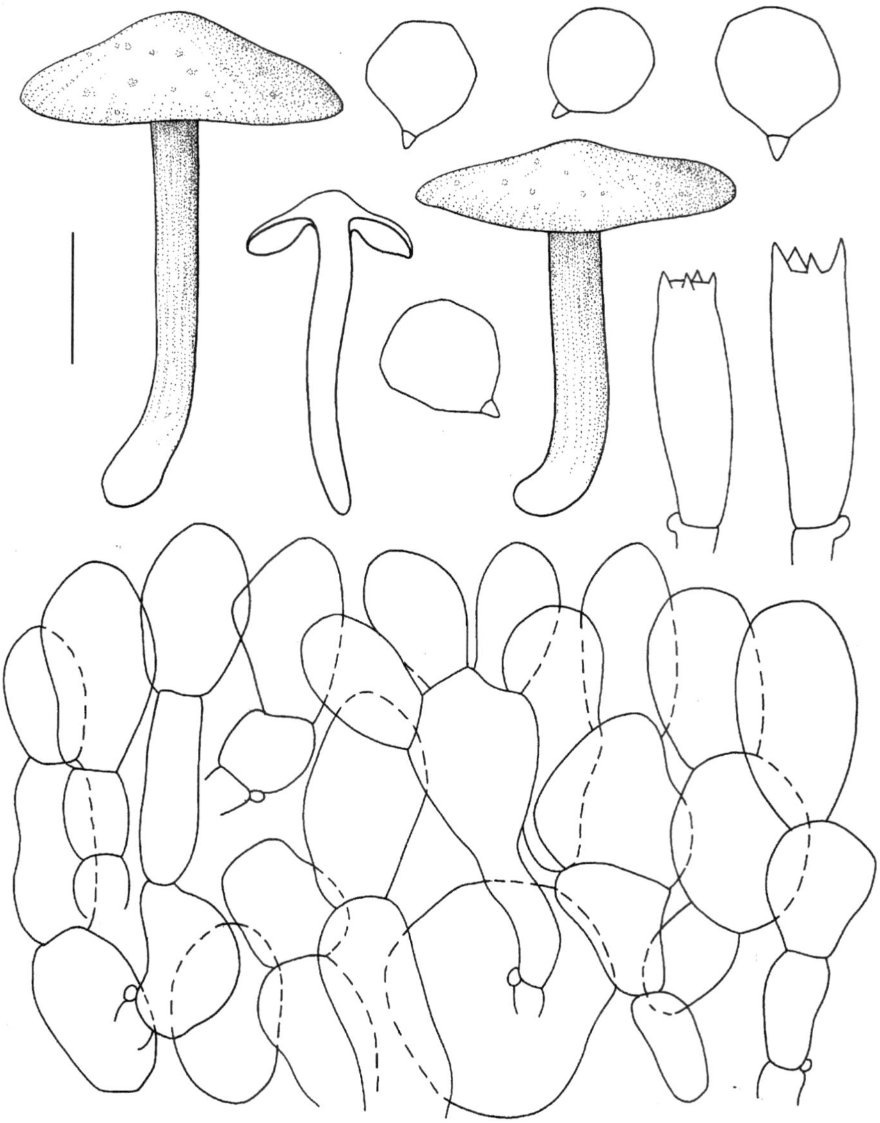

Fig. 3.12 *Entoloma indigoticoumbrinum*. Habit, spores, basidia, and pileipellis. Bar = 10 mm or 10 μm

Habitat & distribution: widespread and abundant in wet sclerophyll forests; main fruiting in two months of autumn (April–May) with remaining records occurring in March and June (Group 3, in Chap. 1, see section "Fruiting Patterns of Species of Entolomataceae").

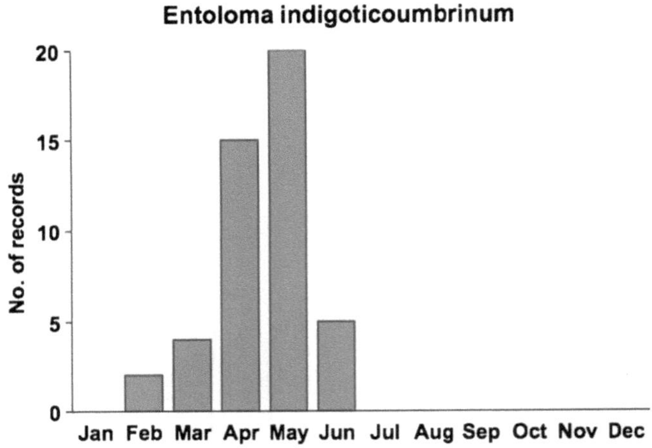

Collections examined. Australia, Tasmania, Adamsons Peak Track, 43° 20′ S, 146° 54′ E, 1 May 1999, G. Gates E 459; Bruny Island, Mt Mangana, 43° 22′ S, 147° 17′ E, 8 April 1999, G. Gates E 374; Chauncy Vale, 42° 37′ S, 147° 16′ E, 24 May 2011, G. Gates E 2297; Clarks Cliffs, 43° 06′ S, 147° 47′ E, 5 June 2001, G. Gates E 1222; Cradle Mtn Lake St Clair NP, 41° 50′ S, 146° 01′ E, 30 April 2004, G. Gates E 1981; Duckhole Lake Track, 43° 22′ S, 146° 53′ E, 4 March 2004, G. Gates E 1936 & E 1937; —, 23 March 2004, G. Gates E 1958; —, 29 May 2010, M.E. Noordeloos 2010064; Evercreech Forest Reserve, 41° 24′ S, 147° 58′ E, 7 June 2004, G. Gates (not conserved); Florentine Rd, Lady Binney Track, 42° 43′ S, 146° 31′ E, 13 May 2003, G. Gates E 1791; Growling Swallet, 42° 41′ S, 146° 30′ E, 22 April 2004, M.E. Noordeloos 2004063 (holotype, HO 543559; isotype, L), —, M.E. Noordeloos 2004064; —, 11 May 2000, G. Gates E 905; —, 1 May 2001, G. Gates E 1152; —, 13 May 2003, G. Gates E 1796; Kermandie Falls, Lower Track, 43° 12′ S, 146° 52′ E, 11 May 1999, G. Gates E 492; —, 19 May 2010, F. Karstedt 1436; —, Upper Track, 26 April 2001, G. Gates E 1135; —, 30 March 1999, G. Gates E 313; Lake St. Clair NP, 42° 04′ S, 146° 09′ E, 22 April 2000, G. Gates E 859; —, 20 April 2002, G. Gates E 1478; Mt Field National Park, Lyrebird Track, 42° 41′ S, 146° 40′ E, 28 Feb. 1999, G. Gates E 157; Mt Wellington, Myrtle Gully, 42° 54′ S, 147° 15′ E, 13 April 1999, G. Gates E 398; North West Bay River, 42° 57′ S, 147° 12′ E, 21 April 2005, G. Gates E 2125; Snug Falls, 43° 05′ S, 147° 13′ E, 24 April 1999, G. Gates E 444; Wielangta, 42° 42′ S, 147° 51′ E, 20 March 1999, G. Gates E 267; —, 13 April 2004, M.E. Noordeloos 2004013.

Entoloma indigoticoumbrinum is a fairly common species in Tasmania, with a rather striking pileus colour, which is a mixture of brown and dark indigo-blue, and dry, pruinose to velutinous-tomentose pileal surface, which tends to become cracked or wrinkled with age. The general aspect is reminiscent of species of the complex of *E. bloxamii*, with which it also shares some microscopic characters, such as the small, isodiametric spores and abundant clamp-connections. The pileipellis structure, however, is strikingly different and would place this species in section *Calliderma* of subgenus *Inocephalus*. Interestingly, the species falls out in the same clade as *E. bloxamii* and *E. prunuloides* in the preliminary phylogenetic analysis based on molecular markers (Morgado-Neves et al. 2012, in prep).

Several similar species have been described in the literature: *Entoloma praestans* Corner & E. Horak differs in the smaller, quadrate spores and *E. burkillii* Massee has a different pileipellis of narrow, cylindrical hyphae and larger spores. *Entoloma coeruleoviride* Corner & E. Horak has narrower spores and pileipellis hyphae, and a fimbriate lamella edge with abundant cheilocystidia.

13. **Entoloma coeruleomagnum** G.M. Gates & Noordel., Persoonia 19:165. 2007.

Original diagnosis: Pileus ad 100 mm latus convexus obscure coeruleus tomentosus versus marginem sulcatus. Lamellae albide violaceo-tinctae. Stipes 60×8–10 mm innate coeruleo-fibrillosus. Sporae 9.0–11.0×8.5–10.0 µm paulisper 5–6-angulatae. Acies lamellarum fertilis. Pileipellis hymenoderma elementis elongates pigmento intracellulari. Fibulae adsunt.

Holotype: Australia, Tasmania, Duckhole Lake Track, 43° 22′ S, 146° 53′ E, 16 April 2005, G. Gates E 2109 (HO 543531; isotype in L).

Etymology: coeruleus (Lat.) = sky blue; magnus (Lat.) = big, referring to the big, blue basidiocarp.

Main characters: robust tricholomatoid basidiocarp, intensely bluish black with velvety cap; pileipellis a palisadoderm; spores small, 5–6-angled (Plate 3.13).

Description:
Pileus 100 mm diam, irregularly convex, with deflexed margin, very dark blackish blue with purple sheen, not hygrophanous, not translucently striate, entirely finely velvety, dry, radially sulcate at margin. Lamellae adnate, ventricose, crowded, up to 18 mm broad, moderately thick, pink-tinged dark violet, with irregular, concolorous edge. Stipe 60×8–10 mm, cylindrical, dark blue with violet tinges, concolorous with pileus, innately fibrous, glabrous, yellowing at base. Odour and taste not distinctive.

Spores 9.0–11.0×8.5–10.0 µm, Q = 1.1–1.25, Qav = 1.15, isodiametric, 5- or 6-angled with rather distinct angles. Basidia 20–40×7–11 µm, 4-spored, clamped. Lamella edge fertile. Cheilocystidia absent. Hymenophoral trama regular, made up of medium-sized, cylindrical to slightly inflated elements, 50–120×5–20 µm. Pileipellis a palisadoderm of erect hyphae with cystidioid terminal elements, 40–120×4–15 µm. Pigment intracellular, brown in terminal elements of pileipellis, purple-blue, and appearing parietal in lower parts of pileipellis. Pileitrama regular, similar to hymenophoral trama. Stipitipellis with numerous cylindrical-flexuous caulocystidia, up to 9 µm in diameter. Clamp-connections frequently observed in pileipellis, stipitipellis and hymenium (Fig. 3.13).

Habitat & distribution: known only from a single collection at the type locality, in a wet mixed forest with *Melaleuca*.

Plate 3.13 *Entoloma coeruleomagnum* (Photos Genevieve Gates)

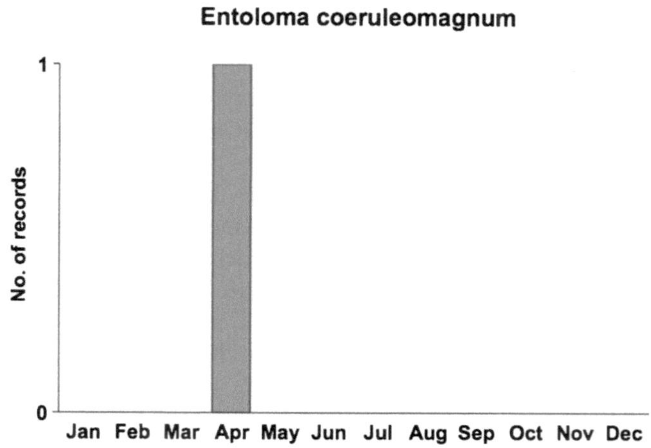

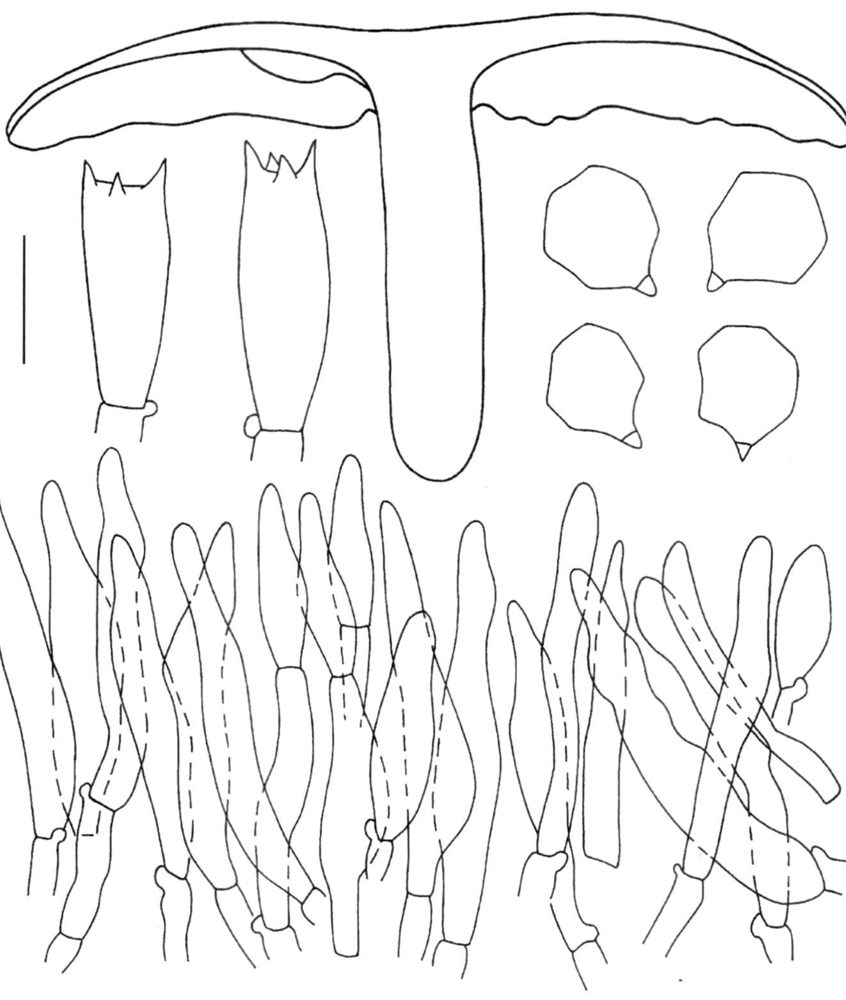

Fig. 3.13 *Entoloma coeruleomagnum.* Habit, spores, basidia, and pileipellis. Bar = 10 mm or 10 µm

Collection examined. Australia, Tasmania, Duckhole Lake Track, 43° 22′ S, 146° 53′ E, 16 April 2005, G. Gates E 2109 (holotype, HO 543531; isotype, L).

At first sight, the description of *E. angulatum* (Cleland) Grgur. might well fit the current species. However, that poorly known species has a simple cutis-like pileipellis (Grgurinovic 1997). There are a few similar species known from Australasia (Horak 1980): *Entoloma praestans* Corner & E. Horak has much smaller, subquadrate spores, *E. burkillii* Massee differs in having much narrower terminal elements in the pileipellis, and *E. coeruleoviride* Corner & E. Horak differs in the green tinges in the pileus, and a sterile lamella edge. *Entoloma jennyae* Noordel. & Cate from Ireland is also very similar, but has a more sky blue pileus and whitish stipe (Noordeloos 1992, 2004).

Section ***Luteifolia*** Noordel., sect. nov.
MycoBank #562978.
Diagnosis: Habit tricholomatoid; lamellae strong yellow; spores small, isodiametric to subisodiametric, thin-walled and weakly angled.
Type species: *Entoloma luridum* Hesler.

The distinctive characters of this section are the combination of a tricholomatoid habit, distinctly yellow lamellae and thin-walled isodiametric spores. Phylogenetic studies by Co-David et al. (2009) have shown that species with these characters end up in a monophyletic clade within the large subgenus *Entoloma*, and deserve therefore to be placed in a separate section. The type species of the section, *Entoloma luridum* Hesler is widespread in North America. There are two similar species in Tasmania, described below.

14. **Entoloma manganaense** G.M. Gates & Noordel., Persoonia 19: 161. 2007.

Original diagnosis: Habitus tricholomatoideus. Pileus 90 mm fuligineus glaber. Lamellae distantes crassae luteae. Stipes 40×25 mm coeruleo-griseus. Sporae 6.0–8.5×5.5–7.5 μm isodiametrales. Cystidia desunt. Pileipellis cutis hyphis 3.5–7.5 μm latis epigmentatis. Fibulae abundantes.

Holotype: Australia, Tasmania, Bruny Island, Mt Mangana, 43° 22′ S, 147° 17′ E, 8 April 1999, G. Gates E 369 (HO 543529; isotype in L).

Etymology: referring to Mt Mangana, the type locality.

Main characters: relatively large tricholomatoid species; pileus brown, glabrous, lubricous; lamellae thickish, bright yellow; stipe blue-grey; spores isodiametric (Plate 3.14).

Description:
Pileus to 90 mm diam, plano-convex with entire, uplifted and undulating margin, not hygrophanous, not translucently striate, red-brown (5F6), more or less uniformly coloured, glabrous, smooth, slightly lubricous. Lamellae sinuate-emarginate, ventricose, moderately crowded, up to 15 mm broad, thickish, cadmium yellow (4A5), with concolorous and entire edge, lamellulae in two tiers. Stipe 40×25 mm, stout, equal, cylindrical, blue-grey (20E4) with slight yellow tinge near base, innately fibrillose, glabrous. Odour none. Taste mild, but not distinctive.

Spores 6.0–8.5×5.5–7.5 μm, on average 7.3×6.9 μm, Q=1.0–1.15, angles weak, 5–7-(or many-)angled, reminiscent of the spores of *Rhodocybe*. Basidia

Plate 3.14 *Entoloma manganaense* (Photos Genevieve Gates)

20–35 × 6–8 μm, 4-spored, clamped. Lamella edge fertile. Cystidia absent. Pileipellis an ixocutis of narrow, cylindrical to inflated hyphae 22–55(–70) × 3.5–7.5 μm; subpellis made up of cylindrical to inflated hyphae to 30 μm in diameter. Pigment pale brown, intracellular in pileipellis. Hymenophoral and pileitrama consisting of short, cylindrical to inflated hyphae, 17–45 × 3–7 μm. Stipitipellis a cutis of narrow, cylindrical hyphae to 3 μm in diameter, with slate blue intracellular pigment. Clamp-connections abundant (Fig. 3.14).

Habitat & distribution: in wet sclerophyll forest litter, only known from the type locality, a small mountain on an island in the D'Entrecasteaux Channel.

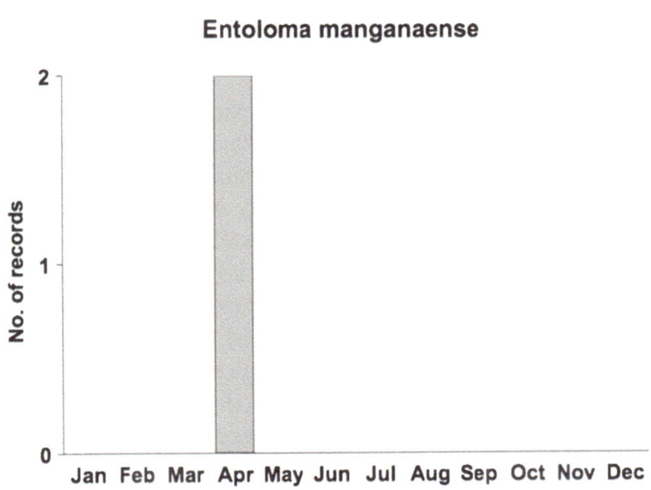

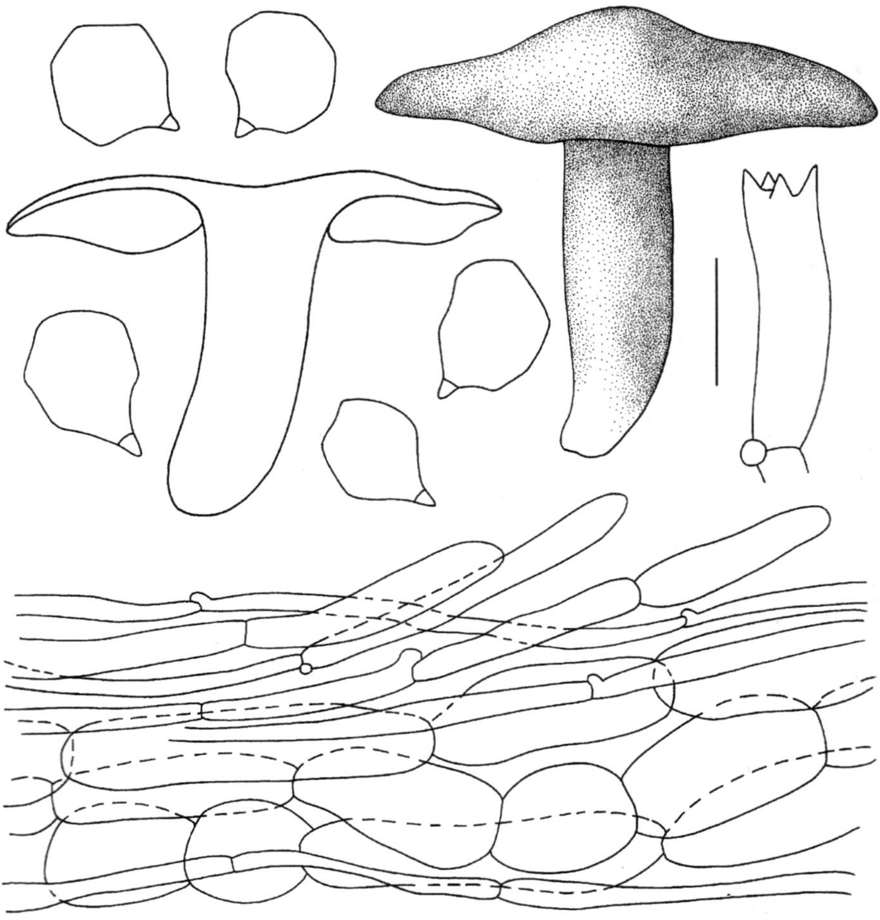

Fig. 3.14 *Entoloma manganaense*. Habit, spores, basidia, and pileipellis. Bar = 10 mm or 10 μm

Collection examined. Australia, Tasmania, Bruny Island, Mt Mangana, 43° 22′ S, 147° 17′ E, 8 April 1999, G. Gates E 369 (holotype, HO 543529; isotype, L); —, 17 April 2011, G. Gates s.n.

This striking species apparently is a rare one. Despite regular collecting trips to the type locality, it has been found only twice in 13 years. *Entoloma manganaense* fits well into subgenus *Entoloma* close to *E. nitidum* Quél., *E. bloxamii* (Berk.) Sacc. and *E. prunuloides* on account of the small, isodiametric, weakly angled spores, the pileipellis in the form of an ixocutis, and abundant clamp-connections. *Entoloma mathinnae* differs mainly in colour of the pileus and stipe. Several similar species are found in South East Asia, viz. *E. assimulatum* Corner & E. Horak from Malaysia, and *E. chalybs* E. Horak from Papua New Guinea (Horak 1980). However, the thick, broad yellow lamellae are distinctive for the present species. *Entoloma cerinum* E. Horak, described from New Zealand, also has yellow lamellae when young, turning pink with age (Horak 2008), but the lamellae in that species are thin and crowded. The European *E. sinuatum* (Bull.: Fr.) P. Kumm. has less brightly yellow coloured, normally thick lamellae and larger, differently shaped spores, and belongs to the distant clade /rhodopolia in the phylogeny (Co-David et al. 2009).

15. **Entoloma mathinnae** G.M. Gates, B.M. Horton & Noordel., Mycotaxon: 107: 176. 2009.

Original diagnosis: Habitus tricholomatoideus. Pileus 40–80 mm, pallide brunneus, glaber. Lamellae distantes, crassae, luteae. Stipes 40×25 mm, albidus demum pallide brunneus, glaber, fibrillosus. Sporae 6.5–8×6–8 μm, isodiametrae. Cystidia absentia. Pileipellis (ixo-)cutis hyphis 2.0–4.0 μm latis pigmento brunneo. Fibulae abundantes.

Holotype: Australia, Tasmania, Mathinna, Ben Ridge Rd, 12 Feb. 2008, G. Gates E 2263 (HO 548298, isotype L).

Etymology: named after the type locality.

Main characters: habit tricholomatoid; pileus light brown; lamellae deep yellow (Plate 3.15).

Description:
Pileus 40–80 mm diam, convex to plano-convex with low umbo, not distinctly hygrophanous, light yellow-brown (5D5), paler towards margin, smooth, glabrous or slightly viscid becoming rimose with age. Lamellae adnate-emarginate, ventricose, moderately crowded, up to 10 mm broad, thickish, bright lemon yellow (3A6–3B6), with entire, concolorous edge, L=ca. 80, l=3–5. Stipe 50–85×15–20 mm, cylindrical, tapering at base, white or pale brown, occasionally with a distinct grey-violet (16E3) hue, innately fibrillose all over, solid then fistulose. Context firm, white. Odour and taste indistinct, fungoid.

Spores 6.5–8×6–8 μm, average 7.3×6.9 μm, Q=1.0–1.2, irregularly 6–8-angled with thin walls and weak angles. Basidia 20–34×7–9 μm, 4-spored, clamped. Lamella edge fertile. Cheilo- and pleurocystidia absent. Hymenophoral trama regular, made up of short, inflated elements, 30–70×8–25 μm with pale yellow, intracellular pigment. Pileipellis a differentiated (ixo-)cutis; suprapellis a slightly gelatinized ixocutis of 2–4 μm wide, cylindrical hyphae; subpellis well differentiated, made up of inflated elements, 35–80×8–25 μm. Pigment brown, intracellular mainly in suprapellis. Pileitrama regular, made up of inflated elements, 50–120×7–27 μm. Clamp-connections abundant (Fig. 3.15).

Habitat & distribution: the habitats of the limited Tasmanian collections range from wet *Eucalyptus delegatensis* high altitude forest with *Leptospermum* understorey to coastal scrub. Known also from Western Australia (K. Syme, pers. comm.) and possibly also New South Wales (see photo of a yellow-gilled Entoloma species on http://www.sydneyfungalstudies.org.au/imagepages/gallery/Page18.htm; webpage viewed February 2012; microscopics of the specimen were not displayed).

Plate 3.15 *Entoloma mathinnae* (Photo Michael Pilkington)

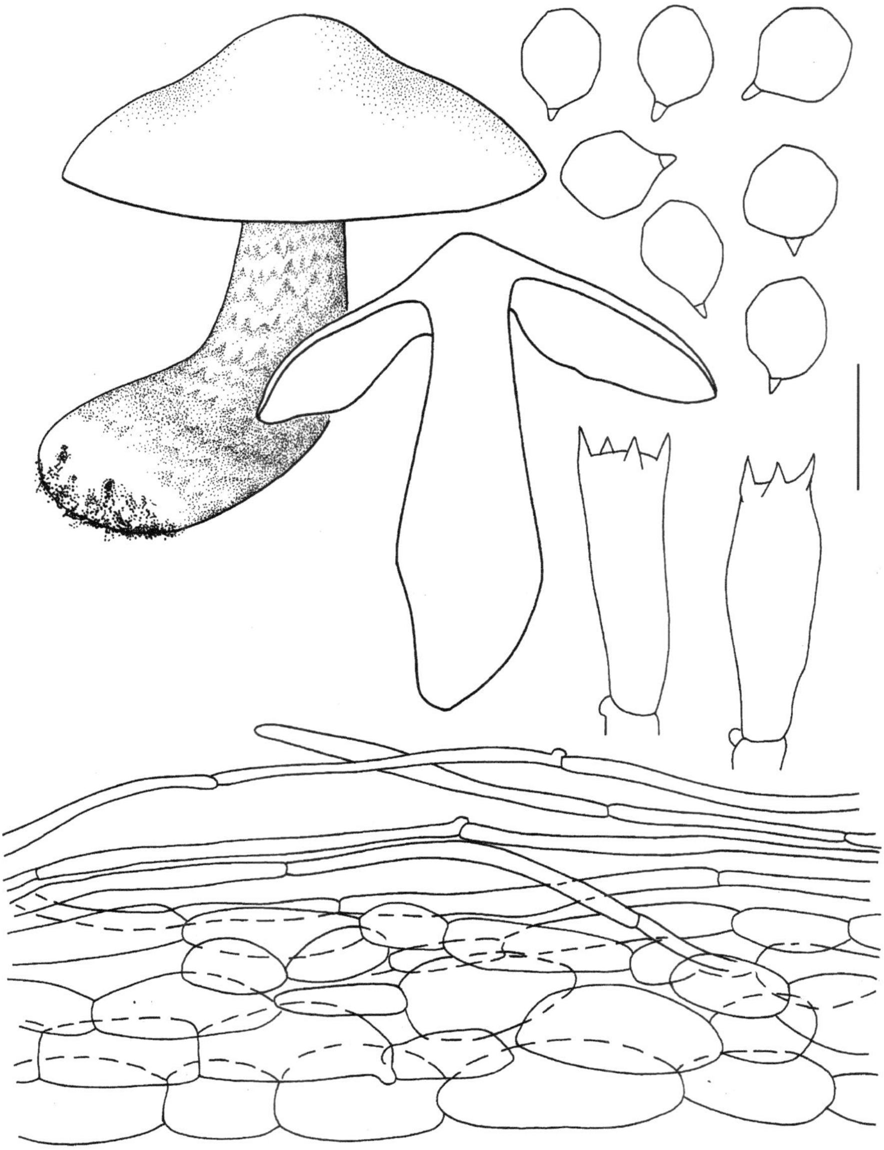

Fig. 3.15 *Entoloma mathinnae.* Habit, spores, basidia, and pileipellis. Bar = 10 mm or 10 μm

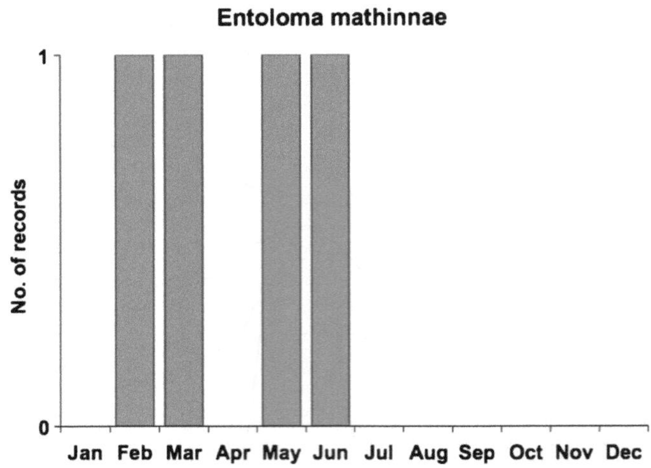

Collections examined. Australia, Tasmania, Mathinna, Ben Ridge Rd, 41° 21′ S, 147° 40′ E, 12 Feb. 2008, G. Gates E 2263 (holotype, HO 548298; isotype, L) & E 2264; —, 25 March 2007, B. Horton s.n. (G. Gates E 2248), HO 548300; Bruny Island, Cape Queen Elizabeth, Track to Mars Beach, 43° 14′ S, 147° 23′ E, 21 June 2009, G. Gates E 2276; Tahune, hanging bridges walk, 43° 06′ S, 146° 43′ E, 20 May 2008, G.Gates E 2267, HO 548299.

Entoloma mathinnae belongs to *Entoloma* subgenus *Entoloma* on account of its tricholomatoid habit, smooth pileus, and small, weakly angled spores. The bright yellow gills without a trace of pink are distinctive. *Entoloma manganaense* G.M. Gates & Noordel. has similar yellow gills, but differs strikingly by the darker red-brown pileus and blue stipe. *Entoloma cerinum* E. Horak from New Zealand also has yellowish lamellae when young, which, however, turn dark pink with age; the pileus is very dark brown, and the hyphae are clampless (Horak 2008). *Entoloma luridum* Hesler from North America has similar yellow lamellae and small spores, but differs in the almost white basidiocarps. *Entoloma sinuatum* (Pers.) P. Kumm. from Europe and North America has a quite different yellow tinge in the lamellae, and much larger, thick-walled and distinctly angled spores, which place it in subgenus *Rhodopolia* (Noordeloos 2004; Co-David et al. 2009). *Entoloma luteifolium* Hesler from Cuba is a small species with a squamulose pileus, typical of subgenus *Cyanula*.

Section ***Nolanidea*** (Fr.) Quél.

Entoloma sect. *Nolanidea* (Fr.) Quél. in Mém. Soc. Émul. Montbéliard, sér. II,5: 118. 1872; *Rhodophyllus* sect. *Nolanidea* (Fr.) Quél. Enchir.: 59. 1886. — *Rhodophyllus* sect. *Clypeati* Romagn. in Bull. trimest. Soc. mycol. Fr. 53: 326. 1937. — *Rhodophyllus* sect. *Apriles* Kühn. & Romagn. ex Romagn. in Bull. Soc. linn. Lyon 43: 332. 1974.

Excl.: *Rhodophyllus* sect. *Nolanidei* sensu Romagn. 1974; Largent 1974 (= *E.* sect. *Rhodopolia*).

Habit tricholomatoid; pileus hygrophanous or not, glabrous or micaceous-fibrillose; spores with well-pronounced angles, thick-walled; pileipellis an (ixo-)cutis to ixotrichoderm of narrow hyphae; pigment intracellular. Clamp-connections abundant in all tissues. — Lectotype (Singer 1951): *Entoloma clypeatum* (L.) Kumm.

Species of this section are all associated with Rosaceae or Ulmaceae, with which they form a special type of ectomycorrhiza (Agerer and Waller 1993). The two species in Tasmania occur with introduced cultivated ornamental plants.

16. Entoloma saundersii (Fr.) Sacc.

Agaricus saundersii Fr., Hymenom. eur.: 192. 1874; *Entoloma saundersii* (Fr.) Sacc., Sylloge fung. 5: 689. 1887; *Rhodophyllus saundersii* (Fr.) Romagn., Bull. trimest. Soc. mycol. Fr. 63: 195. 1947; *Rhodophyllus hiemalis* Lazzari & Blanco, Bol. Gruppo micol. G. Bres. 23: 99. 1980 (invalid); *Entoloma saundersii* var. *hiemalis* [Lazzari & Blanco ex] Bellù in Bol. Gruppo micol. G. Bres. 28: 268. 1985.

Original diagnosis: *A. saundersii* pileo carnoso, e campanulato expanso, obtuso, repando-lobato, adpresse tomentoso, albo, senili fuscescente; stipite solido, aequali, sericeo-fibroso, albo; lamellis leviter adnexis, latis, distantibus, rubellis. *A. majalis* Saund. et Smith t. 46. In terra nuda Angliae, hinc glebulis conspersus. Stipes 3 unc. longus, 1/2 unc. crassus. Pileus 2–3 unc. latus. Species videtur insignis nec hygrophana, ab *A. majali* longe distans; potius affinis *A. sinuato*, prunuloidi vel sequenti.

Holotype: not existing; originally described from England.

Etymology: named after W.W. Saunders.

Selected plates: Bellù in Bol. Gruppo micol. G. Bres. 28: 268. 1985; Cetto, Funghi dal Vero 4, pl. 1426. 1983; Lazzari & Blanco in Bol. Gruppo micol. G. Bres. 23: 99. 1980.

Selected literature: Noordeloos in Persoonia 11: 178–180, Fig. 8. 1981; Noordeloos in Bas et al., Fl. agar. neerl. vol. I: 100–101, fig. 61. 1988.

Main characters: robust species with tricholomatoid habit; pileus sordid white, tinged greyish, micaceous; spores large, isodiametric with blunt angles; associated with Rosaceae and *Ulmus* in parks and gardens (Plate 3.16).

Description:
Pileus 30–65 mm diam, conico-convex to convex with inrolled margin then expanding to irregularly plano-convex with uplifted, undulating marginal zone, sordid white with greyish tinges at centre with age, with silvery fibrillose-micaceous patches, viscid when moist. Lamellae adnexed, ventricose, rather crowded, up to 8 mm broad, thickish, rubbery, cream colour then pale pink with irregular, concolorous edge. Stipe 20–40×8–10 mm, tapering downwards, slightly flexuous, compressed or cylindrical, white turning dingy yellow with age, longitudinally fibrous. Odour and taste farinaceous.

Spores 10–13×9–11 µm, many-angled, isodiametric, Q=1.0–1.2. Basidia 48–56×11–14 µm, 4-spored, clamped. Lamella edge fertile. Cystidia absent. Pileipellis an ixocutis of narrow, cylindrical, 2–5 µm wide hyphae. Hymenophoral and pileitrama regular, made up of short, cylindrical to slightly inflated elements, 20–60×4–12 µm. Clamp-connections abundant in all tissues (Fig. 3.16).

Habitat & distribution: uncommon, occurring in groups, sometimes caespitose, under Rosaceae in gardens, fruiting in spring.

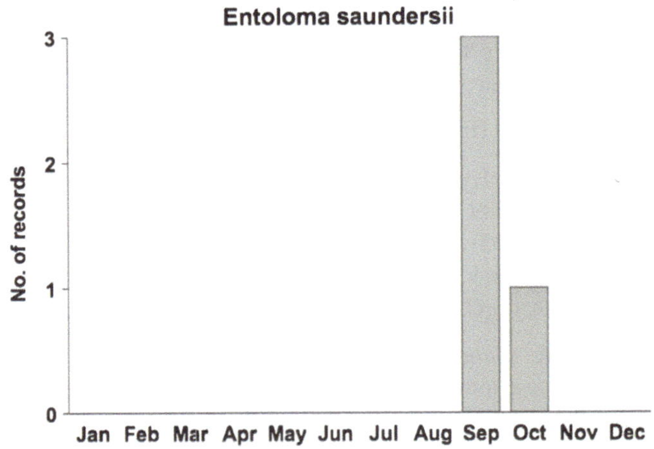

Collection examined. Australia, Tasmania, Taroona, Winmarleigh Ave., 42° 57′ S, 147° 21′ E, 15 Sept. 1998, G. Gates E 55; —, 8 Sept. 2001, G. Gates E 1278; —, 30 Sept. 2004, G. Gates E 2054; —, 3 Oct. 2004, G. Gates E 2060.

The pale pileus with micaceous patches and the large, isodiametric spores are distinctive for *E. saundersii*. This species must have been introduced from Europe, where is frequently found in association with either elms (*Ulmus*) in lanes and gardens, or fruit trees (Rosaceae).

Plate 3.16 *Entoloma saundersii* (Photo Genevieve Gates)

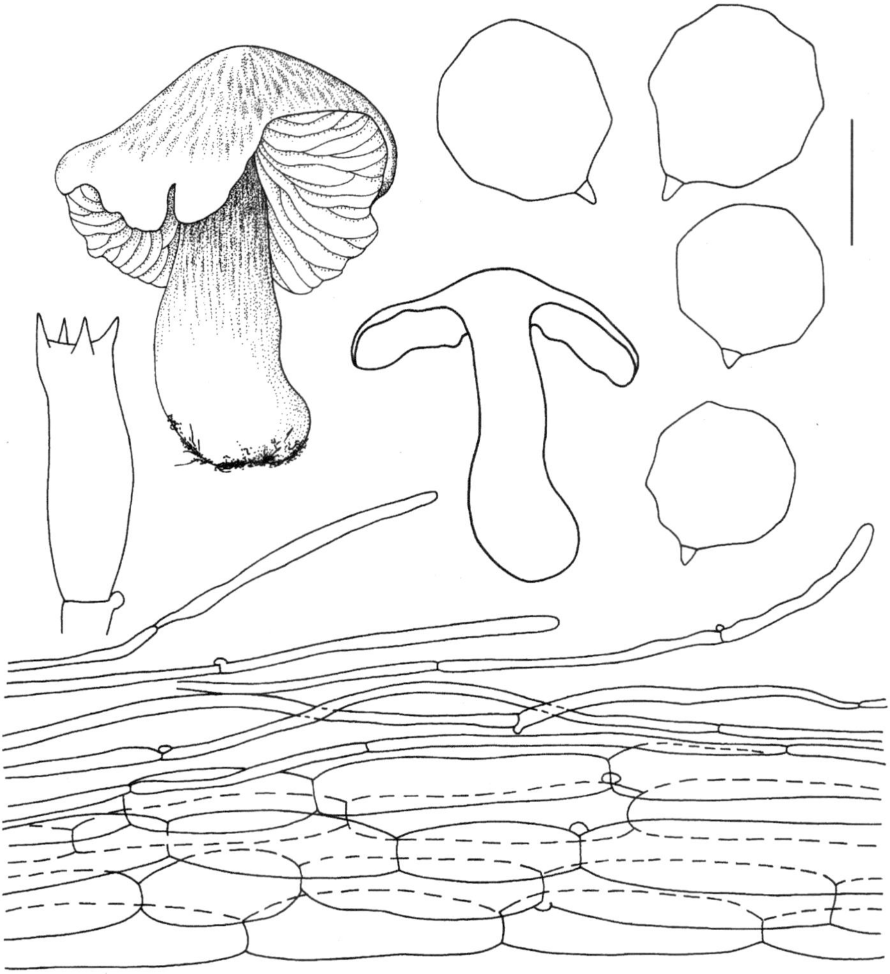

Fig. 3.16 *Entoloma saundersii*. Habit, spores, basidia, and pileipellis. Bar = 10 mm or 10 μm

17. **Entoloma clypeatum** (L.) P. Kumm.

Agaricus clypeatus L., Syst. plant. 2: 1174. 1753; *Entoloma clypeatum* (L.) Kumm., Führ. Pilzk.: 98. 1871; *Rhodophyllus clypeatus* (L.) Quél., Enchir.: 59. 1886; *Hyporrhodius clypeatus* (L.) Schroet. in Cohn, KryptogFl. Schles.: 616. 1889.

Original diagnosis: Agaricus stipatus, pileo hemispherico viscido acuminato, lamellis albis, stipite longo cylindraceis albo. Fungus viscidus papillaris striatus terreis, subtus albus. Habitat in pratis sylvaticis.

Holotype: not existent.

Etymology: clypeus (Lat.) = shield, referring to the shape of the pileus like a Roman shield (Fig. 3.17).

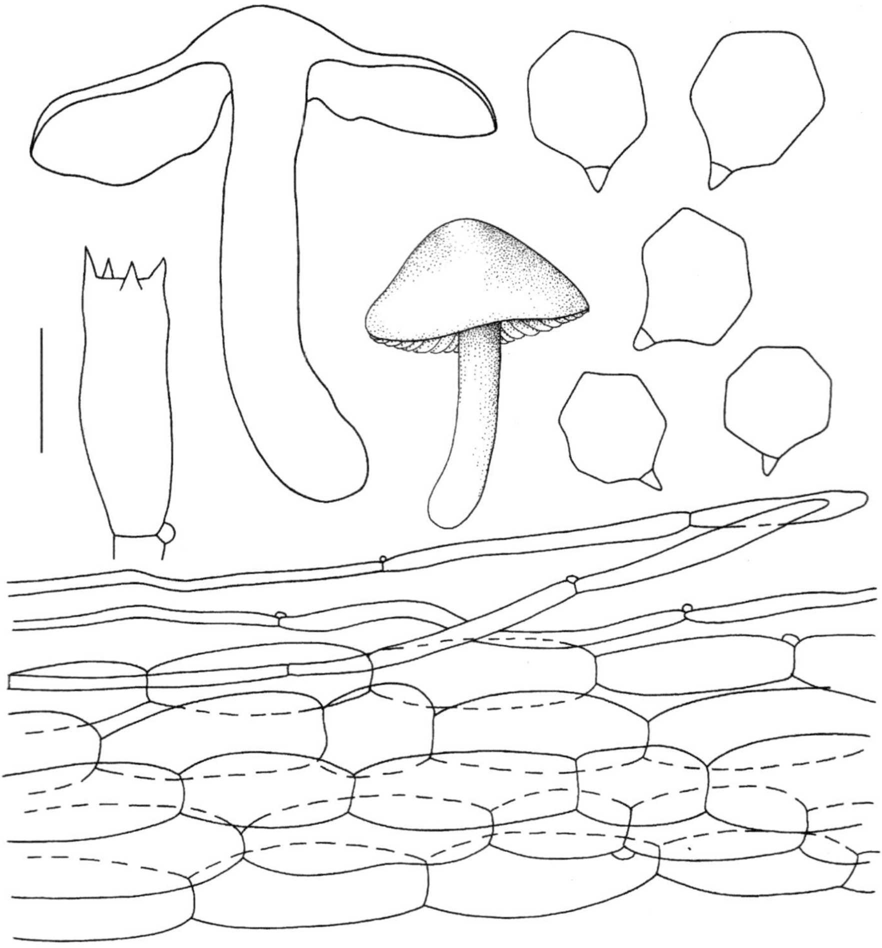

Fig. 3.17 *Entoloma clypeatum*. Habit, spores, basidia, and pileipellis. Bar = 10 mm or 10 μm

Description:
Pileus 20–55 mm diam, conico-convex with obtuse apex, expanding to convex with pronounced umbo with incurved margin, marginal zone undulating-lobed with age, hygrophanous, not translucently striate, blonde-buff to grey-brown (6E4), lubricous when moist, pallescent on drying, becoming radially rugulose. Lamellae sinuate-adnate, segmentiform, crowded, up to 5 mm broad, sordid white then pink often with grey-brown tinge, with irregular, concolorous edge. Stipe 15–46 × 4–11 mm, cylindrical, stout, firm, white to sordid white, longitudinally fibrillose. Odour and taste farinaceous-rancid or cucumber.

Spores 9–11 × 7–9.5 μm, subisodiametric, Q = 1.1–1.3(–1.4) with 5–7 rather pronounced angles, relatively thick-walled. Basidia 23–45 × 7–12 μm, 4-spored, clamped. Lamella edge fertile. Cheilocystidia absent. Pileipellis an ixocutis of narrow, cylindrical hyphae, 2–7 μm wide; subpellis not very distinctly differentiated from pileitrama, which is composed of long chains of short, inflated elements, 30–90×7–20 μm. Pigment brown, intracellular in pileipellis. Clamp-connections abundant in all tissues.

Habitat & distribution: occasionally found in groups, in gardens and public parks, associated with cultivated rosaceous trees, fruiting in spring.

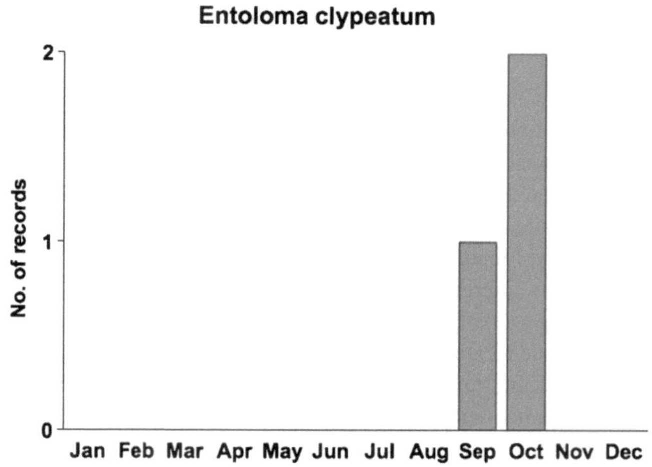

Collections examined. Australia, Tasmania, Sandy Bay, Earl Street, 42° 54′ S, 147° 19′ E, 26 Sept. 2005, G. Gates E 2195; Taroona, Channel Highway, 42° 57′ S, 147° 21′ E, 23 Oct. 2001, coll. Michael Cromer s.n., G. Gates E 1300; —, 13 Oct. 2004, G. Gates E 2060.

The brown-grey, rather smooth pileus without micaceous patches and smaller spores distinguish *E. clypeatum* from *E. saundersii*. It has been introduced from Europe, where it is very common in spring, always in association with rosaceous trees. *Entoloma clypeatum* has also been recorded from New Zealand (Segedin and Pennycook 2001; Pennycook 2004).

Subgenus ***Inocephalus*** Noordel. in Persoonia 11: 145. 1981.

Inocephalus (Noordel.) P.D. Orton in Mycologist 5: 130. 1991.

Habit mycenoid or tricholomatoid; pileus conical or campanulate, expanding with age, fibrillose, tomentose, pruinose, or squamulose; pileipellis a trichoderm or hymeniderm; clamp-connections present or absent.

Type species: *Entoloma inocephalum* (Romagn.) Dennis.

Recent phylogenetic studies by Co-David et al.(2009) and Baroni and Matheny (2011) have shown that *Inocephalus* is polyphyletic.

Section ***Erophila*** (Romagn.) Noordel.

Rhodophyllus sect. *Erophila* Romagn. in Bull. mens. Soc. linn. Lyon 43: 332. 1974; *Entoloma* sect. *Erophila* (Romagn.) Noordel. in Persoonia 11: 86. 1980.

Habit tricholomatoid; pileus radially fibrillose to squamulose; spores many-angled in side-view; lamella edge fertile or heterogeneous; clamps present.

Type species: *Entoloma erophilum* (Fr.) Karsten [= *E. plebejum* (Kalchbr.) Noordel.]

18. **Entoloma plebejum** (Kalchbr.) Noordel.

Agaricus plebejus Kalchbr., Ic. sel. Hymenomyc. Hungariae: 22. 1874; *Entoloma erophilum* var. *plebejum* (Kalchbr.) Sacc., Syll. Fung. 5: 681. 1887; *Entoloma plebejum* (Kalchbr.) Noordel. in Persoonia 12: 462. 1985.

Original diagnosis: In pratis pasciusve, circa Olaszinum. Vernalis, sed etiam sero autumno lectus, in numerosis saepe gregibus. Stipes carnoso-fibrosus, e farcto demum cavus, inaequalis, pro ratione curtis, uncialis, 2‴ et ultra crassus, superficie fibroso striatus, ad basim pube albida obductus, e pallido griseove fuscidulus. Pileus centro tantum caronosus, caeterum tenuis, uncialis et ultra, e convexo planus depressusve, margine inflexus, saepe lobatus, superficie inaequabili, rudi sed non definite squamosa aut fibrillosa, fuligineo-cinereus vel griseo-livescens. Lamellae horizontaliter adnatae, leviter sinuatae, utrinque attenuatae, medio 2‴ latae, subdistantes, firmae, cinerascentes, e sporis demum sordide rubentes. Colores squalidi. Habitus quammodo Ag. Trichol. terrei. amnino e serie Entolomatum Leptonidearum Ag. resuto Fries proximus, ad hoc stipite mode crassiore, haud polito et lamellis subdistantibus nec fere liberis recedens. In centenis exemplaribus vidi semperque constantem reperi.

Holotype: not existent. Type locality: Olaszinum, Hungary.

Etymology: plebejus (Lat.) = common.

Main characters: habit robust mycenoid; pileus greyish brown, fibrillose with micaceous sheen; stipe white; lamellae grey-pink; cheilocystidia scattered, large, up to 100 µm, or absent; spores many-angled.

Description:
Pileus up to 40 mm diam, conical expanding with age, not hygrophanous, not translucently striate, grey-brown, innately fibrillose with micaceous patches. Lamellae adnate-emarginate, moderately distant, grey-pink with entire, concolorous edge. Stipe 30–70 × 3–8 mm, cylindrical, much paler than pileus, almost white, innately fibrillose-striate. Smell and taste farinaceous.

Spores (9–)10.5–14.5(–17) × 7–11 µm, Q = 1.1–1.7, irregularly nodulose-angled in side-view. Basidia 4- rarely also 2-spored, clamped. Lamella edge fertile or heterogeneous. Cheilocystidia, if present, sparse, 35–100 × 5–20 × 2–9 µm, fusiform to slenderly lageniform. Pileipellis a cutis with transitions to a trichoderm, made up of cylindrical to inflated hyphae, 10–20 µm wide, with inflated terminal elements, 55–125 × 15–25 µm. Pigment abundant, brown, intracellular in pileipellis. Clamp-connections present (Fig. 3.18).

Habitat & distribution: uncommon, in parks and gardens under introduced trees.

Fig. 3.18 *Entoloma plebejum.* Habit, spores, basidia, cheilocystidia, and pileipellis. Bar= 10 mm or 10 μm

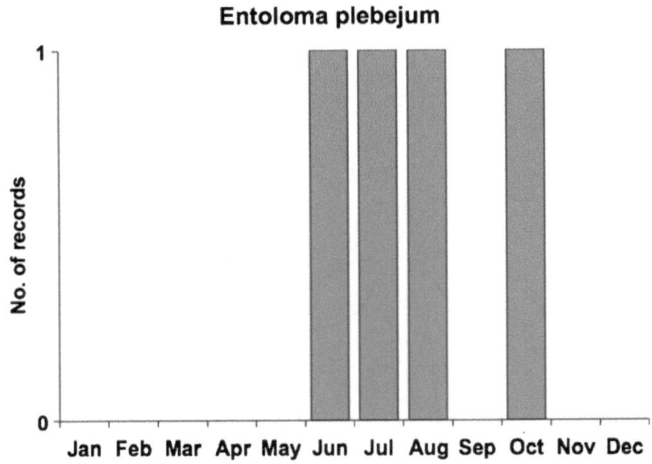

Collections examined. Australia, Tasmania, Marion Bay, 'Marchwiel' property, 42° 48′ S, 147° 52′ E, 5 June 1999, G. Gates E 565; Sandy Bay, suburban garden, 42° 54′ S, 147° 19′ E, 27 Aug. 2000, G. Gates E 1000; —, 1 Oct. 2001, G. Gates E 1295; Taroona, school sports oval, 42° 57′ S, 147° 21′ E, 20 July 1998, G. Gates E 18.

Entoloma plebejum is probably an introduced species, found in urban areas under exotic trees or in disturbed habitats. It originates from Europe, where it is a vernal and occasionally autumnal species. The micaceous pileus, nodulose spores, and large cheilocystidia are distinctive.

Subgenus ***Nolanea*** (Fr.: Fr.) Noordel.

Agaricus tribus Nolanea Fr., Syst. mycol. 1: 10. 1821; *Agaricus* subgenus *Nolanea* (Fr.) Loud., Encl. Pl.: 998. 1929; *Nolanea* (Fr.) Kumm., Führ. Pilzk.: 24. 1871; *Rhodophyllus* subgenus *Nolanea* (Fr.) Quél., Enchir. Fung.: 63. 1886; *Hyporrhodius* subgenus *Nolanea* (Fr.) Schroet. in Cohn, KryptogFl. Schles. 3: 613. 1889. — *Latzinea* O. K., Rev. Gen. Pl. 2: 87. 1891. *Lanolea* Nieuwl. in Amer. Midl. Nat. 4: 381. 1916. — *Arenicola* Velen., Novitates mycologicae novissimae: 62. 1947.

Co-David et al. (2009) made clear that the traditional concept of the (sub)genus *Nolanea* is polyphyletic. More work must be done, however, including many more samples, to sort out the various clades and made a new classification. We therefore maintain the traditional morphological concept here, uniting taxa with a predominantly mycenoid habit and non-cuboid spores.

Section ***Austrofernandae*** Noordel., sect. nov.
MycoBank #562979.
Diagnosis: Habit mycenoid; pileus fibrillose, hygrophanous; pileipellis a differentiated cutis, sometimes with transitions to a trichoderm; pigment of two kinds: encrusting the hyphae of the pileipellis and in addition intracellular, in the form of golden brown, agglutinated clots in the surface hyphae; clamp-connections absent.

Type species: *Entoloma chrysopus* Noordel. & G.M. Gates.

Species of this section resemble those of section *Fernandae* in Europe by the peculiar double pigmentation of the pileipellis, as well as the clampless hyphae.

Plate 3.17 *Entoloma chrysopus* (Photos Michael Pilkington and Machiel Noordeloos)

However, the spores of section *Austrofernandae* are much larger and differently shaped. Two species occur in Tasmania, viz. *Entoloma chrysopus* and *E. maldea*. Similar species also occur in New Zealand.

19. **Entoloma chrysopus** G.M. Gates & Noordel., Persoonia 19: 169. 2007.

Original diagnosis: Pileus 20–35 mm latus conicus vel campanulatus saepe cum papilla ornatus hygrophanus translucido-striatus porphyreobrunneus innate radialiter fibrillosus vel leviter pruinoso-micaceus. Stipes 40–80 × 2–5 mm aureus leviter innate fibrillosus plus minusve politus. Odore aromaticus. Sporae 9.5–14 × 6.5–9.0 µm 5–7-angulatae interdum complexae pseudocruciformae. Acies lamellarum fertilis. Pileipellis cutis differentiatus e hyphis cylindraceis 4–10 µm latis constituis pigmentis leviter incrustantibus vel granuloso-intracellulosus. Fibulae desunt.

Holotype: Australia, Tasmania, Mt Field National Park, Lady Barron Falls, 42° 41′ S, 146° 42′ E, 1 April 1999, G. Gates E 335 (HO 543552; isotype in L).

Etymology: chrysos (Gr.) = yellow; pus (Gr.) = stipe, referring to the yellow stipe.

Main characters: habit mycenoid, often with pronounced conical umbo; pileus purple-brown; stipe golden yellow, subpolished; clamp-connections absent; odour and taste of bubblegum (Plate 3.17).

Description:
Pileus 20–50 mm diam, conical to campanulate, expanding to conico-convex with pronounced small umbo, with deflexed then straight margin, hygrophanous, when moist translucently striate at margin up to half the radius, brown with vinaceous hue or purple-brown, slightly paler and with a slight lilac-pink tinge when fresh near margin (2.5YR 4–3/2, 5YR 2.5–3/2), pallescent along radial streaks when drying, rather pronouncedly innately radially fibrillose or sometimes slightly pruinose-micaceous or finely tomentose overall. Lamellae narrowly adnexed, deeply emarginate, ventricose, moderately crowded, up to 6 mm deep, brown to red-brown with pink hue (7.5YR 6–4/4, 5YR 4/3), sometimes transversely veined, slightly paler towards irregular, concolorous edge, L = 35–50, l = 2–5. Stipe 40–80 × 2–5 mm, cylindrical, gradually broadening towards base, at first appearing rather pale with white pruina overall in youth, glabrescent with age or bruising, then pale yellow-brown to golden yellow (10YR 6–8/6, 10YR 8/4, 2.5Y 8/6–8), slightly fibrillose-striate to almost polished, dry, brittle, with white basal tomentum. Context thin, concolorous with surface. Odour distinctly of bubblegum. Taste of bubblegum.

Spores 9.5–14 × 6.0–9.0 μm, Q = (1.0–)1.2–1.7, rather pronouncedly 5–7-angled in side-view, sometimes complex and somewhat distorted to irregularly 4-angled, almost cruciform. Basidia 18–32 × 6–11 μm, 4-spored, clampless. Lamella edge sterile. Cheilocystidia not observed. Pileipellis a differentiated cutis of septate, cylindrical, 4–10 μm in diameter, hyphae with clavate terminal elements, with parietal and rarely also finely encrusting pigment as well as intracellular pigment both as agglutinated dark brown granules and plasmatic; terminal elements often with a slightly thickened, refringent apex. Pileitrama regular, made up of cylindrical to inflated elements, to 220 × 5–19 μm. Stipitipellis a cutis of narrow, cylindrical hyphae, 3–9 μm in diameter. Clamp-connections absent (Fig. 3.19).

Habitat & distribution: widespread and very common throughout Tasmania in all types of wet forest. Fruiting throughout the year, with a fair number of records in spring and summer months (Group 5, in Chap. 1, see section "Fruiting Patterns of Species of Entolomataceae").

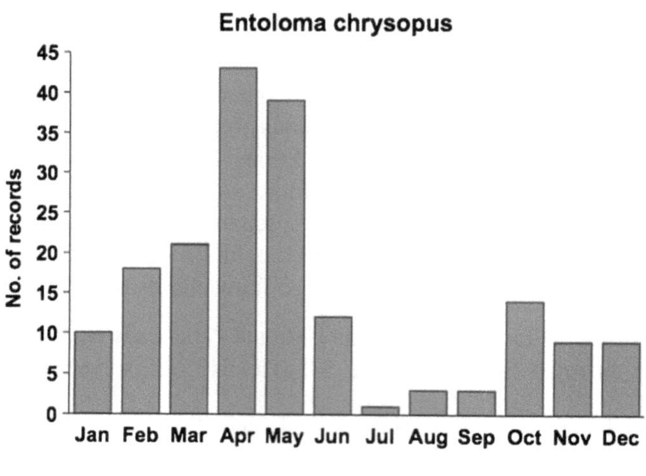

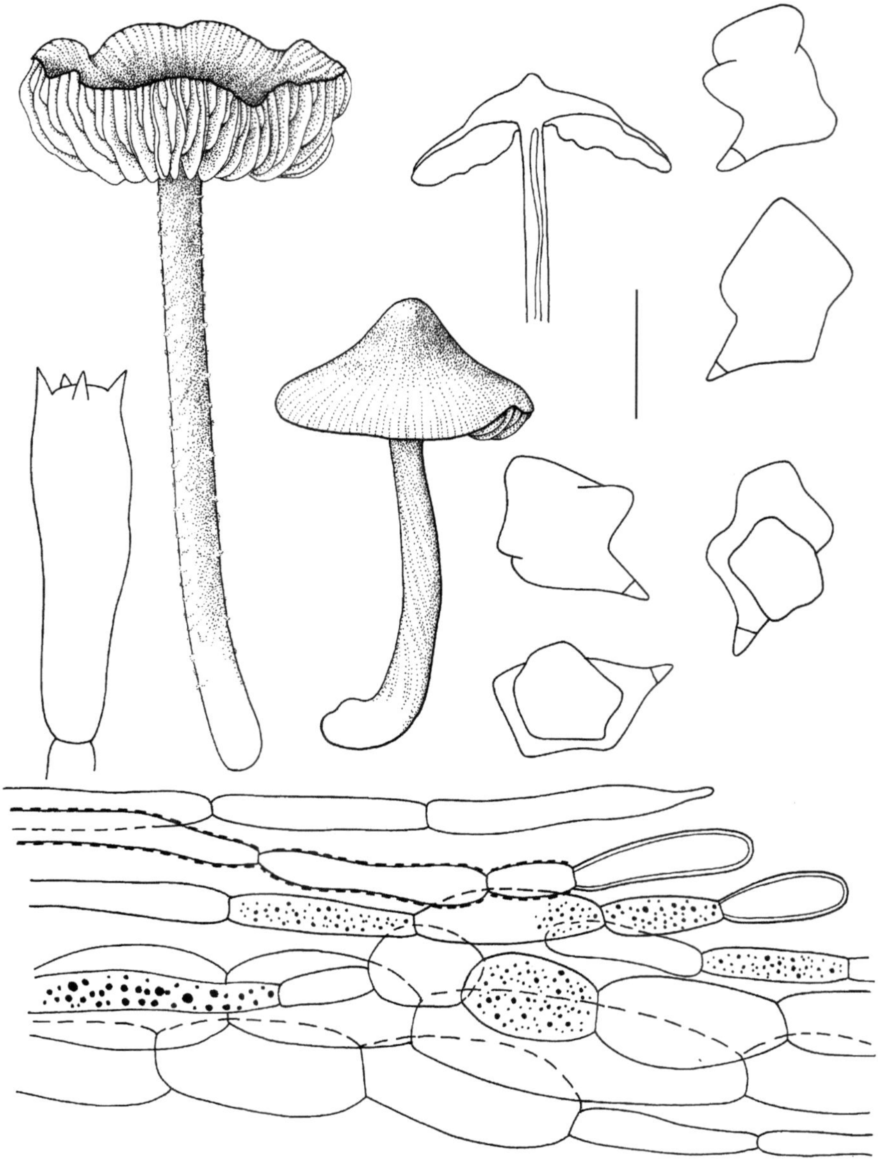

Fig. 3.19 *Entoloma chrysopus*. Habit, spores, basidia, and pileipellis. Bar = 10 mm or 10 μm

Collections examined. Australia, Tasmania, Bermuda Road, 43° 04′ S, 146° 54′ E, 17 April 2003, G. Gates E 1719; —, 16Aug. 2003, G. Gates E 1905; Blue Tier, 41° 11′ S, 148° 00′ E, 22 May 2010, M.E. Noordeloos 2010052; —, Goblin Walk, 14 April 2001, G. Gates E 1094; Cethana, Lacey's property, 43° 22′ S, 147° 17′ E, 2 March 2002, G. Gates E 1446; Clarks Cliffs, 43° 06′ S, 147° 47′ E, 4 May 2002, G. Gates E 1504; Collinsvale, Myrtle Forest, 42° 52′ S, 147° 09′ E, 14 March 2002, G. Gates E 1454; Duckhole Lake Track, 43° 22′ S, 146° 53′ E, 20 April 2004, M.E.

Noordeloos (CTAB134, CTAB137); Growling Swallet, 42° 41′ S, 146° 30′ E, 6 April 2000, G. Gates E 817; —, 22 April 2004, M.E. Noordeloos 2004066; Hellyer Gorge, 41° 16′ S, 145° 37′ E, 26 April 2002, G. Gates E 1490; Lake St. Clair NP, Echo Point to Cynthia Bay, 42° 04′ S, 146° 09′ E, 21 March 2009, M.E. Noordeloos 2009036 (L); Leap Loop, Arve Rd., 43° 09′ S, 146° 47′ E, 31 Aug. 2004, G. Gates E 2050; Liffey Falls, 41° 42′ S, 146° 46′ E, 28 March 1999, G. Gates E 305; —, 25 April 2000, G. Gates E 873; Little Florentine River, F5 Road, 42° 44′ S, 146° 25′ E, 27 Nov. 1999, G. Gates E 767; Marriotts Falls, 42° 43′ S, 146° 39′ E, 7 Oct. 2003, G. Gates E 710, E 711; —, 27 Nov. 1999, G. Gates E 766; Mt Field National Park, Falls Loop, 42° 41′ S, 146° 42′ E, 1 April 1999, G. Gates E 335 (holotype, HO 543552; isotype, L); —, 9 May 1999, G. Gates E 482; —, 4 Nov. 1999, G. Gates E 731; —, 1 June 2010, M.E. Noordeloos 2010069; —, 17 April 2004, M.E. Noordeloos 2004041; Mt Wellington, Betts Vale Track, 42° 55′ S, 147° 15′ E, 24 June 1999, G. Gates E 628; —, Fern Glade, 25 Sept. 1999, G. Gates E 708; —, Jacksons Bend, 14 Oct. 1998, G. Gates E 75; —, 29 Aug. 1999, G. Gates E 687; Old Hartz Mt Track, 43° 12′ S, 146° 52′ E, 19 March 2002, G. Gates E 1461; Reuben Falls, 43° 00′ S, 146° 40′ E, 23 Nov. 1999, G. Gates E 763; —, 4 March 2000, G. Gates E 797; Tahune, Huon Pine Walk, 43°06′ S, 146°44′ E, 18 Oct. 2001, G. Gates E 1294; Warra LTER site, 43° 06′ S, 146° 41′ E, numerous observations between April 2006 and June 2007, G. Gates s.n.

Entoloma chrysopus is a striking *Nolanea* species with its usually prominently umbonate pileus with a distinct brown-purple or vinaceous brown tinge, the often yellowish tinges in the stipe and the odour of bubblegum. Forms are also found with a more greyish stipe and less distinctive odour. However, this species can be easily recognized microscopically by the structure and pigmentation of the pileipellis, with golden granules in some hyphae, as in the closely related *E. maldea*, and the peculiar complex spores, ranging from 5- to 6-angled to almost cruciform. *Entoloma translucidum* E. Horak (= *E. perzonatum* E. Horak) resembles *E. chrysopus* in the pigmentation in the pileipellis and the clampless hyphae, but differs in colour, the often squamulose centre of pileus and fibrillose-twisted stipe surface. Encrusting pigments are absent. Furthermore, it lacks the distinctive odour of *E. chrysopus*.

20. **Entoloma maldea** G.M. Gates & Noordel., Persoonia 19: 171, 2007.

Original diagnosis: Pileus 20–50 mm conicus vel campanulatus saepe cum papilla ornatus hygrophanus translucido-striatus brunneus vel griseobrunneus innate radialiter fibrillosus vel leviter pruinoso-micaceus. Stipes 30–100×2–10 mm pallide griseobrunneus valde fibrilloso-striatus. Odore plus minusve farinaceus. Sapore farinaceus. Sporae 7.0–9.0(–10.0)×6.0–8.0 μm 5–7-angulatae. Cystidia nulla. Pileipellis cutis differentiatus e hyphis cylindraceis 4–15 μm latis constituis pigmentis leviter incrustantibus vel granuloso-intracellulosus, subpellis a elementis inflata constituis. Fibulae desunt.

Holotype: Australia, Tasmania, Chauncy Vale, 42° 37′ S, 147° 16′ E, 29 Sept. 2005, G. Gates E 2197 (HO 543537; isotype in L).

Etymology: maldea (aboriginal) = different.

Main characters: pileus moderately dark brown, silky fibrillose; stipe much paler, greyish brown, silky fibrillose; pigment of three types, viz. intracellular diffuse, granular, and finely encrusting (Plate 3.18).

Plate 3.18 *Entoloma maldea* (Photo Machiel Noordeloos)

Description:
Pileus 20–50 mm diam, campanulate or conico-convex with large umbo, expanding to plano-convex with umbo, with straight margin, hygrophanous, translucently striate up to half the centre or faintly striate at margin only, when moist moderately dark brown with darker greyish brown centre, pallescent to moderately dark or pale reddish brown, innately radially fibrillose or rimose, when wet appearing almost glabrous, fibrils sometimes agglutinated in small adpressed squamules, at centre sometimes slightly villose to very minutely squamulose, becoming micaceous-lustrous particularly on drying. Lamellae narrowly adnate, almost free, ventricose, moderately to very crowded, up to 8 mm broad, thin, white then brownish pink with irregular, concolorous edge, lamellulae in two tiers. Stipe 30–100 × 2–10 mm, cylindrical, often distinctly broadened towards base, pale greyish brown, subglabrous, almost to distinctly polished, entirely densely lustrous-silky fibrillose-striate, sometimes twisted, base whitely tomentose. Odour not distinct to farinaceous. Taste farinaceous.

Spores 7.0–9.0(–10.0) × 6.0–8.0 μm, Q = 1.2–1.6, Qav = 1.3–1.4, regularly to irregularly 5–7-angled in side-view. Basidia 18–30 × 7–11 μm, 4-spored, clampless. Lamella edge fertile. Cheilocystidia absent. Pileipellis a cutis with transition to a trichoderm, made up of cylindrical hyphae, terminal elements sometimes slightly inflated-fusiform, to 15 μm in diameter; subpellis well-differentiated, made up of inflated elements, 20–70 × 4–20 μm. Pigment brown, diffusely intracellular and also, particularly in the narrower hyphae of pileipellis and upper trama, consisting of rather large, clustered golden brown granules, in addition also minutely encrusting, particularly the narrower hyphae in subpellis. Pileitrama regular, made up of cylindrical to fusiform elements, up to 20 μm in diameter. Stipitipellis a cutis of narrow, cylindrical hyphae, up to 12 μm in diameter. Clamp-connections absent (Fig. 3.20).

Habitat & distribution: widespread in wet and dry sclerophyll forest, at a range of altitudes.

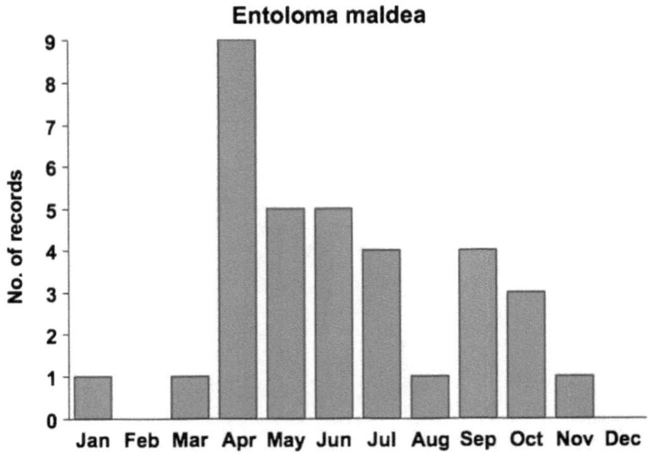

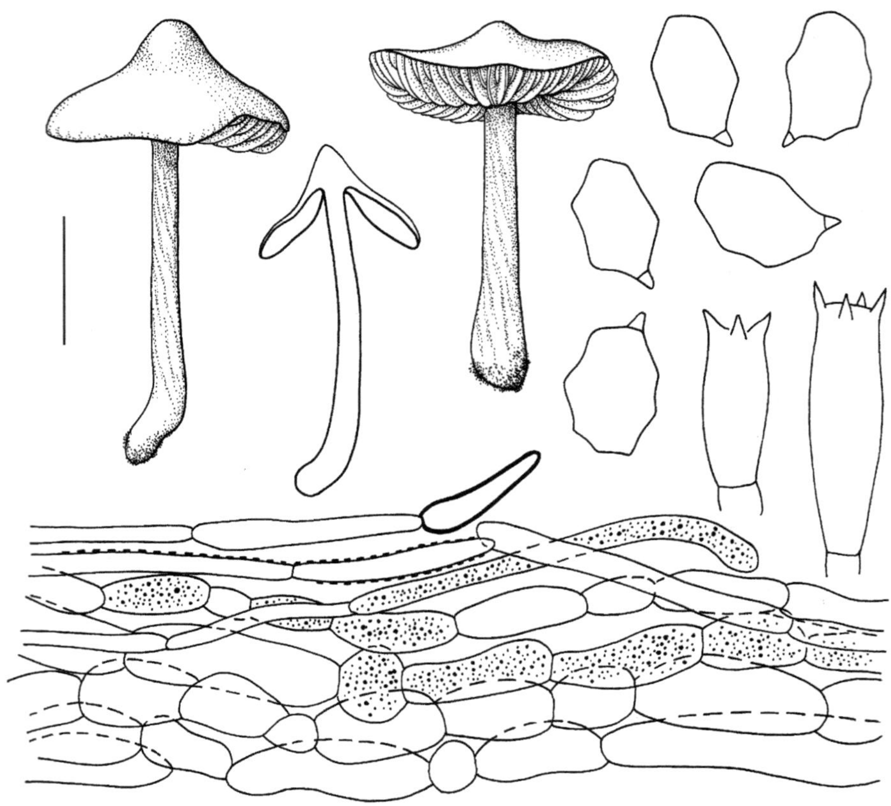

Fig. 3.20 *Entoloma maldea.* Habit, spores, basidia, and pileipellis. Bar = 10 mm or 10 μm

Collections examined. Australia, Tasmania, Blue Tier, 41° 11′ S, 148° 00′ E, 1 April 2000, G. Gates E 808; —, 14 April 2001, G. Gates E 1096; Chauncy Vale, 42° 37′ S, 147° 16′ E, 15 Sept. 2005, G. Gates E 2192; —, 29 Sept. 2005, G. Gates E 2197 (holotype, HO 543537; isotype, L); —, 6 Oct. 2005, G. Gates E 2202; Growling Swallet, 42° 41′ S, 146° 30′ E, 20 April 1999, G. Gates E 437; Kingston, 43° 00′ S, 147° 18′ E, 12 June 2001, coll. J.A. Cooke s.n., G. Gates E 1243; —, 15 July 2002, coll. J.A. Cooke s.n., G. Gates E 1693; —, 21 Sept. 2002, coll. J.A. Cooke s.n., G. Gates E 1341; —, 25 May 2003, coll. J.A. Cooke s.n., G. Gates E 1826; Marriotts Falls, 42° 43′ S, 146° 39′ E, 2 June 2001, G. Gates E 1210; Mt Field, Lady Barron Falls Loop, 42° 41′ S, 146° 42′ E, 1 June 2010, M.E. Noordeloos 2010068; Mt Wellington, Myrtle Gully, 42° 54′ S, 147° 15′ E, 6 Sept. 1999, G. Gates E 692; Scottsdale, Cuckoo Falls Track, 41° 15′ S, 147° 37′ E, 28 April 2006, M.E. Noordeloos 2006033a & b; Styx Valley, Christmas Tree Grove, 42° 50′ S, 146° 41′ E, 13 May 2010, M. E. Noordeloos 2010032; Summerleas Road, Fern Tree, Vrana's property, 42° 56′ S, 147° 16′ E, 11 Nov. 2004, G. Gates E 2062; Tahune, Stringybark Loop, 43° 06′ S, 146° 44′ E, 4 March 2000, G. Gates E 795; Timbs Track, 42° 44′ S, 146° 25′ E, 11 July 2002, G. Gates E 1597.

The distinct features of *E. maldea* are the rather firm basidiocarps with broadly umbonate, very dark brown pilei with contrasting paler stipes. Microscopically it is easily identified by the curious pigment pattern of the pileipellis with hyphae full of dark reddish brown pigment granules, and the brown diffusely intracellular pigment of the differentiated subpellis. This peculiar character is shared with *E. chrysopus*, from which it differs in colour, smell, and in having much smaller spores. Both species apparently belong to the complex of *E. translucidum* E. Horak characterized by a nolaneoid habit, trichodermal aspect of the pileipellis with a distinct subpellis of inflated elements, and clampless hyphae. The New Zealand species lacks encrusting pigment, however, and is described as having a squamulose centre to the pileus. It would be interesting to compare these Southern Hemisphere species at the molecular level with species of the section *Fernandae*, a group of closely related species distributed in the Northern Hemisphere with a similar pigment pattern.

Section ***Papillata*** (Romagn.) Noordel.

Rhodophyllus sect. *Papillata* Romagn. in Bull. mens. Soc. linn. Lyon 43: 330. 1974: *Entoloma* sect. *Papillata* (Romagn.) Noordeloos in Persoonia 10: 246. 1979.

Pigment encrusting at least the narrowest hyphae of pileipellis and pileitrama, sometimes also with intracellular pigment, rarely without encrusting pigment; clamp-connections present, rarely absent; spores isodiametric to heterodiametric, never cuboid or cruciform; cheilocystidia usually absent.

Type species: *Entoloma papillatum* (Bres.) Noordel.

21. **Entoloma obscureotenax** G.M. Gates & Noordel., Persoonia 17: 173. 2007.

Original diagnosis: Pileus 20–40 mm latus obscure brunneus umbonatus hygrophanus pauce translucido-striatus glaber vel innate fibrillosus centro leviter rugulosus. Stipes 25–70 × 1–3 mm obscure brunneus politus. Caro tenax. Odore saporeque forte acidulus. Sporae 10–13 × 8.0–10 μm 5–8-angulatae. Acies lamellarum heterogenea.

Plate 3.19 *Entoloma obscureotenax* (Photo Michael Pilkington)

Cheilocystidia 18–50 × 6–16 µm, late clavata. Pileipellis cutis elementis cylindraceis 2–7 µm latis pigmentis incrustatis constitutis. Fibulae presentes.

Holotype: Australia, Tasmania, Marriotts Falls, 42° 43′ S, 146° 39′ E, 29 May 2003, G. Gates E 1834 (HO 543553; isotype in L).

Etymology: obscureus (Lat.) = dark; tenax (Lat.) = tough, horny, referring to the dark-coloured basidiocarp with horny consistency.

Main characters: basidiocarps mycenoid with a rather tough, cartilaginous context, entirely very dark brown, very hygrophanous from centre to margin, becoming bright yellow; spores heterodiametric; cheilocystidia present; pigment encrusting; clamp-connections present (Plate 3.19).

Description:
Pileus 20–50 mm diam, conico-convex to plano-convex with small pointed umbo, with straight margin, finally irregularly shaped applanate or somewhat concave with uplifted, undulating margin, very hygrophanous, translucently striate at margin or not, very dark red-brown to almost black (10YR 2/2–3, 10YR 3/2–3, 7.5YR 3/2–4, 7.5YR 4/4), sometimes with olivaceous tinges, drying from the centre out in radial streaks to a rather striking golden or olivaceous yellow, glabrous, dry, waxy, cracking with age. Lamellae very narrowly adnate with or without decurrent tooth, ventricose, moderately distant to crowded, up to 8 mm broad, thickish with some

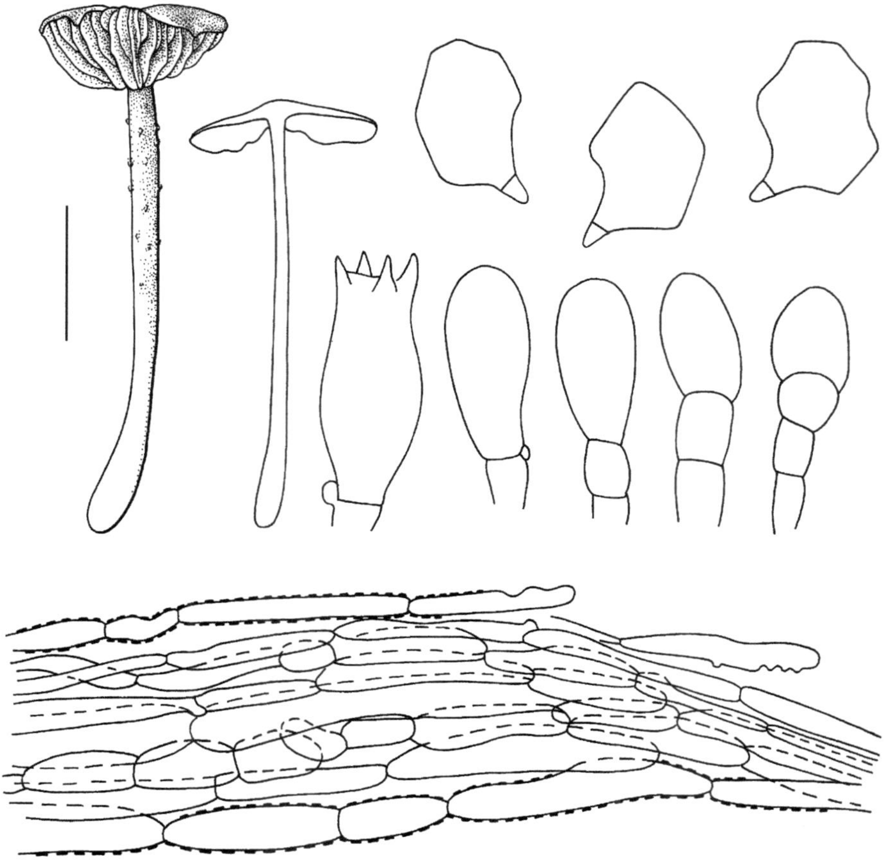

Fig. 3.21 *Entoloma obscureotenax*. Habit, spores, basidia, cheilocystidia, and pileipellis. Bar = 10 mm or 10 μm

intervenosing, very dark yellow-brown to red-brown (7.5YR 3/4–2, 7.5YR 2.5/2), slightly paler towards edge with slightly pruinose, differently coloured, hyaline-brown edge, L = 26–30, l = 3–5. Stipe 25–70 × 1–5 mm, cylindrical or compressed with longitudinal groove, slightly broadened towards base, slender, dry, brittle, very dark red-brown, stuffed brown inner, glabrous, dull, with whitish basal tomentum. Odour spermatic. Taste sweetish, saliva inducing.

Spores 10–13 × 8.0–10 μm, on average 10.5–11 × 8.7–9.2 μm, Q = 1.15–1.4, Qav = 1.2–1.25, heterodiametric, with 5–8 pronounced angles. Basidia 22–35 × 7–11 μm, 4-spored, clamped. Lamella edge sterile (in part) with dense clusters of broadly clavate cheilocystidia, 18–50 × 6–16 μm, often in chains and with refringent, often slightly thickened, walls. Hymenophoral trama regular, made up of long, fusiform elements, 140–220 × 7–19 μm. Pileipellis a cutis of narrow, cylindrical hyphae, 2–7 μm in diameter, gradually passing into pileitrama, which is similar to hymenophoral trama. Pigment brown, parietal and encrusting in pileipellis and pileitrama. Clamp-connections present (Fig. 3.21).

Habitat & distribution: widespread in wet sclerophyll forests, fruiting mainly in autumn (April–May), extending to the winter months July and August (Group 4, in Chap. 1, see section "Fruiting Patterns of Species of Entolomataceae").

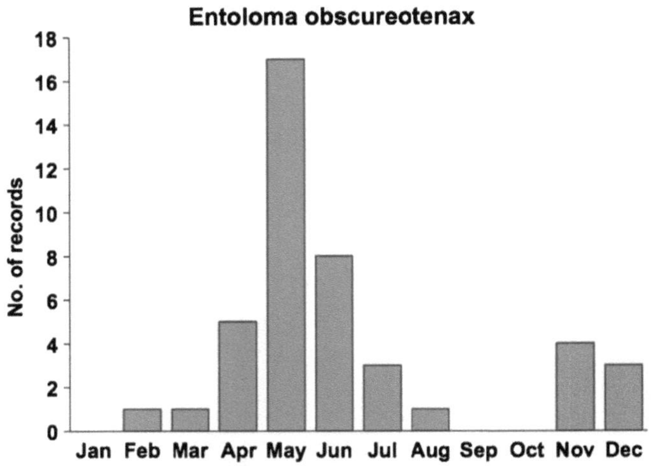

Collections examined. Australia, Tasmania, Arve River Streamside Reserve, 43° 10′ S, 146° 48.5′ E, 2 July 2002, G. Gates E 1592; Evercreech Forest Reserve, 41° 24′ S, 147° 58′ E, 16 June 2003, G. Gates E 1878; Hartz Rd, wildfire plot, 43° 10′ S, 146° 47′ E, 6 May 2010, G. Gates E 2280; Junee Caves State Reserve, 42° 44′ S, 146° 36′ E, 24 May 2003, G. Gates E 1823; Kermandie Falls, Lower Track, 43° 12′ S, 146° 52′ E, 13 June 2005, G. Gates E 2182; —, Upper Track, 11 May 1999, G. Gates E 494; Marriotts Falls, 42° 43′ S, 146° 39′ E, 13 May 2000, G. Gates E 910; —, 29 May 2003, G. Gates E 1834 (holotype, HO 543553; isotype, L); Mt Field National Park, Falls Loop, 42° 41′ S, 146° 42′ E, 1 April 1999, G. Gates E 336; —, 4 Nov. 1999, G. Gates E 729; —, 13 Nov. 1999, G. Gates E 748; —, 17 April 2004, M.E. Noordeloos 2004028; —, 1 June 2010, M.E. Noordeloos 2010072; Scottsdale, Cuckoo Falls Track, 41° 15′ S, 147° 37′ E, 24 May 2010, M.E. Noordeloos 2010057.

Entoloma obscureotenax is well characterized by its overall dark brown-black colour of the basidiocarp, including the lamellae, the striking hygrophanous yellowing of the pileus, and the presence of cheilocystidia. No similar species has been described from New Zealand by Horak (2008). *Entoloma kerocarpus* Hauskn. & Noordel. from Europe is similar, but lacks the yellow tinges in the centre of the pileus, and has differently shaped spores and cystidia.

22. **Entoloma psilocyboides** G.M. Gates & Noordel., Cryptogamie Mycologie 30: 130. 2009.

Original diagnosis: Pileus 7–21 mm latus, conicus demum convexus, umbonatus, hygrophanus, translucido-striatus, obscure brunneus vel griseo-brunneus, pallescens, glabrus. Lamellae adnatae, moderate distantes, pallide brunneo-roseae. Stipes 23–46 × 1–2 mm, cylindraceus, griseo-brunneus, leviter pruinosus. Sporae 8–10 × 6–8 μm, 5–6-angulatae. Acies lamellarum heterogeneae (cheilocystidia et basidia adsunt). Cheilocystidia 30–65 × 5–15 μm, clavata. Pileipellis ex hyphis

Plate 3.20 *Entoloma psilocyboides* (Photo Machiel Noordeloos)

usque ad 5 μm latis cutem formantibus constituta pigmentis intracellularibus. Granula lucentia abundantia. Fibulae absentes.

Holotype: Australia, Tasmania, Growling Swallet, 42° 41′ S, 146° 30′ E, 1 May 2001, G. Gates E 1145 (HO 548338; isotype L).

Etymology: psilocyboides, reminiscent of *Psilocybe montana* (Pers.) P. Kumm. (Plate 3.20).

Main characters: small mycenoid species with distinct small umbo; lamellae dark brown; cheilocystidia large; pigment intracellular; spores simple, 5–6-angled.

Description:
Pileus 7–21 mm diam, conical, expanding to convex with small, acute umbo, with straight margin, hygrophanous, translucently striate, rather dark brown to grey-brown, pallescent on drying to grey-brown, glabrous, dry, sometimes becoming slightly rimose with age. Lamellae adnate, seceding so as to appear adnexed, subventricose, moderately crowded, up to 4 mm broad, pale to dark brown with concolorous edge. Stipe 23–46 × 1–2 mm, cylindrical, slender, stuffed, dark greyish brown, finely pruinose under a hand lens. Odour somewhat spicy or fruity. Taste like fresh grass.

Spores 8–10 × 6–8 μm, Q = (1.1–)1.35–1.6, Qav = 1.45, regularly 5–6-angled in side-view. Basidia 34–44 × 9–12 μm, 4-spored, clamped. Lamella edge heterogeneous. Cheilocystidia 30–65 × 5–15 μm, cylindrical, lageniform or subcapitate, protruding from hymenium. Pileipellis a cutis of narrow, very regularly cylindrical hyphae, ca. 5 μm in diameter. Pigment brown, intracellular and minutely encrusting in

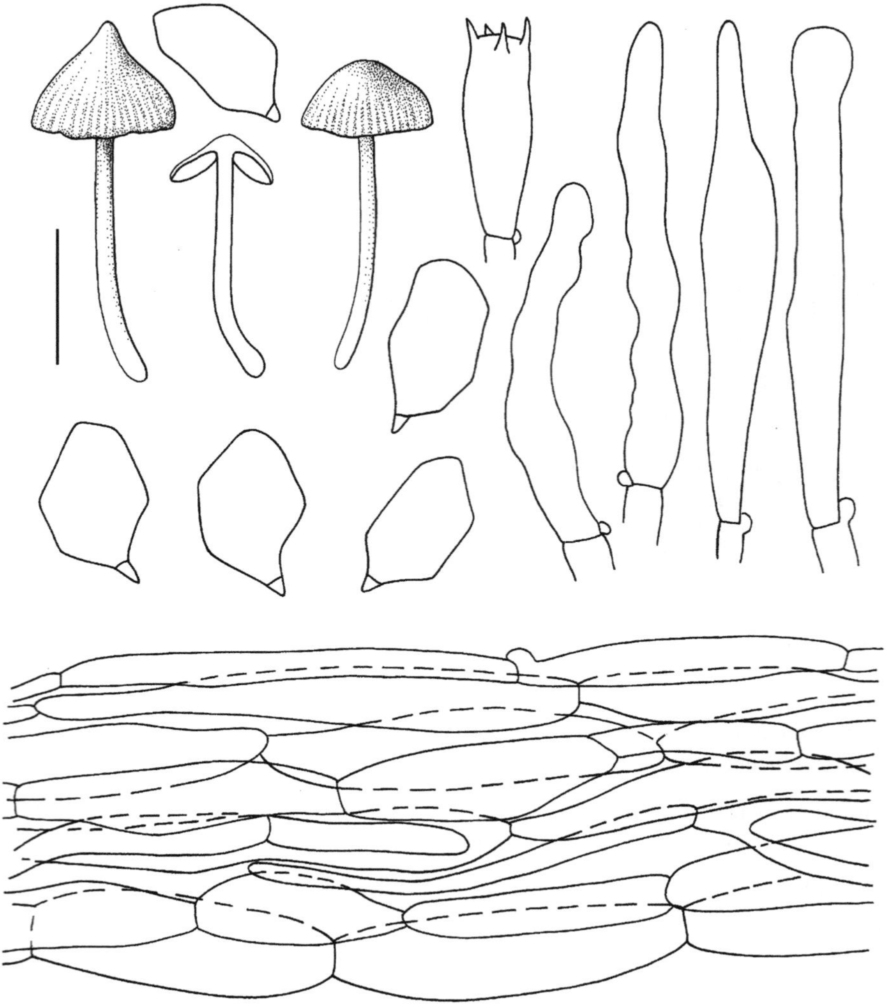

Fig. 3.22 *Entoloma psilocyboides*. Habit, spores, basidia, cheilocystidia, and pileipellis. Bar = 10 mm or 10 μm

pileipellis. Pileitrama regular, made up of inflated, relatively short elements. Clamp-connections observed in pileipellis and hymenium (Fig. 3.22).

Habitat & distribution: terrestrial in small groups in forest litter of wet sclerophyll forest and mixed forest with *Nothofagus cunninghamii* (Hook.) Oerst. and *Atherosperma moschatum* Labill. Uncommon.

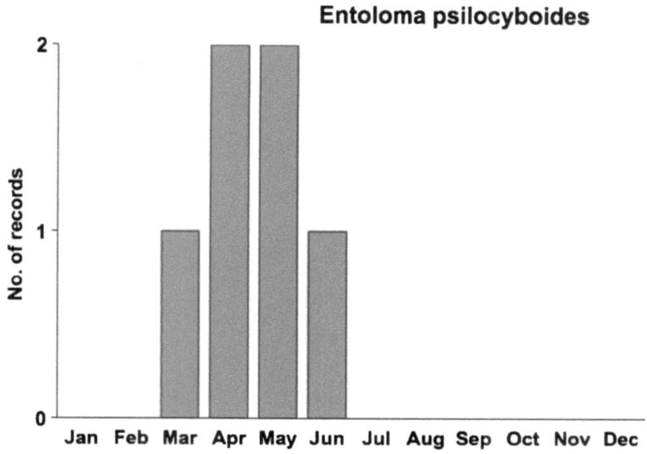

Collections examined. Australia, Tasmania, Growling Swallet, 42° 41′ S, 146° 30′ E, 19 May 1999, G. Gates E 520 (HO 548341); —, 1 May 2001, G. Gates E 1145 (holotype, HO 548338; isotype, L); —, 11 April 2002, G. Gates E 1477 (HO 548339); —, 22 April 2004, G. Gates E 1973 (HO 548340); —, 28 March 2009, M.E. Noordeloos 2009080; Kermandie Falls, Lower Track, 43° 12′ S, 146° 52′ E, 8 June 2002, G. Gates E 1562 (HO 548342).

Entoloma psilocyboides appears in a dark or pale form, but always with distinct brown tinges in the lamellae. Microscopically, the large, protruding cheilocystidia and relatively small, simple spores are distinctive. The dark colours, presence of cheilocystidia, and the relatively short elements of the pileitrama suggest affinity with *E. obscureotenax* G.M. Gates & Noordel., which has much larger spores, differently shaped cheilocystidia, and a predominant heavily encrusting pigment. *Entoloma elegantissimum* E. Horak from New Zealand is a paler species with smaller spores and lacks cheilocystidia. *Entoloma confusum* E. Horak from New Zealand has smaller spores, a rancid smell, and lacks cheilocystidia. *Entoloma atroenigmaticum* Noordel. & Hauskn. from Europe differs by the size and shape of the spores, lacks cheilocystidia and has thickly encrusted hyphae in all parts of the fruit body.

23. **Entoloma phaeophthalmum** Noordel. & G.M. Gates, spec. nov.

MycoBank #564495.

Diagnosis: Habit mycenoid; pileus pale yellow-brown with darker centre; spores 7–9 × 6–7.5 µm, 4–6-angular; cheilocystidia 40–90 × 4–9 µm, cylindrical, fusiform or lageniform, protruding from the hymenium, scattered among basidia; pigment intracellular and encrusting; clamp-connections present.

Holotype: Australia, Tasmania, Mt Wellington, Myrtle Gully, 42° 54′ S, 147° 15′ E, 3 May 2001, G. Gates E 1160 (HO 564349; isotype in L).

Etymology: phaeos (Gr.) = brown; phthalmus (Gr.) = eye, referring to the dark central spot on the pileus.

Plate 3.21 *Entoloma phaeophthalmum* (Photos Genevieve Gates)

Main characters: mycenoid, pallid brown fruit body; stipe slender, polished; cheilocystidia scattered, large, thin-walled (Plate 3.21).

Description:
Pileus 8–25 mm diam, convex, expanding to applanate, with small umbo, with straight margin, hygrophanous, translucently striate, usually distinctly bicoloured, brown-beige to yellow-brown with darker reddish brown umbo, pallescent on drying, glabrous to innately fibrillose, very satin-lustrous when dry. Lamellae adnate with small decurrent tooth, ventricose, moderately distant, up to 4 mm broad, flesh-pink becoming brown-pink with age with entire, concolorous edge, lamellulae in two tiers. Stipe 28–60 × 1–3 mm, cylindrical, pale brown, smooth, polished, base with white mycelium. Odour farinaceous. Taste farinaceous-bitter or soapy.

Spores 7–9 × 6–7.5 µm, Q = 1.1–1.3, Qav = 1.2, 4–6-angled in side-view. Basidia 20–32 × 6–8 µm, 4-spored, clamped. Lamella edge heterogeneous. Cheilocystidia

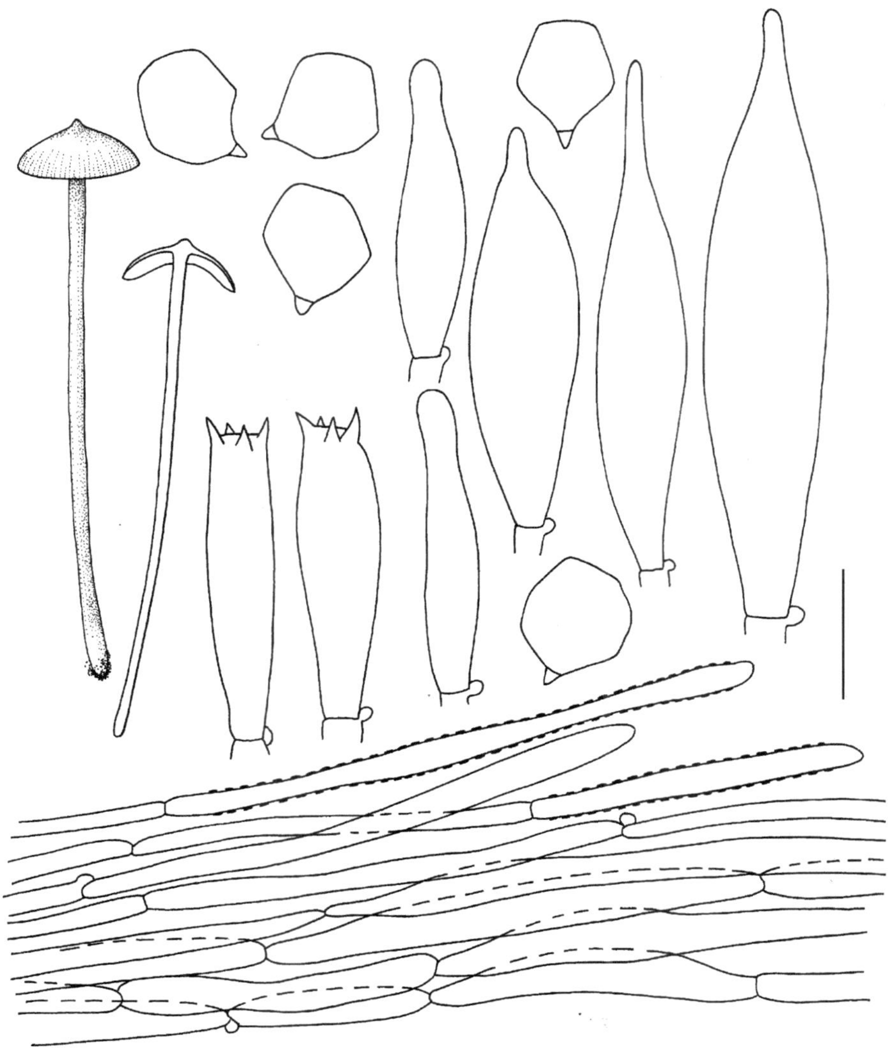

Fig. 3.23 *Entoloma phaeophthalmum*. Habit, spores, basidia, cheilocystidia, and pileipellis. Bar = 10 mm or 10 μm

scattered among basidia, protruding from hymenium, 40–90 × 4–9 μm, very variably shaped from fusiform to clavate, more frequently lageniform with long, tapering or subcapitate, occasionally furcate, neck. Hymenophoral trama regular, made up of medium sized, cylindrical elements, 50–160 × 4–9 μm. Pileipellis a differentiated cutis of cylindrical, septate elements, at centre sometimes forming trichodermal tufts 5–15 μm wide. Pigment pale brown, intracellular, and also minutely encrusting some thinner hyphae of the suprapellis. Pileitrama similar to hymenophoral trama. Clamp-connections fairly abundant (Fig. 3.23).

Habitat & distribution: in wet eucalypt forests, saprotrophic, terrestrial, single or in small groups, rarely subcaespitose. Widespread but infrequently encountered.

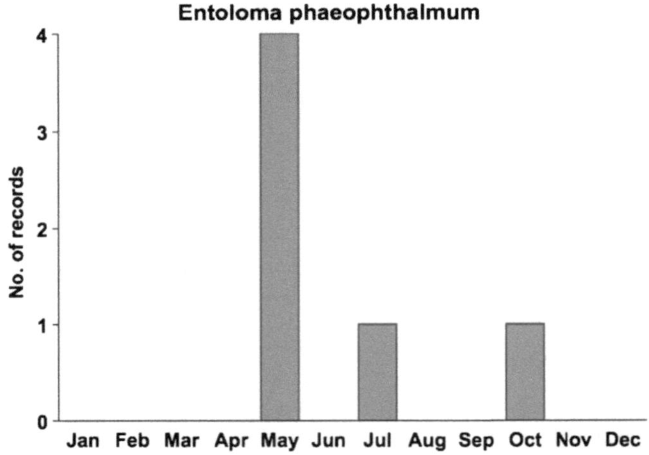

Collections examined. Australia, Tasmania, Bruny Island, Mt. Mangana, 43° 22′ S, 147° 17′ E, 26 May 2001, G. Gates E 1179; Collinsvale, Myrtle Forest, 42° 52′ S, 147° 09′ E, 11 May 2010, M.E. Noordeloos 2010001; Mt Field National Park, Lyre Bird Walk, 42° 41′ S, 146° 40′ E, 27 Oct. 2001, G. Gates E 1304; Mt Wellington, Myrtle Gully, 42° 54′ S, 147° 15′ E, 3 May 2001, G. Gates E 1160 (holotype, HO 564349); North West Bay River, track to Cathedral Rock, 42° 57′ S, 147° 12′ E, 1 May 2003, G. Gates E 1756; Pelverata Falls, 43° 04′ S, 147° 08′ E, 18 July 2002, G. Gates E 1608 & E 1609.

The main distinctive characters of this small *Nolanea* are the pale colours, ranging from almost white to medium brown, the striate, subfibrillose pileus, the polished stipe, a strong farinaceous-rancid smell, and microscopically the irregularly shaped 4–6-angled spores, and scattered, large, very thin-walled cheilocystidia, which are very obvious in fresh material but could be missed in dried material. It has some resemblance to *Entoloma farinolens* E. Horak from New Zealand, which has a similar habit and strong farinaceous smell, but is described as being much darker brown with encrusted pigment only. The spores are more or less similarly 4–6-angled, but the cheilocystidia are differently shaped. *Entoloma deprensum* E. Horak from New Zealand is much darker, has similar cheilocystidia, but differently shaped, 5–6-angled spores, and lacks clamp-connections.

24. **Entoloma fumosopruinosum** G.M. Gates & Noordel., Persoonia 17: 174. 2007.

Original diagnosis: Pileus 15–45 mm latus obtuse conicus vel concico-convexus expansus umbonatus hygrophanus translucido-striatus fumosus minute aerifero-pruinosus. Stipes 20–70 × 2–4 mm fumosus innate fibrillosus vel subpolitus. Odore saporeque nulis. Sporae 10–13(–13.5) × 7.0–10 μm 6–7 vel pluriangulatae-subgibbosae.

Acies lamellarum heterogenea. Cheilocystidia 20–49(–65) × 4–12 µm versiformes lageniformia vel tibiiformia. Pileipellis cutis elementis cylindraceis vel clavatis ad 14 µm latis pigmentis intracellulosis constitutis. Fibulae presentes.

Holotype: Australia, Tasmania, Wielangta, 42° 42′ S, 147° 51′ E, 6 June 2000, G. Gates E 959 (HO 543551; isotype in L).

Etymology: fumosus (Lat.) = smoke; pruinosum (Lat.) = hoary, referring to the smoke grey, hoary aspect of the pileal surface.

Main characters: habit mycenoid; pileus and stipe grey-brown, pileal surface not entirely smooth, but distinctly frosty to fluffy at centre, very hygrophanous; spores variably shaped; cheilocystidia present (Plate 3.22).

Description:
Pileus 10–45 mm diam, truncated conical to convex, expanding to plano-convex with low umbo or with slightly depressed centre, with involute then deflexed margin, slightly to distinctly translucently striate when moist, grey to grey-brown (10YR 3/3–4/4, 10YR 5/4–6/4, 4–5E4–2), paler towards margin (10YR 7/4–6), hygrophanous, noticeably pallescent on drying, entirely densely minutely fibrillose-frosted or with frosted patches, particularly near margin when young, margin then appearing almost white, becoming more pronouncedly frosted-lustrous upon drying, at centre sometimes finely fluffy-tomentose. Lamellae adnate-emarginate with decurrent tooth, segmentiform to subventricose, fairly distant, up to 6 mm broad, often thickish and anastomosing against underside of pileus, sometimes veined on sides, grey with pink tinge (7.5YR 6/4), with irregular, concolorous or slightly paler edge, L = 28–36, l = 3–5. Stipe 20–70 × 2–5(–8) mm, cylindrical, or compressed with longitudinal groove, broadened toward base, grey, concolorous with pileus or much paler grey-brown (10YR 5–6/4 to 10YR 5/3), innately fibrillose, substriate or almost polished, dry, brittle, slender. Odour like cucumber or farinaceous. Taste like cucumber or farinaceous.

Spores 10–13(–13.5) × 7.0–10 µm, on average 11.5–12 × 8.3–9.0 µm, Q = 1.2–1.5, many-angled (6 or 7) in side-view with rather blunt angles, thin-walled. Basidia 20–32 × 7–10 µm, 4-spored, clamped. Lamella edge heterogeneous. Cheilocystidia 20–49(–65) × 4–12 µm, variably shaped, lageniform to tibiiform with round, attenuate or capitate apex, sometimes capitate-mucronate, thin-walled, colourless, scattered amongst basidia, but distinctly protruding from hymenium. Pileipellis a well differentiated cutis of cylindrical hyphae up to 14 µm in diameter with abundant free terminal elements, which may be cylindrical or cystidiform, with intracellular pigment, and resembling the cheilocystidia. Pileitrama regular, made up of long, fusiform elements, 120–230 × 5–20 µm. Clamp-connections present (Fig. 3.24).

Habitat & distribution: widespread in wet sclerophyll forests and rainforests. Fruiting mainly in the peak autumn months April and May, extending to the winter months June-August (Group 4, in Chap. 1, see section "Fruiting Patterns of Species of Entolomataceae").

Plate 3.22 *Entoloma fumosopruinosum* (Photos Michael Pilkington)

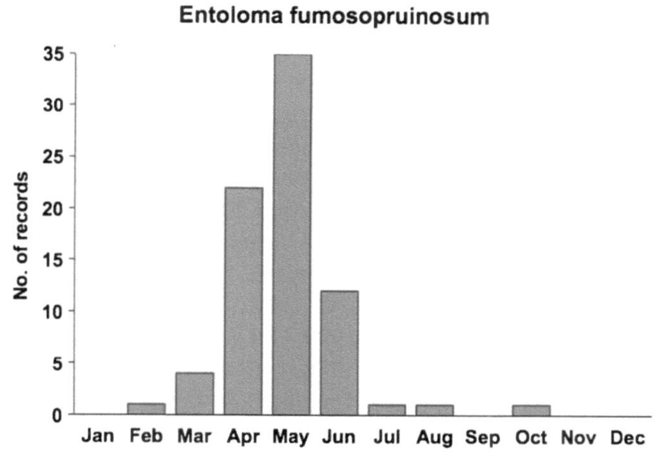

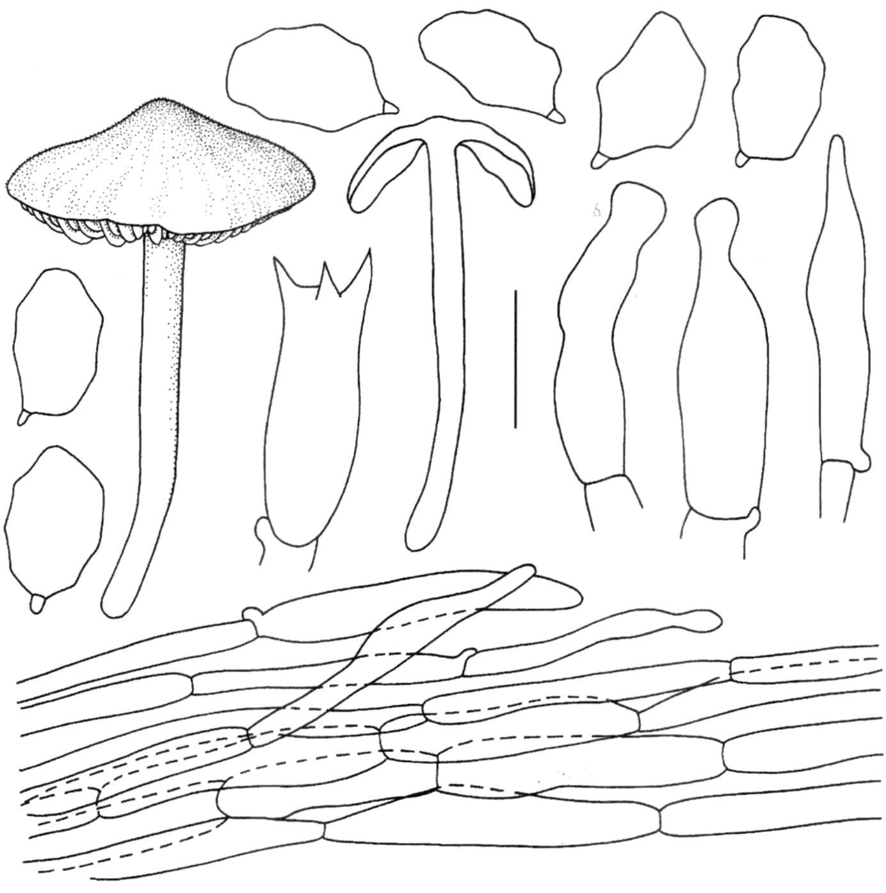

Fig. 3.24 *Entoloma fumosopruinosum*. Habit, spores, basidia, cheilocystidia, and pileipellis. Bar = 10 mm or 10 μm

Collections examined. Australia, Tasmania, Arve River Streamside Reserve, 43° 10′ S, 146° 48.5′ E, 2 July 2002, G. Gates E 1591; Bruny Island, Mt Mangana, 43° 22′ S, 147° 17′ E, 8 April 1999, G. Gates E 388; —, 26 May 2001, G. Gates E 1183; —, 27 May 2004, G. Gates E 2020; Clarks Cliffs, 43° 06′ S, 147° 47′ E, 4 May 2002, G. Gates E 1498; —, 21 April 2003, G. Gates E 1734; Duckhole Lake Track, 43° 22′ S, 146° 53′ E, 8 May 2004, G. Gates E 1992; Growling Swallet, 42° 41′ S, 146° 30′ E, 15 May 1999, G. Gates E 527; —, 6 April 2000, G. Gates E 824; —, 11 May 2000, G. Gates E 903; —, 1 May 2001, G. Gates E 1153; —, 11 April 2002, G. Gates E 1474; —, 13 May 2003, G. Gates E 1794; —, 22 April 2004, M.E. Noordeloos 2004069; —, 8 May 2010, G. Gates E 2281; Kermandie Falls Upper Track, 43° 12′ S, 146° 52′ E, 14 June 1999, G. Gates E 603; —, 30 May 2000, G. Gates E 947; Lake Skinner Track, 42° 57′ S, 146° 44′ E, 11 April 1999, G. Gates E 389a; Lake St. Clair NP, Echo Point to Cynthia Bay, 42° 04′ S, 146° 09′ E, 22 April 2000, G. Gates E 870; Macgregor Peak, 42° 59′ S, 147° 57′ E, 11 April 2004, M.E Noordeloos 2004072a; Mt Field National Park, Lyrebird Walk, 42° 41′ S, 146° 40′ E, 24 April 2001, G. Gates E 1110 & E 1111; —, 2 June 2001, G. Gates E 1207; —, 3 May 2003, G. Gates E 1768; Mt Wellington, Circle Track, 42° 55′ S, 147°15′ E, 11 May 1999, G. Gates E 197; —, Pipeline Track, 17 March 2011, G. Gates E 2287; Myrtle Forest, 42° 52′ S, 147° 09′ E, 19 April 2001, G. Gates E 1103; —, 3 May 2005, G. Gates E 2147; Ralphs Falls, 41° 18′ S, 147° 51′ E, 15 June 2003, G. Gates E 1873; Scottsdale, Forester Road, 41° 06′ S, 147° 37′ E, 29 April 2006, M.E. Noordeloos 2006039; Styx River, Big Tree Reserve, 42° 49′ S, 146° 39′ E, 13 May 2010, M.E. Noordeloos 2010024; Timbs Track, 42° 44′ S, 146° 25′ E, 13 May 2000, G. Gates E 917; Wielangta, 42° 42′ S, 147° 51′ E, 29 April 1999, G. Gates E 448; —, 6 June 2000, G. Gates E 959 (holotype, HO 543551; isotype, L); —, 17 Aug. 2000, G. Gates E 997.

This rather common species can readily be recognized in the field by the frosty, moderately dark brown pileus. The taxonomic position of *E. fumosopruinosum* is unclear because of the combination of the following characters: markedly nodulose-angular, fairly large spores, versiform cheilocystidia and the complex structure of the pileipellis with pileocystidia that are similar to the cheilocystidia. No similar species are described from New Zealand (Horak 2008). It resembles *Entoloma plebejum* (Kalchbr.) Noordel. by the frosted aspect of the pileus, and similar microscopy. The latter species, common and widespread in Europe, and probably introduced to Tasmania, differs by a somewhat stouter habit, paler lamellae, and almost white stipe.

25. **Entoloma fibrosopileatum** G.M. Gates & Noordel., Persoonia 19: 177. 2007.

Original diagnosis: Habitus mycenoideus. Pileus 30–45 mm latus fusce brunneus translucido-striatus fibroso-minute squamulosus. Stipes 55–70 × 3–4 mm argenteo-striatus. Sporae (9.5–)10–14 × 7.0–9.5(–10) μm irregulariter 5–6-angulatae, heterodiametrales, interdum cruciformae. Cystidia desunt. Pileipellis trichoderma elementis ad 20 μm latis. Pigmentum intracellulare. Fibulae adsunt.

Holotype: Australia, Tasmania, Duckhole Lake Track, 43° 22′ S, 146° 53′ E, 20 April 2004, M.E. Noordeloos 2004056 (HO 543558; isotype in L).

Plate 3.23 *Entoloma fibrosopileatum* (Photo Genevieve Gates)

Etymology: fibrosus (Lat.) = fibrous, referring to the fibrous pileus.

Main characters: rather stout *Nolanea* (mycenoid habit) with radially fibrillose to subsquamulose pileus with darker velutinous centre; spores rather large, complex (Plate 3.23).

Description:
Pileus 17–45 mm diam, conical at first, expanding to plano-convex with small umbo, with deflexed then straight margin, hygrophanous, when moist deeply translucently striate up to 2/3 of the radius, dark reddish brown, slightly paler towards margin (7F6–5, 8F6–5, 7D5–4, 7E5–4), pallescent along radial streaks to rather paler yellowish brown, clearly fibrillose, very minutely and darkly velutinous to squamulose at centre, sometimes radially ribbed on margin. Lamellae adnate, deeply emarginate or adnexed to almost free, ventricose, rather crowded, up to 8 mm broad, greyish pink with irregular, concolorous edge, L = 50–60, l = 3–5. Stipe 55–125 × 3–7 mm, cylindrical, gradually broadened towards base, brittle, dry, pale brown (5D5), fluffy-scurfy with white fibrils at apex, longitudinally silvery striate, with white basal tomentum. Context thin, concolorous with surface. Odour rather indistinct, sweetish. Taste mild.

Spores (9.5–)10–14 × 7.0–9.5(–10) µm, on average 11.0–12.5 × 7.1–8.2 µm, Q = 1.2–1.8, very pronouncedly angular, 5- or 6-angled, sometimes twisted and transient to the cruciform type. Basidia 22–40 × 8–12 µm, 4-spored, clamped. Lamella edge fertile. Cystidia absent. Hymenophoral trama regular, made up of short to medium-

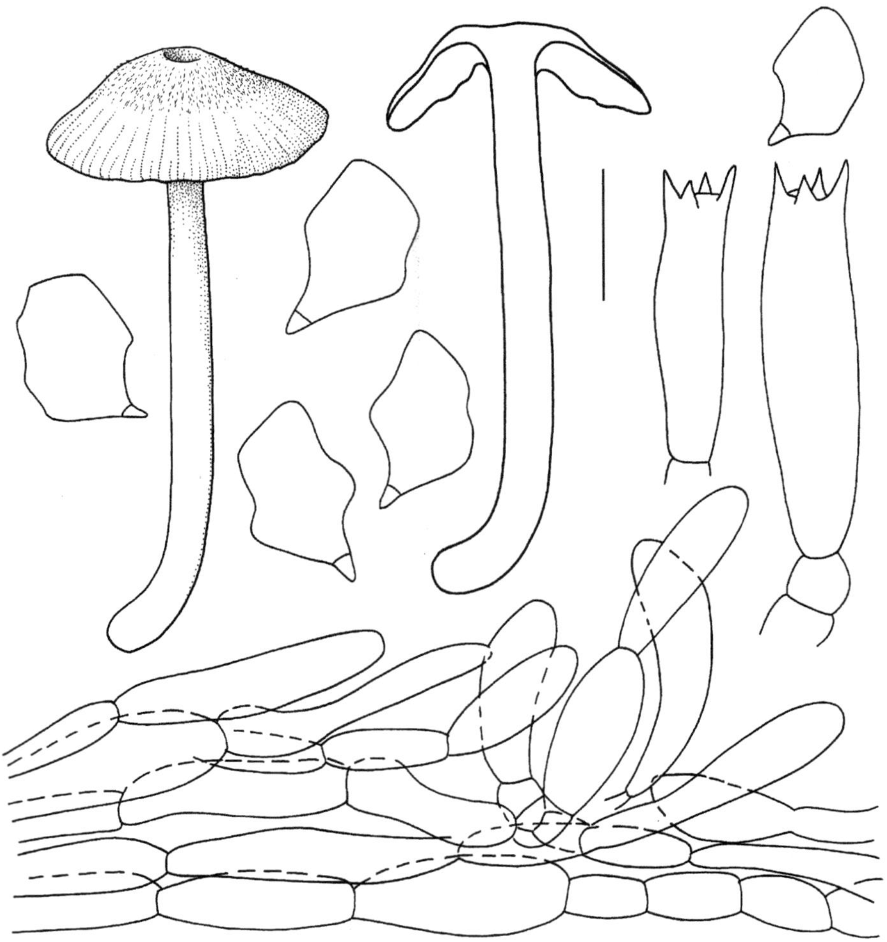

Fig. 3.25 *Entoloma fibrosopileatum*. Habit, spores, basidia, and pileipellis. Bar = 10 mm or 10 μm

sized cylindrical elements, 60–200 × 6–20 μm. Pileipellis a cutis of rather wide, inflated hyphae with trichodermal tufts of clavate, ascending elements, 40–120 × 10–20 μm with very abundant brown intracellular pigment. Pileitrama regular, similar to hymenophoral trama. Clamp-connections rare, but present (Fig. 3.25).

Habitat & distribution: widespread in wet sclerophyll forests, fruiting throughout the year, with a fair number of records in spring and summer months (Group 5, in Chap. 1, see section "Fruiting Patterns of Species of Entolomataceae").

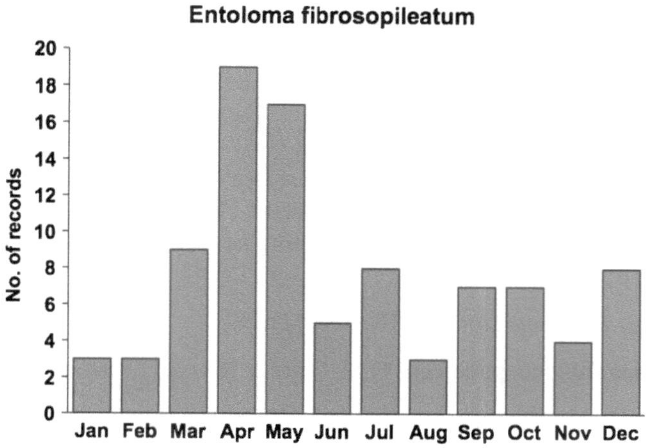

Collections examined. Australia, Tasmania, Arve River Streamside Reserve, 43° 10′ S, 146° 48.5′ E, 7 July 1998, G. Gates E 10; Bermuda Road, 43° 04′ S, 146° 54′ E, 25 July 2002, G. Gates E 1612; Bruny Island, Mt Mangana, 43° 22′ S, 147° 17′ E, 1 June 1999, G. Gates E 557; Cethana, Lacey's property, 41° 31′ S, 146° 04′ E, 27 April 2002, G. Gates E 1489; Darcy Link Rd, Stump Walk, 43° 25′ S, 146° 53′ E, 14 Dec. 2002, G. Gates E 1647; Duckhole Lake Track, 43° 22′ S, 146° 53′ E, 13 July 2002, G. Gates E 1600; —, 14 Dec. 2002, G. Gates E 1648; —, 8 April 2003, G. Gates E 1688; —, 20 April 2004, M.E. Noordeloos 2004056 (holotype, HO 543558; isotype, L); —, 8 May 2004, G. Gates E 1991; —, 16 April 2005, G. Gates E 2115; Ferndene, 41° 10′ S, 146° 20′ E, 28 March 1999, G. Gates E 301; Gowrie Park, O'Neills Creek, 41° 28′ S, 146° 13′ E, 2 May 2005, G. Gates E 2150; Growling Swallet, 42° 41′ S, 146° 30′ E, 6 April 2000, G. Gates E 827; —, 7 Aug. 2001 G. Gates E 2171; —, 22 April 2004, M.E. Noordeloos 2004072; Kermandie Falls, Lower Track, 43° 12′ S, 146° 52′ E, 16 May 2002, G. Gates E 1511; —, 18 March 2004, G. Gates E 1955; Marriotts Falls, 42° 43′ S, 146° 39′ E, 29 May 2003, G. Gates E 1836; Mt Field National Park, Tall Trees Track, 42° 41′ S, 146° 42′ E, 11 Sept. 2001, G. Gates E 1279; —, Lyrebird Track, 42° 41′ S, 146° 40′ E, 3 May 2003, G. Gates E 1763; —, 28 April 2011, M. E. Noordeloos 2011015 (L); Warra LTER site, Bird Track, 43° 05′ S, 146° 39′ E, 27 July 2006, G. Gates E 2246; —, numerous additional observations between April 2006 and June 2007, G. Gates s.n.; Wedge Forest Reserve, 42° 50′ S, 146° 16′ E, 7 March 2004, G. Gates E 1938.

Entoloma fibrosopileatum is a remarkable nolaneoid species with its rather strongly fibrillose, at centre often also minutely squamulose, pileus. In some aspects it is similar to *E. conferendum* (Britzelm.) Noordel., but *E. fibrosopileatum* differs in its pileipellis structure, which is more distinctly trichodermal in places, and the shape of the spores, which are irregularly 5- or 6-angled with pronounced angles, sometimes reminiscent of the cruciform type of *E. conferendum.*

Section **Cosmeoexonema** (Largent & Thiers) Noordel.

Nolanea sect. *Cosmeoexonema* Largent & Thiers in Northwest Sci. 46: 35. 1972; *Rhodophyllus* sect. *Cosmeoexonema* (Largent & Thiers) Romagn. in Bull. mens. Soc. linn. Lyon 43: 331. 1974; *Entoloma* sect. *Cosmeoexonema* (Largent & Thiers) Noordel. in Persoonia 11: 141. 1981.

Pigment encrusting at least the narrowest hyphae of pileipellis and pileitrama, sometimes with additional intracellular pigment; clamp-connections present, rarely absent; spores isodiametric to heterodiametric, never cuboid or cruciform; cheilocystidia usually absent.

Type species: *Entoloma sericeum* [Bull. ex] Quél.

26. **Entoloma fuligineopallescens** G.M. Gates & Noordel., Persoonia 19: 178. 2007.

Original diagnosis: Habitus mycenoideus. Pileus 13–23 mm latus valde hygrophanus griseobrunneus translucidostriatus toto expallens sericeo-fibrillosus interdum paulisper zonatus. Stipes 20–35 × 1–3 mm brunneus vel flavo-brunneus politus. Sporae parvae 6.0–7.0(–7.5) × 5.0–7.0 µm isodiametrales tenuiparietales. Cystidia desunt. Pileipellis cutis hyphis cylindraceis vel inflatis 4.0–9.0 µm latis pigmento valde incrustantibus formata fibulae presentes.

Holotype: Australia, Tasmania, Growling Swallet, 42° 41′ S, 146° 30′ E, 7 Aug. 2001, G. Gates E 1272 (HO 543536; isotype in L).

Etymology: fuligineus (Lat.) = brown; pallescens (Lat.) = pallescent, referring to the basidiocarp becoming pale on drying.

Main characters: small to medium-sized, grey-brown to black *Nolanea* (mycenoid); pileus sericeous, very hygrophanous; stipe polished; spores extremely small, isodiametric (Plate 3.24).

Description:
Pileus 13–25(–47) mm diam, convex with slightly elevated or slightly depressed centre, sometimes truncate, very hygrophanous, when moist translucently striate up to half the radius, dark greyish brown to sepia, distinctly pallescent to pale greyish brown upon drying, often appearing more or less concentrically zoned, glabrous, with aeriferous fibrillose surface when dry. Lamellae adnate-emarginate, ventricose, moderately crowded, up to 3.5 mm broad, brown to brown-red, paler towards entire edge. Stipe 20–35 × 1–3 mm, cylindrical, slender, brittle, brown (5E4) to yellow-brown, glabrous, polished, base with white, mycelial tomentum. Context thin, concolorous with surface. Odour and taste strongly spermatic.

Spores 6.0–7.0(–7.5) × 5.0–7.0 µm, Q = 1.0–1.15, isodiametric, with 6–8 weak angles in side-view, very thin-walled. Basidia 20–30 × 4.0–9.0 µm, 4-spored, clamped. Lamella edge fertile. Cheilocystidia absent. Hymenophoral trama regular, made up of rather long, fusiform, inflated elements, often greater than 120 × 5.0–20 µm, interspersed with narrow cylindrical hyphae 2.0–6.0 µm in diameter with brown, encrusting pigment. Pileipellis a cutis of narrow, cylindrical hyphae, 4.0–9.0 µm wide; subpellis of rather short, inflated hyphae, 20–60 × 5.0–12 µm. Pileitrama regular, made up of long, inflated elements. Pigment abundant, brown,

Plate 3.24 *Entoloma fuligineopallescens* (Photo Genevieve Gates)

parietal-encrusting in pileipellis and upper pileitrama. Clamp-connections abundant in hymenium and also seen frequently in trama (Fig. 3.26).

Habitat & distribution: widespread in wet sclerophyll forests and rainforest, mainly in late winter and early spring (Group 1, in Chap. 1, see section "Fruiting Patterns of Species of Entolomataceae").

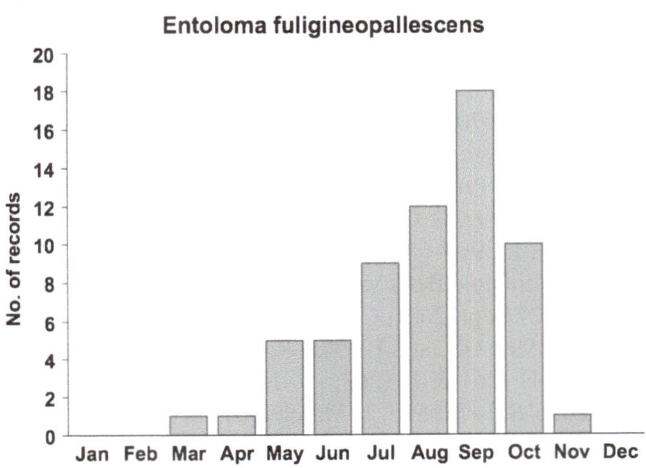

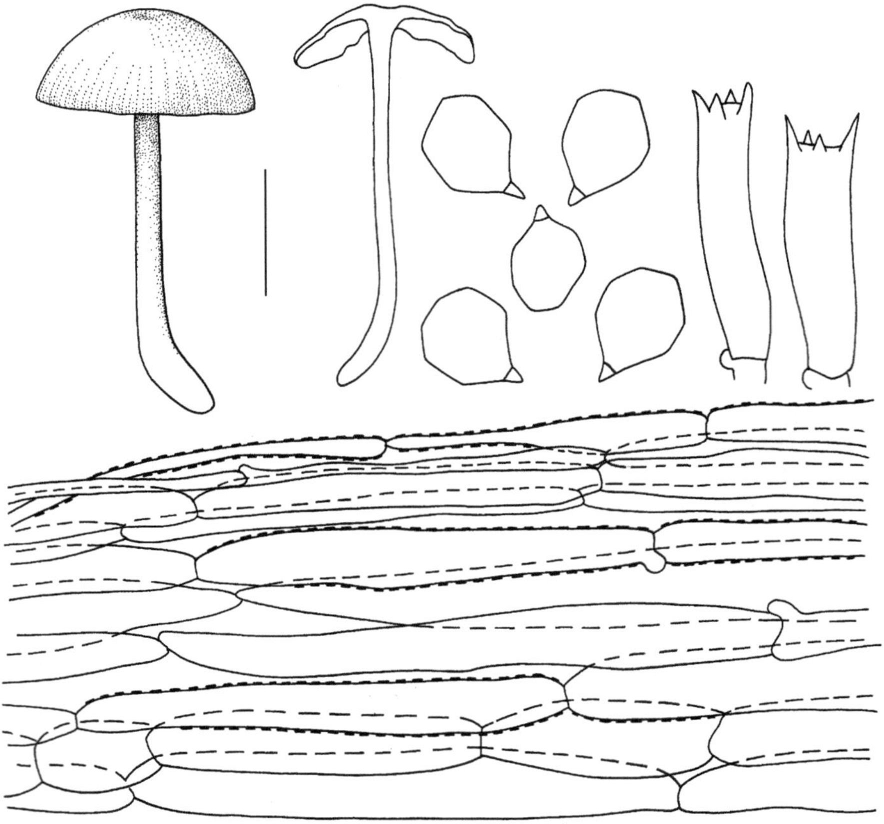

Fig. 3.26 *Entoloma fuligineopallescens*. Habit, spores, basidia, and pileipellis. Bar = 10 mm or 10 μm

Collections examined. Australia, Tasmania, Arm River, 41° 42′ S, 146° 12′ E, 22 Sept. 2001, G. Gates E 1282; Arve Loop Rd, 43° 08′ S, 146° 45′ E, 23 July 1998, G. Gates E 20 & E 23; Bruny Island, Mt Mangana, 43° 22′ S, 147° 17′ E, 9 Sept. 1999, G. Gates E 697; Donnellys Rd, 43° 07′ S, 146° 54′ E, 13 Aug. 1998, G. Gates E 42; Growling Swallet, 42° 41′ S, 146° 30′ E, 1 Oct. 1998, G. Gates E 61; —, 5 Aug. 1999, G. Gates E 670; —, 7 Aug. 2001, G. Gates E 1272 (holotype, HO 543536; isotype, L); Macgregor Peak, 42° 59′ S, 147° 57′ E, 9 Oct. 1999, G. Gates E 713; Mt Field National Park, Falls Loop, 42° 41′ S, 146° 42′ E, 22 Oct. 1998, G. Gates E 92; —, 29 June 1999, G. Gates E 637; —, 2 Sept. 1999, G. Gates E 699; —, 9 June 2001, G. Gates E 1239; —, 8 Oct. 2002, G. Gates E 1641; —, 15 July 2004, G. Gates E 2046; Mt Wellington, Betts Vale Track, 42° 55′ S, 147° 15′ E, 24 June 1999, G. Gates E 627; —, 27 Sept. 2001, G. Gates E 1283; —, Myrtle Gully, 42° 54′ S, 147° 15′ E, 12 Aug. 1999, G. Gates E 674; —, 6 Sept. 1999, G. Gates E 701; —, 31 July 2001, G. Gates E 1261 & E 1268; —, 15 Sept. 2001, G. Gates E 1281; —, Pipeline Track, 42° 55′ S, 147° 15′ E, 23 March 1999, G. Gates E 275; —, Shoobridge

Track, 16 Aug. 1999, G. Gates E 681; —, Silver Falls Track, 2 Oct. 2001, G. Gates E 1240; Timbs Track, 42° 44′ S, 146° 25′ E, 13 May 2000, G. Gates E 918; Warra LTER site, 43° 05′ S, 146° 39′ E, several observations between May and Aug. 2006, G. Gates s.n.; Waterworks Reserve, Hobart, 42° 54′ S, 147° 17′ E, 28 June 1998, G. Gates E 2; —, 30 Aug. 1998, G. Gates E 51.

This species appears to have a unique pattern of maximum fruiting in late winter and early spring, comparable with the Northern Temperate *Entoloma vernum* S. Lundell (Noordeloos 2004).

Entoloma fuligineopallescens is distinctive because of the very small, isodiametric spores. It is a typical *Nolanea*, and may well key out in the group of *E. sericeum* and related species, also when taking into consideration the shape of the pileus, which may be slightly depressed at centre, and the sinuate lamellae. *Entoloma sericeum* Quél and *E. parasericeum* E. Horak differ by their larger spores. These characters indicate that it is not related to the group of *Entoloma undatum* (Fr. ex Gillet) Moser. *Entoloma elaboratum* E. Horak from Argentina is similar, but has a pronouncedly papillate pileus and lacks clamp-connections.

27. **Entoloma austronitens** Noordel. & G.M. Gates, spec. nov.

MycoBank #564496.

Diagnosis: Habit mycenoid; pileus convex, expanding, hygrophanous, brown, deeply translucently striate, shiny; lamellae brown-pink; stipe 40–70×2–3 mm, brown, polished; smell and taste weaky farinaceous; spores 7–10×7–10 μm, isodiametric; cystidia absent; pileipellis cutis on narrow hyphae with encrusting and intracellular pigment; clamp-connections present.

Holotype: Australia, Tasmania, Mt Field National Park, Lyrebird Walk, 42° 41′ S, 146° 40′ E, 28 April 2011, M.E. Noordeloos 2011008 (HO 564350; isotype in L).

Etymology: austro (Lat.) = southern, i.e. the southern counterpart of *Entoloma nitens* (Velen.) Noordel.

Main characters: a brown *Nolanea* (mycenoid) with fibrillose pileus, becoming lustrous with age; stipe long, slender; spores isodiametric; pigments intracellular, encrusting (Plate 3.25).

Description:
Pileus 15–50 mm diam, conical to conic-campanulate at first, then conico-convex to convex expanding with age to plano-convex, with small papilla or very slightly umbilicate, sometimes truncate with small papilla within shallow central depression, with entire, deflexed then straight margin, very hygrophanous, translucently striate at margin or up to centre, dark brown to sepia when moist, pallescent to sordid greyish brown or yellowish brown on drying, innately radially fibrillose, sometimes slightly fibrillose-subsquamulose at centre, becoming very lustrous-shiny on drying. Lamellae adnexed to almost free, ventricose, moderately crowded, up to 4 mm broad, moderately thick, grey-buff with slight pink tinge, with concolorous, entire edge. Stipe 40–70×2–3 mm, cylindrical, slender, fragile, pale brown or yellow-brown, smooth, subpolished, at least

Plate 3.25 *Entoloma austronitens* (Photo Michael Pilkington)

in upper half, below sometimes innately fibrillose. Odour cucumber to subfarinaceous or none. Taste subfarinaceous or indistinct.

Spores 7–10×7–10 μm, isodiametric to subisodiametric, Q=1.0–1.25(–1.5), Qav=1.15, 5–6(–7)-angled. Basidia 30–36×8–11 μm, 4-spored, clamped. Lamella edge fertile. Cystidia absent. Hymenophoral trama regular, made up of inflated elements, 110–200×5–19 μm. Pileipellis a cutis of narrow, cylindrical hyphae, 2–8 μm wide, subpellis not differentiated. Pigment minutely encrusting in pileipellis. Pileitrama regular, made up of long, cylindrical to fusiform hyphae, to 250×5–22 μm, with golden brown intracellular pigment in layer adjacent to pileipellis. Stipitipellis a cutis of narrow, cylindrical hyphae, 2–8 μm wide. Caulocystidia absent. Clamps observed in hymenium, but rare, elsewhere absent or very rare (Fig. 3.27).

Habitat & distribution: common in wet sclerophyll forests, fruiting throughout the year, with a fair number of records in spring and summer months (Group 5, in Chap. 1, see section "Fruiting Patterns of Species of Entolomataceae").

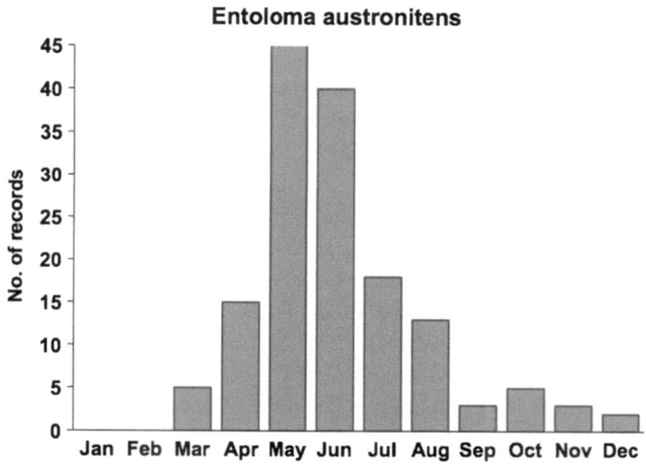

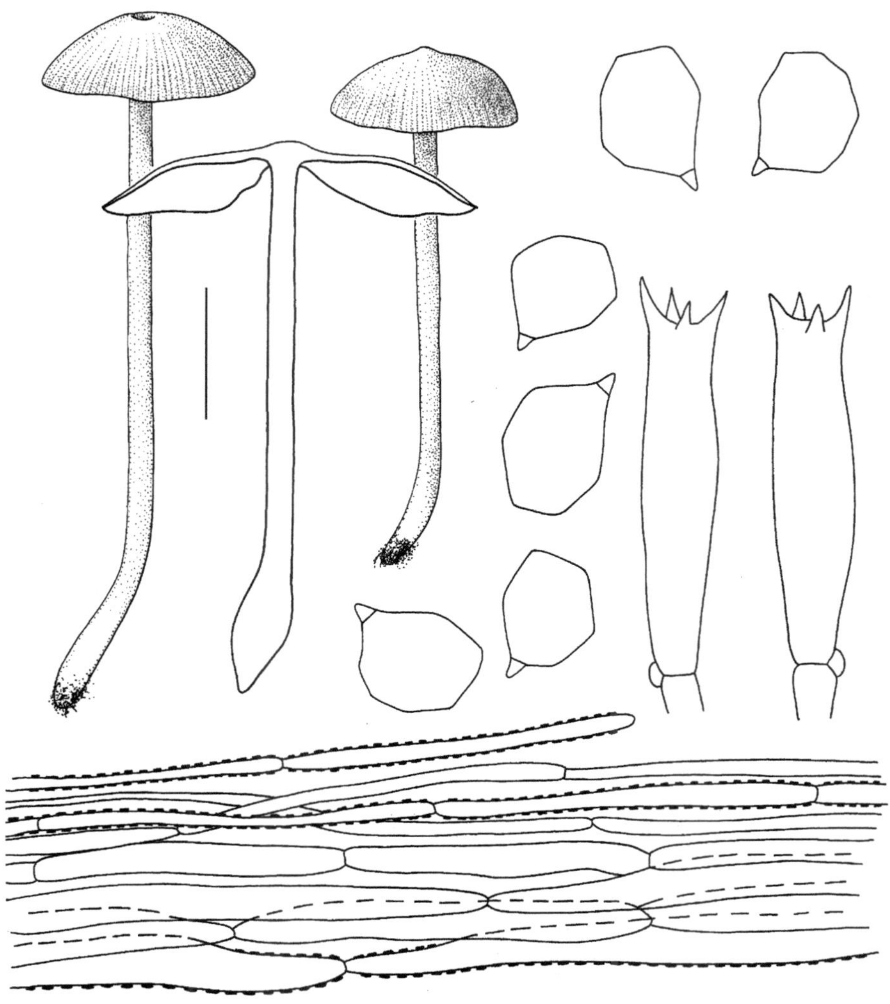

Fig. 3.27 *Entoloma austronitens*. Habit, spores, basidia, and pileipellis. Bar = 10 mm or 10 μm

Collections examined. Australia, Tasmania, Arve River Streamside Reserve, 43° 10′ S, 146° 48.5′ E, 7 July 1998, G. Gates E 5, E 8 & E 9; Birralee, Black Sugar Loaf, 41° 24′ S, 146° 48′ E, 23 Aug. 2001, G. Gates E 1277; Bruny Island, Cape Queen Elizabeth, 43° 14′ S, 147° 23′ E, 27 May 2003, G. Gates E 1830; —, 8 July 2004, G. Gates E 2043; —, Mt Mangana, 43° 22′ S, 147° 17′ E, 1 June 1999, G. Gates E 559; —, 10 June 2000, G. Gates E 967; —, 17 May 2010, M.E. Noordeloos 2010040; —, 26 May 2001, G. Gates E 1186; —, 8 April 1999, G. Gates E 383; —, 9 Sept. 1999, G. Gates E 695; Cape Pillar Track, 43° 09′ S, 147° 56′ E, 27 Aug. 1998; G. Gates E 45; Chauncy Vale, 42° 37′ S, 147° 16′ E, 29 Nov. 2001, G. Gates E 1361; —, 29 July 2003, G. Gates E 1899; Clarks Cliffs, 43° 06′ S, 147° 47′ E, 5 June 2001, G. Gates E 1219; Donnellys Rd, 43° 07′ S, 146° 54′ E, 13 Aug. 1998, G. Gates E 37; Duckhole Lake Track, 43° 22′ S, 146° 53′ E, 15 May 2003, G. Gates E 1808; Fortescue Bay, 43° 09′ S, 147° 57′ E, 27 Aug. 1998, G. Gates E 46; Geeveston, Johns Rd, 43° 16′ S, 146° 55′ E, 17 June 1999, G. Gates E 614; Growling Swallet, 42° 41′ S, 146° 30′ E, 19 May 1999, G. Gates E 523; —, 6 April 2000, G. Gates E815; Hartz Rd, 43° 10′ S, 146° 47′ E, 23 July 1998, G. Gates E 21; Judbury Conservation Area, 43° 10′ S, 146° 47′ E, 20 Aug. 1998, G. Gates E 44; —, 8 July 1999, G. Gates E 657; Junee Caves State Reserve, 42° 44′ S, 146° 36′ E, 24 May 2003, G. Gates E 1825; Kermandie Falls, Lower Track, 43° 12′ S, 146° 52′ E, 23 May 2000, G. Gates E 940; —, 29 May 2001, G. Gates E 1191; —, 31 May 2001, G. Gates E 1202; —, 16 May 2002, G. Gates E 1525; —, 8 June 2002, G. Gates E 1561; —, Upper Track, 43° 12′ S, 146° 52′ E, 12 June 1999, G. Gates E 592; —, 30 May 2000, G. Gates E 957; —, 28 Aug. 2003, G. Gates E 1908; Lake Skinner Track, 42° 57′ S, 146° 44′ E, 11 April 1999, G. Gates E 364; Little Florentine River, F5 Road, 42° 44′ S, 146° 25′ E, 8 May 2003, G. Gates E 1779; Marriotts Falls, 42° 43′ S, 146° 39′ E, 3 Aug. 2000, G. Gates E 994; —, 2 June 2001, G. Gates E 1214; —, 29 May 2003, G. Gates E 1833; Mt Wellington, Betts Vale Track, 42° 55′ S, 147° 15′ E, 26 May 2004, G. Gates E 2016; —, Lower Sawmill Track, 42° 55′ S, 147° 15′ E, 12 June 2001, G. Gates E 1240; —, Myrtle Gully, 42° 54′ S, 147° 15′ E, 8 July 1998, G. Gates E 11 & E 12; —, 13 May 1999, G. Gates E 504; —, Silver Falls, 42° 55′ S, 147° 15′ E, 14 Oct. 1998, G. Gates E 76; Mt. Field National Park, Lady Barron Falls Track, 42° 41′ S, 146°42′ E, 29 June 2000, G. Gates E 980; —, 24 April 2001, G. Gates E 1117; —, 15 July 2004, G. Gates E 2045; —, 1 June 2010, F. Karstedt 1536; —, Tall Trees Track, 42° 41′ S, 146° 42′ E, 4 Nov. 1999, G. Gates E 755; —, 9 May 2000, G. Gates E 900; —, 29 June 2000, G. Gates E 979; —, 19 Aug. 2000, G. Gates E 998; —, 4 Dec. 2001, G. Gates E 1368; —, 8 Oct. 2002, G. Gates E 1642; —, Lyrebird Track, 42° 41′ S, 146° 40′ E, 1 April 1999, G. Gates E 323 & E 328; —, 10 June 1999, G. Gates E 576; —, 31 March 2001, G. Gates E 1061; —, 24 April 2001, G. Gates E 1112; —, 2 June 2001, G. Gates E1209; —, 28 May 2002, G. Gates E 1541; —, 28 April 2011, M.E. Noordeloos 2011008 (holotype, HO 564350); Pelverata Falls, 43° 04′ S, 147° 08′ E, 1 July 1999, G. Gates E 647 & E 648; Ralphs Falls, 41° 18′ S, 147° 51′ E, 23 May 2010, F. Karstedt 1487 and 1491; Sandy Bay, Hytten Hall Gully, 42° 54′ S, 147° 19′ E, 23 April 1999, G. Gates E 440; Snug Falls, 43° 05′ S, 147° 13′ E, 16 July 2002, G. Gates E 1606; St Columba Falls, 41° 19′ S, 147° 55′ E, 21 May 2010, F. Karstedt 1463; Timbs Track, 42° 44′ S, 146°

25′ E, 13 May 2000, G. Gates E 921; —, 21 April 2001, G. Gates E 1107; Warra LTER site, 43° 05′ S, 146° 39′ E, numerous observations between Sept. 2006 and June 2007, G. Gates s.n.; Waterworks, 42° 54′ S, 147° 17′ E, 28 June 1998, G. Gates E 26; —, 30 Aug. 1998, G. Gates E 48, E 50, E 52 & E 53; Wielangta, 42° 42′ S, 147° 51′ E, 3 April 2003, G. Gates E 1856.

Entoloma austronitens is very reminiscent of *E. nitens* (Velen.) Noordel. from Europe with its medium to dark brown, lustrous pileus, subpolished stipe and isodiametric spores. It is also very similar to *Entoloma paraserriceum* E. Horak from New Zealand, which has a darker pileus, a more fibrillose stipe, strong farinaceous smell and taste, slightly more regularly isodiametrial spores, and lacks intracellular pigment. *Entoloma blandiodorum* E. Horak from New Zealand has an aromatic smell and smaller spores.

28. **Entoloma sericeum** (Bull.: Fr.) Quél.

Entoloma sericeum Quél. in Mém. Soc. Émul. Montbéliard, ser. II, 5: 119. 1872 (Champ. Jura Vosges 1); *Rhodophyllus sericeus* (Quél.) Quél., Enchir. Fung.: 59. 1886; *Nolanea sericea* (Quél.) P.D. Orton in Trans. Br. mycol. Soc. 43: 179. 1960. — *Agaricus sericeus* Bull., Herb. France, pl. 413, fig. 2. 1789 non *A. sericeus* Schaeff. 1774. *Rhodophyllus sericeus var. typicus* Kühner & Romagn., Fl. anal.: 193. 1953. *Rhodophyllus sericeus var. nolaniformis* Kühner in Kühn. & Romagn. in Rev. Mycol. 19: 9. 1954; *Entoloma sericeum f. nolaniformis* (Kühner) Noordel. in Persoonia 10: 480. 1980.

Selected literature: Horak, Agaricales New Zealand 1: 252. 2008; Noordeloos, Fungi Europaei 5: 256–259. 1992.

Original diagnosis: Stipe grèle, fibreux-fissile, fistuleux, grisâtre, brillant. Chapeau membraneux, pellucide à stries serrées, hygrophane, brun-fuligineux très brillant par le sec, convexe-plan, mammelonné (3 à 4 cent.). Chair fauve-pâle. Lamelles sinuées, assez serrées, grises puis brunes. Spore sphérique rosée (un 100e de millim.). Été et automne. En troupes dans les prés après les grandes pluies.

Main characters: basidiocarps mycenoid; colour dark brown overall, including the lamellae; pileus glabrous, lustrous; spores isodiametric; clamp-connections present (Plate 3.26).

Pileus 20–40 mm diam, plano-convex to applanate with slight umbo, hygrophanous, very faintly translucently striate, very dark brown, with a satiny sheen. Lamellae adnate, ventricose, crowded, up to 5 mm broad, dark brown with entire, concolorous edge, lamellulae in two tiers. Stipe 15–35 × 2–6 mm, cylindrical, dark brown, subglabrous to innately fibrillose, with white basal mycelium. Odour and taste farinaceous.

Spores 7–9 × 6.5–9, Q = 1.0–1.2, Qav = 1.1, isodiametric to subisodiametric, 5–6-angled in side-view. Basidia 4-spored, clamped. Lamella edge fertile. Cheilocystidia absent. Pileipellis a cutis of narrow, cylindrical hyphae, 2–9 µm wide, with coarsely encrusted walls. Hymenophoral and pileitrama regular, made up of cylindrical to inflated elements, 4–18 µm wide, with encrusting pigment. Clamp-connections present (Fig. 3.28).

Habitat & distribution: known only from a marsupial pasture in one locality.

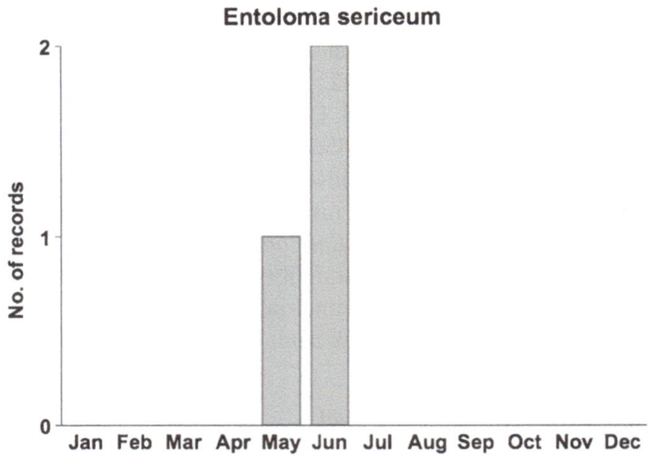

Plate 3.26 *Entoloma sericeum* (Photo Michael Pilkington)

Collection examined: Chauncy Vale, 42° 37′ S, 147° 16′ E, 11 June 2009, E2273a.

This collection fits the descriptions of *Entoloma sericeum* from Europe (Noordeloos 1992, 2004), and it has also been recorded from New Zealand (Horak 2008). The dark brown, silky shining pileus and isodiametric spores are distinctive.

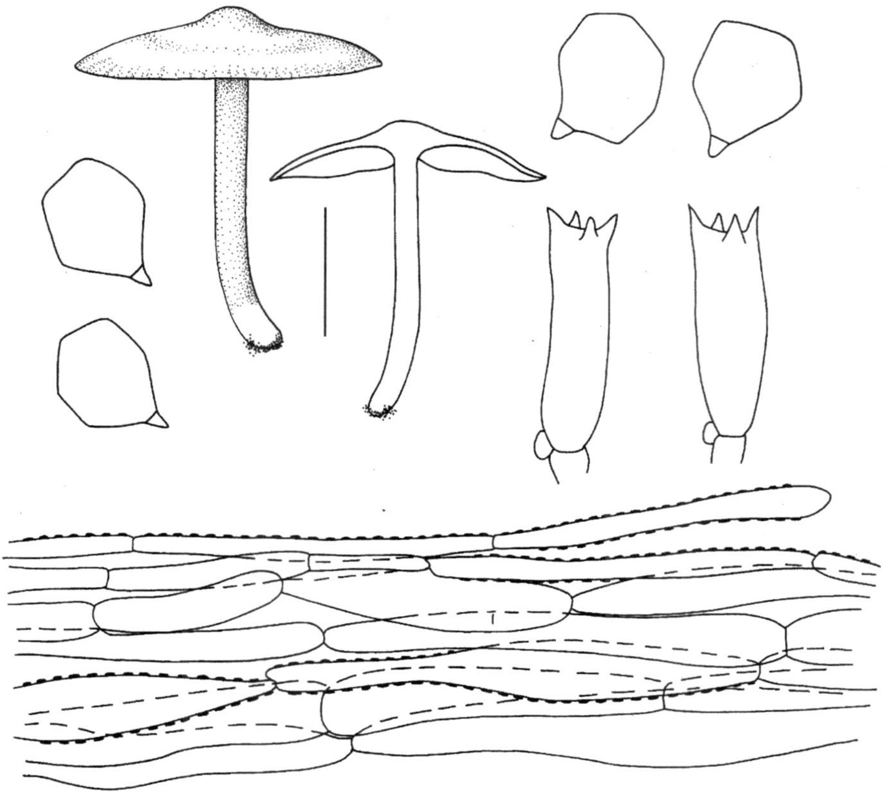

Fig. 3.28 *Entoloma sericeum*. Habit, spores, basidia, and pileipellis. Bar = 10 mm or 10 µm

However, a hitherto unpublished pilot study on the genetic variability of *Entoloma sericeum* in Europe shows that there are several cryptic species within the morphospecies *E. sericeum*. *Entoloma austronitens* Noordel. & G.M. Gates sp. nov., differs by a paler, very hygrophanous pileus, which becomes lustrous on drying, and the presence of intracellular and encrusting pigment. *Entoloma parasericeum* E. Horak from New Zealand is also close, but differs by having a fibrillose pileus.

29. **Entoloma readiae** G. Stev., Kew Bulletin 16: 233. 1962 (as "readii").

Selected literature: Horak, E., Beih. Nova Hedwigia 43: 49–50, fig. 29. 1973; Horak, Agaricales of New Zealand 1: 280, fig. 109, 112. 2008.

Original diagnosis: Pileus 3–4 cm diam, hinnuleus, e convexo plano-depressus, primo humidus impolitusque, siccitate sericeo-nitidus; caro hinnulea, tenuis, sublenta. Lamellae adnatae, subdistantes, ex hinnuleis sordide rubescentes, acieibus aliquantulo pectinatis. Stipes 3–4 cm × 3 mm, hinnuleus, lentus, aequalis, sericeo-striatus, cavus, interdum tortus. Sporae 7–8 × 9–10 µm, angulatae, apiculis prominentibus, in cumulo obscure roseae.

Plate 3.27a *Entoloma readiae* var. *readiae* (Photos Machiel Noordeloos (*left*); Michael Pilkington (*right*))

Holotype: New Zealand, Nelson, Wakefield, Church Valley Road, 17 April 1956, D. Read (K, Stevenson 1038).
Etymology: named after Dorothy Read.

Key to the varieties:

1. Pileus greyish brown with darker central spot — var. ***readiae***
1. Pileus yellow or becoming yellow upon drying — var. ***sulphureum***

29a. **Entoloma readiae var. readiae**

Main characters: small, fragile, collybioid species with greyish brown flattened, often slightly depressed, pileus; stipe concolorous; spores very small (Plate 3.27a).

Description:
Pileus 10–30 mm diam, convex to applanate or almost concave, with slightly to distinctly depressed centre and deflexed to reflexed margin, sometimes infundibuliform, strongly hygrophanous, when moist translucently striate up to centre, pale greyish brown to pinkish grey-brown with darker grey-brown centre (eye), distinctly pallescent along radial streaks from centre becoming very pale yellowish brown, innately radially fibrillose, appearing smooth to the naked eye, but often slightly rugulose at centre (hand lens). Lamellae variably attached from adnate-emarginate to adnate with decurrent tooth, moderately distant, white to creamy white then purely pink with entire or slightly eroded concolorous edge, L = 20–30, l = 1–5. Stipe 30–40 × 1–2 mm, cylindrical, often distinctly broadened at base, concolorous with

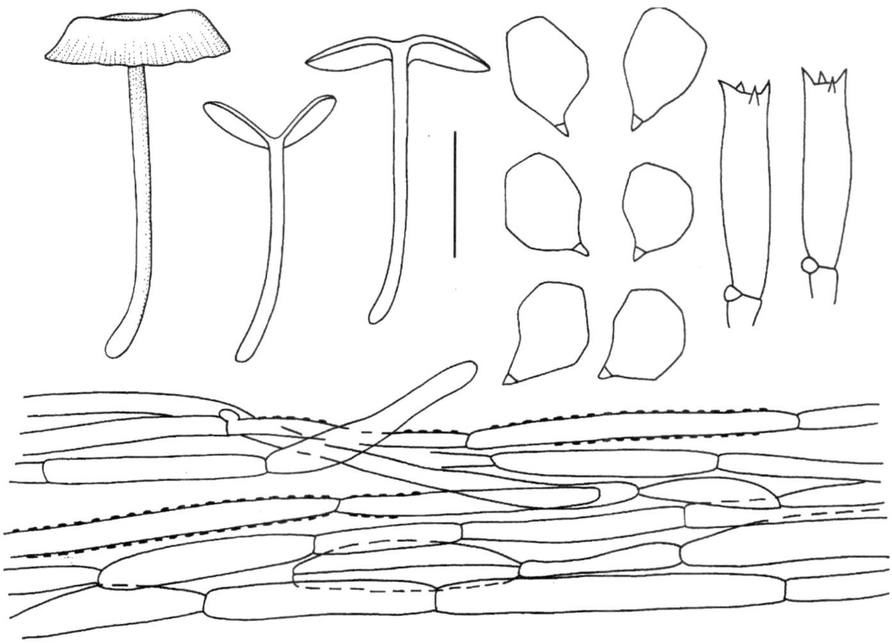

Fig. 3.29 *Entoloma readiae* var. *readiae*. Habit, spores, basidia, and pileipellis. Bar = 10 mm or 10 μm

pileus, sometimes darker red-brown towards base, glabrous, polished. Odour strong, like cucumber or farinaceous. Taste rancid.

Spores 6–8(–9) × 5.5–7.5 μm, 5–6-angled. Basidia 20–36 × 6–10 μm, 4-spored, clamped. Lamella edge fertile. Cheilocystidia absent. Hymenophoral trama regular, made up of medium sized, cylindrical to fusiform elements, 50–150 × 4–20 μm. Pileipellis bilayered: suprapellis a cutis of cylindrical, 4–11 μm wide hyphae, at centre a transition to a trichoderm of up to 15 μm wide, with clavate or fusiform terminal elements; subpellis made up of large, inflated, fusiform hyphae, 40–100 × 8–20 μm, clearly distinct from underlying trama. Pileitrama regular, made up of short, strictly cylindrical elements, 30–70 × 4–15 μm, with non-constricted septae. Pigment brown, finely encrusting in pileipellis and upper pileitrama, in addition sometimes also pale brown, intracellular in subpellis. Stipitipellis a cutis of narrow, cylindrical, 4–9 μm wide, encrusted hyphae. Clamp-connections present in hymenium, rare to frequent in subpellis and rare in other tissues (Fig. 3.29).

Habitat & distribution: in litter, often on very rotten wood, in wet and dry sclerophyll forest and rainforest; very common and widespread in Tasmania, also known from New Zealand (Horak 2008); fruiting throughout the year, with a fair number of records in spring and summer months (Group 5, in Chap. 1, see section "Fruiting Patterns of Species of Entolomataceae").

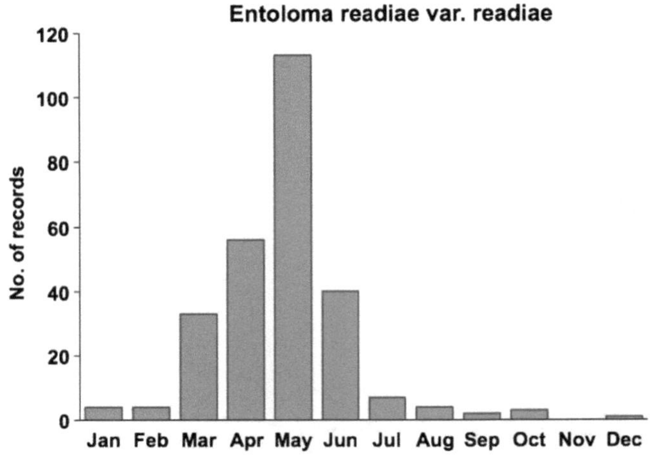

Collections examined. Australia, Tasmania, Bermuda Road, 43° 04′ S, 146° 54′ E, 17 April 2003, G. Gates E 1724; Bluff River State Forest, 42° 31′ S, 147° 40′ E, 20 July 2002, G. Gates E 1610; Bruny Island, Cape Queen Elizabeth, 43° 14′ S, 147° 23′ E, 25 May 2002, G. Gates E 1537; Cape Pillar Track, 43° 09′ S, 147° 56′ E, 1 June 2002, G. Gates E 1544; Chauncy Vale, 42° 37′ S, 147° 16′ E, 10 June 2003, G. Gates E 1862; Clarks Cliffs, 43° 06′ S, 147° 47′ E, 4 May 2002, G. Gates E 1500; Collinsvale, Myrtle Forest, 42° 52′ S, 147° 09′ E, 7 Jan. 1999, G. Gates E 115; —, 18 Feb. 1999; G. Gates E 124; —, 18 March 1999, G. Gates E 243; Creepy Crawly, 42° 50′ S, 146° 22′ E, 15 April 2000, G. Gates E 846; —, 20 March 2001, G. Gates E 1029; Donnellys Rd, 43° 07′ S, 146° 54′ E, 25 May 2000, G. Gates E 942; Duckhole Lake Track, 43° 22′ S, 146° 53′ E, 18 May 2002, G. Gates E 1515; —, 8 April 2003, G. Gates E 1685; —, 20 April 2004, M.E. Noordeloos 2004050; —, 12 April 2008, M.E. Noordeloos 2008017; Ferndene, 41° 10′ S, 146° 20′ E, 28 March 1999, G. Gates E 300; Florentine River Valley, Pagoda Hut, 42° 30′ S, 146° 27′ E, 29 April 2003, G. Gates E 1749; Growling Swallet, 42° 41′ S, 146° 30′ E, 27 March 2001, G. Gates E 1040; —, 4 June 2002, G. Gates E1553; Kermandie Falls, Lower Track, 43° 12′ S, 146° 52′ E, 11 May 1999, G. Gates E 495; —, 23 May 2000, G. Gates E 939; —, 5 April 2001, G. Gates E 1073; —, 29 May 2001, G. Gates E 1194; —, 16 May 2002, G. Gates E 1512; —, 10 April 2003, G. Gates E 1691; —, 24 March 2009, M.E. Noordeloos 2009090; Kermandie Falls, Upper Track, 43° 12′ S, 146° 52′ E, 30 March 1999, G. Gates E 599; —, 30 May 2000, G. Gates E 953; —, 26 April 2001, G. Gates E 1130; Lake St Clair National Park, Cynthia Bay, 42° 04′ S, 146° 09′ E, 8 June 2000, G. Gates E 965; —, Echo Pt to Cynthia Bay, 19 April 2003, G. Gates E 1732; Little Florentine River, F5 Road, 42° 44′ S, 146° 25′ E, 23 May 2002, G. Gates E 1531; Marriotts Falls, 42° 43′ S, 146° 39′ E, 13 May 2000, G. Gates E 916; Mt Field National Park, 42° 41′ S, 146° 42′ E, 29 June 1999, G. Gates E 636; —, 2 May 2000, G. Gates E 886; —, 9 May 2002, G. Gates E 1506; —, 28 May 2002, G. Gates E 1538; —, Falls Loop, 5 April 2000, G. Gates E 812; —, 4 May 2000, G. Gates E 892; —, 9 May 2000, G. Gates E 899; —,

31 March 2001, G. Gates E 1064; —, 25 June 2002, G. Gates E 1588; —, Lyrebird Walk, 42° 41′ S, 146° 40′ E, 1 April 1999, G. Gates E 330; —, 27 Oct. 2001, G. Gates E 1304; Mt Wellington, Betts Vale Track, 42° 55′ S, 147° 15′ E, 20 April 2000, G. Gates E 849; —, Circle Track, 11 March 1999, G. Gates E 206; —, 25 March 1999, G. Gates E 279; —, Fern Glade, 4 March 1999, G. Gates E 163; —, 18 March 2003, G. Gates E 1653; —, Jacksons Bend, 24 June 2000, G. Gates E 977; —, Old Farm Trail, 42° 54′ S, 147° 15′ E, 20 May 1998, G. Gates E 1; —, 8 Aug. 1998, G. Gates E 7; —, 9 March 1999, G. Gates E 176; —, Pillinger Drive, 42° 55′ S, 147° 15′ E, 18 July 2000, G. Gates E 988; —, Reids Track, 30 Aug. 2003, G. Gates E 1910; —, Silver Falls, 4 March 1999, G. Gates E 165 & E 167; —, 16 Sept. 1999, G. Gates E 704; —, 5 May 2001, G. Gates E 1170; —, 27 March 2003, G. Gates E 1660; North West Bay River, 42° 57′ S, 147° 12′ E, 14 Feb. 2002, G. Gates E 1435; —, 31 March 2003, G. Gates E 1662; Peter Murrell Reserve, 43° 00′ S, 147° 18′ E, 23 July 2002, G. Gates E 1611; Ralphs Falls, 41° 18′ S, 147° 51′ E, 15 April 2001, G. Gates E 1869; Styx Valley, 42° 49′ S, 146° 39′ E, 19 March 2009, M.E. Noordeloos 2009023; Timbs Track, 42° 44′ S, 146° 25′ E, 8 May 2003, G. Gates E 1785 & E 1786; —, 2 Oct. 2004, G. Gates E 2055; Warra LTER site, 43° 06′ S, 146° 39′ E, numerous observations between May 2006 and June 2007, G. Gates s.n.; Wedge Forest Reserve, 42° 50′ S, 146° 16′ E, 27 Feb. 2001, G. Gates E 1026; —, 21 April 2001, G. Gates E 1109.

Entoloma readiae is a small but distinct species, easily recognized in the field by its thin, often flattish pileus with distinctly darker, umbilicate centre and pure pink lamellae. The simple, small spores and encrusting pigment as well as the pale colour and polished stipe are reminiscent of the European *Entoloma minutum* (P. Karst.) Noordel., which also frequently has an umbilicate pileus and lamellae with decurrent tooth. *E. readiae* was originally described from New Zealand (Stevenson 1962; Horak 1973, 2008). Horak (1973) did not report clamp-connections from this species, but later Horak (2008) noted the occurrence of clamp-connections at the base of the basidia. This species is fairly common and widespread in Tasmania, in all eucalypt forests and rainforest. Horak (2008) noted a strong resemblance to the European *Entoloma sericeum* (Bull.) Quél, which, although this species occasionally has an umbilicate pileus, it has a less frail stature, darker coloured pileus, fibrillose stipe, strong farinaceous-rancid smell, and coarsely encrusted hyphae in the pileipellis. Yellow-coloured forms of *E. readiae* occur, and are distinguished here as var. *sulphureum* (see below).

29b. **Entoloma readiae var. sulphureum** (E. Horak) G.M. Gates & Noordel., comb. & stat. nov.

Basionym: *Entoloma sulphureum* E. Horak, Beih. Nova Hedwigia 43: 60. 1973.

Original diagnosis: Pileo 10–25 mm lato, primo convexo dein umbilicato vel depresso in centro, ex stramineo vitellino, glabro, sicco, striato. Lamellis adnexo-emarginatis, roseis, ventricosis, acie integra instructis. Stipite 15–40 × 1–2.5 mm, cylindrico, pileo concolori vel luteo-olivaceo, mycelio albo ad basim obtecto, glabro, sicco, fistuloso. Odore saporeque subfarinaceo. Sporis 6–7 × 5–6 µm, 5–6-angulatis. Basidiis 4-sporigeris. Cystidiis nullis. Epicute ex hyphis cylindraceis

Plate 3.27b *Entoloma readiae* var. *sulphureum* (Photo Michael Pilkington)

cutem formantibus, pigmento luteo-brunneo incrustatis. Septis fibulatis. Ad terram in silvis. Novazelandia.

Holotype: New Zealand, Nelson, Golden Bay, Pakawau, 8 May 1968, E. Horak (PDD 27044).

Etymology: sulphureus (Lat.) = sulphur yellow, referring to the colour of the basidiocarp.

Main characters: small, collybioid fruit body with bright yellow colour particularly when drying (Plate 3.27b).

Description:
Pileus 8–25 mm diam, convex then concave with shallowly to distinctly depressed centre, with deflexed then straight margin, hygrophanous, when moist translucently striate up to half the radius, yellowish brown with darker yellow-brown centre (2.5Y 8–7/8, centre 10YR 6–5/6), becoming bright yellow upon drying, glabrous to innately radially fibrillose, slightly rugulose at centre. Lamellae adnate-emarginate, narrowly ventricose, fairly distant, up to 4 mm broad, white then pink, often with yellow tinge, with entire, concolorous edge, L = 20–30, l = 1–3. Stipe 15–30 × 1–2 mm, cylindrical, yellow, concolorous with margin of pileus, slightly darker yellow-brown towards base, glabrous, smooth, polished, with white basal tomentum. Odour strong, farinaceous-rancid. Taste indistinct.

Spores 5.5–7.0×4.5–6.5 μm, Q=1.1–1.4, Qav=1.25, 5–6(–7)-angled in side-view. Basidia 22–34×7–11 μm, 4-spored, clamped. Lamella edge fertile. Cheilocystidia absent. Hymenophoral trama regular, made up of cylindrical to slightly inflated elements, 40–110×6–18 μm. Pileipellis a cutis of narrow, cylindrical, 2–8 μm wide hyphae, gradually passing into pileitrama. Pigment deep yellow, parietal-encrusting and intracellular in pileipellis; also encrusting the narrower hyphae in pileitrama. Pileitrama regular, made up of cylindrical elements 60–125×8–20 μm. Brilliant granules absent. Stipitipellis a cutis of narrow, cylindrical hyphae, 3–8 μm wide with minutely encrusted walls, and yellow intracellular pigment. Clamp-connections present.

Habitat & distribution: in small groups in forest litter and on dead wood; almost all of the fruiting occurs during March–June (Group 3, in Chap. 1, see section "Fruiting Patterns of Species of Entolomataceae").

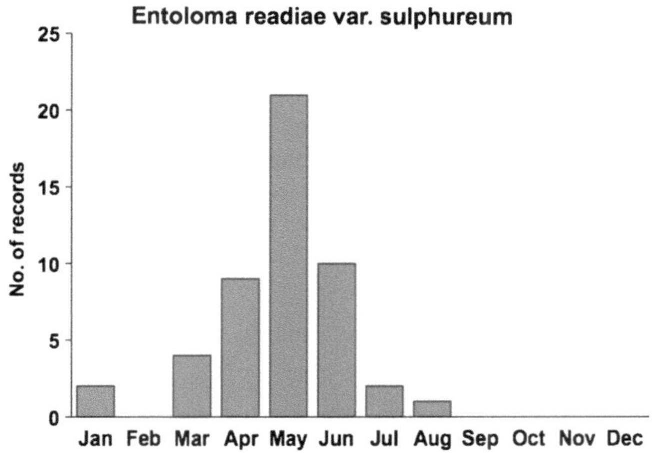

Collections examined. Australia, Tasmania, Bermuda Road, 43° 04′ S, 146° 54′ E, 17 April 2003, G. Gates E 1723; Bruny Island, Mavista Falls Picnic Area, 43° 23′ S, 147° 19′ E, 7 May 2000, G. Gates E 897; Creepy Crawly, 42° 50′ S, 146° 22′ E, 15 April 2000, G. Gates E 842; —, 29 April 2000, G. Gates E 885; Duckhole Lake Track, 43° 22′ S, 146° 53′ E, 18 May 2002, G. Gates E 1518; —, 13 July 2002, G. Gates E 1601; —, 8 April 2003, G. Gates E 1684; Mt Field National Park, Falls Loop, 42° 41′ S, 146° 42′ E, 4 May 2000, G. Gates E 888; —, 29 June 2000, G. Gates E 981; —, 31 March 2001, G. Gates E 1054; —, 24 April 2001, G. Gates E 1124; —, 1 June 2010, M.E. Noordeloos 2010071; North West Bay River, 42° 57′ S, 147° 12′ E, 31 March 2003, G. Gates E 1663; Ralphs Falls, 41° 18′ S, 147° 51′ E, 15 June 2003, G. Gates E 1871; Reuben Falls, 43° 00′ S, 146° 40′ E, 15 May 1999, G. Gates E 513; Styx Valley, Christmas Tree Grove, 42° 50′ S, 146° 41′ E, 13 May 2010, M.E. Noordeloos 2010026; Warra LTER site, 43° 06′ S, 146° 39′ E, numerous observations between June 2006 and June 2007, G. Gates s.n. — New Zealand, Nelson, Golden Bay, Pakawau, 8 May 1968, E. Horak (holotype, PDD 27044).

Horak (1973) described *Entoloma sulphureum* as a very distinctive species with its bright yellow basidiocarps when fresh. In later stages this species gets a brown-yellow pileus, and the stipe an orangy yellow colour. Microscopically it is very similar, however, to *E. readiae* and differs only by the presence of a deep yellow intracellular pigment, which, in mounts in water of fresh material also seems to be diffused among the hyphae of the pileipellis and upper pileitrama. In Tasmania, both taxa have frequently been observed growing on the same decaying log, and intermediate coloured forms also have been seen occasionally. Therefore, it has been decided to merge the two taxa and consider the yellow *E. sulphureum* to be a variety of *E. readiae*. When fresh, this species is barely different from *E. readiae*. It is only as the fruit body begins to hygrophanise and the yellow colour becomes apparent that it can be identified as *E. sulphureum*. This decision is also supported by ongoing molecular phylogenetic studies (Morgado-Neves et al. 2012, in prep.). Both varieties have a wide distribution in Tasmania and New Zealand.

Phylogenetic analyses show that *Entoloma readiae* forms a subclade within the larger */nolanea-claudopus* clade with other Tasmanian taxa like *E. austrorhodocalyx*, in a remote position from *E. sericeum*.

30. **Entoloma convexum** G. Stev., Kew Bulletin 16: 235. 1962.

Original diagnosis: Pileus 1.5–2 cm diam, hemisphaericus, interdum umbone parvo acuto praeditus, badius, humidus striatus, siccus pallide umbrinus et sericeo-fibrillosus. Lamellae liberae vel vix adnexae, subdistantes, latae, sordide roseae. Stipes 3–4 cm × 2 mm, olivaceo-viridis, basi radicante albida, sericeo-striatus vel glaber, cartilagineus, farctus. Sporae 12 × 12 μm, polygonae, apiculis 3 μm longis, in cumulo sordide roseae.

Holotype: New Zealand, Wellington, Butterfly, 7 Aug. 1949, G. Stevenson (K, Stevenson 12).

Etymology: convexus (Lat.) = convex, referring to the shape of the pileus.

Selected literature: Horak, Beihefte Nova Hedwigia 43: 29. 1973; Horak, Beihefte Nova Hedwigia 65: 136. 1980; Horak, Agaricales of New Zealand 1: 246–247. 2008.

Main characters: habit mycenoid; stature medium; pileus moderately dark brown, hygrophanous; lamellae adnexed-emarginate; stipe polished; spores large, isodiametric; cutis differentiated, with encrusting pigment (Plate 3.28).

Description:
Pileus 10–55 mm diam, conico-convex to convex with small umbo, with deflexed then straight margin, expanding to plano-convex or applanate with recurved margin, usually with small papilla, hygrophanous, translucently striate when moist, deep bronze brown (5D6–5E7), paler towards margin, pallescent on drying, becoming a silky pale grey-brown, lubricous when wet, then dry, glabrous. Lamellae adnexed or deeply emarginate, ventricose, up to 8 mm broad, grey-brown when young, then flesh pink becoming deeper pink with maturity, paler towards margin, moderately distant, with two tiers of lamellulae. Stipe (15–) 25–75 × 2–8 mm, cylindrical,

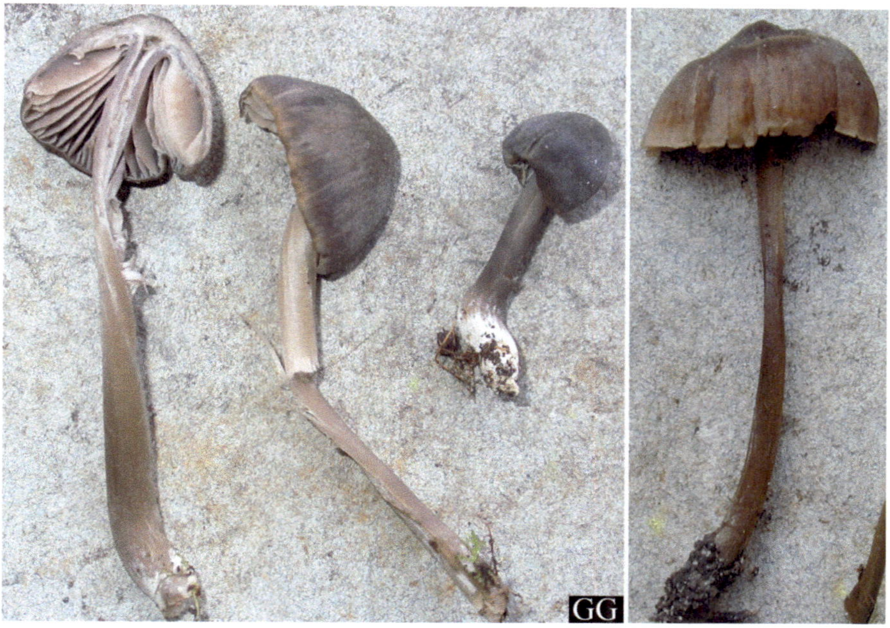

Plate 3.28 *Entoloma convexum* (Photos Genevieve Gates)

slender, slightly swollen at base, brown, glabrous, innately fibrillose, hyaline, subpolished, base with white tomentum. Odour and taste indistinct to farinaceous, often referred to as odour of dirty socks.

Spores (8–)9–13 × 8–12 μm, Q = 1.0–1.2, Qav = 1.1, isodiametric or subisodiametric, 5–6-angled in side-view, with rather thick walls and pronounced angles. Basidia 34–62 × 10–13 μm, 4-spored, clamped. Lamella edge fertile. Cystidia absent. Hymenophoral trama regular, made up of cylindrical to inflated fusiform elements, up to 250 × 8–15 μm, intermingled with narrow, cylindrical hyphae, 2–5 μm wide, with finely encrusted walls. Pileipellis a cutis of repent hyphae, 4–10 μm wide, subpellis not differentiated from underlying pileitrama. Pileitrama regular, made up of cylindrical to fusiform elements, up to 18 μm wide. Hyphae of pileipellis and upper pileitrama with dark brown pigment, sometimes with thickened walls and coarsely encrusted; sometimes a diffuse intracellular pigment is also present. Hyphae of lower pileitrama and hymenophoral trama often finely encrusted. Clamp-connections present (Fig. 3.30).

Habitat & distribution: in wet eucalypt forest and rainforest, widespread in late autumn and in winter.

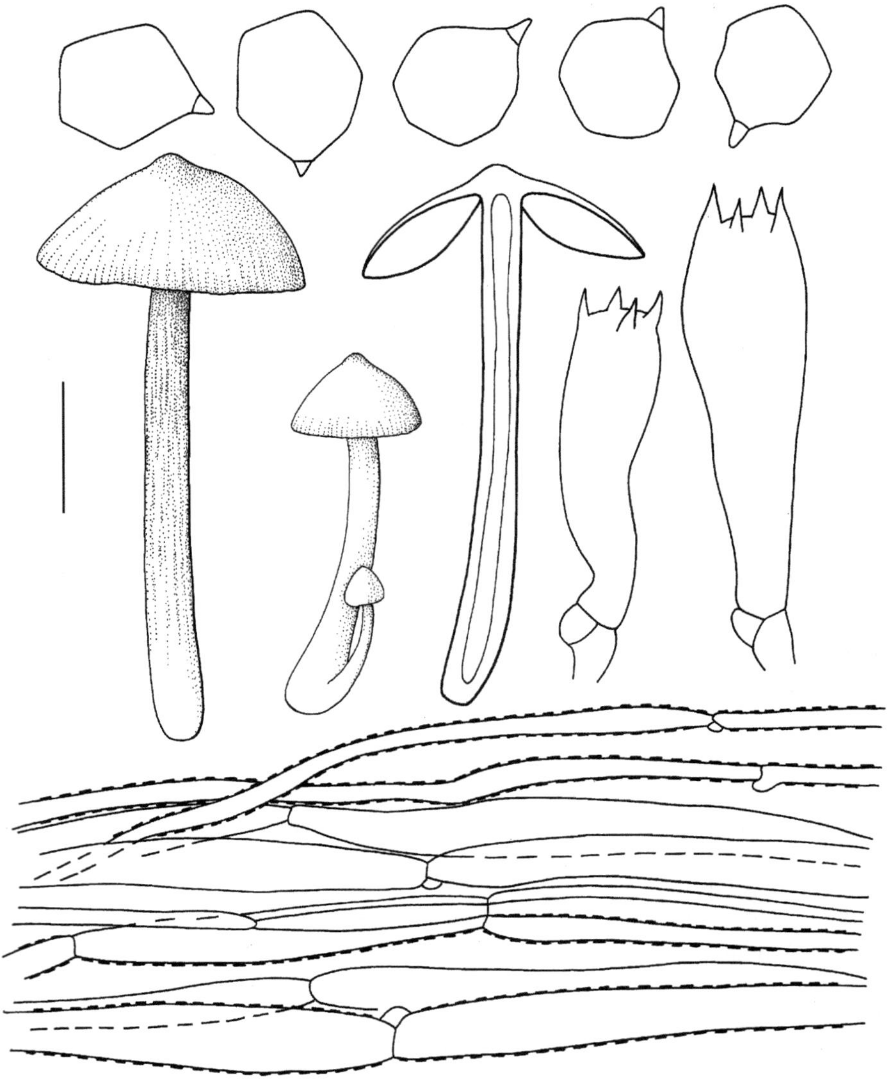

Fig. 3.30 *Entoloma convexum*. Habit, spores, basidia, and pileipellis. Bar = 10 mm or 10 μm

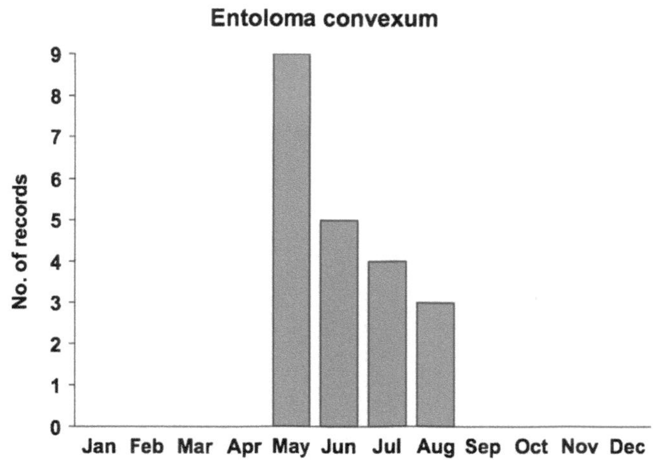

Collections examined. Australia, Tasmania, Bruny Island, Mt Mangana, 43° 22′ S, 147° 17′ E, 26 May 2001, G. Gates E 1188; Julius River, 41° 09′ S, 145° 02′ E, 15 June 2002, G. Gates E 1580; Little Florentine River, F5 Road, 42° 44′ S, 146° 25′ E, 25 May 1999, G. Gates E 540; Marriotts Falls, 42° 43′ S, 146° 39′ E, 2 June 2001, G. Gates E1216; Mt Field National Park, 42° 41′ S, 146° 42′ E, 29 June 1999, G. Gates E 629; —, 25 Aug 2005, G. Gates E 2188; Mt Wellington, Betts Vale Track, 42° 55′ S, 147° 15′ E, 24 June 1999, G. Gates E 626; —, Lenah Valley Track, 9July 2000, G. Gates E 983; —, Lower Sawmill Track, 8 Aug. 1988, G. Gates E 35; —, Shoobridge Track, 25 July 1998, G. Gates E 24; —, 13 May 1999, G. Gates E 505; Lake St Clair National Park, Overland Track, to Windy Ridge, 41° 55′ S, 146° 05′ E, 3 May 2004, G. Gates E 1986; Pelverata Falls, 43° 04′ S, 147° 08′ E, 25 July 2000, G. Gates E 991; —, 26 Aug. 2003, G. Gates E 1907; Timbs Track, 42° 44′ S, 146° 25′ E, 13 May 2005, G. Gates E 919a; —, 23 May 2002, G. Gates E 1533; Warra LTER site, 43° 06′ S, 146° 41′ E, 2 May 2006, M.E. Noordeloos 2006052. — New Zealand, Wellington, Butterfly, 7 Aug. 1949, G. Stevenson 12 (holotype, K).

Entoloma convexum was first described from New Zealand (Stevenson 1962). In Tasmania it is widespread, but uncommon. *Entoloma convexum* is a typical member of subgenus *Nolanea* section *Papillata* on account of the mycenoid habit, glabrous pileus, lack of cystidia, and encrusting pigments. *E. convexum* resembles the European *E. juncinum* (Kühner & Romagn.) Noordel. by the glabrous, almost polished stipe, but that species has a duller, greyish brown overall colour and smaller spores. Some forms of *Entoloma sericeum* (Bull.) Quél. are also superficially similar, but differ by a fibrillose stipe and strong cucumber-like, rancid smell and taste.

31. **Entoloma aromaticum** E. Horak, Beih. Nova Hedwigia 43: 52. 1973.

Original diagnosis: Pileo 10–20 mm lato, primo conico dein convexo-umbonato vel papillato, umbo constanter praedito, griseobrunneo, desiccato pallidiori, glabro vel minute fibrilloso, striato, hygrophano, sicco. Lamellis ex libero adnatis, albidis dein roseis, acie integra instructis. Stipite 20–45 × 1–2 mm, cylindraceo, pileo concolori, fibrillis

Plate 3.29a *Entoloma aromaticum* var. *aromaticum* (Photos Michael Pilkington)

albis basim versus dense obtecto, sicco, fistuloso, fragili. Odore grato vel farinaceo. Sapore rancido. Sporis 9–11×6–7.5 μm, 5–6-angulatis. Basidiis 4-sporigeris. Cystidiis nullis. Epicute ex hyphis cylindraceis cutem formantibus, pigmento brunneo plasmatico vel incrustato instructis. Septis fibulatis defibulatisve. Ad terram in silvis. Novazelandia.

Holotype: New Zealand, Nelson Province, S of Ahaura, Ngahere, 14 March 1968, E. Horak (holotype, PDD 27038).

Etymology: aromaticus (Lat.) = aromatic, sweet, referring to the smell of the basidiocarp.

Selected literature: Horak, Agaricales of New Zealand 1: 236–237. 2008; Wölfel & Hausknecht, Österr. Z. Pilzk. 8: 125. 1999.

Key to the varieties:

1. Spores 9–11×7–8 μm, heterodiametric — var. ***aromaticum***
1. Spores smaller 7–8×6–7 μm, and more isodiameteric — var. ***aromaticellum***

31a. Entoloma aromaticum var. aromaticum

Main characters: small, fragile, mycenoid fruit bodies with yellow-brown to reddish brown colour; odour strong aromatic (Plate 3.29a).

Description:
Pileus 10–30 mm diam, conico-convex or conico-campanulate, quickly expanding to plano-convex or applanate, often with small conical umbo, with straight margin, hygrophanous, when moist translucently striate to centre, medium dark yellow-brown to reddish yellow, often with darker reddish brown centre, paler towards margin, pallescent from centre, becoming a whitish-yellow, glabrous, shining, becoming radially fibrillose upon drying. Lamellae narrowly adnexed to almost free, ventricose, up to 5 mm broad, pale pinkish ochre with entire, concolorous edge, L = 20–32,

l = 1–5. Stipe 30–60 × 2–5 mm, cylindrical or compressed, broadened at base, yellow-brown, concolorous with pileus, glabrous, polished or slightly innately fibrillose in lower part, shining. Context thin, concolorous with surface. Odour strong, aromatic. Taste ranging from farinaceous to rancid, or fish oil.

Spores 9–11 × 7–8 μm, Q = 1.25–1.5, Qav = 1.35–1.4, 6–7(–8)-angled in side-view, with thick walls (usually), and pronounced angles. Basidia 20–43 × 8–11 μm, 4-spored, clampless. Lamella edge fertile. Cheilocystidia absent. Hymenophoral trama regular, made up of fusiform to cylindrical elements, 90–350 × 4–25 μm. Pileipellis a cutis of narrow, cylindrical hyphae, 2–7 μm wide, subpellis gradually passing into pileitrama, made up of long, fusiform or cylindrical elements, 140–320 × 5–20 μm wide. Pileitrama made up of long, fusiform elements, up to 450 × 8–20 μm. Pigment predominantly parietal and intracellular, pallid brown; some narrow hyphae in pileipellis and subpellis minutely encrusted. Stipitipellis a cutis of narrow, cylindrical hyphae, 2–7 μm wide. Caulocystidia absent. Clamp-connections absent (Fig. 3.31).

Habitat & distribution: very common in all forest types in Tasmania. Fruiting throughout the year, with a fair number of records in spring and summer months (Group 5, in Chap. 1, see section "Fruiting Patterns of Species of Entolomataceae"). Originally described from New Zealand, and also recorded from New Caledonia.

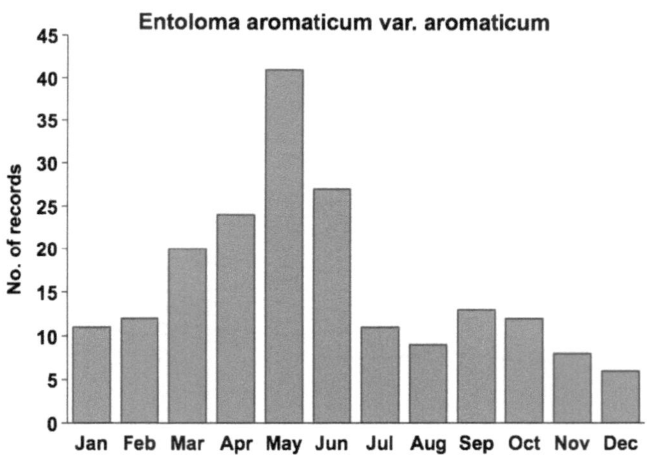

Collections examined. Australia, Tasmania, Bruny Island, Mavista Falls Picnic Area, 43° 23′ S, 147° 19′ E, 8 April 1999, G. Gates E366; —, 1 June 1999, G. Gates E563; —, 9 Sept. 1999, G. Gates E694; —, 17 May 2010, M.E. Noordeloos 2010041 (L); Cape Pillar Track, 43° 09′ S, 147° 56′ E, 29 May 1999, G. Gates E549; —, 28 Oct. 1999, G. Gates E717; —, 1 June 2002, G. Gates E1542; Clarks Cliffs, 43° 06′ S, 147° 47′ E, 21 April 2003, G. Gates E1733; —, 13 March 2004, G. Gates E1942; Collinsvale, Fire Trail, 42° 52′ S, 147° 09′ E, 15 April 1999, G. Gates E417; —, Myrtle Forest, 42° 52′ S, 147° 09′ E, 15 April 1999, G. Gates E428; Donnellys Rd, 43° 07′ S, 146° 54′ E, 17 June 1999, G. Gates E616; —, 23 Sept. 1999, G. Gates E707; —, 19 June 2001, G. Gates E1254; Duckhole Lake Track, 43° 22′ S, 146° 53′ E, 8 April 2003, G. Gates E1680; —, 29 May 2010, M.E. Noordeloos 2001063; Ferndene, 41° 10′ S, 146° 20′ E, 28 March 1999, G. Gates E303; Grassy, King

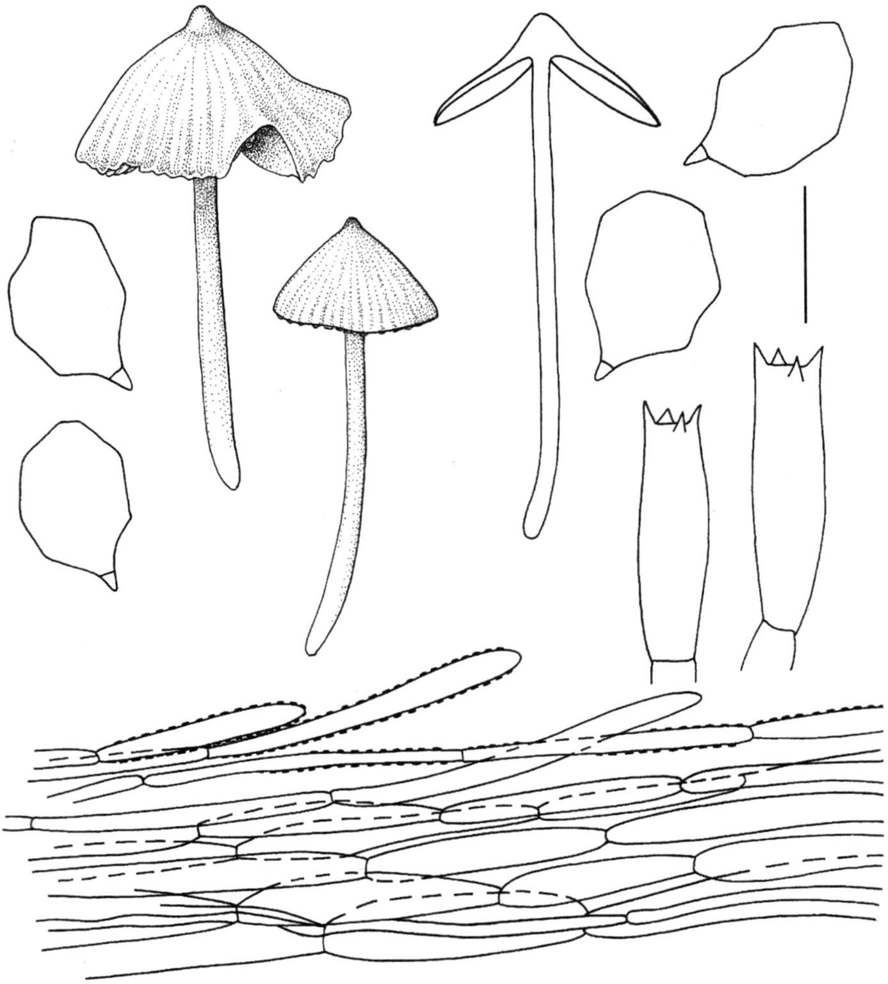

Fig. 3.31 *Entoloma aromaticum* var. *aromaticum*. Habit, spores, basidia, and pileipellis. Bar = 10 mm or 10 μm

Island, 40° 03′ S, 144° 02′ E, 26 Sept. 1998, G. Gates E58; Growling Swallet, 42° 41′ S, 146° 30′ E, 1 Oct. 1998, G. Gates E65; —, 5 Aug. 1999, G. Gates E671; —, 27 March 2001, G. Gates E1044; Hastings Cave, 43° 25′ S, 146° 52′ E, 4 Oct. 1998, G. Gates E66 & E67; Kermandie Falls, Lower Track, 43° 12′ S, 146° 52′ E, 12 Dec. 2002, G. Gates E1646; —, 3 June 2003, G. Gates E1852; —, 24 March 2009, M.E. Noordeloos 2009055 & 2009056 (L); —, Upper Track, 43° 12′ S, 146° 52′ E, 8 Oct. 1998, G. Gates E71 & E73; —, 12 June 1999, G. Gates E587, E589 & E604; —, 26 Aug. 1999, G. Gates E686; Mt Duncan Track, 41° 12′ S, 146° 02′ E, 7 June 1999, G. Gates E569; Mt Wellington, Circle Track, 42° 55′ S, 147° 15′ E, 11

March 1999, G. Gates E202; —, Myrtle Gully, 42° 54′ S, 147° 15′ E, 9 March 1999, G. Gates E177; —, 6 Sept. 1999, G. Gates E702; —, Pipeline Track, 42° 55′ S, 147° 15′ E, 27 April 1999, G. Gates E447; Mt. Field National Park, 42°41′ S, 146°42′ E, 22 Oct. 1998, G. Gates E97; —, 28 Feb. 1999, G. Gates E160; —, 9 May 1999, G. Gates E481; —, 29 June 1999, G. Gates E634; —, 4 Nov. 1999, G. Gates E724; —, Lady Barron Falls Track, 42°41′ × 146°42′, 1 June 2010, M.E. Noordeloos 2010070; —, Lyrebird Walk, 42° 41′ S, 146° 40′ E, 24 April 2001, G. Gates E1113; Myrtle Bank, 'Skemps' property, fern gully, 41° 18′ S, 147° 22′ E, 16 April 2001, G. Gates E1098; North West Bay River, 42° 57′ S, 147° 12′ E, 29 Aug. 2002, G. Gates E1631; Tahune, Huon Pine Walk, 43°06′ S, 146°44′ E, 13 Jan. 2004, G. Gates E1918; Warra LTER site, 43° 06′ S, 146° 41′ E, numerous observations between May 2006 and June 2007, G. Gates s.n.; Waterfall Bay, 43° 04′ S, 147° 57′ E, 13 March 1999, G. Gates E210; Wielangta, 42° 42′ S, 147° 51′ E, 24 Sept. 1998, G. Gates E56; —, 29 Jan. 1999, G. Gates E99; —, 23 Feb. 1999, G. Gates E138; —, 20 March 1999, G. Gates E266. — New Zealand, Nelson Province, S. of Ahaura, Ngahere, 14 March 1968, E. Horak (holotype, PDD 27038).

Entoloma aromaticum can be easily recognized by the fragile, yellowish brown or grey-brown, mycenoid basidiocarps, with polished stipe and strong, fruity aromatic smell. Collections with smaller and more isodiametric spores (7–8 × 6–7 μm, Q = 1.15–1.3) but otherwise identical to the more usual *E. aromaticum* are distinguished here as var. *aromaticellum* (see below). *Entoloma blandiodorum* E. Horak differs by having darker, more robust basidiocarps, smaller, more isodiametric spores, and clamped basidia.

31b. **Entoloma aromaticum** var. **aromaticellum** (E. Horak) G.M. Gates & Noordel. comb. & stat. nov.

Basionym: *Entoloma aromaticellum* E. Horak, Agaricales of New Zealand 1: 234. 2008.

Original diagnosis: Pileus 10–25 mm, primo conicus dein applanatus, papilla obtusa instructus, griseobrunneus vel pallide fuscus, hygrophanus, striatus marginem versus, glaber. Lamellae subliberae vel adnexae, isabelinae dein sordide rosacae, ad aciem concolores. Stipes 25–50 × 1–2.5(–3) mm, cylindricus, aequalis, pileo concolor, glaber vel innate fibrillosus, siccus, fragilis. Odor saporque ope aromatici vel mitissimi. Basidiosporae 7.5–8.5 × (5.5–)6–6.5 μm, 6(–7)-angulatae. Cheilocystidia, pleurocystidia et caulocystidia nulla. Pileipellis ex hyphis cylindraceis cutem formantibus, cellulae terminales cylindricae, 5–7 μm diam, pigmento brunneo incrustatae. Fibulae nullae. Ad terram in silvis mixtis. Nova Zelandia.

Holotype: PDD 72920, isotype ZT 8747.

Differs from the typical variety by the slightly smaller spores, measuring 7–8.5 × 6–7(–7.5) μm, Q = 1.15–1.3 (Plate 3.29b).

Habitat & distribution: widespread in similar habitats as the typical variety.

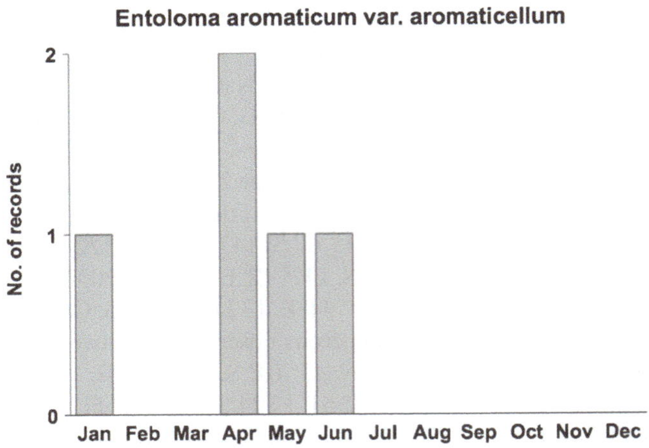

Plate 3.29b *Entoloma aromaticum* var. *aromaticellum* (Photo Fernanda Karstedt)

Collections examined. Australia, Tasmania, Blue Tier, 41° 11′ S, 148° 00′ E, 1 April 2000, G. Gates E806; Creepy Crawly, 42° 50′ S, 146° 22′ E, 6 May 2003, G. Gates E1776; Kermandie Falls Upper Track, 43° 12′ S, 146° 52′ E, 26 April 2001, G. Gates E1138; Mt Field National Park, Lady Barron Falls, 42° 41′ S, 146° 42′ E, 1 June 2010, F. Karstedt 1535. Mt Wellington, Circle Track, 42° 55′ S, 147°15′ E, 26 Jan. 2002, G. Gates E1422. — New Zealand, Nelson, Karamea, Kohaihai River, Nikau Walk, 25 March 2000, E. & A. Horak (holotype PDD 72920).

As pointed out above, some collections of the very common *Entoloma aromaticum* have slightly smaller and more isodiametric spores, which fit the description of *E. aromaticellum* (Horak 2008). Interestingly, the line drawings of the spores by Horak of this species show no differences with those of *E. aromaticum*. Since no other morphological differences can be found, it has been decided to designate these aberrant collections as a variety of *E. aromaticum*.

32. **Entoloma blandiodorum** E. Horak, Agaricales New Zealand 5: 250. 2008.

Original diagnosis: Pileus 5–25(–30) mm, conicus dein hemisphaericus vel convexus, papilla conica distincta instructus, brunneus, hygrophanus, distincte striatus marginem versus, glaber, siccus. Lamellae subliberae vel adnexae, primo isabelinae dein rosaceae, ad aciem concolores. Stipes 25–65 × 1–2.5 mm, cylindricus, aequalis, pileo concolor, glaber, fragilis. Odor saporque mitissimi. Basidiosporae 6.5–8 × 5.5–6.5(–7) µm, hexagonales. Cheilocystidia, pleurocystidia et caulocystidia nulla. Pileipellis ex hyphis cylindraceis cutem formantibus, cellulae terminales cylindraceae, 5–10 µm diam, pigmento brunneo plasmatico et incrustato instructae. Fibulae praesentes. Ad terram in silvis mixtis. Nova Zelandia.

Holotype: New Zealand, Coromandel, Little Barrier Island, Thumb Track, 40 m, on soil among litter under *Leptospermum scoparium* and *Nothofagus truncata*, 12 June 1981, E. Horak (holotype, PDD 74729, isotype, ZT 928).

Main characters: a mycenoid species with strong, aromatic odour; colours dull brown; spores small (Plate 3.30).

Description:
Pileus 30–45 mm diam, conico-convex to convex or hemispherical, slightly expanded with age, with small, acute umbo, with deflexed margin, very hygrophanous, translucently striate at margin, dark grey-brown to sepia brown (6E4–5, 5E3), uniformly coloured, hardly paler towards margin, distinctly pallescent on drying to pale ochre yellow (5C4), shiny with a lustrous sheen. Lamellae deeply emarginate, segmentiform to subventricose, moderately distant, up to 6 mm broad, brown (5C3–4, 5D3–4), with pink tinge when mature, slightly paler towards entire edge, L = 30–40, l = 3–7. Stipe 50–80 × 6–8 mm, compressed with groove, yellow-brown, slightly paler than pileus (5D4–5E4), shiny, polished to the naked eye, but with fine arachnoid fibrils under a hand lens, basal part with white mycelial tomentum.

Plate 3.30 *Entoloma blandiodorum* (Photo Michael Pilkington)

Context thin, subcartilagineous, concolorous with surface. Smell very strong, aromatic. Taste aromatic with nasty aftertaste.

Spores 6.0–8.0 × 5.5–7.0 μm, Q = 1.0–1.3, Qav = 1.2, subisodiametric to shortly heterodiametric, (4–)5–6-angled in side-view. Basidia 22–40 × 6–9 μm, 4-spored, clamped. Lamella edge fertile. Cheilocystidia absent. Hymenophoral trama regular, made up of very long, fusiform elements, 180–350 × 5–20 μm, mixed with narrow connecting hyphae. Pileipellis a thin cutis of narrow, cylindrical hyphae, 2–4 μm wide, subpellis not differentiated from trama. Pileitrama regular, made up of long, fusiform elements, up to 400 × 5–22 μm, mixed with narrow, cylindrical connecting hyphae, 2–5 μm wide. Pigment encrusting the narrow hyphae of pileipellis and the connective hyphae in hymenophoral trama; diffuse intracellular pigment may also be present in upper pileitrama. Clamp-connections abundant in hymenium, rare in other tissues (Fig. 3.32).

Habitat & distribution: widespread in wet sclerophyll forest.

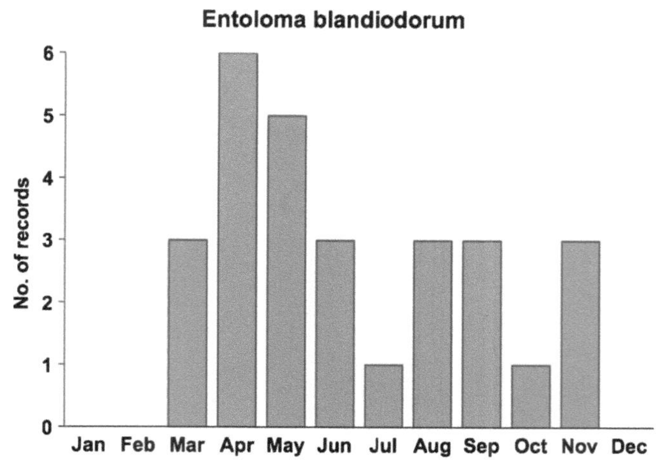

Collections examined. Australia, Tasmania, Arve Rd, Leap Loop, 43° 09′ S, 146° 46′ E, 25 May 2004, G. Gates E 2018 & E 2019; —, 31 Aug. 2004, G. Gates E 2051; Bruny Island, Mt Mangana, 43° 22′ S, 147° 17′ E, 8 April 1999, G. Gates E 381; —, 12 April 2003, G. Gates E 1701; Cethana, Lacey's property, 41° 31′ S, 146° 04′ E, 27 April 2002, G. Gates E 1501; Duckhole Lake Track, 43° 22′ S, 146° 53′ E, 15 May 2003, G. Gates E 1815; —, 23 March 2004, G. Gates E 1959; Ferndene, 41° 10′ S, 146° 20′ E, 28 March 1999, G. Gates E 302; Hobart, Waterworks Reserve, 42° 54′ S, 147° 17′ E, 20 Sept. 2003, G. Gates E 1912; Kermandie Falls, Lower Track, 43° 12′ S, 146° 52′ E, 24 March 2009, M.E. Noordeloos 2009056; Mt Field National Park, 42°41′S, 146°42′E, 28 April 2011, M. E. Noordeloos 2011010 & 2011011 (L); Mt Wellington, Bracken Lane, 42° 55′ S, 147° 15′ E, 25 July 1998, G. Gates E 22; —, Myrtle Gully, 42° 54′ S, 147° 15′ E, 15 June 1999, G. Gates E 609; Ralphs Falls, 41° 18′ S, 147° 51′ E, 8 June 2004, E2033; Wielangta, 42° 42′ S, 147° 51′ E, 5 Sept. 2002, G. Gates E 1634; —, 12 Aug. 2004, G. Gates E 2049; —, 18 Aug. 2005, G. Gates E 2189. — New Zealand, Coromandel, Little Barrier Island, Thumb Track, 40 m, on soil among litter under *Leptospermum scoparium* and *Nothofagus truncata*, 12 June 1981, E. Horak (holotype, PDD 74729).

During the investigation of the Entolomataceae flora of Tasmania, it was noted that the group of nolaneoid species with an aromatic odour varies considerably. In its typical appearance, *Entoloma aromaticum* is a slender, brittle species with relatively pale coloured, deeply translucently striate pileus and rather pale lamellae. Occasionally, darker coloured and more robust "forms" were encountered. With the help of the monograph on the New Zealand taxa (Horak 2008), these forms could be identified as *Entoloma blandiodorum* E. Horak. Apart from the darker and often more robust basidiocarps, this species differs from *E. aromaticum* by the smaller, more isodiametric spores, and abundant clamp-connections in the hymenium. Apparently, *Entoloma gratissimum* E. Horak from New Guinea also belongs to the complex of *E. aromaticum*, with similar microscopics as *E. aromaticum*, but possibly slightly darker colours, more like *E. blandiodorum*.

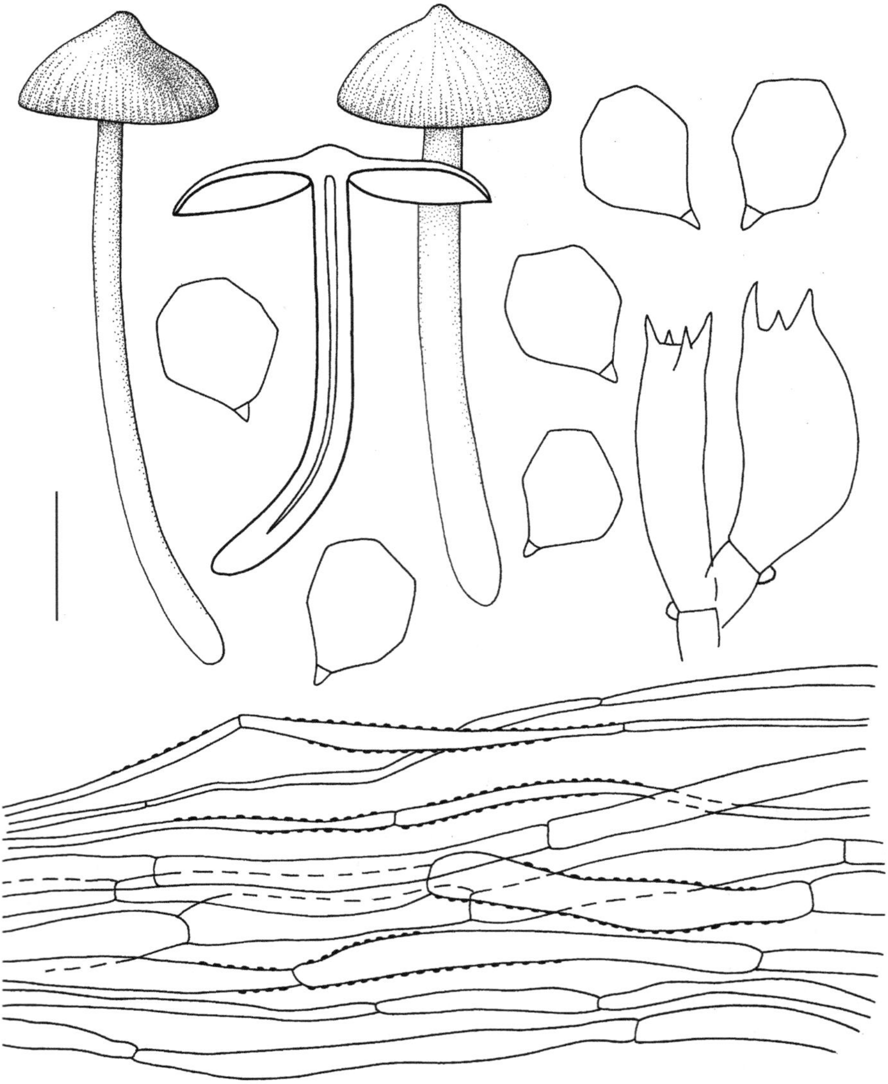

Fig. 3.32 *Entoloma blandiodorum*. Habit, spores, basidia, and pileipellis. Bar = 10 mm or 10 μm

Section ***Staurospora*** (Largent & Thiers) Noordel.

Nolanea sect. *Staurospora* Largent & Thiers in Northwest Sci. 46: 37. 1972; *Rhodophyllus* sect. *Staurospora* (Largent & Thiers) Romagn. in Bull. mens. Soc. linn. Lyon 43: 330. 1974; *Entoloma* sect. *Staurospora* (Largent & Thiers) Noordel. in Persoonia 10: 445. 1980.

Type species: *Entoloma conferendum* (Britzelm.) Noordel.

Habit mycenoid; pileus glabrous or fibrillose; spores cruciform, sometimes in part more cuboid.

33. **Entoloma conferendum** (Britzelm.) Noordel.

Agaricus conferendus Britzelm. in Ber. naturw. Ver. Schwaben, Augsburg 26: 140. 1881; *Nolanea conferenda* (Britzelm.) Sacc., Syll. Fung. 5: 723. 1887; *Entoloma conferendum* (Britzelm.) Noordel. in Persoonia 10: 446. 1980. *Agaricus dissidens* Britzelm. in Ber. naturw. Ver. Schwaben, Augsburg 26: 140. 1881; *Nolanea dissidens* (Britzelm.) Sacc., Syll. Fung. 5 723. 1887. *Agaricus postumus* Britzelm. in Ber. naturw. Ver. Schwaben, Augsburg 26: 140. 1881; *Nolanea subpostuma* (Britzelm.) Sacc., Syll. Fung. 5: 725. 1887. *Nolanea staurospora* Bres., Fungi trident. 1: 18. 1882; *Rhodophyllus staurosporus* (Bres.) J. Lange, Fl. agar. dan. 2: 99. 1937; *Entoloma staurosporum* (Bres.) E. Horak in Sydowia 28: 222. "1974/1975" 1976. *Rhodophyllus staurosporus var. typicus* Kühn. & Romagn., Fl. anal.: 187. 1953. *Rhodophyllus staurosporus var. platyphyllus* Romagn. & Favre in Rev. Mycol. 3: 77. 1938. *Rhodophyllus staurosporus var. obscurior* (Romagn.) Romagn. 1937. *Rhodophyllus staurosporus var. subrugosus* (Romagn.) Romagn. 1937. *Rhodophyllus rickenii* Romagn. in Bull. Soc. mycol. Fr. 48: 320. 1932; *Rhodophyllus staurosporus subsp. rickenii* (Romagn.) Romagn. in Rev. Mycol. 2: 86.1937; *Nolanea rickenii* (Romagn.) Konr. & M. Agaricales: 264. 1948; *Rhodophyllus staurosporus var. rickenii* (Romagn.) Kühner & Romagn., Fl. anal. Champ. super.: 187. 1953. 1948.

Misappl. names: *Nolanea pascua* sensu Britzelm. 1872; Cooke 1884; Ricken 1914. *Nolanea proletaria* sensu Boudier 1906; Rea 1922. *Agaricus cetratus* sensu Britzelm. 1874.

Icon.: Bres. Iconogr. Mycol. 12, pl. 584. 1929 (as *N. staurospora*); Cetto, Funghi Vero 2, pl. 248. 1976; Konrad & Maublanc, Ic. sel. Fung., pl. 178, fig. 1. 1930 (as *N. staurospora*); J. Lange, Fl. agar. dan., pl. 77 A. 1937 (as *R. staurosporus*); Noordeloos, Fungi Europaei vol. 5: pl. 41. 1992; Noordeloos, Fungi Europaei vol. 5a: 1256–1257. 2004.

Original diagnosis: Hut bis über 5 cm breit, unregelmässig glockig, dann flatig, rotlichbraun, bräunlich; gegen den Rand nicht selten zart wellig gefurcht; matt seidig glänzend. Stiel über 1 dm hoch, unter 1 cm, oben halb so breit, weisslich, blassbräunlich, hie und da gedreht, seidig glänzend, hohl, sehr gebrechlich. Lamellen fast frei, blassrot, fleischfarben, ziemlich entfernt; grob, hie und da weit und unregelmässig gekerbt. Geruch stark nach Mehl. Hut und Stielfleisch weisslich, bräunlich. Spores meist mit 4 weit ausgezogenen abgerundeten Enden, 8–10 Mikromill. in Durchmesser. Diese und die Folgende Art werden als nahe verwant zu A. pascuus und zwar hauptsächlich zu jener Form dieses Schwammes zu betrachten sein, welche von Bolton auf t. 35 abgebildet worden ist. Im Sommer in den Bergwälder bei Oberstaufen, einzeln und gesellschaftlich wachsend.

Holotype: not existent.

Selected literature: Horak in Sydowia 28: 222. 1976 (as *E. staurosporum*); Konrad in Bull. trimest. Soc. mycol. Fr. 45: 49–50. 1929; F.H. Möller, Fungi Faeroes 1: 239, fig. 11. 1945; Noordeloos in Persoonia 10: 446–450, fig. 4. 1980; Noordeloos in Bas et al. Fl. agar. neerl., vol. 1: 120–121, fig. 92. 1988; Noordeloos, Entoloma s.l. Fungi Europaei 5: 373–376. 1992.

Plate 3.31 *Entoloma conferendum* (Photos Machiel Noordeloos (*left*), Michael Pilkington (*right*))

Main characters: relatively stout *Nolanea* (mycenoid); pileus greyish brown or reddish brown, very silky-shining; stipe silky-fibrillose; spores cruciform (Plate 3.31).

Description:
Pileus 20–65 mm diam, campanulate to conical, expanding to convex or plano-convex, usually with low, broad umbo, with inflexed then deflexed or straight margin, very hygrophanous, when moist translucently striate at margin, or halfway or fully to the centre in thin-fleshed specimens, dark grey-brown to sepia, only slightly paler towards margin, distinctly pallescent on drying to sordid yellowish or greyish brown, innately radially fibrillose, centre often minutely rugulose-felted, with micaceous sheen, becoming fibrillose-lustrous, shiny on drying. Lamellae emarginate to free, ventricose, moderately distant, up to 11 mm broad, pallid then pink, finally brown-pink, with slightly irregular, concolorous edge, L = 20–50, l = 1–5. Stipe 20–70 × 2–8 mm, cylindrical or compressed with groove, gradually broadened towards base, pale yellowish brown or greyish brown, densely silvery striate with longitudinal fibrils, shiny base with white tomentum. Context more or less concolorous with surface in cortex of pileus and stipe, whitish or greyish in inner parts. Odour farinaceous. Taste slightly rancid, unpleasant.

Spores 7.5–12(–14) × 6–12 μm, Q = 1.0–1.6, cruciform-stellate. Basidia 22–50 × 7–12 μm, 4-spored, clampless. Lamella edge fertile. Cheilocystidia absent. Pileipellis a cutis with transitions to a trichoderm at centre, made up of cylindrical hyphae, 8–14 μm wide, with terminal elements slightly inflated, up to 19 μm at centre. Pigment abundant, brown, intracellular in pileipellis. Clamp-connections absent (Fig. 3.33).

Habitat & distribution: terrestrial in groups, in pastures, paddocks, and along tracks in wet eucalypt forests and rainforest throughout Tasmania. Widely recorded from various parts of the world. Fruiting throughout the year, with a fair number of

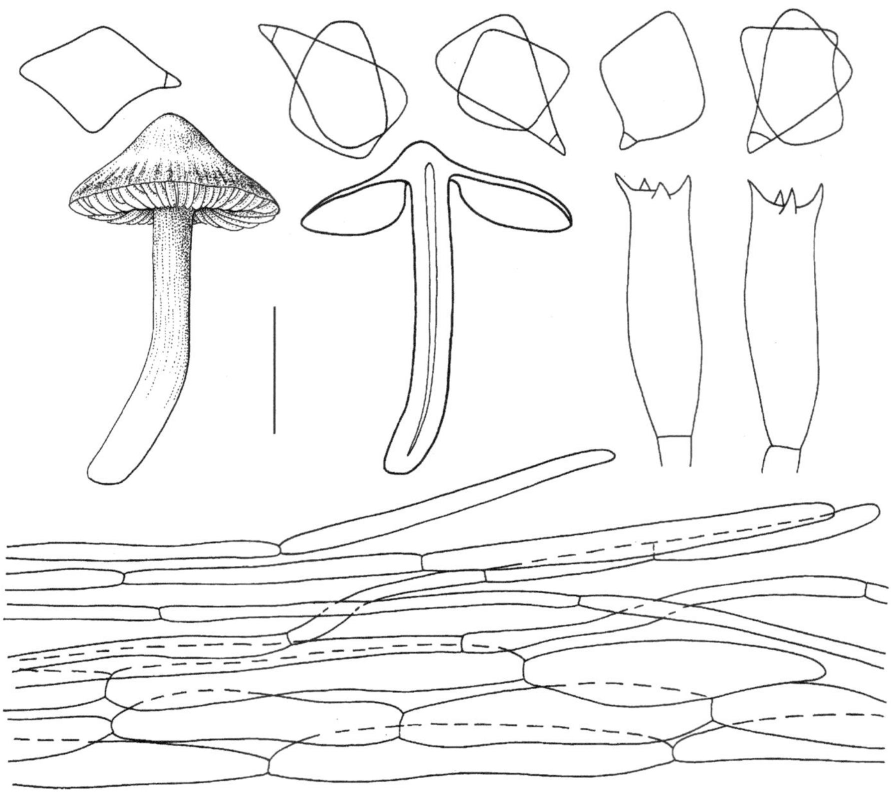

Fig. 3.33 *Entoloma conferendum*. Habit, spores, basidia, and pileipellis. Bar = 10 mm or 10 μm

records in the spring and summer months (Group 5, in Chap. 1, see section "Fruiting Patterns of Species of Entolomataceae").

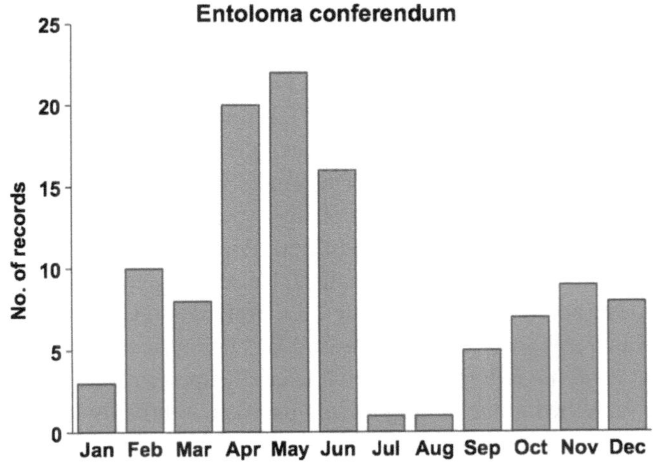

Collections examined. Australia, Tasmania, Adamsons Peak Track, 43° 20′ S, 146° 54′ E, 3 Nov. 2001, G. Gates E 1310; Arve River, Zig-Zag Track, 43° 06′ S, 146° 45′ E, 23 Nov. 1999, G. Gates E 764; Blue Tier, 41° 11′ S, 148° 00′ E, 1 June 2004, G. Gates E 2025; Bluff River State Forest, 42° 31′ S, 147° 40′ E, 9 June 2002, G. Gates E 1156 & E1157; Bruny Island, Mt Mangana, 43° 22′ S, 147° 17′ E, 27 May 2004, G. Gates E 2021; Clarks Cliffs, 43° 06′ S, 147° 47′ E, 29 Sept. 2001, G. Gates E 1285; —, 29 Dec. 2001, G. Gates E 1383; —, 4 May 2002, G. Gates E 1497; Creepy Crawly, 42° 50′ S, 146° 22′ E, 29 April 2000, G. Gates E 879 & E 880; Fortescue Bay, 43° 09′ S, 147° 57′ E, 1 June 2002, G. Gates E 1546; Geeveston, Arve Loop Rd, 43° 08′ S, 146° 45′ E, 23 July 1998, G. Gates E 19; Growling Swallet, 42° 41′ S, 146° 30′ E, 20 April 1999, G. Gates E436; —, 19 May 1999, G. Gates E517; —, 6 April 2000, G. Gates E828; —, 15 June 2000, G. Gates E968; —, 1 May 2001, G. Gates E1150; —, 2 May 2002, G. Gates E1487; —, 4 June 2002, G. Gates E1548; —, 13 May 2003, G. Gates E1803; —, 24 Feb. 2004, G. Gates E1935; Junee Caves State Reserve, 42° 44′ S, 146° 36′ E, 8 April 2000, G. Gates E 832; —, 9 May 2011, M.E. Noordeloos 2011057 (L); Kermandie Falls Upper Track, 43° 12′ S, 146° 52′ E, 6 May 1999, G. Gates E 473; —, 43° 12′ S, 146° 52′ E, 16 Nov. 1999, G. Gates E 757; Lake St Clair National Park, Overland Track, to Windy Ridge, 41° 55′ S, 146° 05′ E, 3 May 2004, G. Gates E 1983; Little Florentine River, F5 Road, 42° 44′ S, 146° 25′ E, 23 May 2002, G. Gates E 1534; Marriotts Falls, 42° 43′ S, 146° 39′ E, 10 June 1999, G. Gates E 581 & E 582; —, 27 Nov. 1999, G. Gates E 768; —, 9 April 2000, G. Gates E 836; —, 13 May 2000, G. Gates E 913 & E 920; Mt Field National Park, Falls Loop, 42° 41′ S, 146° 42′ E, 1 April 1999, G. Gates E 338; —, 4 Nov. 1999, G. Gates E 725 & E 728; —, 13 Nov. 1999, G. Gates E 756; —, 5 April 2000, G. Gates E 813; —, Lyrebird Walk, 42° 41′ S, 146° 40′ E, 27 Oct. 2001, G. Gates E 1303; —, 10 June 1999, G. Gates E 579; Mt Wellington, Fern Glade, 42° 55′ S, 147° 15′ E, 1 Dec. 2005, G. Gates E 2217; —, Myrtle Gully, 42° 54′ S, 147° 15′ E, 9 Dec. 1998, G. Gates E 100; —, 11 Nov. 1999, G. Gates E 745; —, 26 April 2003, G. Gates E 1741; —, Pipeline Track, 42° 55′ S, 147° 15′ E, 16 Sept. 1999, G. Gates E 705; —, Reids Track, 16 Sept. 1999, G. Gates E 703; —, Silver Falls, 22 Dec. 1998, G. Gates E 108 & E 109; —, Sphinx Rock, 24 June 1999, G. Gates E 624; Myrtle Bank, ‘Skemps’ property, fern gully, 41° 18′ S, 147° 22′ E, 15 May 2004, G. Gates E 1998; Myrtle Forest, 42° 52′ S, 147° 09′ E, 10 April 2000, G. Gates E 837; Ralphs Falls, 41° 18′ S, 147° 51′ E, 8 June 2004, G. Gates E 2029; Scottsdale, Cuckoo Falls Track, 41° 15′ S, 147° 37′ E, 24 May 2010, M.E. Noordeloos 2010056; Timbs Track, 42° 44′ S, 146° 25′ E, 20 March 2001, G. Gates E 1027; —, 8 Feb. 2003, G. Gates E 1781; Truganini Track, 42° 56′ S, 147° 21′ E, 30 Sept. 1998, G. Gates E 60; Wielangta, 42° 42′ S, 147° 51′ E, 15 Jan. 2002, G. Gates E 1402. — New Zealand, Nelson, Fringed Hill, 3 June 1956, G. Stevenson 1118 (holotype of *E. nothofagi*, K); Wellington, Catchpole, 3 May 1958, Wellington Botanical Society (G. Stevenson 1322; holotype of *E. botanicum*, K).

Entoloma conferendum is a fairly characteristic species that with some experience can be identified in the field by its campanulate or broadly umbonate, greyish brown pileus with lustrous radial fibrils, the almost free, often broad lamellae, and the obviously fibrillose-striate stipe. In this concept, this species has a very wide geographical distribution, and it has been recorded from both the Northern and Southern Hemisphere. *Entoloma nothofagi* G. Stev. and *E. botanicum* G. Stev. from New Zealand are very similar, and possibly represent synonyms (Noordeloos 1980; Horak 1980). However, Horak (2008) later changed his mind, and considered the New Zealand *E. nothofagi* a good separate species, without offering any explanation why. A study of the holotypes of both New Zealand species in Kew confirms that they probably represent the same species. The differences with *E. conferendum* are minimal. Considering the wide distribution of the conferendum-complex, and its known variability (Noordeloos 1980, 1992, 2004), this species-complex offers an ideal opportunity to do a phylogeographic study with the help of molecular methods.

34. **Entoloma brevispermum** G.M. Gates & Noordel., Persoonia 19: 176. 2007.

Original diagnosis: Pileus 15–35 mm latus conicus demum campanulatus umbonatus hygrophanus translucido-striatus fulvus expallens glaber in sicco leviter fibrillosus. Stipes 40–85 × 2.5–9 mm cylindraceus flavobrunneus politus sub lente leviter fibrillosus. Odore saporeque nullus leviter rancidus sapor farinaceus. Sporae 6.5–9.0 × 5.5–8.5 µm irregulariter 5–6-angulatae interdum subcruciformibus. Basidia tetrasporigera efibulata. Acies lamellarum fertilis. Pileipellis cutis hyphis cylindraceis 4–11 µm latis pigmento incrustantibus formata. Fibulae absentes.

Holotype: Australia, Tasmania, Myrtle Bank, 'Skemps' property, fern gully, 41° 18′ S, 147° 22′ E, 16 April 2001, G. Gates E 1095 (HO 543555; isotype in L).

Etymology: brevis (Lat.) = short; spermus (Lat. & Gr.) = spores, referring to the short spores.

Main characters: often quite a large mycenoid fruit body; pileus yellow-brown, distinctly striate, very hygrophanous; lamellae whitish to pale yellow-pink; stipe subpolished to densely silvery striate; spores small, cruciform (Plate 3.32).

Description:
Pileus 15–50 mm diam, conical to conico-campanulate, usually with pronounced sharp umbo, rarely more or less blunt, conico-convex or convex, with straight margin, with undulating marginal zone, distinctly translucently striate when moist, moderately dark reddish or yellowish brown (5D4), slightly paler at margin, slightly darker on umbo, glabrous to innately radially fibrillose, slightly lubricous at centre when moist, hygrophanous, becoming pale yellow-brown, radially fibrillose-lustrous on drying. Lamellae narrowly adnexed to almost free, ventricose, moderately crowded, up to 6 mm broad, thin, flesh-coloured then pink, with concolorous edge, L = 24–35, l = 1–5. Stipe 40–120 × 2.5–9 mm, cylindrical, slightly broadened towards base, dry, glabrous, longitudinally twisted-striate, brittle, fistulose with age, deep yellow-brown with a white pruina overall which disappears with age or upon bruising. Odour and taste indistinct or of cucumber.

Spores 6.5–9.0 × 5.5–8.5 µm, on average 7.5 × 6.8 µm, Q = 1.0–1.2, very irregularly shaped, subcuboid or cruciform, 5- or 6-angled, with large, blunt angles, often

Plate 3.32 *Entoloma brevispermum* (Photos Machiel Noordeloos (*left*), Michael Pilkington (*right*))

twisted along the central axis. Basidia 24–40 × 7–11 μm, 4-spored, clampless. Lamella edge fertile. Cystidia absent. Hymenophoral trama regular, made up of long, fusiform elements, to 200 × 35 μm. Pileipellis a cutis of cylindrical hyphae 4–11 μm in diameter, subpellis (sometimes distinct) made up of short inflated elements, 30–60 × 7–18 μm, gradually passing into pileitrama. Pileitrama composed of fusiform elements, similar to hymenophoral trama. Pigment minutely encrusting in pileipellis and upper pileitrama. Clamp-connections absent (Fig. 3.34).

Habitat & distribution: widespread in wet sclerophyll forests. Fruiting throughout the year, with a fair number of records in the spring and summer months (Group 5, in Chap. 1, see section "Fruiting Patterns of Species of Entolomataceae").

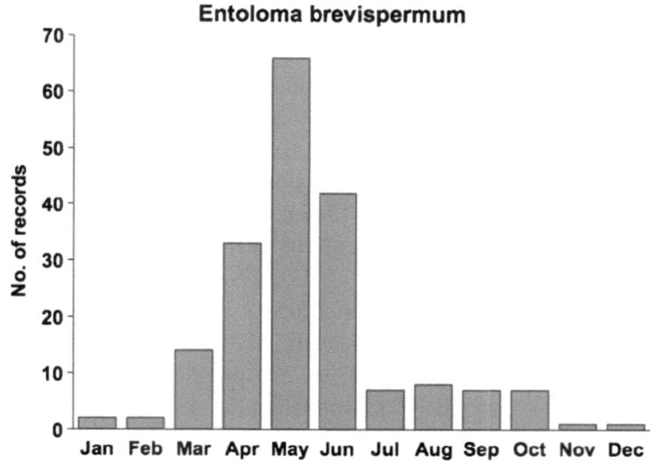

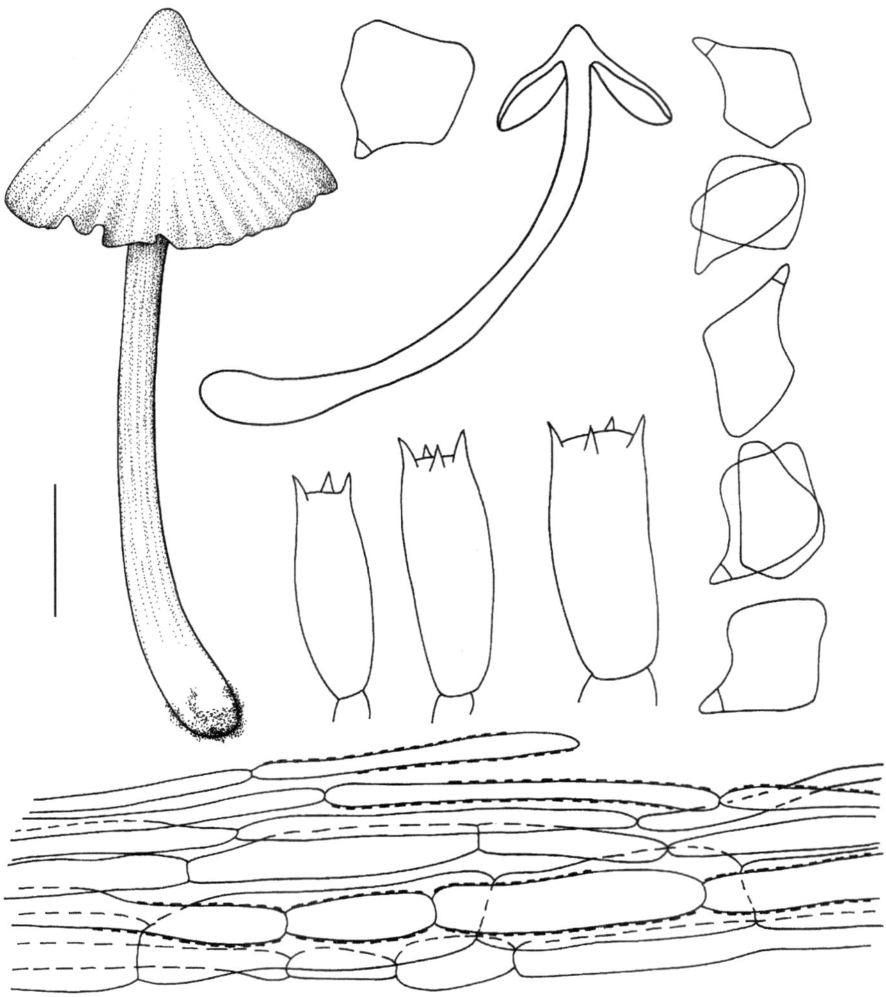

Fig. 3.34 *Entoloma brevispermum*. Habit, spores, basidia, and pileipellis. Bar = 10 mm or 10 µm

Collections examined. Australia, Tasmania, Arve Rd, Keoghs Walk, 43° 08′ S, 146° 47′ E, 10 April 2001, G. Gates E 1089 & E 1093; Bermuda Road, 43° 04′ S, 146° 54′ E, 17 April 2003, G. Gates E 1718; Blue Tier, 41° 11′ S, 148° 00′ E, 1 April 2000, G. Gates E 807; Bluff River State Forest, 42° 31′ S, 147° 40′ E, 9 June 2002, G. Gates E 1568; Bruny Island, Fluted Cape, 43° 22′ S, 147° 22′ E, 9 Sept. 1999, G. Gates E 698; Calverts Lagoon, 43° 01′ S, 147° 29′ E, 3 Sept. 2005, G. Gates E 2190; Chauncy Vale, 42° 37′ S, 147° 16′ E, 20 Nov. 2001, G. Gates E 1323; —, 5 Oct. 2004, G. Gates E 2058; —, 15 Sept. 2005, G. Gates E 2194; —, 29 Sept. 2005, G. Gates E 2198; —, 6 Oct. 2005, G. Gates E 2203; Creepy Crawly, 42° 50′ S, 146° 22′ E, —, 15 April 2000, G. Gates E 845; —, 29 April 2000, G. Gates E 897 & E 878; —, 20 March 2001, G. Gates E 1031; Donnellys Rd, 43° 07′ S, 146° 54′ E, 25 May 2000, G. Gates E 943; Duckhole Lake Track, 43° 22′ S, 146° 53′ E, 18 May 2002, G. Gates E 1516, E 1517 & E 1524; —, 6 June 2002, G. Gates E 1559; —, 13 July 2002, G. Gates E 1598; 22 Oct. 2002, G. Gates E 1643; —, 8 April 2003, G. Gates E 1687; —, 15 May 2003, G. Gates E 1806 & E 1814; —, 15 Jan. 2004, G. Gates E 1922; —, 12 Feb. 2004, G. Gates E 1933; —, 29 May 2010, F. Karstedt 1436; —, Evercreech Forest Reserve, 41° 24′ S, 147° 58′ E, 16 June 2003, G. Gates E 1879; Florentine River Valley, Pagoda Hut, 42° 30′ S, 146° 27′ E, 29 April 2003, G. Gates E 1747; Geeveston, Rutherfords Road, 43° 16′ S, 146° 55′ E, 23 June 1998, G. Gates E 3; Growling Swallet, 42° 41′ S, 146° 30′ E, 14 Oct. 1999, G. Gates E 714; —, 11 May 2000, G. Gates E 906; —, 27 March 2001, G. Gates E 1045; —, 1 May 2001, G. Gates E 1141; —, 12 March 2005, G. Gates E 2090; Hellyer Gorge, 41° 16′ S, 145° 37′ E, 26 April 2002, G. Gates E 1490; Julius River, 41° 09′ S, 145° 02′ E, 15 June 2002, G. Gates E 1581 & E 1584; Junee Caves State Reserve, 42° 44′ S, 146° 36′ E, 24 May 2003, G. Gates E 1822; —, 28 March 2009, M.E. Noordeloos 2009070; Lake St Clair National Park, Overland Track, to Windy Ridge, 41° 55′ S, 146° 05′ E, 3 May 2004, G. Gates E 1988 & E 1989; Little Florentine River, F5 Road, 42° 44′ S, 146° 25′ E, 8 April 2000, G. Gates E 829; Marriotts Falls, 42° 43′ S, 146° 39′ E, 29 May 2003, G. Gates E 1835; Milkshake Hills, 41° 06′ S, 145° 10′ E, 14 June 2002, G. Gates E 1575; Mt Field National Park, Falls Loop, 42° 41′ S, 146° 42′ E, —, 1 April 1999, G. Gates E 337; —, 9 May 2000, G. Gates E 899; —, 29 June 2000, G. Gates E 982; —, 24 July 2001, G. Gates E 1264; —, 8 Aug. 2002, G. Gates E 1623; —, 19 July 2003, G. Gates E 1895; —, 17 April 2004, M.E. Noordeloos 2004089 (CTAB108); —, 7 May 2005, G. Gates E 2161; —, 8 Sept. 2005, G. Gates E 2191; Mt Wellington, Betts Vale, 42° 55′ S, 147°15′ E, 13 Feb. 1999, G. Gates E 121; —, 24 June 2000, G. Gates E 976; —, Circle Track, 25 March 1999, G. Gates E 290; —, Fern Glade, 8 May 1999, G. Gates E 476; —, 24 June 1999, G. Gates E 625; —, 5 May 2001, G. Gates E 1169; —, 17 April 2005, G. Gates E 2117; —, 1 Dec. 2005, G. Gates E 2218; —, Jacksons Bend, 4 May 1999, G. Gates E 467; —, 7 June 1999, G. Gates E 573, E 574, & E 575; —, 24 June 2000, G. Gates E 974 & E 975; —, 17 May 2001, G. Gates E 1177; —, 27 Sept. 2001, G. Gates E 1284; —, Myrtle Gully, 42° 54′ S, 147° 15′ E, 9 July 1998, G. Gates E 1a; —, 13 May 1999, G. Gates E 503; —, 12 Aug. 1999, G. Gates E 675; —, 22

June 2000, G. Gates E 973; —, 19 Aug. 2003, G. Gates E 1906; —, 7 Oct. 2003, G. Gates E 1913; —, Pillinger Drive, 42° 55′ S, 147° 15′ E, 4 March 1999, G. Gates E 162; —, Silver Falls Service Track, 4 March 1999, G. Gates E 172; Myrtle Bank, 'Skemps' property, fern gully, 41° 18′ S, 147° 22′ E, 16 April 2001, G. Gates E 1095 (holotype, HO 543555; isotype, L); Myrtle Forest, 42° 52′ S, 147° 09′ E, 28 May 1999, G. Gates E 547; —, 14 April 2003, G. Gates E 1706; Needles Picnic Area, 42° 46′ S, 146° 24′ E, 8 April 2000, G. Gates E 831; Notley Gorge, 41° 21′ S, 146° 55′ E, 25 April 2000, G. Gates E 974; —, 25 April 2003, G. Gates E 1787; Oldina Forest Reserve, Noel Jago Track, 41° 00′ S, 145° 40′ E, 16 June 2002, G. Gates E 1583; Ralphs Falls, 41° 18′ S, 147° 51′ E, 15 June 2003, G. Gates E 1874; Reuben Falls, 43° 00′ S, 146° 40′ E, 15 May 1999, G. Gates E 514; Snug Falls, 43° 05′ S, 147° 13′ E, 23 Aug. 1999, G. Gates E 684; 16 July 2002, G. Gates E 1605; St Columba Falls, 41°19′, 147°55′, 15 April 2001, G. Gates E 1099 & E 1099a; Timbs Track, 42° 44′ S, 146° 25′ E, 29 April 2000, G. Gates E 881; 23 May 2002, G. Gates E 1530; 8 May 2003, G. Gates E 1782; Warra LTER site, 43° 06′ S, 146° 39′ E, numerous observations between May 2006 and June 2007; Wedge Forest Reserve, 42° 50′ S, 146° 16′ E, 6 May 2003, G. Gates E 1771; Wielangta, 42° 42′ S, 147° 51′ E, 14 July 1998, G. Gates E 13; —, 20 March 1999, G. Gates E 261; —, 29 March 2001, G. Gates E 1047; —, 14 June 2001, G. Gates E 1245; —, 3 April 2003, G. Gates E 1671.

This remarkable *Nolanea* has very small and rather irregularly shaped spores, which are similar to those of *E. conferendum* (Britzelm.) Noordel. but smaller and less pronouncedly cruciform. *Entoloma cucurbita* E. Horak differs by larger spores and darker colours.

35. **Entoloma stellatum** G.M. Gates & Noordel., Persoonia 19: 180. 2007.

Original diagnosis: Habitus mycenoideus. Pileus 10–25 mm latus campanulatus vel conicus demum convexus valde hygrophanus brunneus translucido-striatus toto expallens sericeo-fibrillosus. Stipes 30–50 × 1–3 mm pallide flavo-brunneus politus. Sporae parvae 8.0–11 × 7.0–10 μm irregulariter cruciformibus. Cystidia lageniformia. Pileipellis cutis hyphis cylindraceis vel inflatis 2–5 μm latis pigmento intracellulosa formata. Fibulae presentes.

Holotype: Australia, Tasmania, Duckhole Lake Track, 43° 22′ S, 146° 53′ E, 6 June 2002, G. Gates E 1558 (HO 543535; isotype in L).

Etymology: stella (Lat.) = star, referring to the star-shaped spores.

Main characters: small mycenoid species with a brown hygrophanous pileus; stipe yellow-brown, polished; spores very irregularly shaped, cruciform; cheilocystidia large (Plate 3.33).

Description:
Pileus 10–25 mm diam, campanulate or conico-convex, expanding to convex or applanate, usually with acute to minutely umbonate centre, rarely truncate with slight

Plate 3.33 *Entoloma stellatum* (Photos Michael Pilkington)

central depression, with straight margin, when moist translucently striate up to half the radius, brown or ochre brown (5 C4–5), hygrophanous, pallescent on drying to silky ochraceous buff, glabrous, fibrillose-shiny when dry. Lamellae adnate with decurrent tooth, ascending, segmentiform to ventricose, moderately distant, up to 6 mm broad, moderately thick, pink, becoming ochraceous pink with age, with entire, concolorous edge, lamellulae in two tiers. Stipe 30–50 × 1–3 mm, cylindrical, often broadened to subbulbous at base, brittle, pallid at apex, yellow-brown, ochre or brown towards base, glabrous, polished, base with white mycelial tomentum. Odour faintly spermatic. Taste slightly bitter.

Spores 8.0–11 × 7.0–10 μm, Q = 1.0–1.2, very irregularly cruciform. Basidia 20–38 × 9–11 μm, 4-spored, clamped. Lamella edge heterogeneous. Cheilocystidia scattered among basidia, 30–90 × 6–15 × 2–4 μm, lageniform, sometimes slender, sometimes with very broad basal part, with tapering neck, often ending up in a slightly broadened, spathula-like apex, sometimes with moniliform neck ending bluntly, thin walled. Hymenophoral trama regular, made up of inflated elements, 90–120 × 9–20 μm. Pileipellis a cutis of cylindrical hyphae 2–5 μm in diameter, overlaying a subpellis of rather short, inflated elements, 30–70 × 8–18 μm. Pileitrama regular, made up of inflated elements. Pigment plasmatic, brown. Clamp-connections rare, at base of basidia and in pileipellis (Fig. 3.35).

Habitat & distribution: widespread but relatively uncommon, in wet sclerophyll forests and rainforests.

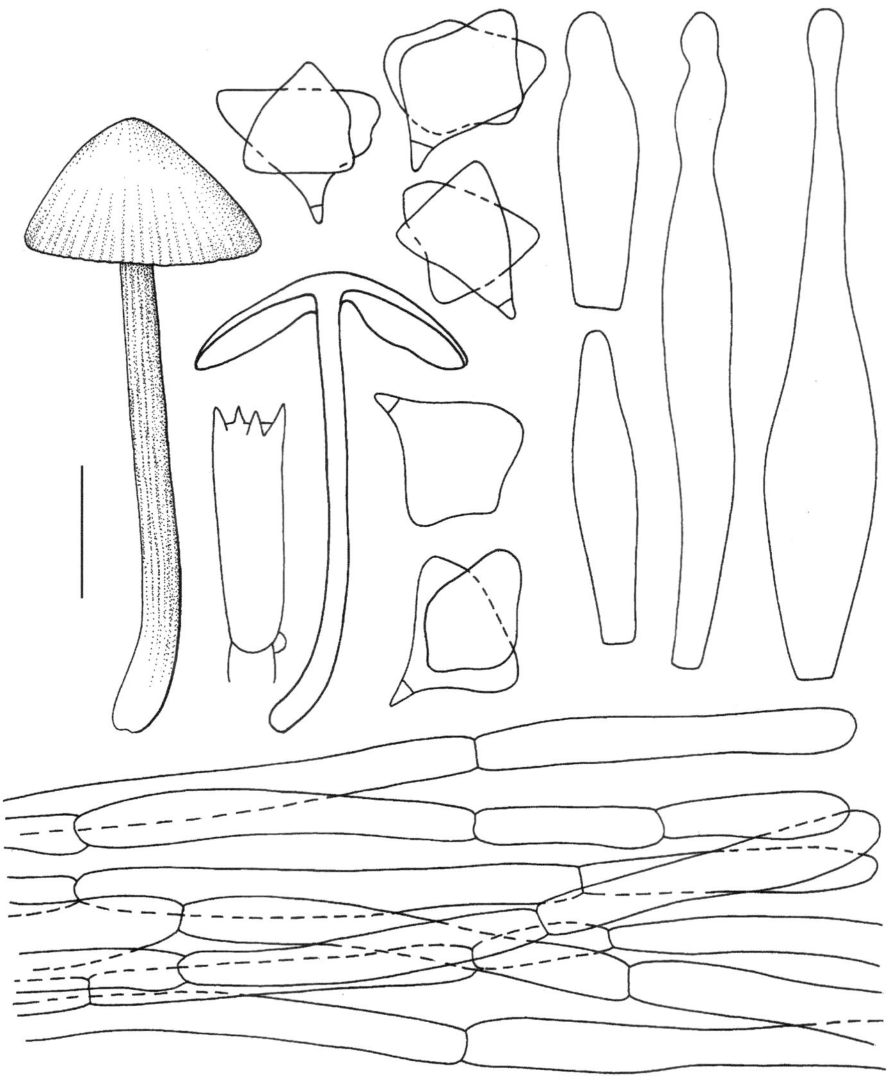

Fig. 3.35 *Entoloma stellatum*. Habit, spores, basidia, cheilocystidia, and pileipellis. Bar = 10 mm or 10 μm

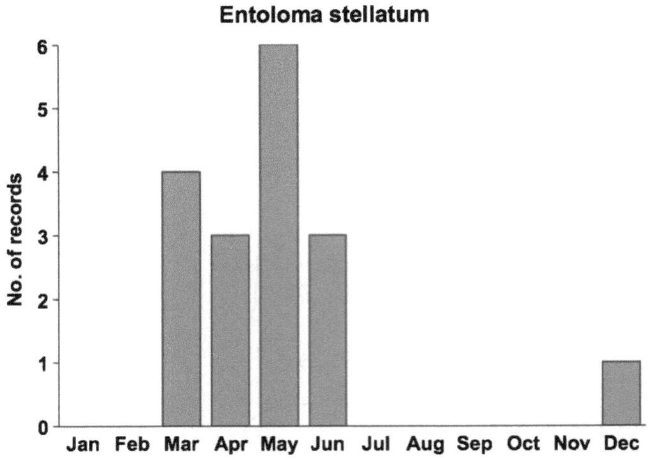

Collections examined. Australia, Tasmania, Creepy Crawly, 42° 50′ S, 146° 22′ E, 23 May 2002, G. Gates E 1529; Dora Falls, 42° 58′ S, 146° 55′ E, 16 Dec. 2004, G. Gates E 2068; Duckhole Lake Track, 43° 22′ S, 146° 53′ E, 18 May 2002, G. Gates E 1523; —, 6 June 2002, G. Gates E 1558 (holotype, HO 543535; isotype, L); Lake St Clair National Park, Echo Point to Cynthia Bay, 42° 04′ S, 146° 09′ E, 26 March 2005, G. Gates E 2103; Mt Cripps Karst Area, 41° 36′ S, 145° 46′ E, 24 March 2005, G. Gates E 2101; Myrtle Forest, 42° 52′ S, 147° 09′ E, 10 March 2005, G. Gates E 2057; Timbs Track, 42° 44′ S, 146° 25′ E, 8 May 2003, G. Gates E 1784; Warra LTER, 43° 06′ S, 146° 41′ E, 27 April 2004, G. Gates E 1978; —, 24 May 2005, G. Gates E 2172; —, Bird Track, 43° 06′ S, 146° 39′ E, 8 June 2006, G. Gates E 2243.

Entoloma stellatum is widespread in Tasmania. None of the known species with cuboid or cruciform spores matches this species (Horak 1976, 1977). It resembles *E. conferendum* (Britzelm.) Noordel. on account of the rather smooth pileus and cruciform spores, but that species lacks cheilocystidia.

36. **Entoloma procerum** G. Stev., Kew Bull. 16: 233. 1962.

Syn. *Entoloma inconspicuum* G. Stev., Kew Bull. 16: 236. 1962.

Original diagnosis: Pileus 1.5–4.5 cm diam, campanulatus, umbonatus vel centro gelatinosus, margine striato et subplicato, udus ceraceus, humidus, siccitate sericeus, margine ultra lamellas excedente; caro tenuis, umbrina. Lamellae adnexae dein liberae, subconfertae, roseolae demum obscuriores, aciebus undulatis vel serrulatis. Stipes 7–15 cm × 4–8 mm, cremeicolor, basi albo-tomentosus, laevis, sericeostriatus, cavus, fragilis, saepe tortus, fractus sucum aquosum exsudans. Sporae 10 × 13 µm, valde lobatae, apiculis prominentibus, in cumulo testaceae.

Holotype: New Zealand, Taupo, 12 March 1958, G. Stevenson 1260 (K).

Selected literature: Horak, Agaricales of New Zealand 1: 210–212, Figs 90, 2008.

Main characters: habit fairly stout mycenoid; pileus very dark brown; stipe pale; spores complex, cruciform to 5–6-angled (Plate 3.34).

Plate 3.34 *Entoloma procerum* (Photo Michael Pilkington)

Description:

Pileus 10–30 mm diam, conico-convex to convex, with umbo, blunt to slightly depressed centre, with deflexed margin, slightly hygrophanous, when moist obscurely translucently striate at margin, dark yellowish brown to very dark brown, distinctly pallescent along radial streaks to medium brown, innately radially fibrillose, at centre sometimes minutely squamulose. Lamellae adnate with decurrent tooth, subventricose, moderately crowded, up to 6 mm broad, sordid pinkish brown, with concolorous or dark brown fimbriate edge. Stipe 60–80×5–9 mm, cylindrical or compressed with groove, pale horn brown, innately fibrillose, almost polished, basal part with white mycelial tomentum, apex finely pruinose. Context brittle. Odour and taste not distinct.

Spores 9.5–13×8.5–12 µm, Q=1.0–1.2, mainly cruciform, also 5–6-angled or cuboid, very irregularly shaped. Basidia 30–45×9–14 µm, 4-spored, clamped. Lamella edge sterile. Cheilocystidia clavate, sometimes septate, 42–130×10–19 µm, often filled with brown, intracellular pigment. Hymenophoral trama regular, made up of long, inflated elements. Pileipellis a differentiated cutis of repent to slightly ascending hyphae, 5–13 µm wide with clavate terminal elements, 40–120×5–17 µm, with brown intracellular and also encrusting pigment. Pileitrama regular, made up of inflated hyphae, 7–19 µm wide. Brilliant granules absent. Stipitipellis a cutis of cylindrical hyphae with scattered distinct terminal elements (caulocystidia), 42–110×6–19 µm, with brown, intracellular pigment. Clamp-connections present (Fig. 3.36).

Habitat & distribution: widespread in Tasmanian wet eucalypt forests; originally described from New Zealand.

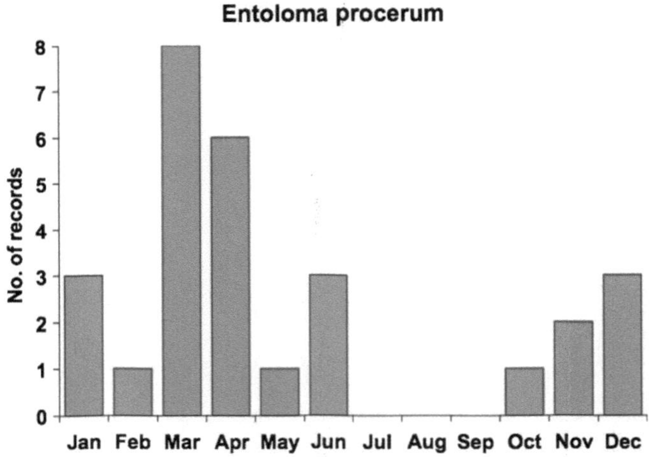

Collections examined. Australia, Tasmania, Bruny Island, Mt Mangana, 43° 22′ S, 147° 17′ E, 8 April 1999, G. Gates E 377; Clarks Cliffs, 43° 06′ S, 147° 47′ E, 29 Dec. 2001, G. Gates E 1382; Kermandie Falls, Lower Track, 43° 12′ S, 146° 52′ E, 22 Dec. 2005, G. Gates E 2227; —, Upper Track, 30 March 1999, G. Gates E 315; —, 26 April 2001, G. Gates E 1137; Mt Wellington, Myrtle Gully, 42° 54′ S, 147° 15′ E, 9 March 1999, G. Gates E 175; —, 13 April 1999, G. Gates E 402; —, 15 June 1999, G. Gates E 608; —, 27 Nov. 2001, G. Gates E 1347; North West Bay River, 42° 57′ S, 147° 12′ E, 16 June 2001, G. Gates E 1252; —, 17 Nov. 2001, G. Gates E 1318; —, 21 April 2005, G. Gates E 2131.— New Zealand, Wellington, Keith George Park, 30 May 1949, G. Stevenson 599 (holotype, K); Taupo, 12 March 1958, G. Stevenson 1260 (holotype of *E. inconspicuum*, K).

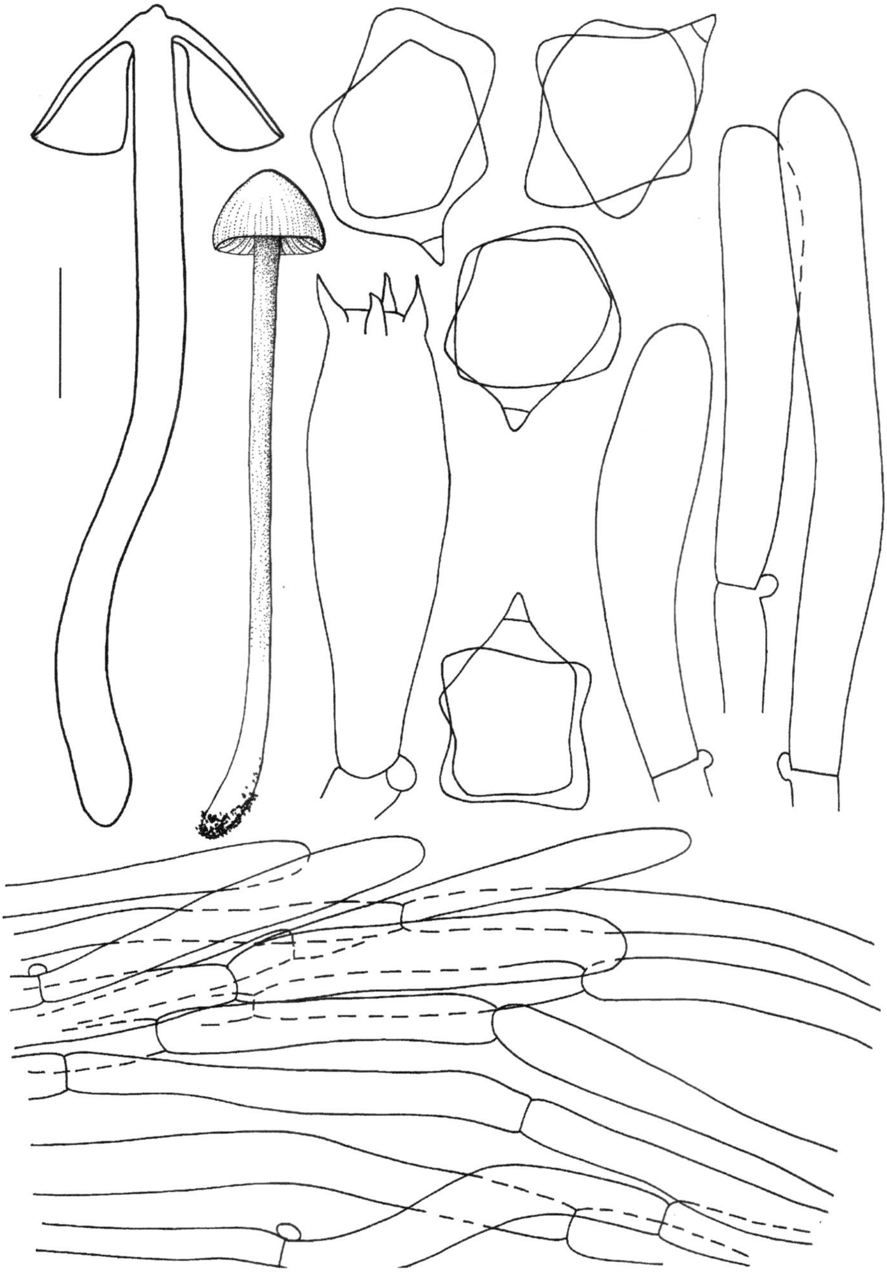

Fig. 3.36 *Entoloma procerum*. Habit, spores, basidia, cheilocystidia, and pileipellis. Bar = 10 mm or 10 μm

Entoloma procerum is distinctive on account of the very dark brown pileus, which is definitely not glabrous but covered in coarse, radially arranged fibrils with a very slightly and minutely squamulose centre, the paler, almost polished stipe, the very irregular more or less cruciform spores and the very long, vesiculose-clavate cystidia with dark brown intracellular pigment. Horak (2008) suggested that *E. procerum* is endemic to New Zealand, but, like several other so-called New Zealand endemics, this fungus appears to be quite widespread in Tasmania. The complete lack of brilliant granules and the presence of encrusting pigment suggest that *E. procerum* is more likely related to *E. conferendum* (Britzelm.) Noordel. in subgenus *Nolanea* than to the true *Inocephalus* species such as *Entoloma canoconicum* E. Horak from New Zealand, which has more regularly shaped cuboid spores, and very abundant brilliant granules which turn the trama completely milky under the microscope. This could be confirmed by ongoing molecular studies.

Section ***Lepiotoidei*** Noordel. & G.M. Gates, sect. nov.
MycoBank #562980.
Diagnosis: Habit mycenoid; pileus radially fibrillose to squamulose; pigment intracellular; clamp-connections present.
Type species: *Entoloma lepiotoides* G.M. Gates & Noordel.

37. **Entoloma lepiotoides** G.M. Gates & Noordel., Persoonia 19: 182. 2007.

Original diagnosis: Habitus lepiotoideus. Pileus 8–25 mm latus toto rufobrunneus tomentosus demum squamulosus. Stipes 10–40 × 1–4 mm pallidus plus minusve politus. Sporae parvae 7.0–9.0 × 7.0–8.5 μm (sub)isodiametrales. Cystidia nulla. Pileipellis trichoderma elementis fusoideus 40–120 × 15–25 μm pigmento intracellulosa formata. Fibulae presentes.

Holotype: Australia, Tasmania, Mt Field National Park, Falls Loop, 42° 41′ S, 146° 42′ E, 31 May 2003, G. Gates E 1847 (HO 543534; isotype in L).

Etymology: resembling a *Lepiota* species.

Main characters: habit mycenoid; basidiocarps small and velutinous, soft, reminiscent of a brown *Lepiota* species; spores small; clamp-connections present (Plate 3.35).

Description:
Pileus 8–25 mm diam, conical to conico-convex with or without small umbo, expanding to plano-convex with small umbo, with deflexed then straight margin, uniformly very dark reddish brown to almost black, burnt umber or chocolate brown, not hygrophanous, not translucently striate, densely woolly-tomentose all over, often becoming cracked and more or less concentrically scaly at margin, showing the slightly paler background, centre often remaining non-squamulose, dull. Lamellae adnate with small decurrent tooth or emarginate, segmentiform to

Plate 3.35 *Entoloma lepiotoides* (Photo Genevieve Gates)

ventricose, moderately crowded, up to 6 mm broad, moderately thin, pale grey, almost white at first then pallid brown, with faint pink tinge and always consistently very pale, with eroded, concolorous edge, lamellulae in two tiers. Stipe 10–40 × 1–4 mm, cylindrical, curved or straight, pale grey-brown to whitish grey, contrastingly paler than pileus, glabrous or innately fibrillose, not truly polished, with white basal tomentum. Odour and taste indistinct.

Spores 7.0–9.0 × 7.0–8.5 μm, Q = 1.0–1.3, Qav = 1.2, isodiametric to subisodiametric, 5–7-angled in side-view. Basidia 18–32 × 7–9 μm, 4-spored, clamped. Lamella edge fertile. Cystidia absent. Hymenophoral trama regular, made up of cylindrical to inflated elements, 40–160 × 5–20 μm. Pileipellis a true trichoderm, made up of triangular bundles of fusiform to lageniform elements, 40–120 × 15–25 μm. Subpellis made up of rather short elements, 20–60 × 12–40 μm, gradually passing into pileitrama, made up of cylindrical to inflated elements, 70–200 × 12–36 μm. Pigment abundant, brown, intracellular in pileipellis. Brilliant granules present but not abundant. Clamp-connections present in hymenium and in pileipellis (Fig. 3.37).

Habitat & distribution: widespread in wet sclerophyll forests.

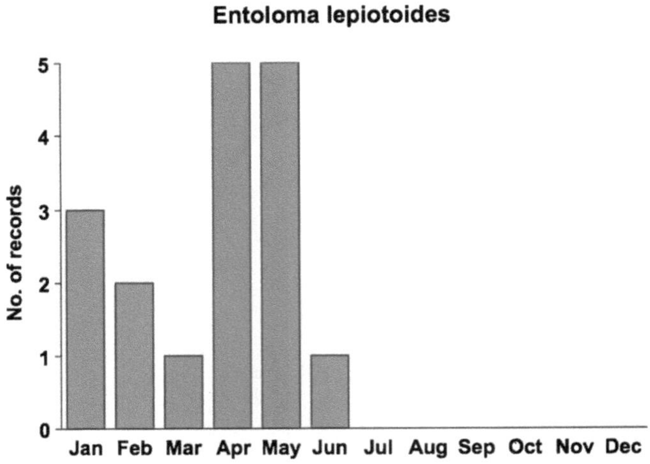

Collections examined. Australia, Tasmania, Donnellys Rd, 43° 07′ S, 146° 54′ E, 23 May 2000, G. Gates E 932; Forester Road, Scottsdale, 41° 06′ S, 147° 37′ E, 29 April 2006, M.E. Noordeloos 2006035; Kermandie Falls, Lower Track, 43° 12′ S, 146° 52′ E, 5 April 2001, G. Gates E 1079; —, 5 Jan. 2002, G. Gates E 1386; Mt Field National Park, Falls Loop, 42° 41′ S, 146° 42′ E, 31 May 2003, G. Gates E 1847 (holotype, HO 543534; isotype, L); Wielangta, 42° 42′ S, 147° 51′ E, 23 Feb. 1999, G. Gates E 130; —, 3 April 2003, G. Gates E 1674.

Entoloma lepiotoides is a soft, delicate species with a brown, woolly then distinctly regularly squamulose pileus, resembling a *Lepiota* species. The small spores, lack of cheilocystidia, and presence of clamp-connections are distinctive. Only a few species in the literature are similar. *Entoloma pervelutinum* E. Horak from Papua New Guinea has a very dark brown, concentrically squamulose pileus, and more or less similar spores and pileipellis, but clearly differs in the dark stipe with lilac or porphyry tinges, and the presence of clavate cheilocystidia.

38. **Entoloma sepiaceovelutinum** G.M. Gates & Noordel., Persoonia 19: 183. 2007.

Original diagnosis: Habitus mycenoideus vel collybioideus. Pileus 12–32 mm latus expansus haud hygrophanus haud vel paulisper translucido-striatus toto intense sepiaceus tomentosus demum squamulosus. Stipes 12–43 × 2–3 mm pallid griseo-brunneus politus. Odore spermaticis. Sporae 10–14(–15) × 7.0–9.0(–10) µm hetero-diametrales 6–8-angulatae. Basidia 4-sporigera fibulata. Cheilocystidia sparsa lageniformia. Pileipellis trichoderma elementis cylindraceus vel clavatis 60–160 × 7–22 µm pigmento intracellulosa formata. Fibulae presentes rarae.

Holotype: Australia, Tasmania, Mt Field National Park, Lady Barron Falls, 42° 41′ S, 146° 42′ E, 7 Dec. 2004, G. Gates E 2065 (HO 543533; isotype in L).

Etymology: sepia = dark brown; velutinus (Lat.) = velvety, referring to the sepia coloured, velvety pileus.

Main characters: habit mycenoid; pileus very dark sepia brown, entirely velutinous; lamellae broadly adnate with decurrent tooth; stipe pallid, polished; cheilocystidia very prominent, lageniform; spores large, heterodiametric (Plate 3.36).

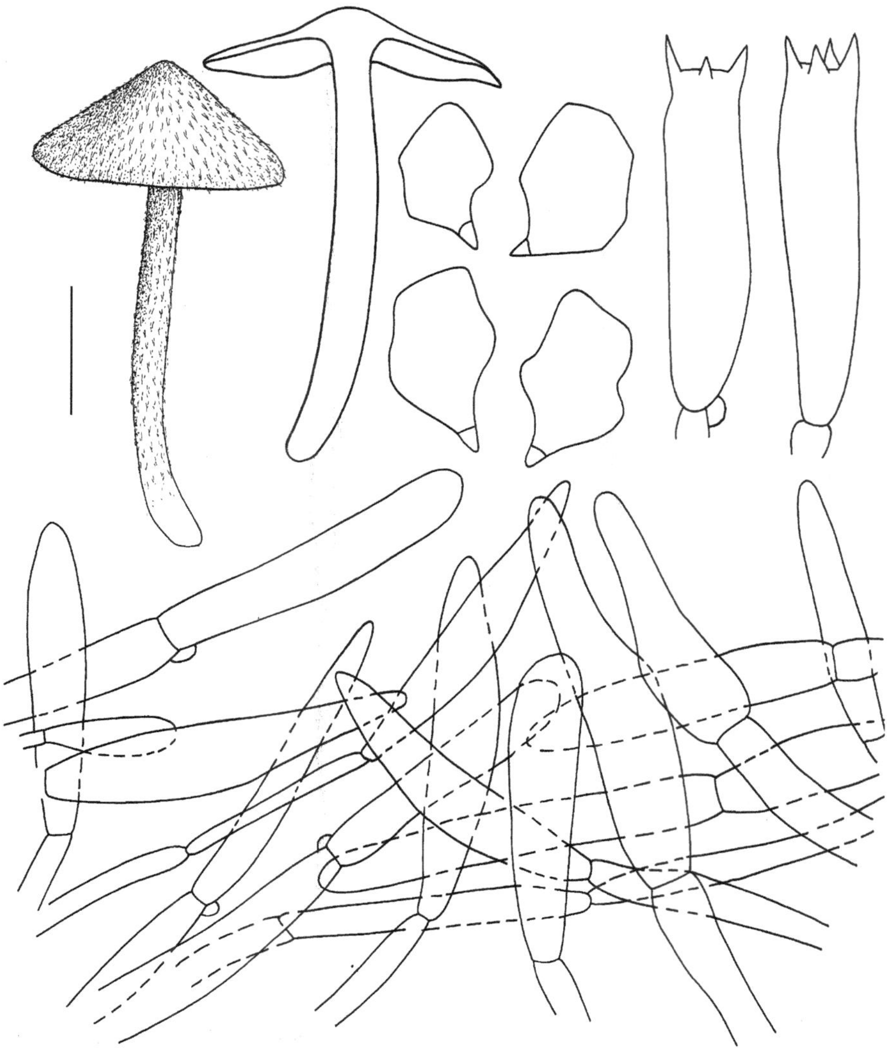

Fig. 3.37 *Entoloma lepiotoides*. Habit, spores, basidia, and pileipellis. Bar = 10 mm or 10 μm

Description:
Pileus 12–32 mm diam, expanded plano-convex, almost flat with deflexed to straight margin, not hygrophanous, not or indistinctly translucently striate, entirely very dark sepia brown, dry, velutinous or breaking up into small pointed squamules at centre and imbricate scales towards margin. Lamellae broadly adnate with small decurrent tooth, segmentiform to ventricose, moderately crowded, up to 6 mm broad, pale pinkish brown with concolorous edge, lamellulae in two tiers. Stipe 12–43 × 2–3 mm, cylindrical, straight, pale grey-brown (5D3), glabrous, polished, with white basal tomentum. Odour spermatic. Taste slightly saliva inducing.

Plate 3.36 *Entoloma sepiaceovelutinum* (Photos Genevieve Gates (*upper left*), Michael Pilkington)

Spores 10–14(–15) × 7.0–9.0(–10) µm, heterodiametric with 6–8 pronounced, rounded angles. Basidia 24–40 × 9–14 µm, 4-spored, clamped. Lamella edge heterogeneous. Cheilocystidia scattered, long, protruding from hymenium, lageniform with gradually tapering neck or with moniliform neck and slightly widened, blunt apex. Hymenophoral trama regular, made up of cylindrical to inflated elements, 50–190 × 6–23 µm. Pileipellis a cutis with transition to a trichoderm, particularly at centre, made up of long, cylindrical to clavate terminal elements, 30–70 × 6–20 µm. Pileitrama regular, made up of cylindrical to slightly inflated elements, 60–160 × 7–22 µm. Pigment brown, intracellular in pileipellis. Brilliant granules absent. Clamp-connections very rare, only seen in hymenium at base of basidia (Fig. 3.38).

Habitat & distribution: widespread but infrequently collected in wet sclerophyll forests.

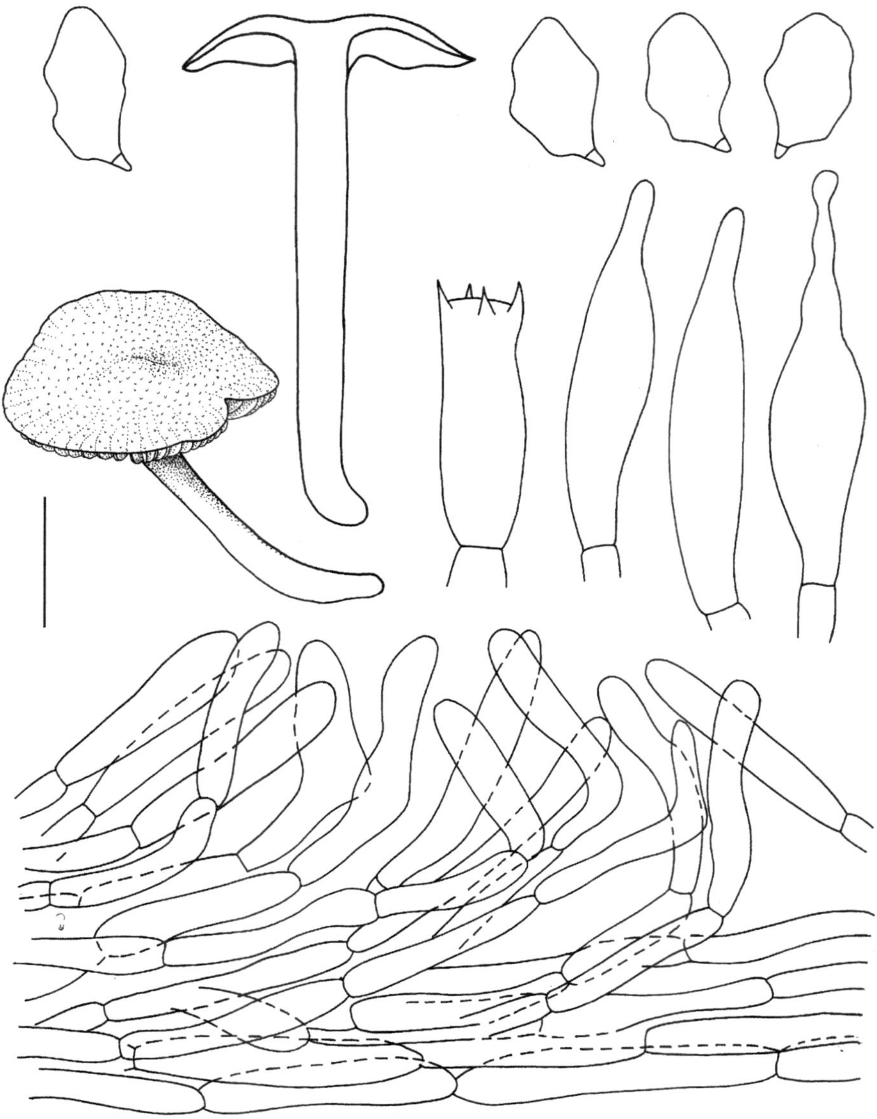

Fig. 3.38 *Entoloma sepiaceovelutinum*. Habit, spores, basidia, cheilocystidia, and pileipellis. Bar = 10 mm or 10 μm

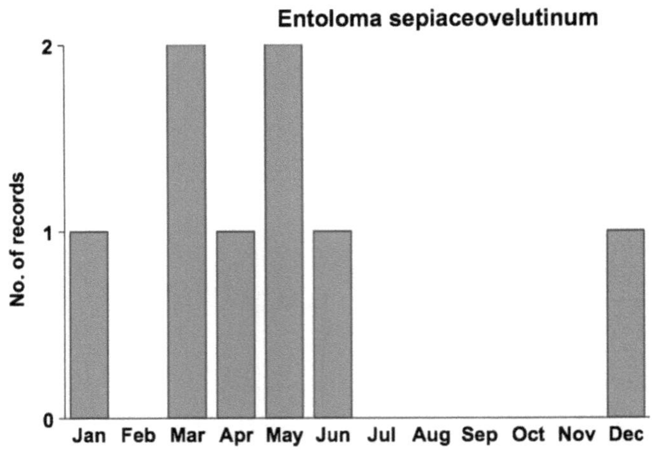

Collections examined. Australia, Tasmania, Growling Swallet, 42° 41′ S, 146° 30′ E, 12 March 2005, G. Gates E 2091; —, 28 March 2009, M.E. Noordeloos 2009083 (L); Kermandie Falls Upper Track, 43° 12′ S, 146° 52′ E, 5 Jan. 2006, G. Gates E 2235; Mt Field National Park, Lady Barron Falls, 42° 41′ S, 146° 42′ E, 7 Dec. 2004, G. Gates E 2065 (holotype, HO 543533; isotype, L); St Columba Falls, 41° 19′ S × 147° 55′ E, 21 May 2010, F. Karstedt 1456; Warra LTER, Bird Track, 43° 06′ S, 146° 39′ E, 19 June 2007, G. Gates E 2261.

Entoloma sepiaceovelutinus superficially resembles *E. lepiotoides*, but differs strikingly by the much larger spores and presence of large cheilocystidia. *Entoloma corneum* E. Horak from New Zealand is much smaller, with a translucently striate pileus, less distinctly angled spores, and clampless basidia.

39. **Entoloma strigosum** G.M. Gates & Noordel., Cryptogamie Mycologie 30: 132. 2009.

Original diagnosis: Pileus 8–20 mm latus, hemisphaericus vel convexus, haud hygrophanus, haud translucido-striatus, obscure brunneus, strigosus. Lamellae adnatae, distantes, obscure brunneus. Stipes 10–15 × 1–2 mm, cylindraceus, brunneus, fibrillosus. Sporae 10–12 × (6–)7–8 μm, angulate nodulosae. Acies lamellarum fertilis. Cheilocystidia desunt. Pileipellis ex hyphis usque ad 5 μm latis, cutem formantibus constituta cum fasciculis erectis ex elementius terminalibus fusiformibus, 90–150 × 10–17 μm, pigmentis intracellularibus. Granula lucentia absentia. Fibulae presentes.

Holotype: Australia, Tasmania, Duckhole Lake Track, 43° 22′ S, 146° 53′ E, 20 April 2004, E 1972 (HO 548343; isotype L).

Etymology: strigosum (Lat.) = hairy, referring to the strigose tufts on the pileal surface.

Main characters: tiny mycenoid species with strigose pileus; stipe fibrous; pigment intracellular; hyphae clamped (Plate 3.37).

Description:
Pileus 8–20 mm diam, convex-hemispherical, with incurved margin, not hygrophanous, not translucently striate, very dark brown (7F8), entirely covered with minute

Plate 3.37 *Entoloma strigosum* (Photos Machiel Noordeloos (*left*), Genevieve Gates (*right*))

strigose tufts. Lamellae adnate, distant, up to 2.5 mm broad, thick, dark brown with entire margin. Stipe 10–15 × 1.5–2 mm, tapering downwards, cylindrical, dark brown, fibrous. Odour none. Taste unknown.

Spores 10–12 × (6–)7–8 μm, Q = 1.3–1.4(–1.7), irregularly angled-nodulose. Basidia 38–49 × 11–14 μm, broadly clavate, 4-spored, clamped. Lamella edge fertile. Cheilocystidia absent. Hymenophoral trama consisting of cylindrical hyphae to 9 μm wide, with pale brown, intracellular pigment. Pileipellis a cutis of repent hyphae with trichodermal fascicles of fusiform terminal elements, 90–150 × 10–17 μm, with abundant, brown, intracellular pigment. Pileitrama regular, made up of cylindrical hyphae, ca. 10 μm wide. Caulocystidia narrowly cylindrical to fusiform, ca.

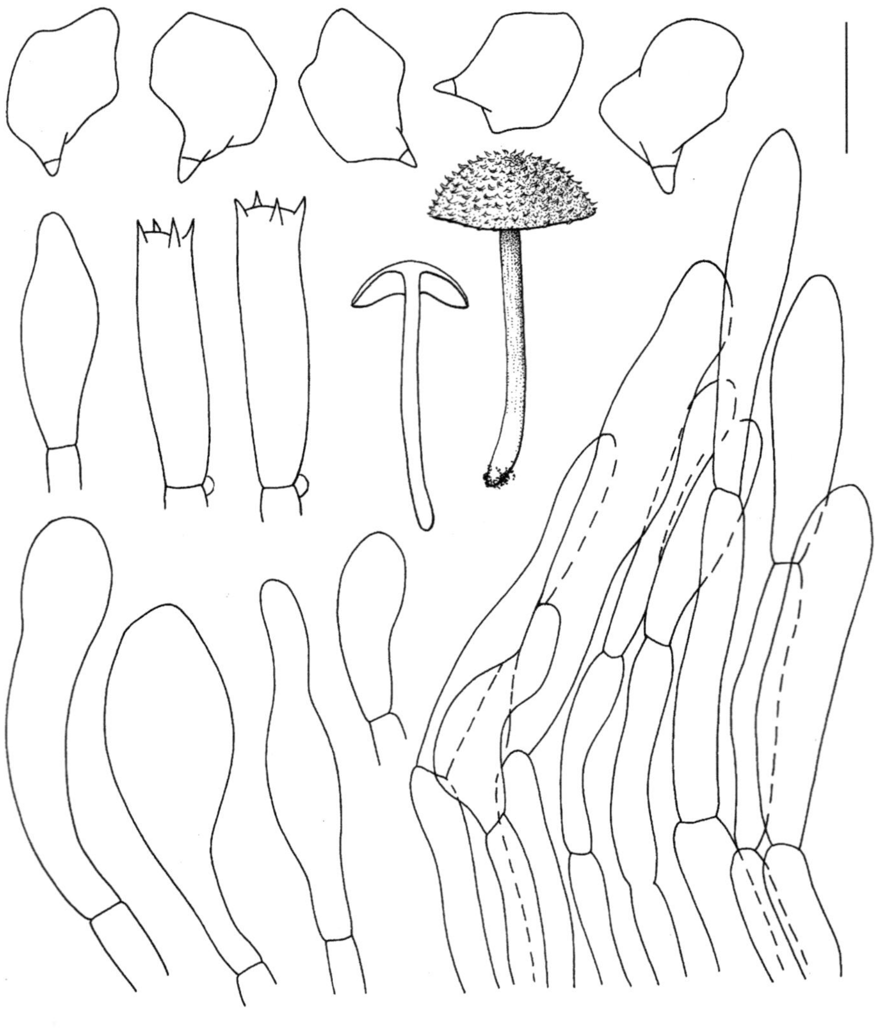

Fig. 3.39 *Entoloma strigosum*. Habit, spores, basidia, cheilocystidia, and pileipellis. Bar = 10 mm or 10 μm

50 × 5 μm with intracellular pigment. Clamp-connections observed in hymenium and hymenophoral trama (Fig. 3.39).

Habitat & distribution: terrestrial in small groups in forest litter in wet sclerophyll forest and mixed forest, the latter with a wide range of rainforest species including *Nothofagus cunninghamii* (Hook.) Oerst. (at Duckhole Lake), and *Atherosperma moschatum* Labill. (at Wielangta). Very rare, only known from a few localities.

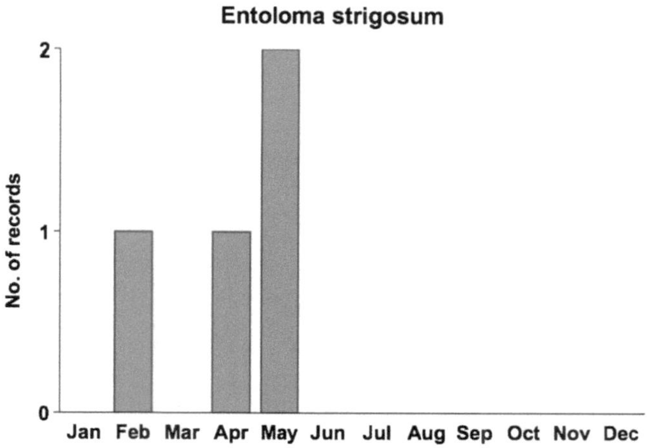

Collections examined. Australia, Tasmania, Dismal Swamp, 40° 57′ S, 144° 51′ E, 5 May 2011, M.E. Noordeloos 2011039 (L); Duckhole Lake Track, 43° 22′ S, 146° 53′ E, 15 May 2003, E 1811 (HO 548344); —, 20 April 2004, E 1972 (holotype, HO 548343; isotype, L); Wielangta Forest Reserve, 42° 42′ S, 147° 51′ E, 23 Feb. 1999, G. Gates E 126 (HO 548345).

In the field, *Entoloma strigosum* could be taken for a member of subgenus *Pouzarella*, but the lack of encrusting pigment and the presence of clamp-connections suggest inclusion in subgenus *Inocephalus*. *Entoloma densisquamosum* E. Horak from Papua New Guinea is similar in having a brown, densely squamulose-strigose pileus and dark lamellae, but differs by having much simpler, smaller spores, and by the lack of clamp-connections. This is also the case for *E. villosulum* Corner & E. Horak from Malaysia, which in addition has paler lamellae.

Subgenus ***Leptonia*** (Fr.: Fr.) Noordel. in Persoonia 11: 146. 1981.

Agaricus tribus *Leptonia* Fr.: Fr., Syst. mycol. 1: 10. 1821; *Agaricus* subgen. *Leptonia* (Fr.) Loudon, Encycl. Pl.: 998. 1827; *Leptonia* (Fr.: Fr.) P. Kumm., Führ. Pilzk.: 24. 1871.

Habit mycenoid or collybioid, rarely omphalioid; pileus not or weakly hygrophanous, fibrillose, tomentose, or squamulose; pileus varying from a cutis to a trichoderm or hymeniderm of inflated elements with intracellular, rarely with additional encrusting, pigment; clamp-connections present or absent.

Type species: *Entoloma euchroum* (Pers.: Fr.) Donk.

Section **Leptonia** (Fr.: Fr.) Noordel. in Persoonia 11: 146. 1981.
Habit mycenoid or slenderly tricholomatoid; stipe often fibrillose or squamulose; pileipellis a trichoderm of inflated elements with intracellular, and often additionally encrusting, pigment; clamp-connections present.

Type species: *Entoloma euchroum* (Pers.: Fr.) Donk.

Plate 3.38 *Entoloma panniculus* (Photo Michael Pilkington)

40. **Entoloma panniculus** (Berk.) Sacc., Sylloge Fungorum 5: 692. 1887.

Agaricus (Entoloma) panniculus Berk. in Hooker, Flora Tasmanica, vol. 2: 245. 1859. — *Entoloma scabripes* E. Horak, Agaricales New Zealand 5: 175. 2008.

Original diagnosis: Pileo tenui campanulato obtuso flocculoso stipiteque deorsum incrassato fibrilloso basi albo-tomentoso atro-violaceis, lamellis adnatis secedentibus.

Holotype: Australia, Tasmania, nr. Deloraine, Cheshunt, Quarry Hill, 22 March 1856, W. Archer (K).

Etymology: panniculus (Lat.) = little basket.

Main characters: mycenoid habit; basidiocarps deep blue; pileus and stipe squamulose; pigment in pileus purple-brown, intracellular and blue encrusting (Plate 3.38).

Description:
Pileus 20–30 mm diam, conical to conico-convex finally convex, with or without small umbo, with deflexed then straight margin, not hygrophanous, not translucently striate, uniformly deep blue (19E8–19F8, 20E8–20F8), entirely woolly-tomentose to warted-squamulose with prominent, neat, erect squamules, dull. Lamellae very narrowly adnexed to almost free, ventricose, moderately distant, up to 5 mm broad,

off-white then with pink hue, brown-pink when old, with somewhat irregular, concolorous edge, L = 30–40, l = 3–5. Stipe 40–70 × 2–4 mm, cylindrical, often distinctly broadened at base or apex, deep blue, entirely fibrillose-squamulose with deep blue fibrils (19E8–19F8, 20E8–20F8) on slightly paler background (16–17D4, 16–17E4). Odour farinaceous or mealy. Taste oily.

Spores irregularly 9–12 × 6–8 μm, Q = 1.2–1.7, Qav = 1.5, 5–7-angled, heterodiametric with pronounced angles. Basidia 20–40 × 6–9 μm, 4-spored, clamped. Lamella edge fertile. Cheilocystidia absent. Pileipellis an entangled trichoderm of cylindrical, 8–12 μm wide hyphae with irregularly shaped, cystidiform, cylindrico-clavate, sometimes subcapitate terminal elements, 25–7 × 8–13 μm; subpellis rather well differentiated, made up of cylindrical elements. Pigment brown-purple, intracellular in suprapellis, and blue parietal and granular in subpellis. Pileitrama regular, made up of cylindrical to inflated elements, up to 200 × 5–19 μm. Stipitipellis an entangled trichoderm of cylindrical hyphae, 6–12 μm wide with cystidioid, irregularly shaped flexuous, clavate, or lageniform terminal elements ("caulocystidia"), 20–60 × 4–12 μm, with deep blue intracellular pigment, and sometimes somewhat thickened, encrusted walls. Brilliant granules absent. Clamp-connections abundant (Fig. 3.40).

Habitat & distribution: single or in small groups on soil especially under *Blechnum wattsii* mossy trunks and very decayed logs. Main fruiting in two months of autumn (April–May) with most of the remaining records occurring in March and June (Group 3, in Chap. 1, see section "Fruiting Patterns of Species of Entolomataceae").

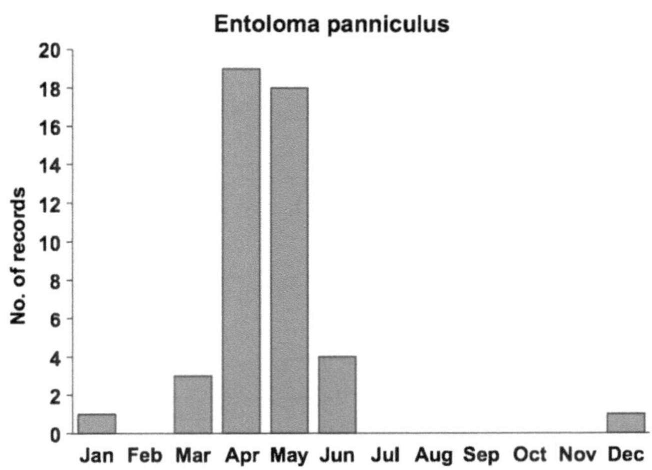

Collections examined. Australia, Tasmania, Bermuda Road, 43° 04′ S, 146° 54′ E, 17 April 2003, G. Gates E 1714; Collinsvale, Myrtle Forest, 42° 52′ S, 147° 09′ E, 15 April 1999, G. Gates E 425; —, 14 March 2002, G. Gates E 1456; Dismal Swamp (nr. Smithton), 40° 57′ S, 144° 51′ E, 5 May 2011, G. Gates (M.E. Noordeloos 2011036); Evercreech Forest Reserve, 41° 24′ S, 147° 58′ E, 16 June 2003, G. Gates E 1876; Growling Swallet, 42° 41′ S, 146° 30′ E, 11 May 2000, G. Gates E 902; Junee Caves State Reserve, 42° 44′ S, 146° 36′ E, 15 April 2000, G. Gates E 848;

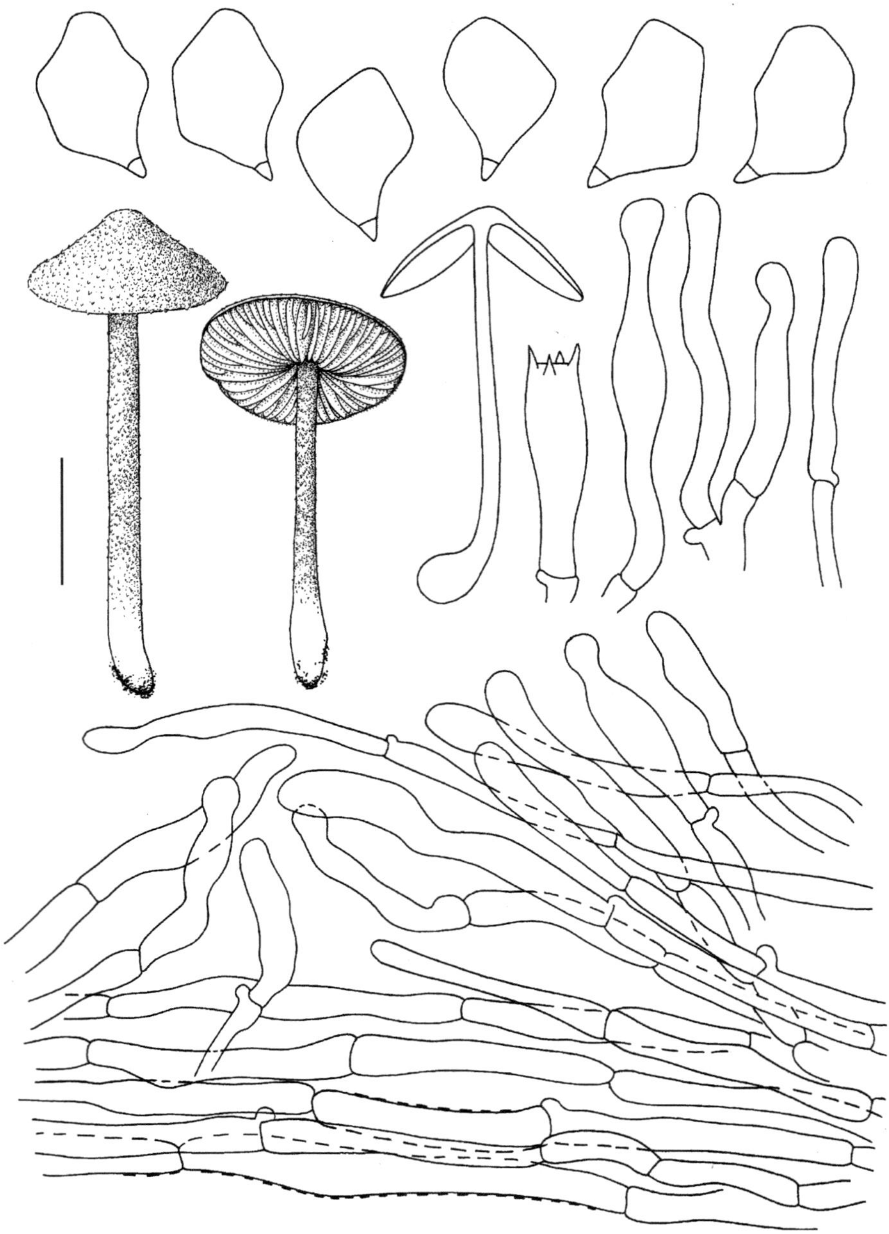

Fig. 3.40 *Entoloma panniculus*. Habit, spores, basidia, cheilocystidia, and pileipellis. Bar = 10 mm or 10 μm

Kermandie Falls Lower Track, 43° 12′ S, 146° 52′ E, 5 April 2001, G. Gates E 1068; Meander, nr. Deloraine, Cheshunt, Quarry Hill, 22 March 1856, W. Archer (holotype, K); Mt Field National Park, Falls Loop, 42° 41′ S, 146° 42′ E, 31 May 2003, G. Gates E 1843; —, Lady Barron Falls Track, 1 April 1999, G. Gates E 334; —, 24 April 2001, G. Gates E 1129; —, Russell Falls, 17 April 2004, M. E. Noordeloos 2004030; Mt Wellington, Betts Vale Track, 42° 55′ S, 147° 15′ E, 25 March 1999, G. Gates E 277; —, Fern Glade, 17 April 2005, G. Gates E 2118; —, Myrtle Gully, 42° 54′ S, 147° 15′ E, 13 April 1999, G. Gates E 399; Styx Valley, Big Tree Reserve, 42° 49′ S, 146° 39′ E, 13 May 2010, F. Karstedt 1392; Wielangta, Rainforest Loop Walk, 42° 42′ S, 147° 51′ E, 29 April 1999, G. Gates E 455.

Entoloma panniculus is very distinctive with its deep blue fruit bodies and squamulose pileus and stipe. The clamped hyphae, lack of brilliant granules, and trichodermal pileipellis place it in subgenus *Leptonia*. *Entoloma endotum* Noordel. & G.M. Gates has similar blue basidiocarps, but differs by the somewhat more robust fruit bodies, less conspicuously squamulose pileus, deeper, dull blue colour, orange-brown tinges in the lamellae, and larger spores. *Entoloma purpureum* Petch, as described and depicted by Horak (1980), is very similar and might differ only in the less obviously squamulose stipe. *Entoloma scabripes* E. Horak, recently described from New Zealand (Horak 2008), must be considered synonymous with *E. panniculus*. The differences given (habit, pileipellis structure, lack of cheilocystidia, and presence of caulocystidia) do not stand. Horak also refers to the dark blue pigment which turns brown instead of vividly green-blue as a distinguishing character, but does not indicate precisely whether the latter character refers to *E. panniculus*. We have never found such a discolouration in *E. panniculus*.

41. **Entoloma tomentosolilacinum** G.M. Gates & Noordel., Persoonia 19: 217. 2007.

Original diagnosis: Pileus 5–15 mm latus convexus demum expansus lilacino-griseus minute tomentosus. Stipes15–35 × 1.5–4 mm pallide lilacino-griseus vel coeruleo-griseus fibrillosus. Sporae (5.5–)6.0–7.0 × 5.5–7.0 μm (sub)isodiametricae. Cystidia desunt. Pileipellis trichoderma elementis septatis inflates 10–15 μm latis pigmentis intracellulosis vel parietalis constituis. Fibulae rarae.

Holotype: Australia, Tasmania, Mt Field National Park, Lady Barron Falls Track, 42° 41′ S, 146° 42′ E, 31 March 2001, G. Gates E 1058 (HO 543549; isotype in L).

Etymology: tomentosus (Lat.) = tomentose; lilacinus (Lat.) = lilac, referring to the lilac, tomentose pileal surface.

Main characters: small, collybioid or mycenoid species with a remarkable lilac, metallic shiny pileus and bluish lilac stipe; spores very small, isodiametric to sub-isodiametric; pileipellis differentiated (Plate 3.39).

Description:
Pileus 5–15 mm diam, convex to plano-convex, with blunt centre, not umbilicate, with involute then deflexed margin, not hygrophanous, not translucently striate,

Plate 3.39 *Entoloma tomentosolilacinum* (Photos Genevieve Gates)

lilac-grey to greyish ruby (12E3), with metallic sheen, entirely tomentose, breaking up into very minute, imbricate squamules. Lamellae adnate, segmentiform, moderately crowded, up to 3 mm broad, sordid white, with entire, concolorous edge, lamellulae in two tiers. Stipe 15–35 × 1.5–4 mm, cylindrical, straight or flexuous, sometimes slightly broadened towards base, more or less concolorous with pileus or more blue-grey, innately silky fibrillose, shiny, with white basal tomentum. Odour farinaceous or spermatic. Taste salty to farinaceous.

Spores (5.5–)6.0–7.0 × 5.5–7.0 µm, Q = 1.0–1.2, isodiametric to subisodiametric, with very thin walls, slightly but distinctly 7–many-angled in side-view. Basidia

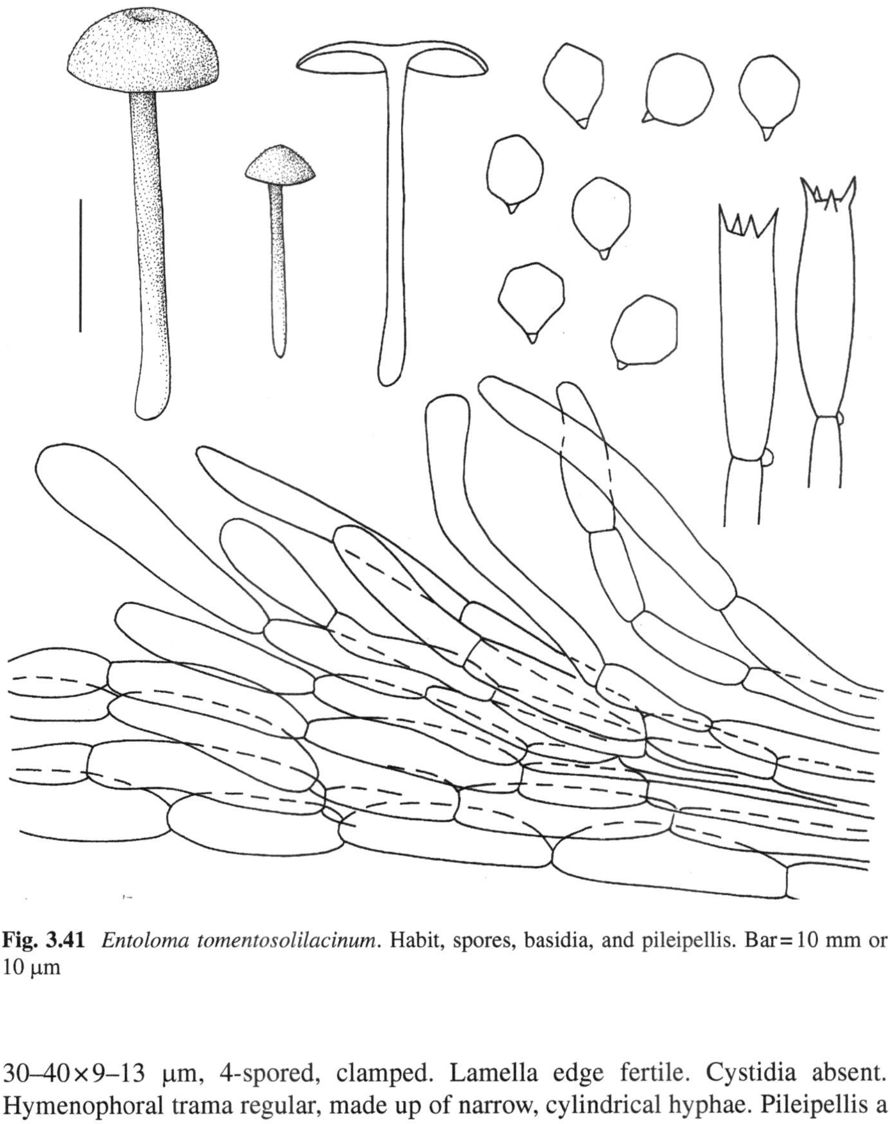

Fig. 3.41 *Entoloma tomentosolilacinum*. Habit, spores, basidia, and pileipellis. Bar = 10 mm or 10 μm

30–40 × 9–13 μm, 4-spored, clamped. Lamella edge fertile. Cystidia absent. Hymenophoral trama regular, made up of narrow, cylindrical hyphae. Pileipellis a differentiated cutis made up of septate hyphae 10–15 μm in diameter with trichodermal clusters of (semi)erect septate terminal endings to 20 μm in diameter. Pigment brown-lilac, intracellular in pileipellis and in upper pileitrama, also deposited at inner side of the hyphal walls as internal encrustations, blue, parietal in lower part of pileipellis and in upper pileitrama. Crystals frequent in pileitrama. Brilliant granules absent. Clamp-connections present but very rare (Fig. 3.41).

Habitat & distribution: widespread but uncommon, on the ground and on very rotten wood in wet sclerophyll forests.

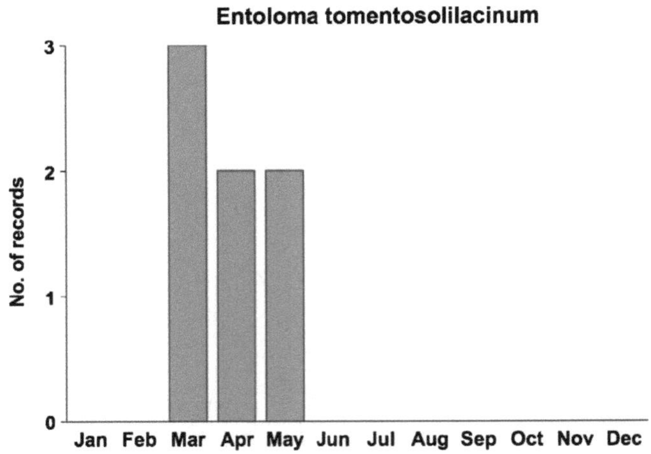

Collections examined. Australia, Tasmania, Liffey Falls, 41° 42′ S, 146° 46′ E, 28 March 1999, G. Gates E 296; Mt Cripps Karst Area, Philrod Track, 41° 36′ S, 145° 46′ E, 24 March 2005, G. Gates E 2100; Mt Field National Park, Lady Barron Falls Track, 42° 41′ S, 146° 42′ E, 5 April 2000, G. Gates E 811; —, 31 March 2001, G. Gates E 1058 (holotype, HO 543549; isotype, L); Scottsdale, Forester Road, 41° 06′ S, 147° 37′ E, 31 May 2005, G. Gates E 2175; Warra LTER site, Bird Track, 43° 06′ S, 146° 39′ E, 19 April 2007, G. Gates E 2250.

Entoloma tomentosolilacinum is a small member of section *Leptonia*, characterized by its grey-lilac colour, small spores, and scarcity of clamp-connections. The small spores are distinctive and the only other species in the section sharing this character, viz. *E. coelestinum* (Fr.) Hesler from Europe, differs clearly in the overall blue colour of the basidiocarp. The woody habitat is not unusual for members of section *Leptonia*.

42. **Entoloma violascens** G.M. Gates & Noordel., Cryptogamie Mycologie 30: 121. 2009.

Original diagnosis: Pileus 6–21 mm latus, conicus vel convexus, haud hygrophanus, haud translucido-striatus, violaceo-brunneus, minute squamulosus. Lamellae adnexae, moderate distantes, sordide albae demum roseae. Stipes 15–55 × 1–2 mm, cylindraceus, violaceus, politus. Sporae 8–12 × 7–9.5 µm, 5–6-angulatae. Acies lamellarum fertilis. Cheilocystidia absentes. Pileipellis cutis vel trichoderma, ex elementis inflatis, ad 16 µm latis, constituta pigmentis intracellularibus. Granula lucentia absentia. Fibulae presentes.

Holotype: Australia, Tasmania, Mt. Wellington, Fern Glade, 42° 55′ S, 147° 15′ E, 5 May 2001, G. Gates E 1166 (HO 548320; isotype L).

Etymology: violascens = turning violaceous, referring to the colour of the basidiocarps.

Main characters: habit mycenoid; pileus and stipe with distinct violaceous tinges; cheilocystidia absent; clamp-connections present (Plate 3.40).

Plate 3.40 *Entoloma violascens* (Photo Genevieve Gates)

Description:
Pileus 6–21 mm diam, conical to convex, not hygrophanous, not translucently striate, moderately dark to fairly dark violet-brown (9–10D3), dry, finely squamulose all over. Lamellae adnexed to (almost) free, ventricose, moderately distant, up to 4 mm broad, sordid white to flesh pink (6A2) with entire, concolorous edge. Stipe 15–55 × 1–2 mm, cylindrical, slender, pliant, violet (18A4) becoming violet-grey in older specimens, dry, glabrous, polished, with white mycelium at base. Odour not distinct. Taste not distinct.

Spores 8–12 × 7–9.5 μm, Q = (1.1–)1.2–1.5, Qav = 1.35, heterodiametric, irregularly 5–6-angled, sometimes more or less nodulose. Basidia 40–60 × 10–16 μm, 4-spored, clamped. Lamella edge fertile. Cystidia absent. Hymenophoral trama regular, made up of cylindrical 6–15 μm wide hyphae. Pileipellis a cutis with transition to a trichoderm of inflated elements 10–16 μm wide, gradually passing into the pileitrama of similar, cylindrical to slightly inflated hyphae. Pigment golden brown, intracellular in pileipellis and upper pileitrama. Brilliant granules absent. Clamp-connections present (Fig. 3.42).

Habitat & distribution: very rare, in wet sclerophyll forest.

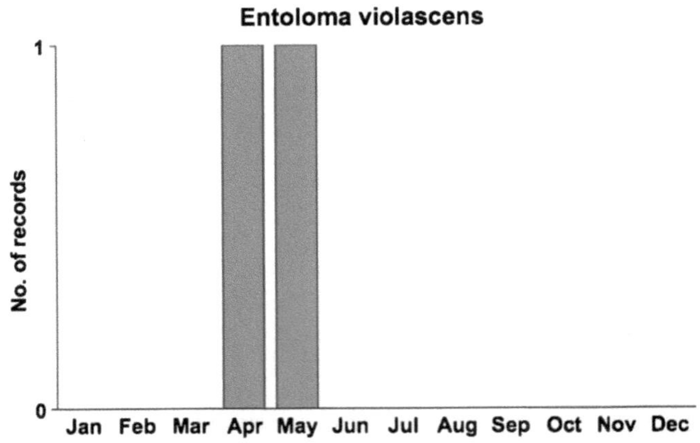

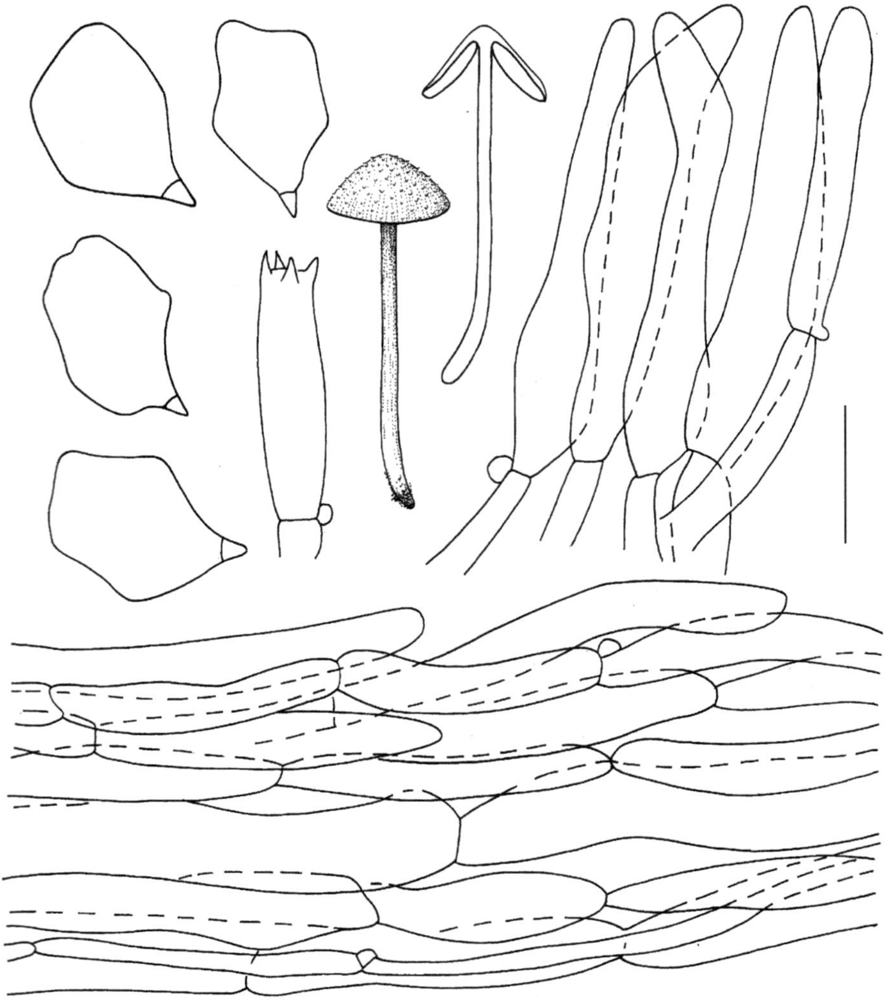

Fig. 3.42 *Entoloma violascens*. Habit, spores, basidia, cheilocystidia, and pileipellis. Bar = 10 mm or 10 μm

Collections examined. Australia, Tasmania, Collinsvale, Myrtle Forest, 42° 52′ S, 147° 09′ E, 14 April 2003, G. Gates E 1704 (HO 548321); Mt Wellington, Fern Glade, 42° 55′ S, 147° 15′ E, 5 May 2001, G. Gates E 1166 (holotype, HO 548320; isotype, L).

Entoloma violascens belongs to subgenus *Leptonia* section *Leptonia* on account of its trichodermal pileipellis structure and clamped hyphae (Noordeloos 2004). *Entoloma panniculus* (Berk.) Sacc. differs by the overall persistent blue colour, and a scabrous stipe. *Entoloma peraffine* E. Horak from New Zealand is superficially similar, but differs strikingly by the blue-amethyst tinge in the lamellae, smaller, less angular spores, and the absence of clamp-connections. *Entoloma gentile* E. Horak from Papua New Guinea has purplish tinges in the pileus and stipe, but a fibrillose surface of both pileus and stipe. *Entoloma perfidum* E. Horak, also from Papua New Guinea, has more bluish tinges and larger hyphae in the pileipellis.

43. **Entoloma endotum** Noordel. & G.M. Gates, spec. nov.

MycoBank #564497.

Diagnosis: Habit tricholomatoid; pileus dark sky blue, tomentose then squamulose; lamellae and context with yellowish tinges; stipe violet-blue; spores 11.5–14 × 7.5–9.0 μm, heterodiametric, rather pronouncedly 5–7-angular in side-view; clamp-connections present.

Holotype: Tasmania, Hobart, Mt. Wellington, Circle Track, 42° 55′ S, 147°15′ E, 11 March 1999, G. Gates E 198 (HO 562483; isotype in L).

Etymology: endota (aboriginal) = beautiful.

Main characters: habit robust, almost tricholomatoid; pileus dark blue-violet, dry, adpressedly squamulose; lamellae golden brown or orange-brown; stipe with fibrillose-squamulose covering, context of stipe with yellow tinges; spores 10–14 μm long; clamp-connections present (Plate 3.41).

Description:
Pileus 40–60 mm diam, conico-convex, hemispherical to convex, with or without low, broad umbo, with deflexed then straight margin, not hygrophanous, not translucently striate, Prussian blue (21F8) or dark violet-blue all over, dry, tomentose all over at first, then with adpressed triangular squamules in central part and radially fibrillose towards margin. Lamellae adnexed or adnate-emarginate, subventricose to ventricose, moderately distant, up to 9 mm broad, moderately thick, creamy pink when young then pale golden brown or orange-brown, with slight pink tinge, with subentire, concolorous margin, L = 36–44, l = 3–5. Stipe 50–80 × 6–8 (apex) × 10–13 (base) mm, cylindrical, gradually broadening towards base, solid, rather firm, longitudinally fibrillose-floccose or girdled with dark blue-violet fibrils on white to pale yellow background, whitely tomentose at base. Context firm, yellow in stipe, whitish in pileus. Odour none. Taste salty or bitter to acrid.

Spores 10–14 × 7.5–9.0 μm, av. 10.5–12.7 × 7.9–8.5 μm, Q = 1.25–1.45(–1.8), heterodiametric, 5–7-angled in side-view with pronounced angles. Basidia

Plate 3.41 *Entoloma endotum* (Photo Michael Pilkington)

30–50 × 10–12 μm, 4-spored, clavate, clamped. Lamella edge fertile, cystidia absent. Hymenophoral trama regular, made up of 10–20 μm wide hyphae. Pileipellis an entangled trichoderm of semi-erect to erect, cylindrical hyphae with cylindrical to narrowly clavate, or cystidiform terminal elements, 20–90 × 5–20 μm, with abundant clamp-connections, subpellis poorly to distinctly differentiated, made up of relatively short, cylindrical to inflated elements, 50–120 × 8.0–20 μm, gradually passing into trama. Pileitrama regular, comprised of cylindrical to inflated hyphae 40–125 × 10–28 μm wide. Brilliant granules absent. Stipitipellis a cutis of loosely arranged, cylindrical 5–12 μm wide hyphae, with scattered trichodermal tufts of clustered terminal elements (caulocystidia) similar to the terminal elements in pileipellis. Pigment abundant, dark blue, intracellular in pileipellis, upper pileitrama and stipitipellis, encrusting pigments not observed. Clamp-connections present on almost all septae (Fig. 3.43).

Habitat & distribution: found in wet sclerophyll with or without *Melaleuca* understorey. Uncommon.

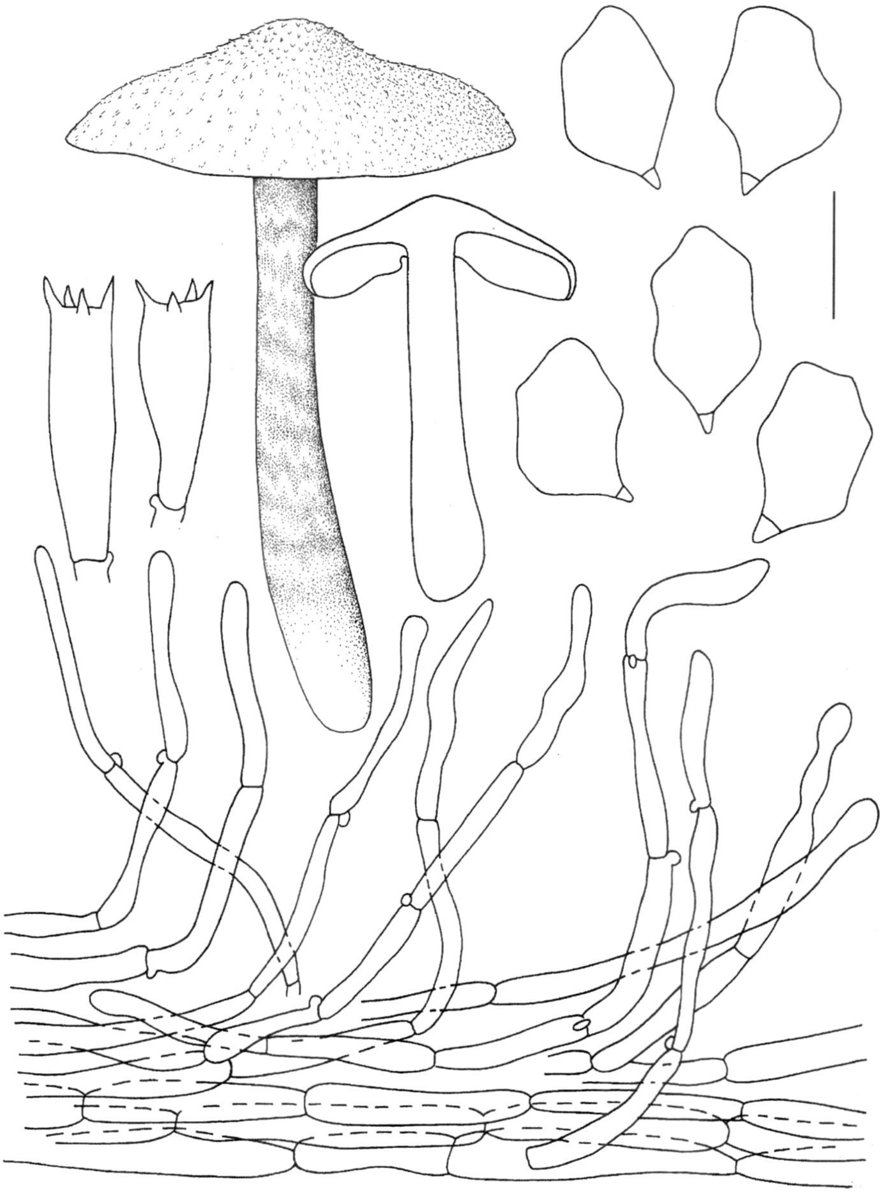

Fig. 3.43 *Entoloma endotum*. Habit, spores, basidia, and pileipellis. Bar = 10 mm or 10 μm

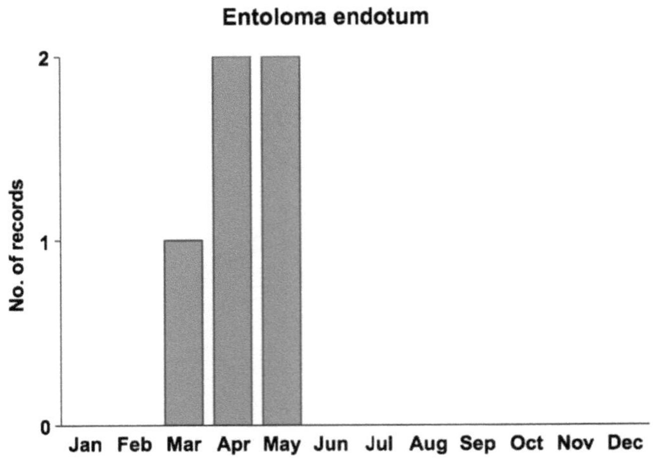

Collections examined. Australia, Tasmania, Bruny Island, Mt Mangana, 43° 22′ S, 147° 17′ E, 17 April 2011, G. Gates E 2294; —, 30 April 2011, M.E. Noordeloos 2011019 (L); —, 17 May 2010, M.E. Noordeloos 2010038; Mt Wellington, Circle Track, 42° 55′ S, 147° 15′ E, 11 March 1999, G. Gates E 198 (HO 562483, holotype).

Entoloma endotum is placed in subgenus *Leptonia* on account of the habit, clamped basidia and hyphal septae, and distinctly angled, heterodiametric spores. Macroscopically, the golden brown to orange-brown tinge in the mature lamellae and the yellow tinges in the context of the stipe are distinctive. *Entoloma panniculus* is a less robust species without the yellow tinges in the context, with white then purely pink lamellae, and smaller spores. *E. panniculus* also differs microscopically by having encrusting pigments in the lower part of the pileipellis.

Subgenus ***Trichopilus*** (Romagn.) Noordel.

Rhodophyllus subgenus *Trichopilus* Romagn. in Beih. Nova Hedwigia 59: 50. 1978; *Entoloma* subgenus *Trichopilus* (Romagn.) Noordel. in Persoonia 11: 86. 1980. *Rhodophyllus* subgenus *Leptonidium* Kühner in Bull. trimest. Soc. mycol. Fr. 93: 446. 1977 (nom. nud., no Latin description).

Habit tricholomatoid; pileus usually umbonate, rarely applanate, not or weakly hygrophanous, sericeous, radially fibrillose, velutinous, tomentose or squamulose; pileipellis a cutis with transitions to a trichoderm or a trichoderm of inflated hyphae with intracellular pigment; hymenophoral trama regular, made up of inflated elements up to 500 µm long; lamella edge heterogeneous or sterile with fusiform or clavate, often mucronate or lageniform to lecythiform cheilocystidia; clamp-connections usually present in all tissues.

Type species: *Entoloma jubatum* (Fr.: Fr.) P. Karst.

44. **Entoloma porphyrescens** E. Horak, Beih. Nova Hedwigia 43: 37. 1973.

Original diagnosis: Pileo 25–65 mm lato, umbonato, obscure brunneo vel fuligineo reflexo porphyreo instructo, radialiter squamoso vel squarroso, haud striato, sicco.

Lamellis ex adnato emarginatis, brunneo-roseis, primo purpureo vel lilacino tinctu, acie fimbriata albidaque instructis. Stipite 30–70×3–6 mm, cylindraceo vel attenuato apicem versus, pileo concolori vel pallidiori, dense fibrillis fuligineis, lilacinis vel purpureis obtecto, ad basim frequ. albido, sicco, fistuloso, fragili. Caro in stipite lilacina. Odore saporeque acidulo. Sporis 7.5–8.5×5–6 μm, 5(6)-angulatis. Basidiis 4-sporigeris. Cheilocystidiis 30–65×8–20 μm, polymorphis (lecythiformibus, clavatis vel uteriformibus), tenui-tunicatis, hyalinis. Pleuro- et caulocystidiis nullis. Epicute ex hyphis cylindraceis cutem vel trichodermium formantibus, pigmento brunneo plasmatico instructis. Septis defibulatis. Ad terram in silvis. Novazelandia.

Holotype. New Zealand, Fiordland, Lake Te Anau, Mt Luxmore, Kepler Track, 26 March 1969, E. Horak (PDD 27027).

Etymology: porphyrescens (Lat.) = becoming porphyry red, referring to the colour of the basidiocarps.

Selected literature: Horak, Beihefte Nova Hedwigia 65: 139. 1980.

Main characters: habit tricholomatoid; pileus and stipe deep porphyry red to purple-brown, pileus finely radially fibrillose-squamulose; cheilocystidia with moniliform neck (Plate 3.42).

Description:
Pileus up to 80 mm diam, campanulate-conical at first, expanding to convex or plano-convex, always with pronounced umbo, with deflexed then straight margin, not hygrophanous, not translucently striate, dry, deep red-brown with an almost deep ruby, but not truly violet, sheen ("porphyry red", 5YR 3/3–4), very finely velutinous or minutely squamulose at first then more or less radially fibrillose-minutely squamulose, with fine, radially arranged fibrils that agglutinate in small squamules, showing paler pinkish brown flesh between the fibrils. Lamellae narrowly adnate-emarginate or almost free, sometimes with short decurrent tooth, ventricose, moderately crowded, up to 8 mm broad, greyish porphyraceous when young, then greyish pink, with pallid sometimes deep carmine-red tinged, fimbriate edge, L = 60–80, l = 3–7. Stipe 45–90×5–10 mm, often gradually broadening towards base (8–15 mm), longitudinally fibrous to minutely squamulose, more or less similar to pileipellis, but with slightly more violet-red tinge (9C5–4, 9D5–4, when young 9E5–4), base whitely tomentose. Context concolorous with surface in cortex, pallid pinkish in inner parts. Odour and taste indistinct.

Spores 6.0–8.5(–10)×5.0–6.0(–8.0) μm, Q = 1.2–1.55, Qav. = 1.35, 6–7-angled. Basidia 24–40×7–12 μm, 4-spored, clamped. Lamella edge heterogeneous with abundant cheilocystidia, 20–70×2–6 (neck)×6–12 μm (middle part), lecythiform to lageniform, often with moniliform neck and conspicuous head. Hymenophoral trama regular, made up of long, cylindrical to inflated elements, to 20 μm wide. Pileipellis a cutis with transitions to a trichoderm, made up of very long, fusiform elements, 40–120×6–18 μm. Pileitrama regular, made up of inflated elements, to 200×5–20 μm. Pigment brown, intracellular. Brilliant granules absent. Stipitipellis

Plate 3.42 *Entoloma porphyrescens* (Photos Machiel Noordeloos (*top*), Michael Pilkington (*bottom*))

Fig. 3.44 *Entoloma porphyrescens*. Habit, spores, basidia, cheilocystidia, and pileipellis. Bar = 10 mm or 10 μm

a cutis with transitions to a trichoderm, made up of long, cylindrical elements, 3–14 μm wide with brownish purple intracellular pigment. Clamp-connections abundant (Fig. 3.44).

Habitat & distribution: wet eucalypt forests and rainforest. Widely distributed including at high altitude (700 m) where it was found in a very stunted form. Also known from New Zealand, where it was first described. Fruiting spread out from late spring to early winter (Group 2, in Chap. 1, see section "Fruiting Patterns of Species of Entolomataceae").

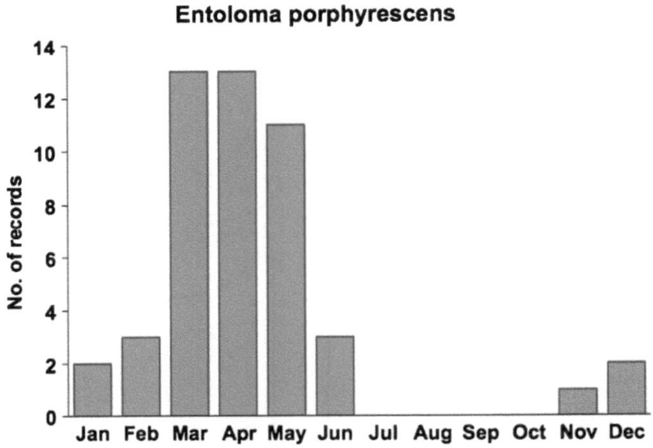

Collections examined. Australia, Tasmania, Blue Tier, Goblin Walk, 41° 11′ S, 148° 00′ E, 1 April 2000, G. Gates E 803; —, 14 April 2001, G. Gates E 1095a; —, 1 June 2004, G. Gates E 2026; Creepy Crawly, 42° 50′ S, 146° 22′ E, 20 March 2001, G. Gates E 1032; Florentine Rd, Lady Binney Track, 42° 43′ S, 146° 31′ E, 30 Dec. 2003, G. Gates E 1917; Franklin River Nature Trail, 42° 13′ S, 146° 01′ E, 15 April 2008, M.E. Noordeloos 2008023 (L); Growling Swallet, 42° 41′ S, 146° 30′ E, 6 April 2000, G. Gates E 820; —, 27 March 2001, G. Gates E 1042; —, 28 March 2009, G. Gates s.n.; Kermandie Falls Lower Track, 43° 12′ S, 146° 52′ E, 2 Dec. 1999, G. Gates E 773; Liffey Falls, 41° 42′ S, 146° 46′ E, 28 March 1999, G. Gates E 297; Little Florentine River, F5 Road, 42° 44′ S, 146° 25′ E, 8 May 2003, G. Gates E 1783; Marriotts Falls, 42° 43′ S, 146° 39′ E, 13 May 2000, G. Gates E 911; Mt Field National Park, Lyre Bird Walk, 42° 41′ S, 146° 40′ E, 1 April 1999, G. Gates E 325; —, 42° 41′ S, 146° 42′ E, 17 April 2004, M.E. Noordeloos 2004030 (L); Mt Wellington, Circle Track, 42°55′ S, 147° 15′ E, 8 Jan. 2000, G. Gates E 786; —, Myrtle Gully, 42° 54′ S, 147° 15′ E, 9 March 1999, G. Gates E 186; —, 18 May 2000, G. Gates E 924; —, Radfords Track, 42° 55′ S, 147° 15′ E, 18 Nov. 2000, G. Gates E 1017; Tahune, Huon Pine Walk, 43° 06′ S, 146° 44′ E, 14 March 2009, M.E. Noordeloos 2009082; Wielangta, Rainforest Loop Walk, 42° 42′ S, 147° 51′ E, 23 Feb. 1999, G. Gates E 133; —, 6 June 2000, G. Gates E 985; —, 29 March 2001, G. Gates E 1046; —, 2 April 2003, G. Gates E 1670. — New Zealand, Otago Province, Fjordland N.P., Track to Mt Luxmore, Lake Te Anau, 26 March 1969, E. Horak (holotype, PDD 27027).

Entoloma porphyrescens is fairly common in Tasmania and New Zealand. Its fruit bodies are easily identifiable in the field by the typical porphyry colour (a deep red that tends to violet) and radially fibrillose, umbonate pileus. Microscopically, the capitate cheilocystidia are the most striking features of this species. It is very similar to the similarly coloured Northern Hemisphere *Entoloma porphyrophaeum*

(Fr.) P. Karst., from which it differs mainly by the size and shape of the spores (8–12 × 6–8 µm for the latter) and cheilocystidia (Noordeloos 1992, 2004). The Tasmanian collections fit well with the holotype. Our type-study shows that, contrary to the description of Horak (1973, 2008), the spores are not 5–6 angled, however, but 6–7-angled in side-view.

Entoloma subgenus ***Cyanula*** (Romagn.) Noordel., comb. & stat. nov.
MycoBank #562982.

Basionym: *Rhodophyllus* sect. *Cyanula* Romagn. in Bull. mens. Soc. linn. Lyon 43: 328. 1974; *Entoloma* sect. *Cyanula* (Romagn.) Noordel. in Persoonia 11: 452. 1982.

Syn.: *Leptonia* sect. *Chalybaeae* Konr. & M., Agaricales: 261. 1948 (lacking a Latin description). *Rhodophyllus* sect. *Fragiles* Romagn. in Bull. mens. Soc. linn. Lyon 43: 329. 1974. *Rhodophyllus* sect. *Rufocarnei* Romagn. in Bull. mens. Soc. linn. Lyon 43: 329. 1974. *Rhodophyllus* sect. *Fuliginosa* Romagn. in Bull. mens. Soc. linn. Lyon 43: 329. 1974. *Leptonia* sect. *Albidicaules* Larg. in Mycologia 66: 1012. 1974. *Leptonia* sect. *Roseicaules* Larg. in Mycologia 66: 1013. 1974. *Leptonia* sect. *Viridicaules* Larg. in Mycologia 66: 1013. 1974. *Leptonia* sect. *Cereicaules* Larg. in Mycologia 66: 1014. 1974. *Leptonia* sect. *Chromocystotae* Larg. in Bibl. mycol. 55: 130. 1977.

Interpr. err.: *Leptonia* sect. *Paludocybe* sensu Larg. in Mycologia 66: 1013. 1974.
Basidiocarps collybioid or omphalioid; pileus usually distinctly depressed to umbilicate, with involute margin when young, white, or with pink, blue, violaceous, red, green, yellow or brown tinges, minutely squamulose at least at centre when mature; stipe white, yellow, brown, blue, violaceous or grey, often smooth and polished, less often fibrillose striate to flocculose; basidia always clampless; cheilocystidia, if present, always more or less cylindrical to clavate; pileipellis more or less trichodermal, at least at centre, made up of inflated elements, up to 45 µm wide. Pigment exclusively intracellular in pileipellis and stipitipellis; trama often with brilliant granules and/or oily substances, making it difficult to observe the hyphae; clamp-connections absent from all tissues.

Type species: *Entoloma serrulatum* (Fr.: Fr.) Hesler.
Members of subgenus *Cyanula* occur all over the world. They often have striking colours viz. blue, violet, green, yellow, or any tinge of yellow-brown or brown. Species concepts are mainly based on colour, surface structure of pileus and stipe, spore size and shape and nature of the lamella edge. Species groups with similar morphological characters occur in various parts of the world, which makes it sometimes hard to tell whether a certain species found in Tasmania is conspecific with one of the European, North American of Asian species described (Noordeloos 2004; Largent 1977, 1994; Horak 1980, 2008). However, often subtle differences have been found to support our species concepts. Preliminary results of our molecular studies confirm the Tasmanian species often cluster together in their own clades, distinct from those of other continents.

Cyanula is the most species rich subgenus in Tasmania with about 40 species. To conform with Noordeloos (1992, 2004) the species are arranged in 'stirps', an informal grouping based on similarity of macroscopic characters. Despite the fact that most species are very attractive, being often vividly coloured, identification is often difficult. This is mainly due to the fact that colours may change upon aging and/or upon drying, and in the variability in size and shape of the spores, and in the pigmentation of the cheilocystidia. We think we have a good impression of the variability of some of the species described below, but others are based on single or few collections. More work is needed to establish sound species concepts, and we hope that with our results published, we will stimulate further studies of this interesting group.

Stirps ***Serrulatum***

Lamella edge of the serrulatum-type with brownish black, blue or purple-blue intracellular pigment.

45. **Entoloma tasmanicum** G.M. Gates & Noordel., spec. nov.

MycoBank #564498.

Diagnosis: Habit collybioid; pileus blackish brown, entirely tomentose-minutely squamulose, not translucently striate; lamellae sordid brown-pink with blackish-brown, fimbriate edge; stipe same colour as pileus, towards base often with yellowish tinge, fibrillose; spores 8–10×5.5–7 µm, 5–6(–7)-angular; lamella edge of serrulatum type with blackish-brown pigment; pileipellis trichodermal with brown, intracellular pigment; clamp-connections absent.

Holotype: Australia, Tasmania, North West Bay River, 42° 57′ S, 147° 12′ E, 21 April 2005, G. Gates E 2133 (HO 561501; isotype in L).

Etymology: named after Tasmania.

Main characters: basidiocarps collybioid, blackish brown; pileus tomentose, not striate; lamellae blue in youth with dark brown to black fimbriate edge; stipe fibrillose, grey-blue with yellow base; spores small, 8–10×5–7 µm (Plate 3.43).

Description:

Pileus 20–40 mm diam, truncate-conical, convex to plano-convex, shallowly depressed to umbilicate, sometimes cup-shaped, with deflexed then straight margin, not hygrophanous, not translucently striate, black to very dark brown, tomentose all over, sometimes becoming minutely squamulose or rimose with age. Lamellae adnate with small decurrent tooth, arcuate, segmentiform to narrowly ventricose, moderately distant, up to 4 mm broad, dark grey to greyish blue or grey-violet sometimes with distinct yellow-green hues, with heavily pigmented very dark brown to black serrulate edge, L=25–40, l=3–5. Stipe 20–40×2–4 mm, cylindrical, dark to moderately dark grey-blue or violet-blue all over with yellow to yellow-green spot at base, dry, subpolished or with scattered, loose, grey or silvery fibrils, and finely

Plate 3.43 *Entoloma tasmanicum* (Photos Genevieve Gates (*inset*), Machiel Noordeloos)

squamulose at apex. Context thin, concolorous with surface, often tinged yellow or yellow-green in inner part of basal half of stipe, which sometimes extends over the whole stipe up into the context of the pileus. Odour none. Taste none or soapy or rancid.

Spores 8–10 × 5.5–7 μm, Q = 1.2–1.7, Qav = 1.35–1.45, heterodiametric, 5–6(–7)-angled in side-view. Basidia 20–42 × 8–12 μm, 4-spored, clampless. Lamella edge entirely sterile, consisting of a broad band of hyphae running parallel to the edge, with dense clusters of cheilocystidia at close intervals. Cheilocystidia 20–70 × 4–9 μm, cylindrical to narrowly clavate, sometimes subcapitate, with very dark brown intracellular pigment. Hymenophoral trama regular, made up of cylindrical hyphae 5–19 μm wide. Pileipellis a cutis with transitions to a trichoderm, made up of cylindrical hyphae 8–18 μm wide, with cylindrical to clavate terminal elements, 50–120 × 9–20 μm. Pigment dark brown, intracellular, diffuse and granular. Pileitrama regular, made up of cylindrical elements 4–28 μm wide. Brilliant granules present. Clamp-connections absent (Fig. 3.45).

Habitat & distribution: wet eucalypt forests and rainforests. Widespread. Fruiting extends from late spring to early winter (Group 2, in Chap. 1, see section "Fruiting Patterns of Species of Entolomataceae").

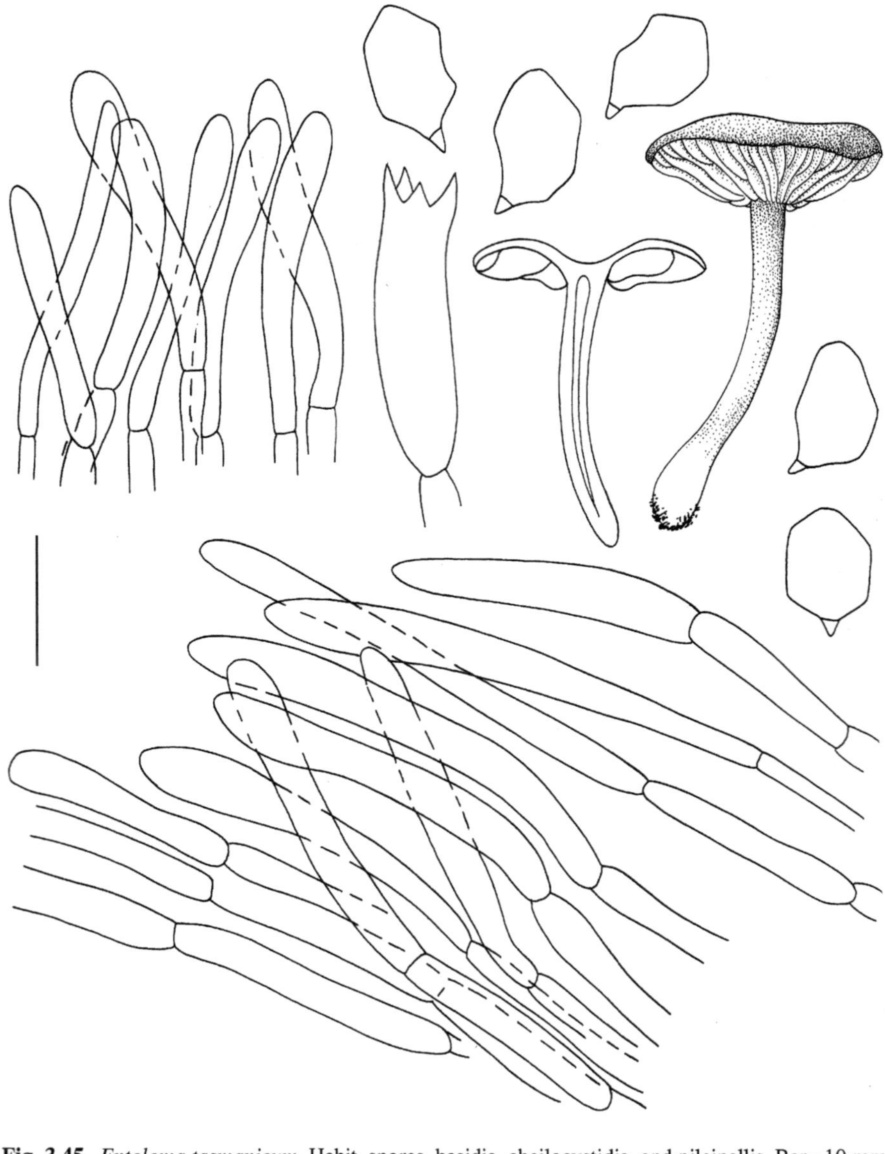

Fig. 3.45 *Entoloma tasmanicum*. Habit, spores, basidia, cheilocystidia, and pileipellis. Bar = 10 mm or 10 μm

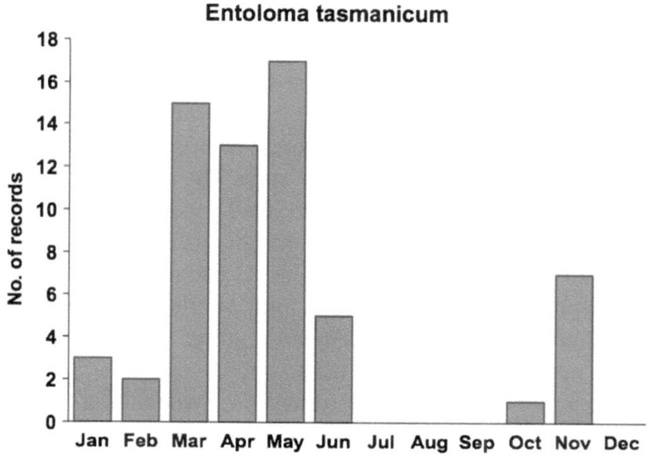

Collections examined. Australia, Tasmania, Arthur & Hellyer Rivers Junction, 41° 13′ S, 145° 31′ E, 5 June 1999, G. Gates E 568; Bruny Island, Mt Mangana, 43° 22′ S, 147° 17′ E, 8 April 1999, G. Gates E 373, E 386; —, 1 June 1999, G. Gates E 558; —, 26 May 2001, G. Gates E 1180, E 1189; —, 15 Nov. 2001, G. Gates E 1314; —, 27 May 2003, G. Gates E 1828; —, 27 May 2004, G. Gates E 2024; —, 29 April 2011, M.E. Noordeloos 2011020 (L); Chauncy Vale, 42° 37′ S, 147° 16′ E, 13 Oct. 2001, G. Gates E 1292; —, 22 Nov. 2001, G. Gates E 1327; —, 29 Nov. 2004, G. Gates E 1354; Collinsvale, Myrtle Forest, 42° 52′ S, 147° 09′ E, 18 March 1999, G. Gates E 259; —, 2 March 2000, G. Gates E 791; Dismal Swamp, 40° 57′ S, 144° 51′ E, 5 May 2011, G. Gates (M.E. Noordeloos 2011037, L); Evercreech Forest Reserve, 41° 24′ S, 147° 58′ E, 16 June 2003, G. Gates E 1881; Franklin River, 42° 13′ S, 146° 01′ E, 22 March 2005, G. Gates E 2096; Growling Swallet, 42° 41′ S, 146° 30′ E, 6 April 2000, G. Gates E 819; —, 1 May 2001, G. Gates E 1148; —, 2 May 2002, G. Gates E 1492; —, 13 May 2003, G. Gates E 1798; —, 17 Feb. 2005, G. Gates E 2079; Kermandie Falls Upper Track, 43° 12′ S, 146° 52′ E, 5 Nov. 1998, G. Gates E 103; —, 30 March 1999, G. Gates E 320; —, 16 Nov. 1999, G. Gates E 760; Lake St Clair National Park, Echo Point to Cynthia Bay, 42° 04′ S, 146° 09′ E, 22 April 2000, G. Gates E 952; Liffey Falls, 41° 42′ S, 146° 46′ E, 20 Jan. 2001, G. Gates E 1024; Mt Field National Park, 42° 41′ S, 146° 42′ E, 9 May 1999, G. Gates E 484; —, 29 June 1999, G. Gates E 631; Mt Wellington, Circle Track, 42° 55′ S, 147° 15′ E, 11 March 1999, G. Gates E 200; —, Jacksons Track, 25 March 1999, G. Gates E 287; —, Pipeline Track, 14 March 2011, G. Gates E 2286; North West Bay River, 42° 57′ S, 147° 12′ E, 17 Nov. 2001, G. Gates E 1316; —, 24 June 2003, G. Gates E 1884; —, 18 May 2004, G. Gates E 2008; —, 21 April 2005, G. Gates E 2133; Reuben Falls, 43° 00′ S, 146° 40′ E, 4 March 2000, G. Gates E 793; Warra LTER site, at roadside, 43° 06′ S, 146° 41′ E, 27 April 2004, G. Gates E 1975; —, 26 April 2005, G. Gates E 2139; —, Manuka Rd., 2 Nov. 2000, G. Gates E 1004; Waterfall Bay, 43° 04′ S, 147° 57′ E, 13 March 1999, G. Gates E 216; Wielangta, Rainforest Loop Walk, 42° 42′ S, 147° 51′ E, 23 Feb.1999, G. Gates E 142.

Entoloma tasmanicum belongs to the polymorphic group of *Entoloma serrulatum* (Fr.: Fr.) Hesler, characterized by a special structure of the lamellae edge which is completely sterile, made up of a strand of hyphae which runs parallel to the edge, with irregularly spaced dense bundles of cylindrical, clavate to fusiform terminal elements (cheilocystidia), which usually are heavily blue or brown pigmented with intracellular pigment (Noordeloos 2004). The distinguishing characters of *Entoloma tasmanicum* are the very dark, almost black basidiocarps, with often very distinct yellow to yellow-green tinges in the stipe base, which sometimes are also present in the lamellae and in the context of pileus and stipe. The stipe surface is not polished but distinctly striate, and sometimes also with small darker dots or minute squamules in the upper part. The terminal elements of the lamella edge are long and fusiform, filled with brownish black intracellular pigment. The trama often has deeply pigmented brown hyphae, the content of which turns purple in 5 % KOH. Furthermore, the spores are relatively small for the serrulatum-group. *E. tasmanicum* is one of the most common *Cyanula* species in Tasmania.

Entoloma *waikaremoana* E. Horak from New Zealand is a tiny species from the same group, with turquoise green or glaucous green fruit bodies, and somewhat smaller spores (Horak 2008). *Entoloma querquedula* (Romagn.) Noordeloos from Europe also has greenish yellow or olivaceous tinges, but is much paler than *E. tasmanicum*, and has larger spores (Noordeloos 2004).

46. **Entoloma splendidum** Noordel. & G.M. Gates, spec. nov.

MycoBank #564499.

Diagnosis: Habit collybioid, entirely sky blue; pileus 10–25 mm, deeply translucently striate; lamellae with blue fringed edge; stipe 10–36 × 1–2 mm, polished; spores 8–12 × 6–8 µm, 5–6- angular; lamella edge of serrulatum type; pileipellis a cutis with transitions to a trichoderm of inflated hyphae, 9–20 µm wide; pigment blue, intracellular; clamp-connections absent.

Holotype: Australia, Tasmania, Growling Swallet, 42° 41′ S, 146° 30′ E, 21 April 2007, G. Gates E 2252 (HO 561522; isotype in L).

Etymology: splendidus (Lat.) = splendid.

Main characters: delicate collybioid species with bright blue, translucent striate pileus, blue fimbriate lamella edge and blue polished stipe (Plate 3.44).

Description:
Pileus 10–25 mm diam, convex with slight depression, with deflexed, often slightly crenate margin, not hygrophanous, indistinctly to distinctly translucent-striate up to centre, deep blue (19E3–20E4), darker at centre, minutely squamulose at centre, radially fibrillose towards margin. Lamellae sinuate or adnate with short decurrent tooth, segmentiform, moderately distant, up to 3 mm broad, white to pale blue-grey or dark grey-blue with pink hue, with slightly serrulate to subentire, blue-black edge, lamellulae in two tiers. Stipe 10–36 × 1–2 mm, cylindrical, fragile, sky blue, same colour or (much) paler than pileus, glabrous, polished, with white basal tomentum. Odour spermatic. Taste saliva inducing.

Plate 3.44 *Entoloma splendidum* (Photos Michael Pilkington)

Spores 8–12×6–8 μm, Q=1.2–1.5, Qav=1.3, heterodiametric, 5–6 angled in side-view. Basidia 30–40×8–11 μm, 4-spored, clampless. Lamella edge of serrulatum-type with long, cylindrical to clavate cheilocystidia, 20–50×4–9 μm, in clusters with blue intracellular pigment. Pileipellis a cutis with transitions to a trichoderm, made up of cylindrical hyphae, 8–15 μm wide with clavate terminal elements, 40–90×9–20 μm. Pigment intracellular, blue, diffuse and granular in pileipellis. Pileitrama regular, made up of cylindrical hyphae, 5–15 μm wide. Brilliant granules present. Clamp-connections absent (Fig. 3.46).

Habitat & distribution: saprotrophic, among mosses in wet sclerophyll forest; rarely found, known from only two localities.

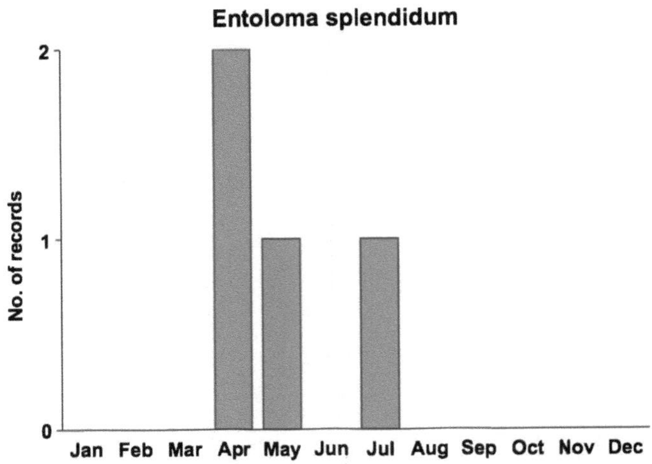

Fig. 3.46 *Entoloma splendidum*. Habit, spores, basidia, cheilocystidia, and pileipellis. Bar = 10 mm or 10 μm

Collections examined. Australia, Tasmania, Collinsvale, Myrtle Forest 42° 52′ S, 147° 09′ E, 28 April 2005, G. Gates E 2143; —, 11 May 2010, G. Gates & M.E. Noordeloos s.n.; Growling Swallet, 42° 41′ S, 146° 30′ E, 10 July 2001, G. Gates E 1259; —, 21 April 2007, G. Gates E 2252 (holotype, HO 561522).

Entoloma splendidum is one of the most striking species of *Entoloma* in Tasmania, with its dark blue, translucently striate pileus, contrasting with the pale lamellae with beautiful dark blue serrulate lamella edges. This species must be placed in the serrulatum-group on account of the structure and colour of the lamella edge. It resembles *Entoloma gomerense* Noordeloos & Wölfel, originally described from the Island of Gomera, Canary Islands, Spain, but now known from a few other localities in Europe (Noordeloos 2004).

47. **Entoloma violaceocoeruleum** Noordel. & G.M. Gates, spec. nov.

MycoBank #564500.

Diagnosis: Habit collybioid; pileus 5–25 mm, blue-violet then violet-pink with minute blackish squamules; lamellae violet-blue with violet-red, fimbriate edge; stipe concolorous with pileus, polished; spores 8–10×6–7 μm, 5–6-angular; lamella edge sterile, of serrulatum type with pink-violet intracellular pigment; pileipellis cutis to trichoderm of clavate elements with blue-violet intracellular pigment; clamp-connections absent.

Holotype: Australia, Tasmania, Growling Swallet, 42° 41′ S, 146° 30′ E; 28 March 2009, M.E. Noordeloos 2009086 (HO 564351; isotype in L).

Etymology: violaceus (Lat.) = violet; coeruleus (Lat.) = blue.

Main characters: habit collybioid, a member of the serrulatum group with delicate violaceous blue colours; lamella edge violet-red (Plate 3.45).

Description:
Pileus 5–25 mm diam, convex to plano-convex with deflexed to straight margin, not hygrophanous, deeply translucently striate, blue-violet at first, soon violaceous pink, covered with very minute blackish blue squamules at centre spreading towards margin, fibrillose at margin. Lamellae adnexed, ventricose, moderately distant, up to 4 mm broad, blue-violet at first, soon violet-pink with finely violet-red, serrulate edge, lamellulae in two tiers. Stipe 20–30×0.5–1 mm, filiform-cylindrical, deep blue then fading to violet-pink, polished. Odour none. Taste not tried.

Spores 8–10×6–7 μm, 5–6-angled in side-view. Basidia 22–40×6–10 μm, 4-spored. Lamella edge sterile, of serrulatum-type. Cheilocystidia in dense clusters, clavate to broadly clavate or sphaeropedunculate, 30–60×12–25 μm, with purple-violet intracellular pigment. Hymenophoral trama regular, made up of more or less cylindrical, 4–13 μm wide hyphae. Pileipellis a trichoderm of broadly clavate elements, 30–90×18–30 μm, with violaceous pigment. Pileitrama similar to hymenophoral trama. Brilliant granules very abundant. Stipitipellis a cutis of narrow, cylindrical hyphae, 3–7 μm wide. Caulocystidia absent. Clamp-connections absent (Fig. 3.47).

Habitat & distribution: widespread in wet sclerophyll forest. Uncommon.

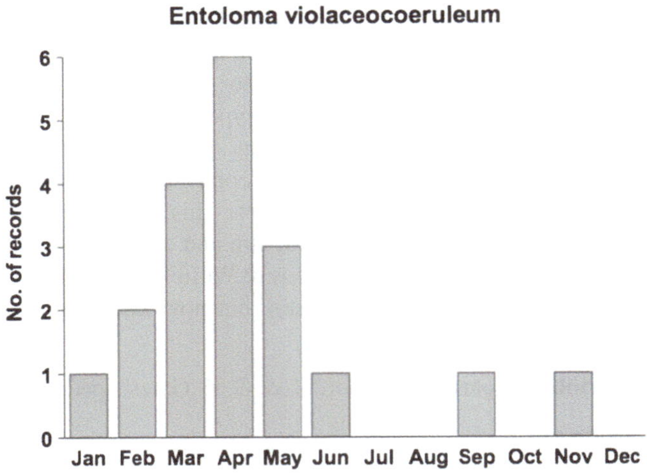

Plate 3.45 *Entoloma violaceocoeruleum* (Photo Machiel Noordeloos)

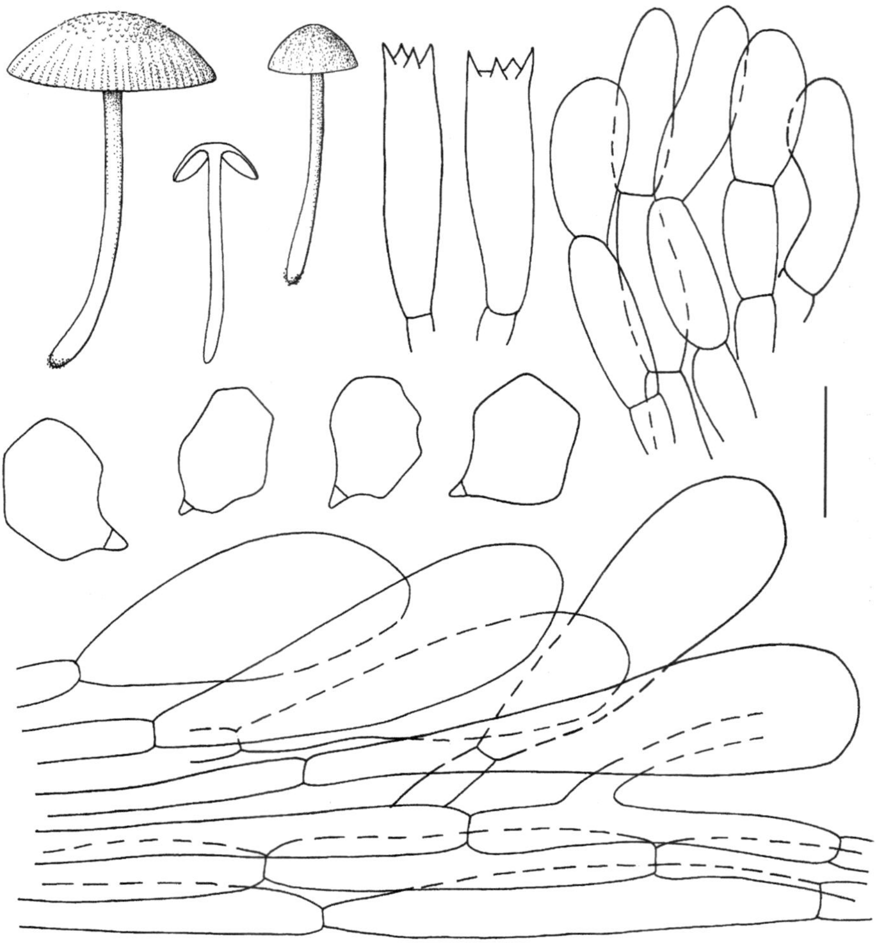

Fig. 3.47 *Entoloma violaceocoeruleum*. Habit, spores, basidia, cheilocystidia, and pileipellis. Bar = 10 mm or 10 μm

Collections examined. Australia, Tasmania, Bruny Island, Mt Mangana, 43° 22′ S, 147° 17′ E, 31 March 2009, M.E. Noordeloos 2009087; Clarks Cliffs, 43° 06′ S, 147° 47′ E, 13 March 2004, G. Gates E 1943; Collinsvale, Myrtle Forest 42° 52′ S, 147° 09′ E, 14 April 2003, G. Gates E 1707; —, 10 March 2005, G. Gates E 2086; —, 3 May 2005, G. Gates E 2149; Duckhole Lake Track, 43° 22′ S, 146° 53′ E, 18 May 2002, G. Gates E 1519; —, 6 June 2002, G. Gates E 1557; —, 15 May 2003, G. Gates E 1809; Growling Swallet, 42° 41′ S, 146° 30′ E, 28 March 2009, M.E. Noordeloos 2009086 (holotype, HO 564351); Kermandie Falls Lower Track, 43° 12′ S, 146° 52′ E, 10 April 2003, G. Gates E 1690; Mt Wellington, Myrtle Gully, 42° 54′ S, 147° 15′ E, 12 Sept. 2002, G. Gates E 1637; —, Reids Track, 42° 55′ S, 147° 15′ E, 27 Feb. 1999, G. Gates E 150; North West Bay River, 42° 57′ S, 147°

12′ E, 21 April 2005, G. Gates E 2137; —, 18 April 2007, G. Gates E 2249; Styx Valley, Christmas Tree Grove, 42° 50′ S, 146° 41′ E, 26 April 2011, M.E. Noordeloos 2011001 (L); Tahune, Huon Pine Walk, 43° 06′ S, 146° 44′ E, 22 Feb. 2005, G. Gates E 2084; Warra, Manuka Road, 43° 06′ S, 146° 41′ E, 2 Nov. 2000, G. Gates E 1005; —, 15 April 2004, G. Gates E 1971.

Entoloma violaceocoeruleum is a very beautiful delicate *Cyanula* with blue-violet colours, which change considerably with age as the dark blue colour slowly disappears and the violet-pink colour remains in most of the pileus and stipe. The strikingly serrulate lamella edge suggests affinity with the group of *Entoloma serrulatum*. No similar species could be found in the relevant literature.

48. **Entoloma natalis-domini** G.M. Gates & Noordel., Cryptogamie Mycologie 30: 112. 2009.

Etymology: referring to natalis-domini = nativity, in association with the name of the type locality.

Original diagnosis: Pileus 8–27 mm latus, conico-convexus, acute umbonatus, haud hygrophanus, haud translucido-striatus, obscure brunneus, glabrus. Lamellae adnatae, confertae, griseo-albae, demum roseotinctae, acies lamellarum caesiotinctae. Stipes 25–37 × 1.2–2 mm, cylindraceus, coeruleo-griseus, politus. Sporae 8–10(–10.5) × 6–8 μm, (4–)5(–6)-angulatae. Acies lamellarum sterilis. Cheilocystidia 32–55 × 6–12 μm contento caesio. Pileipellis cutis vel trichoderma, ex elementis cylindraceis, 5–14 μm latis, constituta pigmentis intracellularibus. Granula lucentia abundantia. Fibulae desunt.

Holotype: Australia, Tasmania, Styx Valley, Christmas Tree Grove, 42° 50′ S, 146° 41′ E, 9 June 2007, G. Gates E 2260 (HO 548304; isotype L).

Main characters: habit mycenoid; pileus acute, very dark brown; lamella edge dark blue, fimbriate; stipe grey-blue, polished (Plate 3.46).

Description:
Pileus 8–27 mm diam, conico-convex with sharp umbo, with somewhat terraced surface, with straight, entire to crenulate margin, not hygrophanous, slightly translucently striate, very dark brown, velutinous at centre, glabrous at margin. Lamellae adnate with short decurrent tooth, seceding, ventricose, crowded, up to 4 mm broad, greyish blue-pink, paler towards the dark blue, fimbriate edge. Stipe 25–37 × 1.2–2 mm, slightly broadened towards base, cylindrical, dark blue-grey, paler at the base, polished with pinkish basal mycelium. Context white. Odour not distinctive. Taste not distinctive.

Spores 8–10(–10.5) × 6–8 μm, average 9–9.5 × 7–7.5 μm, Q = 1.2–1.5, Qav = 1.35, heterodiametric, (4–)5(–6)-angled in side-view. Basidia 17–25 × 7–9 μm, 4-spored, clampless. Lamella edge sterile, of serrulatum-type, rarely fertile. Cheilocystidia 32–55 × 6–12 μm, cylindrical to narrowly clavate, with blue intracellular pigment. Hymenophoral trama regular, made up of long, cylindrical elements, 80–150 × 4–14 μm. Pileipellis a cutis with transition to a trichoderm at centre, made up of 9–12(–14) μm wide hyphae, with clavate terminal elements at centre, 11–20 μm

Plate 3.46 *Entoloma natalis-domini* (Photos Michael Pilkington)

wide, with abundant bright blue pigment; subpellis made up of long, cylindrical elements, 5–14 μm wide with brown intracellular pigment. Pileitrama regular, as hymenophoral trama. Brilliant granules very abundant in all tissues. Stipitipellis a cutis of cylindrical hyphae, 5–15 μm wide with blue, intracellular pigment. Caulocystidia absent. Clamp-connections absent (Fig. 3.48).

Habitat & distribution: in small groups, terrestrial in wet sclerophyll forest and mixed forest. Known only from a few localities.

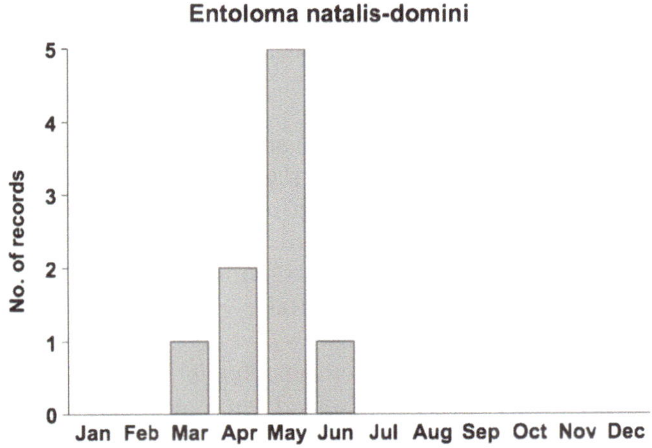

Fig. 3.48 *Entoloma natalis-domini*. Habit, spores, basidia, cheilocystidia, and pileipellis. Bar = 10 mm or 10 μm

Collections examined. Australia, Tasmania, Kermandie Falls Lower Track, 43° 12′ S, 146° 52′ E, 19 May 2010, F. Karstedt 1446; Styx Valley, Big Tree Reserve, 42° 49′ S, 146° 39′ E, 13 May 2010, F. Karstedt 1391; —, Christmas Tree Grove, 42° 50′ S, 146° 41′ E, 24 April 2001, G. Gates E 1115; —, 9 June 2007, G. Gates E 2260 (holotype, HO 548304; isotype, L); —, 13 May 2010, M.E. Noordeloos 2010009; Wedge Forest Reserve, 42° 50′ S, 146° 16′ E, 28 May 2011, G. Gates s.n.

The blue fringed lamella edge suggests that *Entoloma natalis-domini* is related to the serrulatum group of subgenus *Cyanula*, but its dark brown to almost black, conical, papillate, and often terraced pileus and the small (4–)5-angled spores place it in a rather isolated position there. One collection (Karstedt 1391) is very similar to those of the type locality, with the exception that some specimens in that collection have a fertile lamella edge, without cheilocystidia. *Entoloma sassafras* G.M. Gates & Noordel. is another dark-coloured species with a pale to dark carmine to violet-black lamella edge, but it lacks blue tinges in the stipe, and has differently shaped spores. *Entoloma asprelloides* G. Stev. from New Zealand differs not only by a collybioid habit with depressed pileus and lack of blue tinges in the stipe, but also by the larger spores (Horak 2008). *Entoloma glaucopus* E. Horak, although macroscopically different in stature and colour, appears somewhat similar in microscopy, but has slightly larger spores. The European *Entoloma linkii* (Fr.) Noordel. has a collybioid habit, with similar dark brown pileus, but lacks blue tinges in the stipe, and has larger spores. *Entoloma caesiocinctum* (Kühner) Noordel. another species from the serrulatum-group with a brown pileus, is a widespread species from Europe and North America, but differs by a paler, usually deeply translucently striate, umbilicate pileus, and larger spores (Noordeloos 2004).

49. **Entoloma sassafras** G.M. Gates & Noordel., Cryptogamie Mycologie 30: 114. 2009.

Original diagnosis: Pileus 11–30 mm latus, plano-convexus, centro depressus, haud hygrophanus, haud translucido-striatus, flavo-brunneus regulariter minute squamulosus. Lamellae adnatae, confertae, griseae vel flavo-griseae, demum roseotinctae, acies lamellarum violaceotinctae. Stipes 25–45×3–4 mm, cylindraceus, obscure griseus, politus sed apicem squamulosus. Sporae 9–10.5×7–8 μm, 5–7-angulatae. Acies lamellarum sterilis. Cheilocystidia 32–60×5–11 μm, contento caesio. Pileipellis cutis vel trichoderma ex elementis fusiformis, 25–80×7–25 μm latis, constituta pigmentis intracellaribus. Granula lucentia abundantia. Fibulae desunt.

Holotype: Australia, Tasmania, Warra LTER site, Bird Track, 43° 06′ S, 146° 39′ E, 5 June 2007, G. Gates E 2256 (HO 548308; isotype L).

Etymology: named after sassafras, the common name of the tree *Atherosperma moschatum*, with which it is often associated.

Main characters: habit collybioid; pileus very dark, almost black with neat, regularly arranged squamules; lamella edge deep carmine to violet-black; stipe dark grey, apex squamulose (Plate 3.47).

Description:
Pileus 11–30 mm diam, convex to plano-convex with depressed centre and deflexed to straight, black-rimmed margin, not hygrophanous, not translucently striate, entirely black in youth, then dark yellow-brown, overlain by radiating, thick, black squamules, which are more concentrated on the black, scaly disc, and regularly spread towards the margin, sometimes a distinct violet hue may be present near the margin. Lamellae adnate-emarginate to adnexed, segmentiform to subventricose, crowded, up to 4.5 mm broad, dark grey or yellow-grey, sometimes more yellow near pileus, with subentire, pale to deep carmine to violet-black edge. Stipe

Plate 3.47 *Entoloma sassafras* (Photos Michael Pilkington)

25–45 × 3–4 mm, cylindrical, dark grey, finely squamulose at apex, glabrous or with some sparse fibrils towards base, with white to pinkish basal tomentum. Odour not distinct. Taste not distinct or somewhat bitter.

Spores 9–10.5 × 7–8 µm, average 9.5 × 7.5 µm, Q = 1.2–1.5, heterodiametric, 5–7-angled in side-view with pronounced angles, thick-walled. Basidia 20–37 × 7–11 µm, 4-spored, clampless. Lamella edge sterile with dense clusters of clavate to utriform cheilocystidia, 32–60×5–11 µm, with granular, blue pigment. Hymenophoral trama regular, made up of cylindrical hyphae, 4–20 µm wide. Pileipellis a trichoderm of cylindrical to fusiform terminal elements, 25–80×7–25 µm with brown, clotted-granular intracellular pigment. Pileitrama regular, similar to hymenophoral trama. Brilliant granules abundant. Caulocystidia absent. Clamp-connections absent (Fig. 3.49).

Habitat & distribution: terrestrial and on living and dead trunks of sassafras in wet sclerophyll and mixed forests, not common.

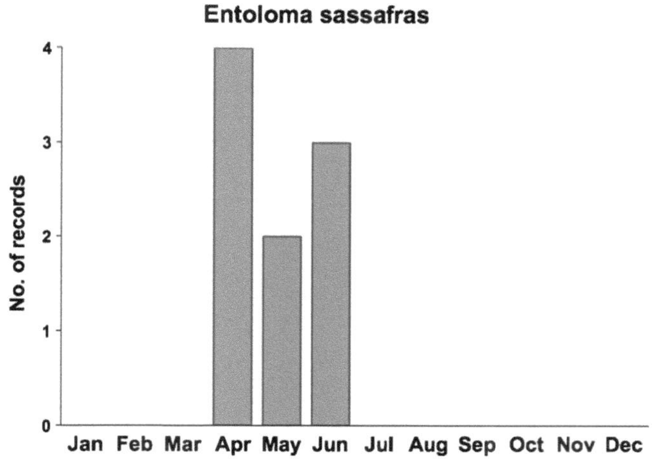

Collections examined. Australia, Tasmania, Duckhole Lake Track, 43° 22′ S, 146° 53′ E, 15 May 2003, G. Gates E 1804 (HO 548312); Kermandie Falls Lower Track, 43° 12′ S, 146° 52′ E, 19 May 2010, F. Karstedt 1442; Lake Skinner Track, 42° 57′ S, 146° 44′ E, 11 April 1999, G. Gates E 368 (HO 548311); Lake St Clair National Park, Echo Point to Cynthia Bay, 42° 04′ S, 146° 09′ E, 20 April 2002, G. Gates E 1481 (HO 548310); Mt Field National Park, Lyrebird Walk, 42° 41′ S, 146° 40′ E, 1 April 1999, G. Gates E 322 (HO 548307); —, 2 June 2001, G. Gates E 1208 (HO 548306); —, Russell Falls Loop, 42° 41′ S, 146° 42′ E, 29 June 1999, G. Gates E 639 (HO 548309); Warra LTER site, Bird Track, 43° 06′ S, 146° 39′ E, 5 June 2007, G. Gates E 2256 (holotype, HO 548308; isotype, L); —, coupe WR008J, 43° 06′ S, 146° 41′ E, 27 April 2004, G. Gates E 1976 (HO 548305).

Entoloma sassafras is a striking species with its very dark, densely squamulose pileus, violet-blue lamella edge and dark grey stipe. *Entoloma atrellum* E. Horak from New Zealand is similar, but differs by having a blue pileus, blue lamellae edge, and lageniform cheilocystidia. *Entoloma melanocephalum* G. Stev. from New Zealand has more predominantly blue-green colours in the pileus, lamellae and stipe, a blue lamella edge, and larger spores.

50. **Entoloma viridomarginatum** (Cleland) E. Horak, Beih. Nova Hedwigia 65: 301. 1980.

Leptonia viridomarginata Cleland, Trans. Proc. Royal Soc. South Australia 51: 303. 1927.— *Entoloma caesiomarginatum* E. Horak, Beih. Nova Hedwigia 43: 70.

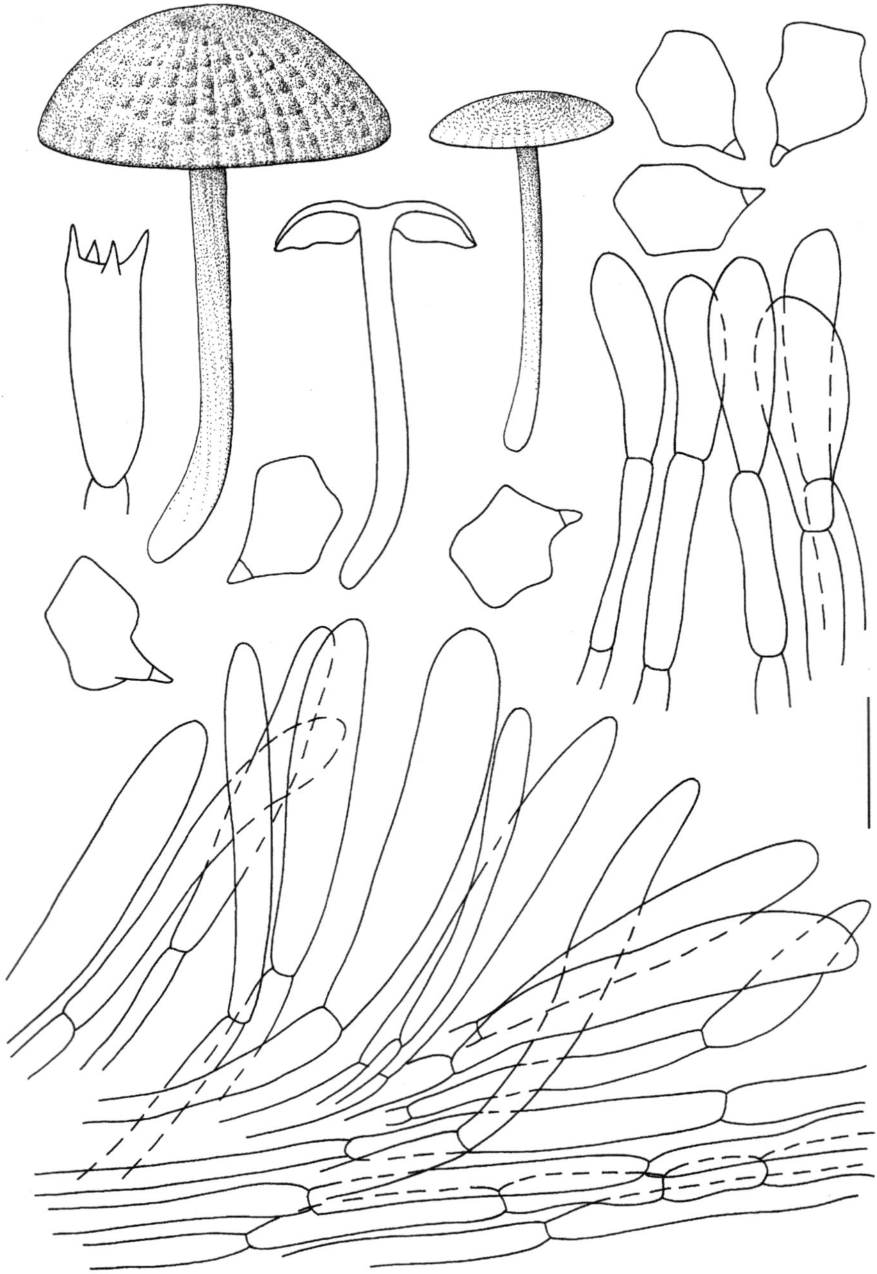

Fig. 3.49 *Entoloma sassafras*. Habit, spores, basidia, cheilocystidia, and pileipellis. Bar = 10 mm or 10 μm

1973.— *Entoloma viridomarginatum* var. *milfordense* Wölfel & Hausknecht, Österr. Z. Pilzk. 8:136. 1999.

Original diagnosis: Pileus ¾ in. (18 mm) in diameter, slightly convex with the centre dimpled, clad with small fibrillose scales, edge slightly sulcate, dark green (near Dusky Olive Green). Gills slightly sinuate with a decurrent tooth, moderately distant, alternate ones short, of a Light Pinkish Cinnamon tint passing into dark green which edges the gills. Stem 2½ ins. (6.2 cm) high, slender, flexuous, twisted, shining, finely punctate above, rather tough, flesh heterogeneous from that of the pileus, hollow, dark green. Shed spores of salmony tint, microscopically angular pear-shaped, 11–11.5 × 7.5 μm, cystidia not seen. On the ground. June.

Holotype. Australia, South Australia, Adelaide, Mt. Lofty, 13 June 1925, Cleland 10850 (ADW 13536).

Etymology: viridis (Lat.) = green; marginatus (Lat.) = of margin, referring to the green lamella edge.

Selected literature: Grgurinovic, Larger fungi of South Australia: 378. 1997; Grgurinovic & Simpson, Fungal Diversity 8: 105. 2001; Horak, Agaricales of New Zealand I: 146. 2008; Wölfel & Hausknecht, Österr. Z. Pilzk. 8: 136. 1999.

Main characters: small collybioid *Cyanula*; fruit body predominantly green-yellow; lamella edge green (Plate 3.48).

Description:
Pileus 6–20 mm diam, convex to plano-convex with depressed to umbilicate centre, not truly hygrophanous, obscurely translucently striate at margin, at first very dark viridian green, green-blue, olivaceous brown or brown, only slightly paler towards margin, becoming paler green-yellow to green-brown to ochre orange with age and broadly translucently striate, minutely squamulose all over at first, densely squamulose at centre and distinctly radially fibrillose with squamules spreading to margin. Lamellae adnate-decurrent, segmentiform to subventricose, moderately crowded, up to 4 mm broad, sordid pink with contrasting green, minutely serrulate edge. Stipe 20–40 × 2–3 mm, cylindrical or compressed with groove, sometimes slightly broadening towards base, blue, green-blue or deep green with contrasting yellow to yellow-orange tinged base, fading to yellowish or yellow-orange with age, polished, glabrous. Context orangy pink in stipe. Odour and taste indistinct.

Spores 9.0–12 × 5.5–8 μm, Q = 1.2–1.6, Qav = 1.4, heterodiametric, 5–7-angled in side-view. Basidia 20–32 × 8–11 μm, 4-spored, clampless. Lamella edge sterile of serrulatum-type. Cheilocystidia 20–60 × 7–15 μm, subcylindrical to clavate, sometimes septate, filled with blue-green or brown-green intracellular pigment. Hymenophoral trama regular, made up of cylindrical to inflated, up to 20 μm wide hyphae. Pileipellis a cutis with transition to a trichoderm, made up of inflated, clavate terminal elements, 5–15 μm wide. Pigment brown-green, sometimes also bluish in part of the hyphae, intracellular. Pileitrama regular, made up of cylindrical elements, 5–14 μm. Brilliant granules abundant. Caulocystidia absent. Clamp-connections absent (Fig. 3.50).

Plate 3.48 *Entoloma viridomarginatum* (Photos Michael Pilkington and Machiel Noordeloos)

Habitat & distribution: found in wet and dry eucalypt forests and rainforest in Tasmania; also found in South Australia. Fruiting throughout the year, with a fair number of records in spring and summer months (Group 5, in Chap. 1, see section "Fruiting Patterns of Species of Entolomataceae").

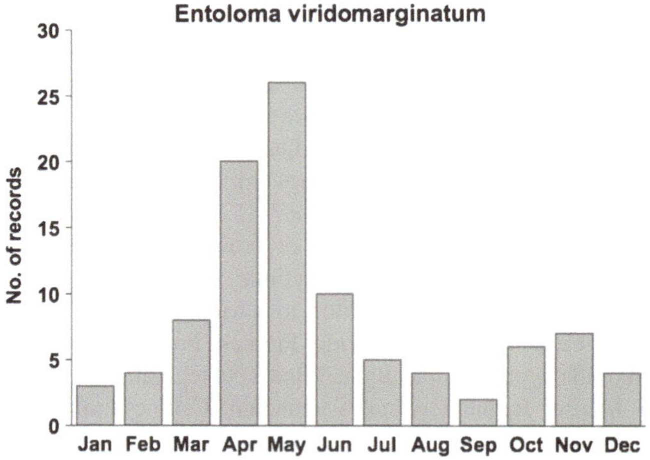

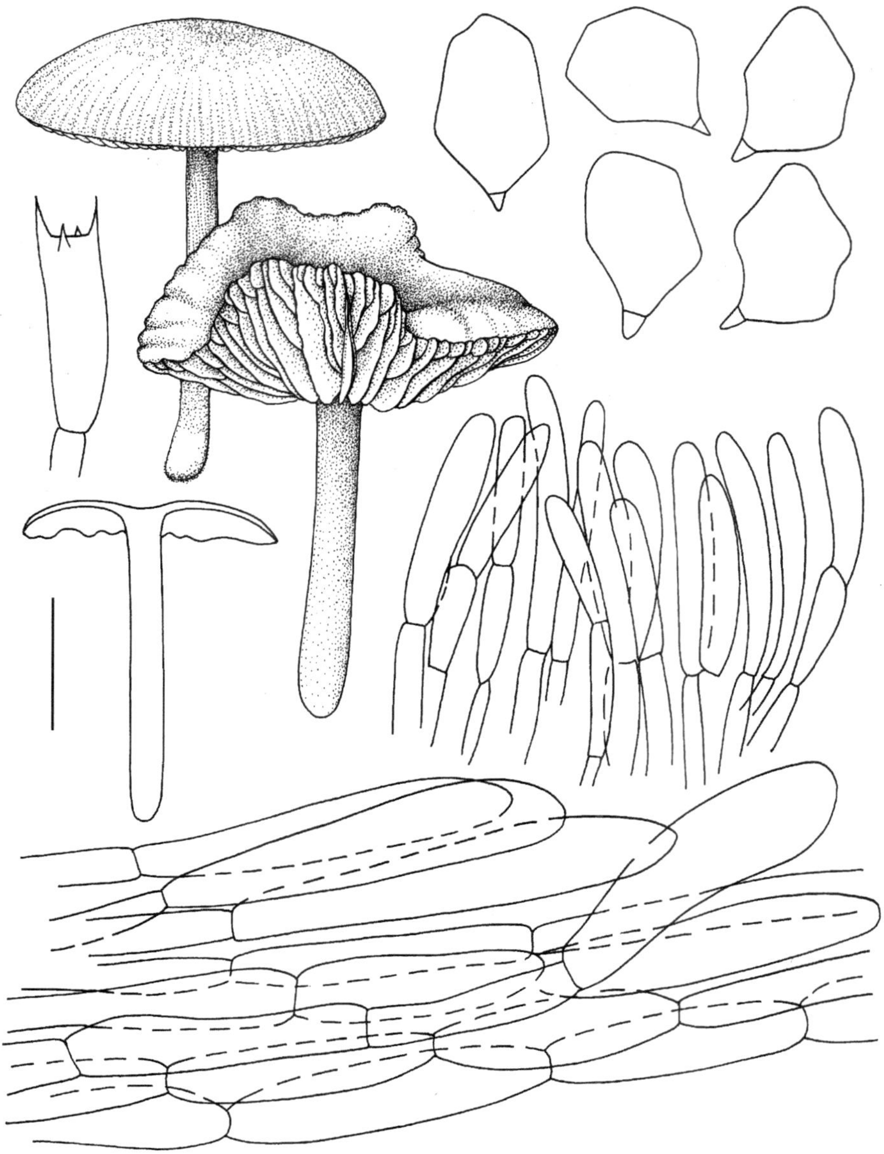

Fig. 3.50 *Entoloma viridomarginatum*. Habit, spores, basidia, cheilocystidia, and pileipellis. Bar = 10 mm or 10 μm

Collections examined. Australia, South Australia, Adelaide, Mt. Lofty, 13 June 1925, Cleland 10850 (ADW 13536, holotype). — Tasmania, Chauncy Vale, 42° 37′ S, 147° 16′ E, 13 Oct. 2001, G. Gates E 1288; Clarks Cliffs, 43° 06′ S, 147° 47′ E, 13 March 2004, G. Gates E 1951; Collinsvale, Fire Trail, 42° 52′ S, 147° 09′ E, 15 April 1999, G. Gates E 421; —, Myrtle Forest, 25 Nov. 1998, G. Gates E 105; —,

15 April 1999, G. Gates E 426; Douglas Apsley National Park, 2 May 2011, G. Gates (M.E. Noordeloos 2011028); Duckhole Lake Track, 43° 22′ S, 146° 53′ E, 16 April 2005, G. Gates E 2113; Geeveston, Donnellys Rd, 43° 07′ S, 146° 54′ E, 13 Aug. 1998, G. Gates E 43; —, 12 June 1999, G. Gates E 595 & E 597a; —, 30 July 2002, G. Gates E 1614; —, 10 April 2003, G. Gates E 1699; Grassy, King Island, 40° 03′ S, 144° 02′ E, 25 Sept. 1998, coll. D.A. Ratkowsky s.n., G. Gates E 59; Growling Swallet, 42° 41′ S, 146° 30′ E, 1 May 2001, G. Gates E 1156; —, 28 March 2009, M.E. Noordeloos 2009097; Lake St. Clair National Park, Echo Point to Cynthia Bay, 42° 04′ S, 146° 09′ E, 21 March 2009, M.E. Noordeloos 2009033 (L); Marriotts Falls, 42° 43′ S, 146° 39′ E, 24 May 2003, G. Gates E 1819; Mt Field National Park, Russell Falls, 42° 41′ S, 146° 42′ E, 29 June 1999, G. Gates E 641; Mt Wellington, Myrtle Gully, 42° 54′ S, 147° 15′ E, 15 June 1999, G. Gates E 612; —, 3 May 2001, G. Gates E 1158; —, Silver Falls, 42° 55′ S, 147° 15′ E, 27 Feb. 1999, G. Gates E 149 & E 151; Notley Gorge, 41° 21′ S, 146° 55′ E, 23 Oct. 1999, G. Gates E 716; Wielangta, 42° 42′ S, 147° 51′ E, 23 Feb.1999, G. Gates E 128.

Entoloma viridomarginatum is the chameleon among the *Entoloma* species of Tasmania, with a wide range of colours from deep blue-green, viridian green to yellowish orange, depending on age of fruit body and weather conditions. For this reason we do not consider var. *milfordense* worth separating, in agreement with Horak (2008).

Stirps ***Transmutans***

Pileus and stipe blue or violet mixed with pink; pileus not or faintly translucently striate, entirely finely squamulose; stipe polished.

51. **Entoloma transmutans** G.M. Gates & Noordel., Cryptogamie Mycologie 30: 123. 2009.

Etymology: transmutans (Lat.) = changing, referring to the colour changes in the developing pileus.

Original diagnosis: Pileus 4–20 mm latus, convexus vel plano-convexus, haud hygrophanus, haud translucido-striatus, obscure atro-coeruleus vel coeruleo-griseus, minute squamulosus, versus marginem virgatus. Lamellae adnatae, distantes, griseo-roseae, acies lamellarum brunneo-tinctae. Stipes 7–20 × 1–2 mm, cylindraceus, violaceus, politus. Sporae 8–10 × 6–9 μm, 6–8-angulatae. Acies lamellarum sterilis. Cheilocystidia 20–42 × 9–25 μm, cylindracea vel clavata. Pileipellis cutis vel trichoderma, ex elementis inflatis, 6–25 μm latis, constituta pigmentis intracellularibus. Granula lucentia abundantia. Fibulae absentes.

Holotype: Australia, Tasmania, Mt. Field National Park, Lyrebird Walk, 42° 41′ S, 146° 43′ E, 1 April 1999, G. Gates E 331 (HO 548322, isotype L).

Main characters: small, collybioid species with combinations of blue, grey and pink colours in the basidiocarp (Plate 3.49).

Plate 3.49 *Entoloma transmutans* (Photos Michael Pilkington)

Description:
Pileus 5–30 mm diam, convex then plano-convex with rounded to umbilicate centre, not hygrophanous, not translucently striate, dark blue-grey at first, then breaking up into minute violet-grey (11D4–5, 11E4–5), or blackish blue squamules at centre and radially fibrillose-virgate towards margin, showing delicate pinkish purple background between the squamules and fibrils, becoming a violet-pink-blue colour upon drying. Lamellae adnate-emarginate, moderately distant, up to 4.5 mm broad, white to pale blue-grey at first then greyish pink with fimbriate, concolorous or pale pinkish brown edge, L = ca. 40, l = 3–5. Stipe 7–20 × 1–2 mm, cylindrical, bluish at first then purplish pink, concolorous with background of pileus, shiny, polished or innately fibrillose. Odour spermatic. Taste saliva inducing.

Spores 8–10 × 6–9 μm, Q = 1.2–1.5, heterodiametric with 6–8 angles. Basidia 28–35 × 9–12 μm, 4-spored, clampless. Lamella edge completely sterile with broadly clavate, vesiculose or sphaeropedunculate cheilocystidia, 20–42 × 9–25 μm, with or without pale brown intracellular pigment, with a hint of violet. Hymenophoral trama regular, made up of inflated hyphae to 20 μm wide. Brilliant granules abundant. Pileipellis a cutis of 7–12 μm wide hyphae, with transitions to a trichoderm, made up of clavate terminal elements, 24–90 × 6–25 μm; subpellis filamentous, made up of narrow, cylindrical hyphae, 5–12 μm wide. Pigment abundant, brown in suprapellis, greyish blue in subpellis. Pileitrama regular, made up of inflated elements to 25 μm wide. Stipitipellis a cutis of cylindrical hyphae, 6–12 μm wide. Caulocystidia present at apex of stipe, 30–70 × 5–10 μm, cylindrical or strangulate, sometimes subcapitate. Clamp-connections absent (Fig. 3.51).

Habitat & distribution: distributed widely in forest litter of wet and dry sclerophyll forest and in mixed forest. Fruiting throughout the year, with a fair number of records in spring and summer months (Group 5, in Chap. 1, see section "Fruiting Patterns of Species of Entolomataceae").

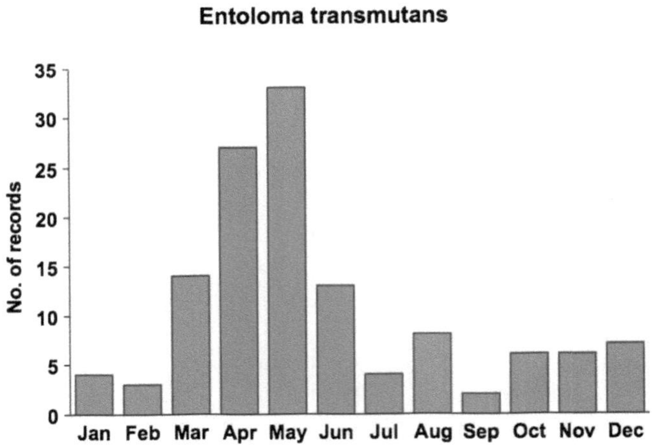

Collections examined. Australia, Tasmania, Adamsons Peak Track, 43° 20′ S, 146° 54′ E, 1 May 1999, G. Gates E 464; Arve Road, Zig-Zag Track, 43° 06′ S, 146° 45′ E, 23 Nov. 1999, G. Gates E 761; Bruny Island, Mt Mangana, 43° 22′ S, 147° 17′ E, 8 April 1999, G. Gates E 365 & E 375; —, 1 June 1999, G. Gates E 552; —, 7 April 2001, G. Gates E 1081; 26 May 2001, G. Gates E 1182; 30 April 2011, M.E. Noordeloos 2011021 (L); Chauncy Vale, 42° 37′ S, 147° 16′ E, 13 Oct. 2001, G. Gates E 1291; —, 28 June 2003, G. Gates E 1891 (HO 548324); Clarks Cliffs, 43° 06′ S, 147° 47′ E, 4 May 2002, G. Gates E 1503; Collinsvale, Myrtle Forest, 42° 52′ S, 147° 09′ E, 18 March 1999, G. Gates E 251; —, 15 April 1999, G. Gates E 424; Duckhole Lake Track, 43° 22′ S, 146° 53′ E, 18 May 2002, G. Gates E 1521; Geeveston, Donnellys Rd, 43° 07′ S, 146° 54′ E, 23 May 2000, G. Gates E 931; —, 24 Oct. 2000, G. Gates E 1003; —, 29 Oct. 2001, G. Gates E 1306; —, 30 July 2002, G. Gates E 1615; Growling Swallet, 42° 41′ S, 146° 30′ E, 20 April 1999, G. Gates E 434; —, 6 April 2000, G. Gates E 825; —, 11 May 2000, G. Gates E 908a; —, 1 May 2001, G. Gates E 1149; 22 April 2004, M.E. Noordeloos 2004007; —, 28 March 2009, M.E. Noordeloos 2009081; Junee Caves State Reserve, 42° 44′ S, 146° 36′ E, 24 May 2003, G. Gates E 1820; Kermandie Falls Upper Track, 43° 12′ S, 146° 52′ E, 11 May 1999, G. Gates E 499; —, 26 Aug. 1999, G. Gates E 685; —, 30 May 2000, G. Gates E 954; —, 26 April 2001, G. Gates E 1136; Lake Skinner Track, 42° 57′ S, 146° 44′ E, 11 April 2004, G. Gates E 392; Lake St Clair National Park, Echo Point to Cynthia Bay, 42° 04′ S, 146° 09′ E, 24 March 2001, G. Gates E 1039; Marriotts Falls, 42° 43′ S, 146° 39′ E, 13 May 2000, G. Gates E 912; Mt Field National Park, Lady Barron Falls Track 42° 41′ S, 146° 42′ E, 22 Oct. 1998, G. Gates E 90; —, 1 April 1999, G. Gates E 349; —, 9 Nov. 1999, G. Gates E 736, G. Gates E 778; 19 July 2001, G. Gates E 1282; —, Lyrebird Walk, 42° 41′ S, 146° 40′ E, 1 April 1999, G. Gates E 331 (holotype, HO 548322; isotype, L); —, Tall Trees Track, 42° 41′ S, 146° 42′ E, 4 Nov. 1999, G. Gates E 723, G. Gates E 726; —, 13 Nov. 1999, G. Gates E 747; Sandy Bay, gully near Hytten Hall, 42° 54′ S, 147° 19′ E, 2 Aug. 1999, G. Gates E 699; Styx Valley, Big Tree, 42° 49′ S, 146° 39′ E, 19 March 2009, M.E. Noordeloos 200900 1; Mt. Wellington, Circle Track, 42° 55′ S, 147° 15′ E, 11 March 1999, G. Gates E 203; —, 30 Nov. 1999, G. Gates E 771 (HO 548323); —, Jacksons Bend, 25

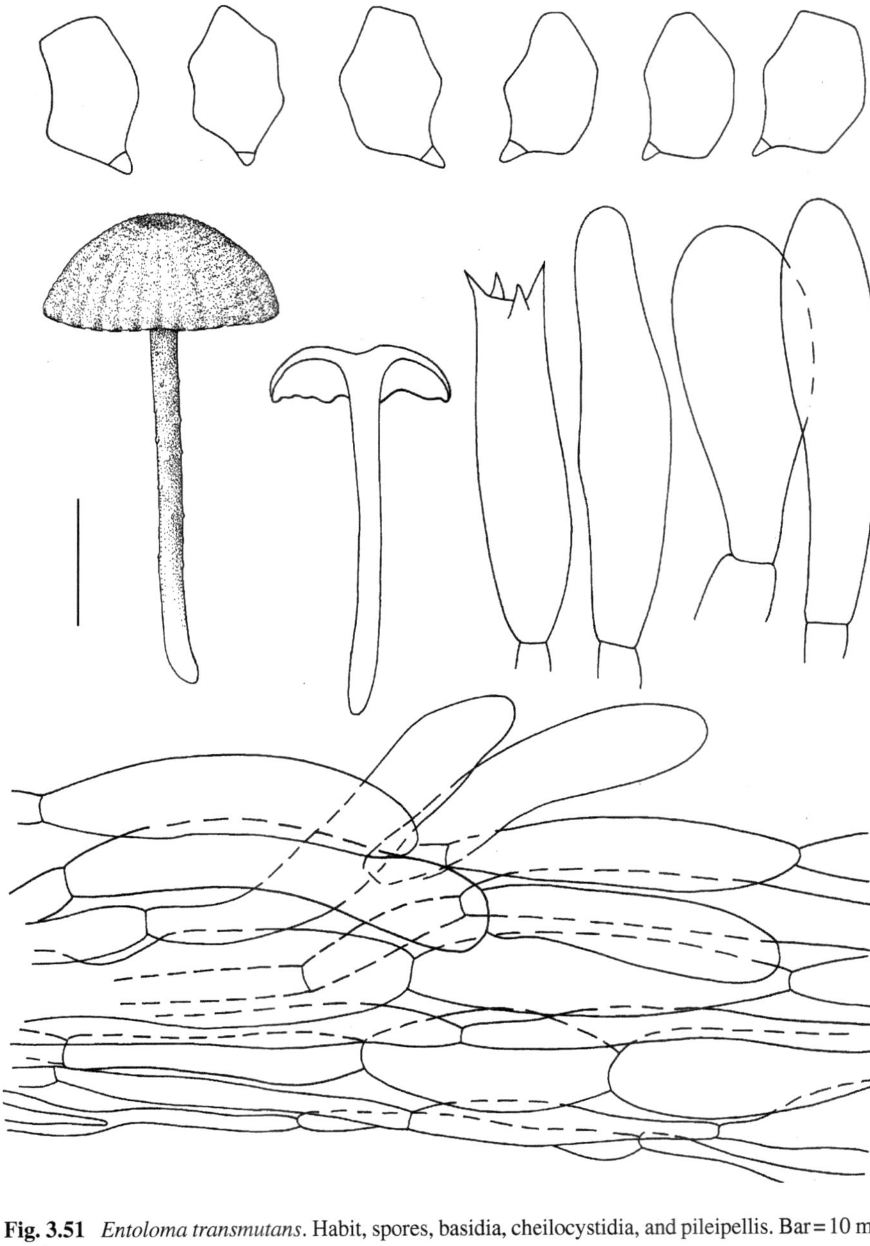

Fig. 3.51 *Entoloma transmutans*. Habit, spores, basidia, cheilocystidia, and pileipellis. Bar = 10 mm or 10 μm

March 1999, G. Gates E 291; —, Myrtle Gully, 42° 54′ S, 147° 15′ E, 9 March 1999, G. Gates E 178; —, 12 Aug. 1999, G. Gates E 678; —, 31 July 2001, G. Gates E 1267; —, 17 Jan. 2002, G. Gates E 1412; —, Old Farm Rd, 13 April 1999, G. Gates E 406; —, 6 Sept. 1999, G. Gates E 691; Silver Falls, 42° 55′ S, 147° 15′ E, 27 Feb. 1999, G. Gates E 155; —, 4 March 1999, G. Gates E 171; —, 5 May 2001, G. Gates

E 1167; —, 27 March 2003, G. Gates E 1659; North West Bay River, 42° 57′ S, 147° 12′ E, 14 Aug. 2001, G. Gates E 1274; —, 29 Aug. 2002, G. Gates E 1632; —, 1 May 2003, G. Gates E 1757; —, 18 May 2004, G. Gates E 2009; —, 21 April 2005, G. Gates E 2130; Wielangta, 42° 42′ S, 147° 51′ E, 15 Jan. 2002, G. Gates E 1401.

Entoloma transmutans is a remarkable little species in subgenus *Cyanula* with the delicate purple-pink background colour of the pileus and stipe and often pinkish brown lamella edge. It finds its place in stirps *Corvinum* (Noordeloos 1992, 2004), but differs from all species in that group in colour. The New Zealand species, *Entoloma melanocephalum* G. Stev. and *E. asprelloides* G. Stev., differ in colour, spore size and spore shape.

Stirps ***Austroprunicolor***

Pileus delicate violaceous pink or blue, contrasting with the pallid, whitish, polished stipe.

52. **Entoloma austroprunicolor** G.M. Gates & Noordel., Persoonia 19: 185. 2007.

Original diagnosis: Pileus 10–50 mm latus convexus demum expansus interdum umbonatus obscure coeruleo-violaceus interdum roseo-tinctus fibrilloso-velutinus demum minute squamulosus. Stipes 30–75×2–6 mm initio albus demum pallide griseo tinctus glabrus pseudopolitus vel leviter fibrillosus. Sporae 10–13(–16)×6.5–9.0(–10) μm. Acies lamellarum sterilis. Cheilocystidia 20–30×5–9 μm cylindracea vel clavata. Pileipellis trichoderma elementis usque ad 20 μm latis pigmentis intracellulosis. Fibulae absentes.

Holotype: Australia, Tasmania, Kermandie Falls, Lower Track, 43° 12′ S, 146° 52′ E, 5 Jan. 2002, G. Gates E 1389 (HO 543543; isotype in L).

Etymology: austro (Lat.) = southern; prunicolor (Lat.) = plum-coloured.

Main characters: habit collybioid; pileus reddish purple, fibrillose-velutinous then minutely squamulose all over, markedly contrasting with the pallid, almost white, glabrous or innately fibrillose stipe; spores large; cheilocystidia present; clamp-connections absent (Plate 3.50).

Description:
Pileus 10–50 mm diam, hemispheric to convex, expanding with age, often with slight umbo, margin deflexed, not hygrophanous, not translucently striate, dark blue to purple-blue (21E4) at first, then with reddish purple tinges, purplish grey or pinkish purple-grey (13E3), entirely fibrillose-velutinous at first, breaking up into small radially arranged fibrillose squamules. Lamellae adnate, segmentiform, rather crowded, up to 6 mm broad, moderately thick, white then pale pink with entire, concolorous edge. Stipe 30–75×2–6 mm, cylindrical with slight broadening at base, dry, brittle, fistulose with age, white or pale grey-violet, glabrous, innately fibrillose. Context purple in cortex of pileus, white in stipe and inner parts of pileus. Odour not distinctive. Taste usually none but sometimes slightly radishy or peppery.

Spores 10–13(–16)×6.5–9.0(–10) μm, Q = 1.3–1.8, Qav = 1.50, heterodiametric, 6–8-angled in side-view with pronounced angles. Basidia 33–40×9–14 μm, 4-spored, clampless. Lamella edge heterogeneous. Cheilocystidia 20–30×5–9 μm, irregularly

Plate 3.50 *Entoloma austroprunicolor* (Photo Michael Pilkington)

cylindrical, narrowly clavate or lageniform, rather thin-walled and inconspicuous. Pileipellis a cutis with transitions to a trichoderm, made up of cylindrical to inflated hyphae up to 20 μm in diameter, with clavate terminal elements, 40–125 × 11–25 μm. Pileitrama regular, consisting of rather narrow cylindrical hyphae 4.5–9 μm in diameter, with abundant brilliant granules. Pigment purple-brown, diffusely intracellular and in the form of agglutinated granules. Stipitipellis a cutis of loosely arranged, cylindrical hyphae, 2.0–7.0 μm in diameter. Caulocystidia absent. Clamp-connections absent (Fig. 3.52).

Habitat & distribution: widespread and common in wet sclerophyll forests. Fruiting spread from late spring to early winter (Group 2, in Chap. 1, see section "Fruiting Patterns of Species of Entolomataceae").

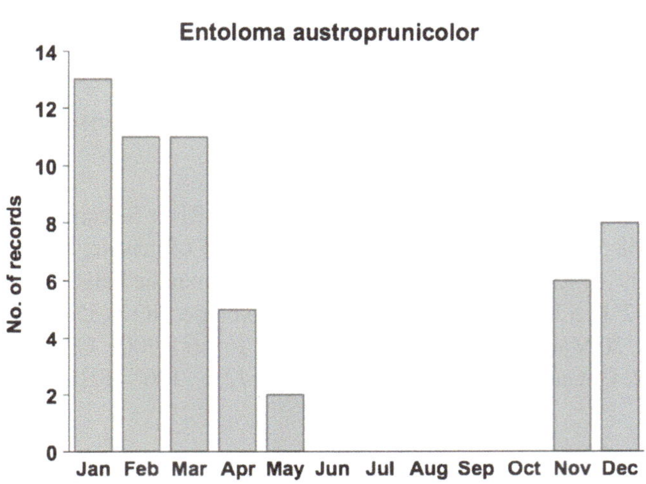

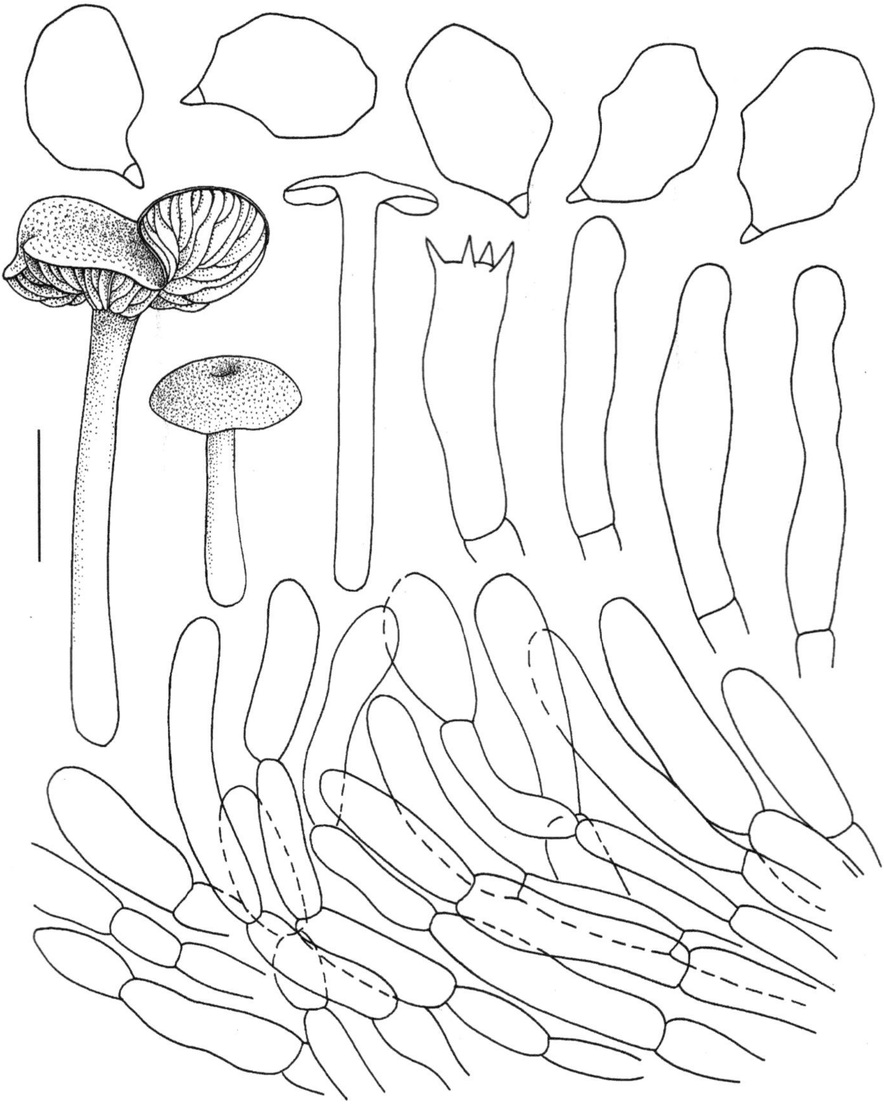

Fig. 3.52 *Entoloma austroprunicolor*. Habit, spores, basidia, cheilocystidia, and pileipellis. Bar = 10 mm or 10 μm

Collections examined. Australia, Tasmania, Arve River Streamside Reserve, 43° 10′ S, 146° 48.5′ E, 22 Dec. 2001, G. Gates E 1375; Collinsvale, Myrtle Forest, 42° 52′ S, 147° 09′ E, 15 April 1999, G. Gates E 427; Kermandie Falls, Lower Track, 43° 12′ S, 146° 52′ E, 5 Jan. 2002, G. Gates E 1389 (holotype, HO 543543; isotype, L); —, Upper Track, 30 March 1999, G. Gates E 316; —, 26 April 2001, G. Gates E 1134; —, 10 Jan. 2002, G. Gates E 1392; Liffey Falls, 41° 42′ S, 146° 46′ E, 10 Feb. 2004, G.

Gates E 1930; Mt Field National Park, Lady Barron Falls Track 42° 41′ S, 146° 42′ E, 1 April 1999, G. Gates E 339; —, 9 Nov. 1999, G. Gates E 738; Mt Wellington, Jacksons Track, 42° 55′ S, 147° 15′ E, 19 Feb. 2002, G. Gates E 1440; —, Myrtle Gully, 42° 54′ S, 147° 15′ E, 11 Jan. 1999, G. Gates E 118; —, 9 March 1999, G. Gates E 127; —, 17 Jan. 2002, G. Gates E 1415; Warra LTER site, 43° 06′ S, 146° 41′ E, 3 Feb. 2004, G. Gates E 1925; Wielangta, 42° 42′ S, 147° 51′ E, 20 March 1999, G. Gates E 118a; —, rainforest loop, 15 Jan. 2002, G. Gates E 1398; —, 26 Feb. 2002, G. Gates E 1445.

Entoloma austroprunicolor represents a species of subgenus *Cyanula* because of the general stature of the basidiocarp, clampless hyphae, and abundant brilliant granules. It has some similarity with *Entoloma queletii* (Boud.) Noordel. from Europe, which has a differently coloured vinaceous pink pileus tending to ochre when old, a white, fibrillose stipe and well differentiated cheilocystidia on a sterile lamella edge. In Largent (1994), *Entoloma austroprunicolor* would fit section *Albidicaules*, but it does not match any of the North American species.

Stirps ***Roseum***

Basidiocarps with delicate pink or deep carmine colour; blue or violaceous tinges absent.

53. **Entoloma austroroseum** G.M. Gates & Noordel., Persoonia 19: 187. 2007.

Original diagnosis: Pileus 10–26 mm latus convexus demum expansus leviter umbilicatus margine inflexus haud vel paulisper translucido-striatus intense roseus centro obcurioir glaber vel centro minute squamulosus. Stipes 11–20 × 1.5–3 mm pileo concolorus glabrus politus. Sporae 7.5–10 × 6.0–8.0 μm 5–7-angulatae heterodiametricae. Acies lamellarum fertilis. Cheilocystidia absentes. Pileipellis cutis vel trichoderma elementis latis pigmentis intracellulosis. Fibulae absentes.

Holotype: Australia, Tasmania, Collinsvale, Myrtle Forest, 42° 52′ S, 147° 09′ E, 19 April 2001, G. Gates E 1101 (HO 543542; isotype in L).

Etymology: austro (Lat.) = southern; roseum (Lat.) = pink.

Main characters: habit collybioid, basidiocarp delicate pink (Plate 3.51).

Description:

Pileus 10–26 mm diam, shallowly convex (to 3 mm high) to plano-convex with slight umbilicus, margin straight and entire, not hygrophanous, faintly translucently striate, pink (11B5) with darker centre (11B7), entirely finely squamulose. Lamellae adnate with small decurrent tooth, ventricose, moderately crowded, up to 3 mm broad, moderately thick, whitish or pale pink, sometimes with a pink edge. Stipe 11–20 × 1.5–3 mm, cylindrical, pale pink, dry, polished, with white basal tomentum. Odour spermatic. Taste sweetish.

Spores 7.5–10 × 6.0–8.0 μm, on average 8.5–9.3 × 6.1–7.3 μm, Q = 1.2–1.6, heterodiametric, 5–7-angled in side-view. Basidia 32–55(–65) × 6–12(–15) μm, 4-spored, clampless. Lamella edge fertile. Hymenophoral trama regular, made up of cylindrical elements, 60–100 × 5–9 μm, with abundant brilliant granules.

Plate 3.51 *Entoloma austroroseum* (Photos Michael Pilkington)

Pileipellis a trichoderm to a hymeniderm at centre, consisting of cylindrical to clavate terminal elements, 30–70 × 8–23 μm, with abundant pink intracellular pigment. Pileitrama regular, made up of cylindrical hyphae, 7–24 μm in diameter. Stipitipellis a cutis of narrow, cylindrical hyphae, 3–9 μm wide. Caulocystidia absent. Brilliant granules abundant. Clamp-connections absent (Fig. 3.53).

Habitat & distribution: known only from the type locality, a gully within a wet sclerophyll forest, from which collections were made on 11 different visits.

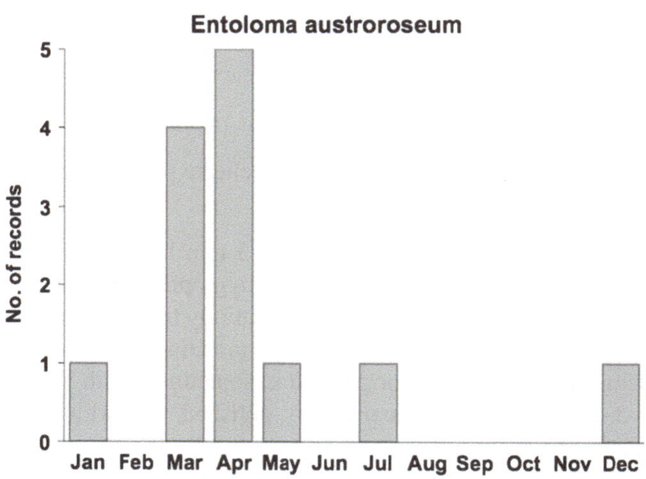

Collections examined. Australia, Tasmania, Collinsvale, Myrtle Forest 42° 52′ S, 147° 09′ E, 18 March 1999, G. Gates E 239; —, 15 April 1999, G. Gates E 423; —, 19 April 2001, G. Gates E 1101 (holotype, L; isotype, HO 543542); —, 14 April

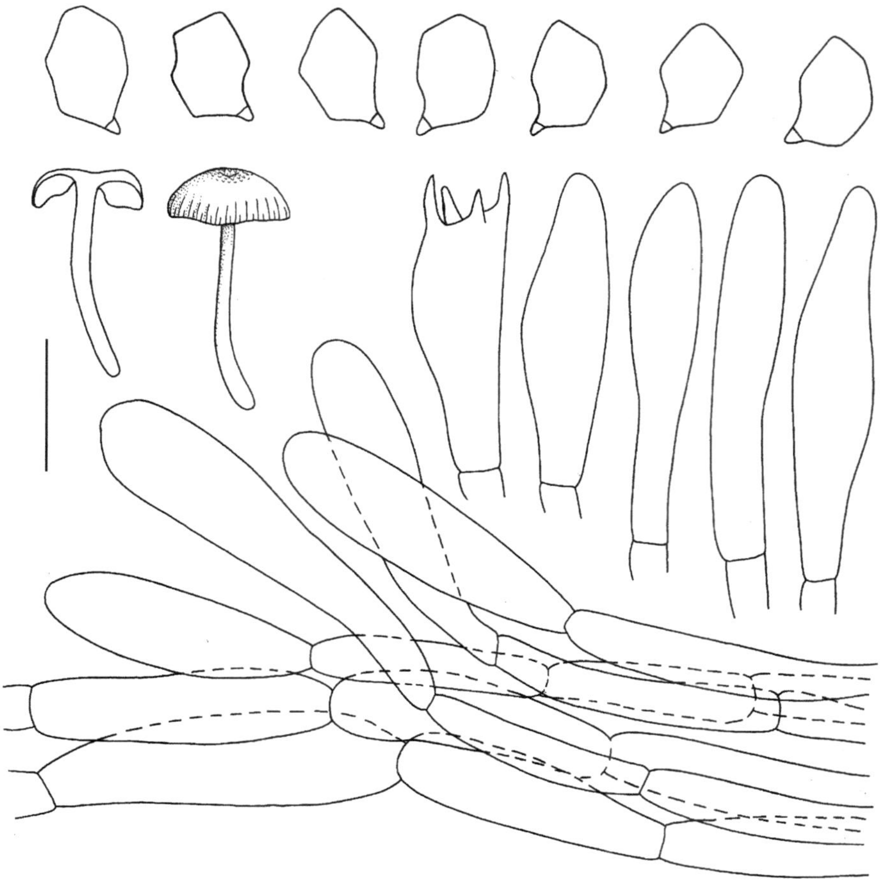

Fig. 3.53 *Entoloma austroroseum*. Habit, spores, basidia, cheilocystidia, and pileipellis. Bar = 10 mm or 10 μm

2003, G. Gates E 1705; —, 10 March 2005, G. Gates E 2085; —, 29 March 2005, G. Gates E 2107; —, 3 Jan. 2006, G. Gates E 2232.

Entoloma austroroseum is very similar to *E. roseum* (Longyear) Hesler from the Northern Hemisphere, but the cheilocystidia are differently shaped. No similar species is described in the works of Horak (1973, 1980, 2008) from Australasia and New Zealand.

54. **Entoloma carminicolor** G.M. Gates & Noordel., Persoonia 19: 188. 2007.

Original diagnosis: Pileus 12–20 mm latus convexus demum expansus umbilicatus margine inflexus obscure rufocarmineus minute squamulosus. Lamellae adnexae roseae rufo margine fimbriato. Stipes 30–35 × 3–5 mm pileo concolorus glabrus politus. Sporae 7.5–10 × (5–)5.5–7(–7.5) μm 5–7-angulatae heterodiametricae. Acies lamellarum sterilis. Cheilocystidia 20–43 × 8–14 μm cylindracea vel clavata.

Plate 3.52 *Entoloma carminicolor* (Photo Genevieve Gates)

Pileipellis cutis vel trichoderma elementis latis pigmentis intracellulosis. Fibulae absentes.

Holotype: Australia, Tasmania, Donnellys Road, 43° 07′ S, 146° 54′ E, 3 April 1999, G. Gates E 354 (HO 543541; isotype in L).

Etymology: carminicolor (Lat.) = carmine-coloured.

Main characters: basidiocarps collybioid; fruit body carmine red; lamella edge sterile, carmine red (Plate 3.52).

Description:
Pileus 12–20 mm diam, 3–4 mm high, plano-convex with umbilicate centre, with deflexed margin, not hygrophanous, not translucently striate, uniformly deep carmine red to ruby pink (12F6), dry, finely squamulose at centre only. Lamellae adnate with decurrent tooth, segmentiform, moderately crowded, up to 3 mm broad, narrow, grey-pink with carmine red flocculose edge. Stipe 30–35 × 3–5 mm, cylindrical, or laterally flattened and longitudinally grooved, paler than pileus but with similar carmine colour, glabrous, dry, brittle, polished, base with white tomentum. Odour spermatic. Taste sweetish, saliva inducing.

Spores 7.5–10 × (5.0–)5.5–7(–7.5) µm, on average 8.5–9.3 × 6.1–6.5 µm, Q = 1.2–1.6, usually 5–7-angled in side-view with pronounced angles. Basidia 22–34 × 8–11 µm,

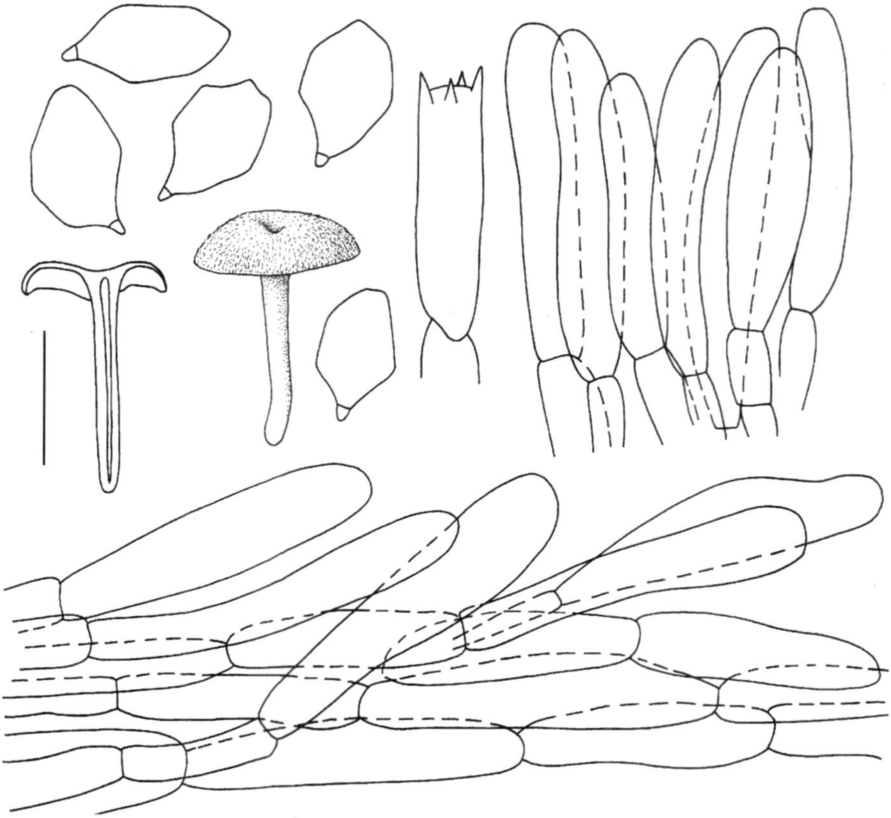

Fig. 3.54 *Entoloma carminicolor*. Habit, spores, basidia, cheilocystidia, and pileipellis. Bar = 10 mm or 10 μm

4-spored, clampless. Lamella edge entirely sterile, of the serrulatum-type, composed of a dense strand of hyphae running along the edge with, at irregular intervals, dense bundles of clavate, cylindrical, or lageniform cheilocystidia 20–43 × 8–14 μm, thin-walled, except for the apical part, which is slightly thickened and refringent, with granulose, pale carmine, intracellular pigment. Hymenophoral trama regular, made up of cylindrical elements, 60–120 × 5–15 μm, often with some pale brownish intracellular pigment. Pileipellis a cutis with transitions to a trichoderm, made up of repent, cylindrical hyphae, 4–14 μm in diameter, with clavate or subclavate terminal elements, 34–70 × 5–16 μm. Pigment pinkish red, intracellular in pileipellis and upper pileitrama. Stipitipellis a cutis of narrow, cylindrical hyphae, 4–9 μm wide. Caulocystidia absent. Brilliant granules present in trama. Clamp-connections absent (Fig. 3.54).

Habitat & distribution: known only from a single site, where a marsupial lawn borders a wet sclerophyll forest near Geeveston, Tasmania. Collections were made during six separate visits.

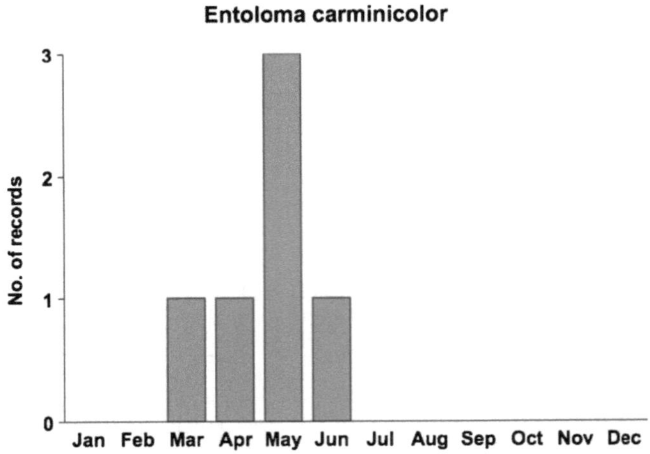

Collections examined. Australia, Tasmania, Donnellys Road, 43° 07′ S, 146° 54′ E, 16 March 1999, G. Gates E 236; —, 3 April 1999, G. Gates E 354 (holotype, HO 543541; isotype, L); —, 30 May 2001, G. Gates E 930; —, 19 May 2005, G. Gates E 2168.

Entoloma carminicolor is a characteristic species with its deep carmine red basidiocarps and flocculose (in part), coloured lamella edge. Similar species from other regions are *E. puroides* E. Horak, described from Papua New Guinea, with a smooth pileus and brown lamella edge, and the European *E. rufocarneum* (Berk.) Noordel., which differs, among other things, by having larger spores and a concolorous, fertile lamella edge.

Stirps ***Obscureovirens***

Basidiocarps with dark green to olivaceous tinges, particularly when fresh.

55. **Entoloma obscureovirens** G.M. Gates & Noordel., Persoonia 19: 206. 2007.

Original diagnosis: Pileus 20–48 mm latus convexus demum expansus umbilicatus margine inflexus obscure griseobrunneus vel griseoater interdum virido tinctus centro velutino demum squamuloso versus margine virgatus. Lamellae roseae griseo-tinctae margine fimbriato. Stipes 30–60 × 3–10 mm pileo griseovirens fibrillosus. Sporae 8.5–12 × 6.0–8.0 µm 6-angulatae heterodiametricae. Acies lamellarum sterilis. Cheilocystidia 40–85 × 9–20 µm cylindracea vel clavata. Pileipellis cutis vel trichoderma elementis 45–90 × 11–22 µm pigmentis intracellulosis. Fibulae absentes.

Holotype: Australia, Tasmania, Wielangta, 42° 42′ S, 147° 51′ E, 5 May 2005, G. Gates E 2157 (HO 543540; isotype in L).

Etymology: obscureus (Lat.) = dark; virens (Lat.) = becoming green.

Main characters: rather stout, dark coloured *Cyanula* (collybioid); pileus opaque, virgate, grey-brown; stipe greenish grey, fibrillose; taste soapy-acrid (Plate 3.53).

Plate 3.53 *Entoloma obscureovirens* (Photo Michael Pilkington)

Description:
Pileus 20–50 mm diam, plano-convex when young, becoming applanate with age with a slightly depressed centre, finally somewhat concave, with straight margin, not hygrophanous, not translucently striate, uniformly very dark charcoal grey or grey-brown, often with green hue when young, becoming olivaceous grey-brown with brownish grey centre when old, radially virgate-fibrillose with velutinous then minutely squamulose centre, becoming somewhat paler with age (5E6). Lamellae adnexed to free, segmentiform, crowded, up to 9.5 mm broad, thin, flesh-coloured pink with grey tinge, becoming yellowish with age, with entire, concolorous edge, lamellulae in two tiers. Stipe 30–60 × 3–10 mm, cylindrical or slightly broadened towards base, or laterally compressed with longitudinal groove, dry, stuffed or fistulose, grey-green (1D3–2), often bright yellow at base, usually rather distinctly longitudinally fibrillose with loose fibrils, base whitely tomentose. Odour indistinct. Taste rancid, soapy, burning.

Spores 8.5–11 × 6.0–8.0 μm, Q = 1.2–1.7, heterodiametric with (5–)6–7(–8) angles in side-view. Basidia 20–40 × 8.0–12 μm, 4-spored, clampless. Lamella edge sterile, composed of dense clusters of cylindrical to clavate or fusiform cheilocystidia, 40–85 × 9–20 μm, thin-walled, originating from a dense band of hyphae running along the lamella edge (=serrulatum-type). Hymenophoral trama regular, made up of cylindrical hyphae, 4–20 μm in diameter, with brilliant granules. Pileipellis a cutis with transitions to a trichoderm, made up of clavate terminal elements,

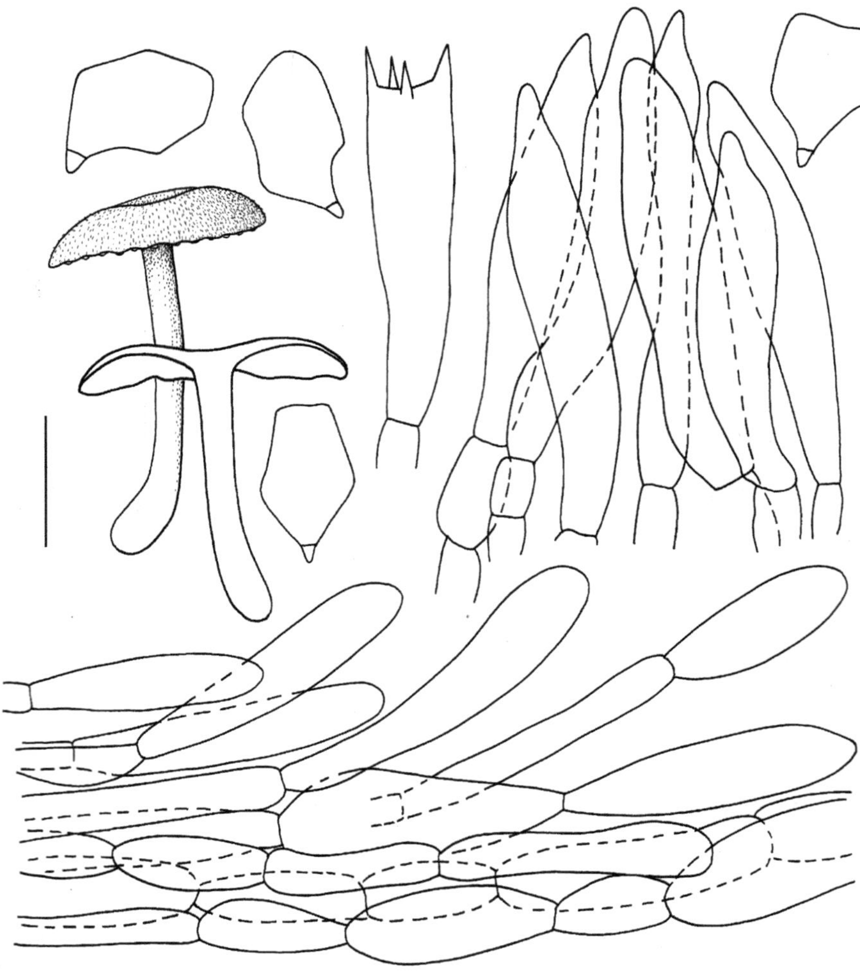

Fig. 3.55 *Entoloma obscureovirens*. Habit, spores, basidia, cheilocystidia, and pileipellis. Bar = 10 mm or 10 μm

45–90 × 11–22 μm, with abundant, brown, intracellular pigment. Pileitrama regular, made up of cylindrical to slightly inflated hyphae, to 30 μm in diameter, with brilliant granules and scattered vascular hyphae. Stipitipellis a cutis with transitions to a trichoderm, made up of clavate to fusiform terminal elements (caulocystidia), 30–80 × 6.0–19 μm, with dark grey-brown, intracellular pigment. Clamp-connections absent (Fig. 3.55).

Habitat & distribution: widespread but infrequently encountered in wet sclerophyll forests.

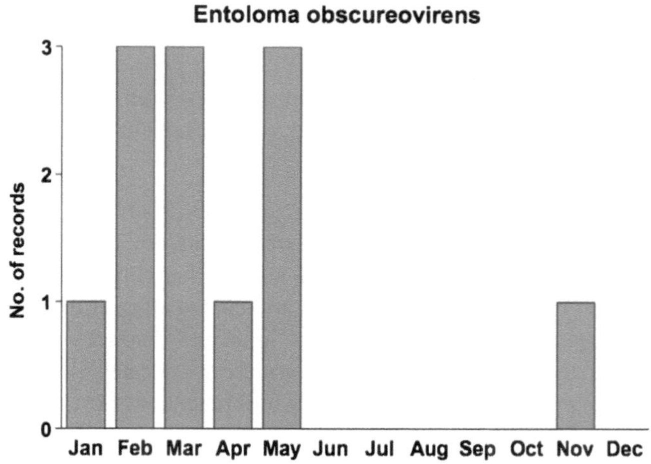

Collections examined. Australia, Tasmania, Dismal Swamp, 40° 57′ S, 144° 51′ E, 5 May 2011, G. Gates (M.E. Noordeloos 2011035); Taranna Forest Walk, 43° 04′ S, 147° 53′ E, 10 May 2003, G. Gates E 1787; Waterfall Bay, 43° 04′ S, 147° 57′ E, 13 March 1999, G. Gates E 209; Wielangta, 42° 42′ S, 147° 51′ E, 23 Feb.1999, G. Gates E 132; —, 20 March 1999, G. Gates E 268; —, 23 Nov. 2000, G. Gates E 1022;—, 15 Jan. 2002, G. Gates E 1399; —, 5 May 2005, G. Gates E 2157 (holotype, HO 543540; isotype, L).

Entoloma obscureovirens has been observed at the type locality for many years, and its variability is therefore well known. The stout habit, dark pileus, greenish tinges (particularly in the stipe), fibrillose stipe surface and hot burning taste are distinctive. It belongs to section *Cyanula*, and keys out near the European species *E. griseoviridulum* Courtec. (Noordeloos 2004), which differs in having a less fibrillose, slightly differently coloured stipe. The dark colours, especially when the stipe base is yellowish, could confuse *E. obscureovirens* with *E. tasmanicum*, which can readily be distinguished by the bluish lamellae with dark, serrulate edge. No similar species could be found that were described by Horak (1973, 1978, 1980, 2008).

Stirps ***Albidae***

Basidiocarps pure white, at least when fresh, sometimes becoming yellowish ochre with age.

56. **Entoloma albidosimulans** G.M. Gates & Noordel., Persoonia 19: 208. 2007.

Original diagnosis: Habitus collybioideus. Pileus 10–35 mm latus albus tomentosus. Stipes 20–30×2–3 mm albus politus. Sporae 8.5–11×7.0–8.5 µm 6–7-angulatae. Acies lamellarum sterilis. Cheilocystidia 20–50×5.0–9.5 µm cylindracea vel clavata. Pileipellis trichodermalis elementis 4.0–14 µm latis elementis terminals ad 20 µm latis, epigmentatis. Fibulae adsunt.

Holotype: Australia, Tasmania, Growling Swallet, 42° 41′ S, 146° 30′ E, 22 April 2004, M.E. Noordeloos 2004065 (CTAB145) (HO 543556; isotype in L).

Etymology: albidus (Lat.) = whitish; simulans (Lat.) = resembling, referring to the likeness with *Entoloma sericellum.*

Main characters: small, white collybioid species, resembling *E. sericellum* (Fr.) P. Kumm. but with 6- or 7-angled spores and a sterile lamella edge (Plate 3.54).

Description:
Pileus 10–35 mm diam, convex, becoming applanate with age, with or without slight central depression, with deflexed then straight margin, not hygrophanous, not translucently striate, pure white when fresh, slowly changing to pale ochre yellow or pink with age (10YR 8/4; 4A2–3), finely tomentose all over, becoming radially fibrillose with age. Lamellae adnate, often slightly to distinctly emarginate with distinct decurrent tooth, segmentiform to ventricose, moderately distant, up to 6 mm broad, white then pink, with entire or slightly fimbriate, concolorous edge, L = up to 40, l = 3–7. Stipe 20–30 × 2–3 mm, cylindrical, often distinctly broadened towards base, white to yellowish, innately fibrillose appearing almost smooth and polished, glabrous. Odour indistinct or pleasant. Taste mild.

Spores 8.5–11 × 7.0–8.5 μm, on average 10.0–10.6 × 7.5–7.7 μm, 6- or 7-angled with rather simple angles. Basidia 18–30 × 7.0–11 μm, 4-spored, clampless. Lamella edge sterile, with dense clusters of fusiform, lageniform, cylindrical to narrowly clavate cheilocystidia, 20–50 × 5.0–9.5 μm, interspersed infrequently with scattered basidia. Brilliant granules not observed or sparse. Pileipellis a cutis with transitions to a trichoderm, made up of cylindrical hyphae, 4.0–14 μm in diameter, with clavate terminal elements up to 20 μm in diameter. Pigment absent. Clamp-connections absent or very rarely present in hymenium (Fig. 3.56).

Habitat & distribution: widespread in wet sclerophyll forests and rainforests. Fruiting throughout the year, with a fair number of records in the spring and summer months (Group 5, in Chap. 1, see section "Fruiting Patterns of Species of Entolomataceae").

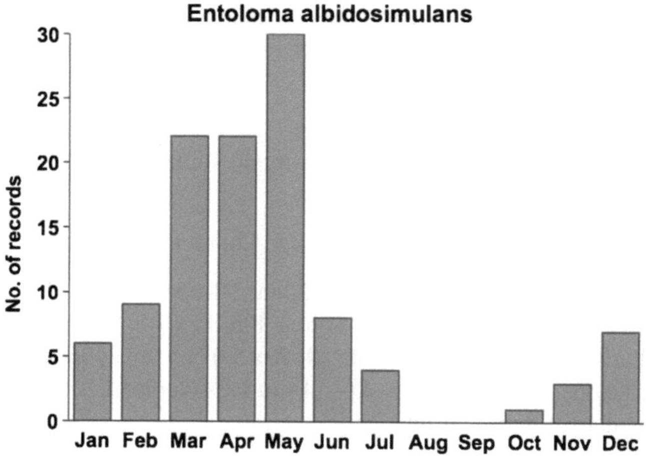

Plate 3.54 *Entoloma albidosimulans* (Photos Michael Pilkington (*top*), Machiel Noordeloos (*bottom*))

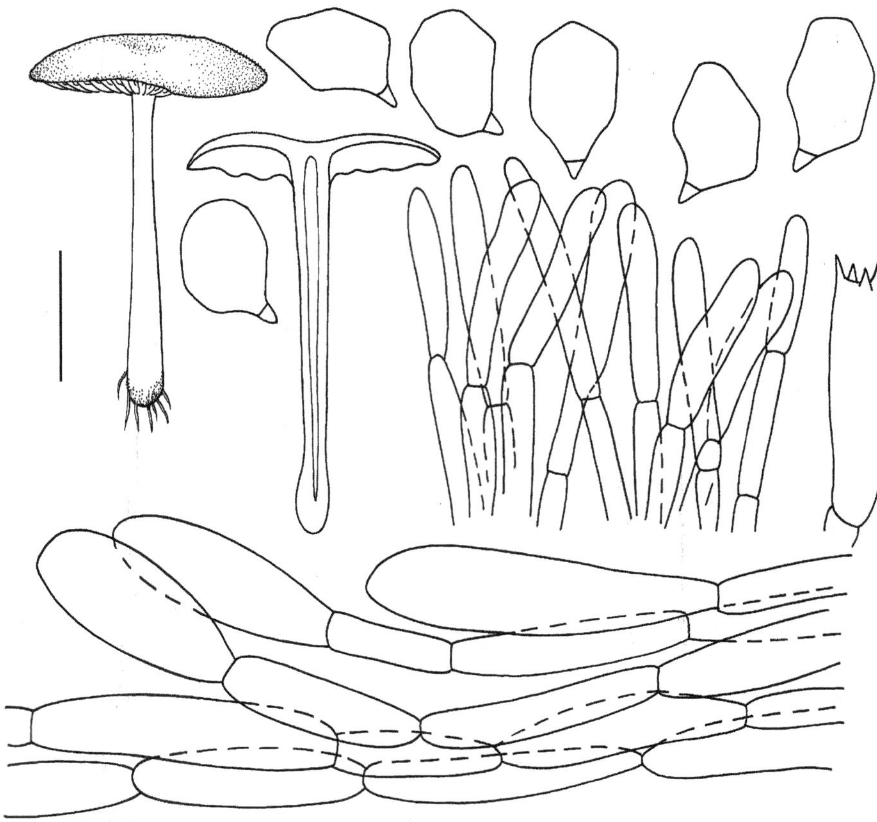

Fig. 3.56 *Entoloma albidosimulans*. Habit, spores, basidia, cheilocystidia, and pileipellis. Bar = 10 mm or 10 μm

Collections examined. Australia, Tasmania, Arve River Streamside Reserve, 43° 10′ S, 146° 48.5′ E, 7 July 1998, G. Gates E 6; Arve River, Zig-Zag Track, 43° 06′ S, 146° 45′ E, 15 May 1999, G. Gates E 512; —, 22 Dec. 2001, G. Gates E 1372; Collinsvale, Myrtle Forest, 42° 52′ S, 147° 09′ E, 18 March 1999, G. Gates E 258; —, 15 April 1999, G. Gates E 430; Creepy Crawly, 42° 50′ S, 146° 22′ E, 15 April 2000, G. Gates E 850; Dalco Creek, 43° 23′ S, 147° 01′ E, 15 March 2009, M.E. Noordeloos 2009010; Dismal Swamp, 40° 57′ S, 144° 51′ E, 5 May 2011, M.E. Noordeloos 2011041 (L); Donnellys Rd. Geeveston, 43° 07′ S, 146° 54′ E, 16 March 1999, G. Gates E 237; Evercreech Forest Reserve, 41° 24′ S, 147° 58′ E, 16 June 2003, G. Gates E 1880; Growling Swallet, 42° 41′ S, 146° 30′ E, 19 May 1999, G. Gates E 521; —, 6 April 2000, G. Gates E 822; —, 11 May 2000, G. Gates E 908; —, 1 May 2001, G. Gates E 1154; —, 11 April 2002, G. Gates E 1473; —, 13 May 2003, G. Gates E 1793; —, 22 April 2004, M.E. Noordeloos 2004065 (CTAB145) (holotype, HO 543556; isotype, L); Kermandie Falls, Upper Track, 43° 12′ S, 146° 52′ E, 6 May 1999, G. Gates E 317; —, 6 May 1999, G. Gates E 475; Liffey Falls,

41° 42′ S, 146° 46′ E, 28 March 1999, G. Gates E 299; Marriotts Falls, 42° 43′ S, 146° 39′ E, 29 May 2003, G. Gates E 1838; Mt Field National Park, Falls Loop, 42° 41′ S, 146° 42′ E, 31 May 2003, G. Gates E 1848; —, Lyrebird Walk, 42° 41′ S, 146° 40′ E, 1 April 1999, G. Gates E 329; —, Tall Trees Track, 42° 41′ S, 146° 42′ E, 24 Nov. 2005, G. Gates E 2215; Mt Wellington, Jacksons Track, 42° 55′ S, 147° 15′ E, 14 Oct. 1998, G. Gates E 74; —, Myrtle Gully, 42° 54′ S, 147° 15′ E, 8 July 1998, G. Gates E 29; —, 13 April 1999, G. Gates E 400; Scottsdale, Forester Road, 41° 06′ S, 147° 37′ E, 29 April 2006, M.E. Noordeloos 2006034; Styx Valley, Big Tree, 42° 49′ S, 146° 39′ E, 19 March 2009, M.E. Noordeloos 2009021; Timbs Track, 42° 44′ S, 146° 25′ E, 27 March 2004, G. Gates E 1962; Wielangta, 42° 42′ S, 147° 51′ E, 23 Feb. 1999, G. Gates E 139; —, 18 Dec. 1999, G. Gates E 782; —, 15 Jan. 2002, G. Gates E 1396.

The present species was originally identified as *E. sericellum* in the field because of the general colour and habit of the basidiocarps. Microscopically, however, there are quite striking differences, particularly with regard to the sterile lamella edge and clampless hyphae. Also, molecular data show that it has a distant position with regard to *E. sericellum* (Co-David et al. 2009). The following New Zealand species are similar (Horak 1973, 2008): *Entoloma niveum* G. Stev. differs strikingly in the fibrillose, translucently striate pileus, smaller and narrower spores, capitate cheilocystidia, and encrusting pigment in the pileipellis (type studied, K). *Entoloma peralbidum* E. Horak has abundant clamp-connections, encrusting pigment, and somewhat slenderer spores. *Entoloma parasericellum* Corner & E. Horak from Sabah has larger spores, a simple cutis-like pileipellis, and very long, cylindrical cheilocystidia. *Entoloma neosericellum* E. Horak (2008) from New Zealand is very similar, but differs by the presence of lageniform pleurocystidia and absence of clamp-connections.

57. **Entoloma totialbum** G.M. Gates & Noordeloos, Cryptogamie Mycologie 30: 120. 2009.

Original diagnosis: Pileus 11–34 mm latus, convexus umbilicatus, haud hygrophanus, haud translucido-striatus, albus demum leviter roseotinctus, glabrus. Lamellae adnatae, confertae, albae demum roseae. Stipes 27–40 × 2–5 mm, cylindraceus, albus, glaber sed non politus. Sporae 8–11 × 6–8 μm, polyangulatae vel nodulosae. Acies lamellarum sterilis. Cheilocystidia 27–75 × 4–11 μm, lageniformia. Pileipellis cutis ex elementis cylindraceis vel leviter inflatis, 10–15 μm latis, constituta pigmentis nullis. Granula lucentia absentia. Fibulae abundantes.

Holotype: Australia, Tasmania, University of Tasmania, gully near Hytten Hall, 42° 54′ S, 147° 19′ E, 15 May 2006, G. Gates E 2239 (HO 548319; isotype L).

Etymology: totus (Lat.) = totally; albus (Lat.) = white, referring to the totally white basidiocarps.

Main characters: all white, collybioid species with large, nodulose spores and long, protruding cheilocystidia (Plate 3.55).

Plate 3.55 *Entoloma totialbum* (Photo Genevieve Gates)

Description:
Pileus 11–34 mm diam, convex, expanding with age, centrally depressed, with straight, entire margin, not hygrophanous, not translucently striate, white becoming pink-tinged, glabrous, dry, slightly aeriferous. Lamellae adnate with decurrent tooth, segmentiform to subarcuate, crowded, up to 5 mm broad, moderately thick, white then pink, with serrulate edge. Stipe 27–40 × 2–5 mm, equal, cylindrical, white, slightly yellowing at base, apex pruinose, downwardly glabrous but not polished, fistulose. Odour fruity then slightly farinaceous when cut. Taste not known.

Spores 8–11 × 6–8 μm, Q = 1.2–1.7, Qav = 1.45, irregularly nodulose-angled, with 7–9 blunt angles in outline. Basidia 4-spored, clamped. Lamella edge heterogeneous with abundant, large, protruding cheilocystidia, 27–75 × 4–11 μm, narrowly to broadly lageniform. Hymenophoral trama regular, made up of inflated elements, 40–200 × 5–10 μm. Brilliant granules absent. Pileipellis a differentiated cutis of cylindrical hyphae, 10–15 μm wide. Stipitipellis a cutis of narrow, cylindrical hyphae 5–12 μm wide. Caulocystidia absent. Clamp-connections abundant (Fig. 3.57).

Habitat & distribution: widespread in a range of habitats, but infrequently encountered.

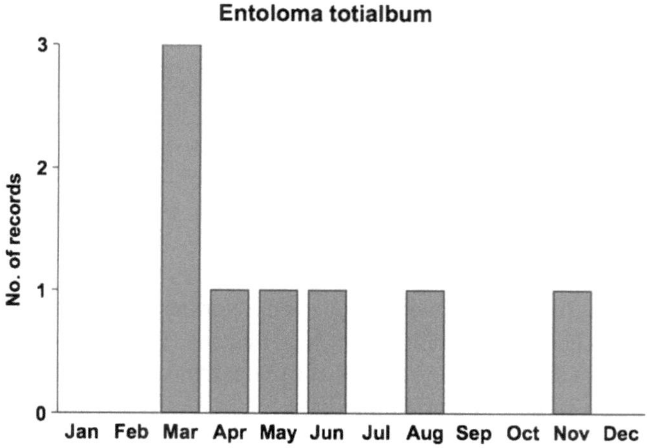

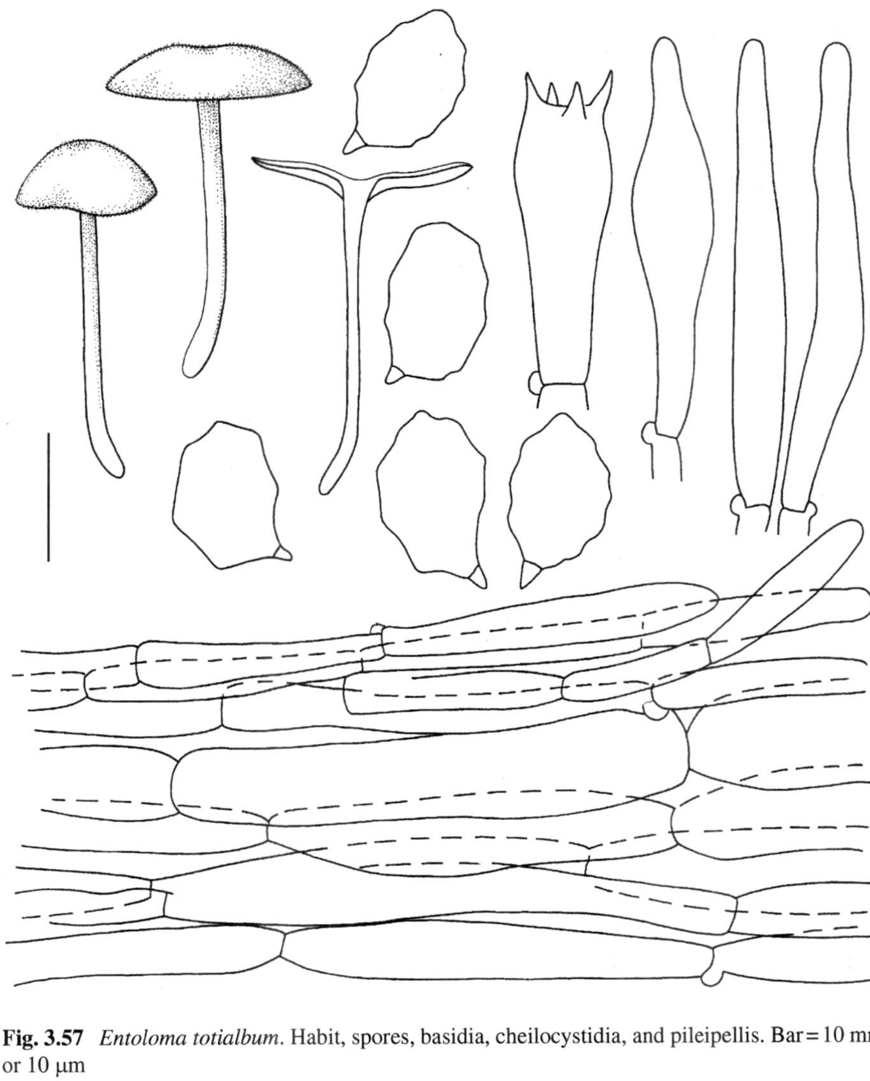

Fig. 3.57 *Entoloma totialbum*. Habit, spores, basidia, cheilocystidia, and pileipellis. Bar = 10 mm or 10 μm

Collections examined: Australia, Tasmania, Collinsvale, Myrtle Forest, 42° 52′ S, 147° 09′ E, 26 March 2011, G. Gates E 2288; Mt Field NP, Lady Barron Falls Track, 42° 41′ S, 146° 42′ E, 1 April 1999, G. Gates E 352; North West Bay River, 42° 57′ S, 147° 12′ E, 16 June 2001, G. Gates E 1250; —, 29 Aug. 2002, G. Gates E 1630; Sandy Bay, University of Tasmania, gully near Hytten Hall, 42° 54′ S, 147° 19′ E, 15 May 2006, G. Gates E 2239 (holotype, HO 548319; isotype, L).

At first sight *Entoloma totialbum* is reminiscent of *E. sericellum* (Fr.) P. Kumm., a common species from the Temperate Zone of the Northern Hemisphere (Noordeloos 2004), which has also been recorded from New Zealand (Horak 2008). However, that species often has a distinct yellow tinge in the pileus and stipe, and differently shaped spores. *Entoloma albidosimulans* G.M. Gates & Noordel. from Tasmania (Gates and Noordeloos 2007) has a completely sterile lamellae edge, regularly shaped spores with 6–7 angles, and clamp-connections are lacking or very scarce. *Entoloma peralbidum* E. Horak from New Zealand differs by the shape of the spores and presence of encrusting pigment in the hyphae of the pileipellis. *Entoloma niveum* G. Stev. from New Zealand has differently shaped, larger spores, less abundant clamp-connections, and strangulated cheilocystidia. The macroscopically similar white species from Australasia in Horak (1980) lack the above combination of characters. Manimohan et al. (1995, 2006) described a number of white species from Kerala State, India, which differ consistently from *Entoloma totialbum*, particularly with regard to the pink tinge in the mature pileus and size and shape of the spores.

Stirps ***Roseoluteum***

Basidiocarps with pale pink-violet pileus and pale yellow stipe.

58. **Entoloma roseoluteolum** G.M. Gates & Noordel., Persoonia 19: 209. 2007

Original diagnosis: Habitus collybioideus. Pileus 25–35 mm latus roseus leviter violaceo-tinctus leviter translucidostriatus, radialiter rimoso-fibrillosus centro squamulosus. Stipes 60–65 × 4.0–5.5 mm albido-flavus politus. Sporae 10–12 × (6.0–)7.0–8.0 µm (5–)6–7-angulatae. Acies lamellarum fertilis. Pileipellis trichodermalis elementis 11–22 µm latis pigmentis intracellulosis. Fibulae adsunt.

Holotype: Australia, Tasmania, Kermandie Falls, Lower Track, 43° 12′ S, 146° 52′ E, 9 Jan. 2005, G. Gates E 2074 (HO 543539; isotype in L).

Etymology: roseus (Lat.) = pink; luteolum (Lat.) = yellowish, referring to the colours of the basidiocarp.

Main characters: slender *Leptonia*; pileus pink, tinged with violet, finely squamulose at centre; stipe polished, yellowish; spores large; cheilocystidia absent (Plate 3.56).

Description:

Pileus 25–35 mm diam, plano-convex with depressed to umbilicate centre, with straight margin, not hygrophanous, slightly translucently striate at margin, pink (7A3) with pale violet (18A4) margin, centre darker reddish pink or Venetian red

Plate 3.56 *Entoloma roseoluteolum* (Photos Genevieve Gates)

(8E6), breaking up into small squamules at centre, radially virgate-fibrillose towards margin. Lamellae adnate, moderately crowded, segmentiform, up to 4 mm broad, thin, pink, with concolorous, entire edge, lamellulae in two tiers. Stipe 60–65 × 4.0–5.5 mm, cylindrical with slightly enlarged base, slender, equal or flexuous, very pale yellowish or whitish (3A2), polished. Context thin, pinkish in pileus, pale yellow in stipe, more intensely yellowish in stipe base. Odour not distinctive. Taste saliva inducing.

Spores 10–12 × (6.0–)7.0–8.0 μm, (5–)6–7-angled in side-view. Basidia 15–22 × 7.0–11 μm. Lamella edge fertile. Cheilocystidia absent. Hymenophoral trama regular, made up of cylindrical to slightly inflated elements to 18 μm in diameter with abundant brilliant granules. Pileipellis a cutis with transitions to a trichoderm of cylindrical to inflated hyphae, with clavate terminal elements, 30–70 × 11–20 μm, containing pale pinkish brown intracellular pigment. Pileitrama regular, made up of inflated elements, 40–120 × 6.0–20 μm, with abundant brilliant granules. Vascular hyphae abundant. Stipitipellis a cutis of cylindrical hyphae 5.0–9.0 μm in diameter. Clamp-connections absent (Fig. 3.58).

Habitat & distribution: known only from the type locality, a wet sclerophyll forest along a river.

Fig. 3.58 *Entoloma roseoluteolum*. Habit, spores, basidia, and pileipellis. Bar = 10 mm or 10 μm

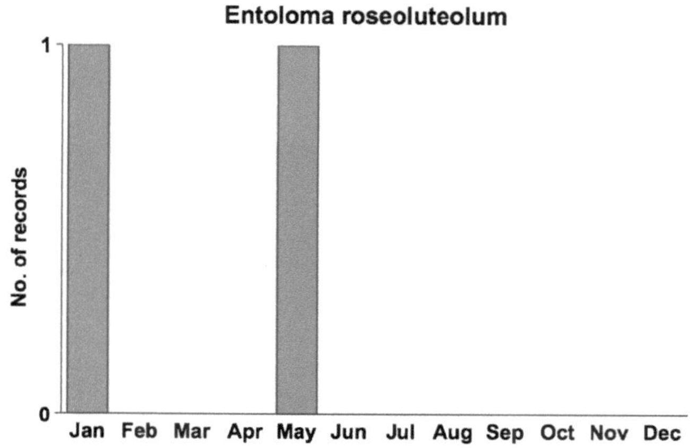

Collection examined. Australia, Tasmania, Kermandie Falls, Lower Track, 43° 12′ S, 146° 52′ E, 23 May 2000, G. Gates E 935; —, 9 Jan. 2005, G. Gates E 2074 (holotype, HO 543539; isotype, L).

Entoloma roseoluteolum is distinctive because of the pinkish pileus with violaceous tinges and pallid, polished stipe. It clearly belongs to section *Cyanula* on account of the brilliant granules in the trama and clampless hyphae. There are no similar species known from the Southern Hemisphere. *Entoloma ritae* Noordel. & Wölfel from Europe is somewhat similar, but differs by the lack of violaceous tinges in the pileus and the abundance of clamp-connections.

Stirps ***Sarcitulum***

Basidiocarps yellow to yellow-brown, pileus translucently striate.

Morphospecies with yellowish to yellow-brown pileus occur widely all over the globe, and show a remarkable variability (Largent 1977, 1994; Noordeloos 2004; Horak 1980; Romagnesi and Gilles, 1979). They show a rather big variation, both in macroscopical characters (striation of the pileus, colour of the lamellae and lamella edge, colour of stipe), and microscopy (size and shape of spores, presence and shape of cheilocystidia). During this study, numerous collections have been studied, and we tried to distinguish clear-cut taxa. However, some question marks remain, and some collections could not be accommodated satisfactorily, needing more material to study.

59. **Entoloma aurantiolabes** G.M. Gates & Noordel., Persoonia 19: 211. 2007.

Original diagnosis: Habitus collybioideus. Pileus usque ad 30 mm latis brunneo violaceo-tinctus demum ochraceus translucido-striatus radialiter fibrillosus centro squamulosus. Stipes 20–50 × 2–4 mm initio violaceogriseo tinctus demum pallide brunneus politus. Toto aurantio labes. Sporae 10–12 × 7.5–9.0 μm, 5–7-angulatae. Cheilocystidia 19–40 × 6.0–13 μm cylindracea vel clavata. Pileipellis trichodermalis elementis 8–15 μm latis pigmentis intracellulosis. Fibulae desunt.

Holotype: Australia, Tasmania, North West Bay River, 42° 57′ S, 147° 12′ E, 18 May 2004, G. Gates E 2000 (HO 543544; isotype in L).

Plate 3.57 *Entoloma aurantiolabes* (Photos Michael Pilkington and Machiel Noordeloos)

Etymology: aurantius (Lat.) = orange; labes (Lat.) = spotted.

Main characters: typical *Cyanula* (collybioid) with violet-blue tinges in pileus and stipe, which fade with age; all parts stain with orange splotches when bruised or when old (Plate 3.57).

Description:
Pileus 10–40 mm diam, conico-campanulate, sometimes truncate expanding to convex or plano-convex with slightly to distinctly depressed centre, slightly to distinctly hygrophanous, translucently striate when moist to half the radius, very variable in colour, violet-brown, sepia-brown, ochre brown or pallid brown, purple-violet hue disappearing with age, yellow-brown towards margin, pallescent and becoming

lustrous on drying, developing bright orange splotches with age or when bruised, centre minutely tomentose to squamulose, innately radially fibrillose towards margin. Lamellae adnate to subdecurrent, segmentiform to subventricose, rather crowded, up to 4 mm broad, moderately thin, pallid grey-pink or pale cream-coloured, with entire, concolorous, rarely partly stained brown edge, staining mottled orange with age or when bruised, with two tiers of lamellulae. Stipe 20–50×2–4 mm, cylindrical or laterally compressed and longitudinally grooved, blue-grey or violet-grey at first, especially in the upper part and in young, fresh specimens, and in the basal part more yellowish grey, losing the blue colour with age then grey or grey-brown, dry, glabrous, polished, with white basal tomentum, with orange splotch at base, particularly when old or bruised. Odour indistinct. Taste musty, earthy.

Spores 10–12×7.5–9.0 μm, Q=1.2–1.6, rather irregularly 5–7-angled in side-view, Basidia 19–32×8.0–11 μm, 4-spored, clampless. Lamella edge sterile. Cheilocystidia cylindrical-clavate, 19–40×6.0–13 μm, usually colourless, occasionally with brown intracellular pigment. Hymenophoral trama regular, made up of cylindrical to inflated elements to 20 μm in diameter. Pileipellis a transition between a cutis and a trichoderm, made up of cylindrical to inflated hyphae, 8–15 μm in diameter, with clavate terminal elements to 26 μm in diameter. Pigment brownish grey, intracellular in pileipellis. Brilliant granules present in trama. Clamp-connections absent (Fig. 3.59).

Habitat & distribution: common in dry and wet sclerophyll forest, preferably in litter in open, more or less sun-lit places among tree ferns and in more open situations. Fruiting throughout the year, with a fair number of records in spring and summer months (Group 5, in Chap. 1, see section "Fruiting Patterns of Species of Entolomataceae").

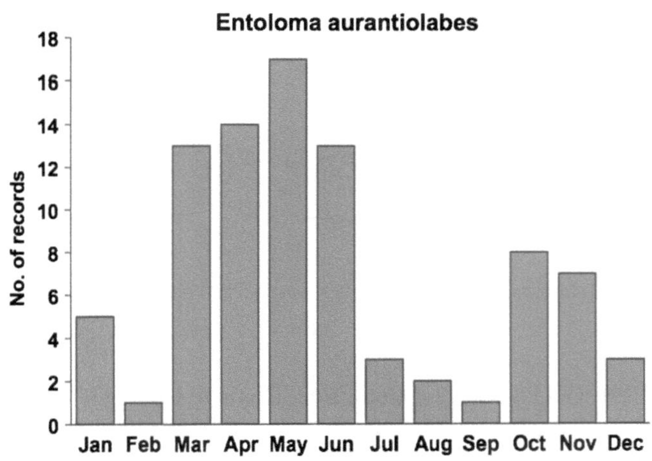

Collections examined. Australia, Tasmania, Arthur & Hellyer Rivers Junction, 41° 13′ S, 145° 31′ E, 5 June 1999, G. Gates E 567; Arve River Streamside Reserve, 43° 10′ S, 146° 48.5′ E, 2 July 2002, G. Gates E 1593; Bruny Island, Mt Mangana, 43° 22′ S, 147° 17′ E, 8 April 1999, G. Gates E 372; Clarks Cliffs, 43° 06′ S, 147° 47′ E, 4 May 2002, G. Gates E 1496; Collinsvale, Myrtle Forest, 42° 52′ S, 147° 09′

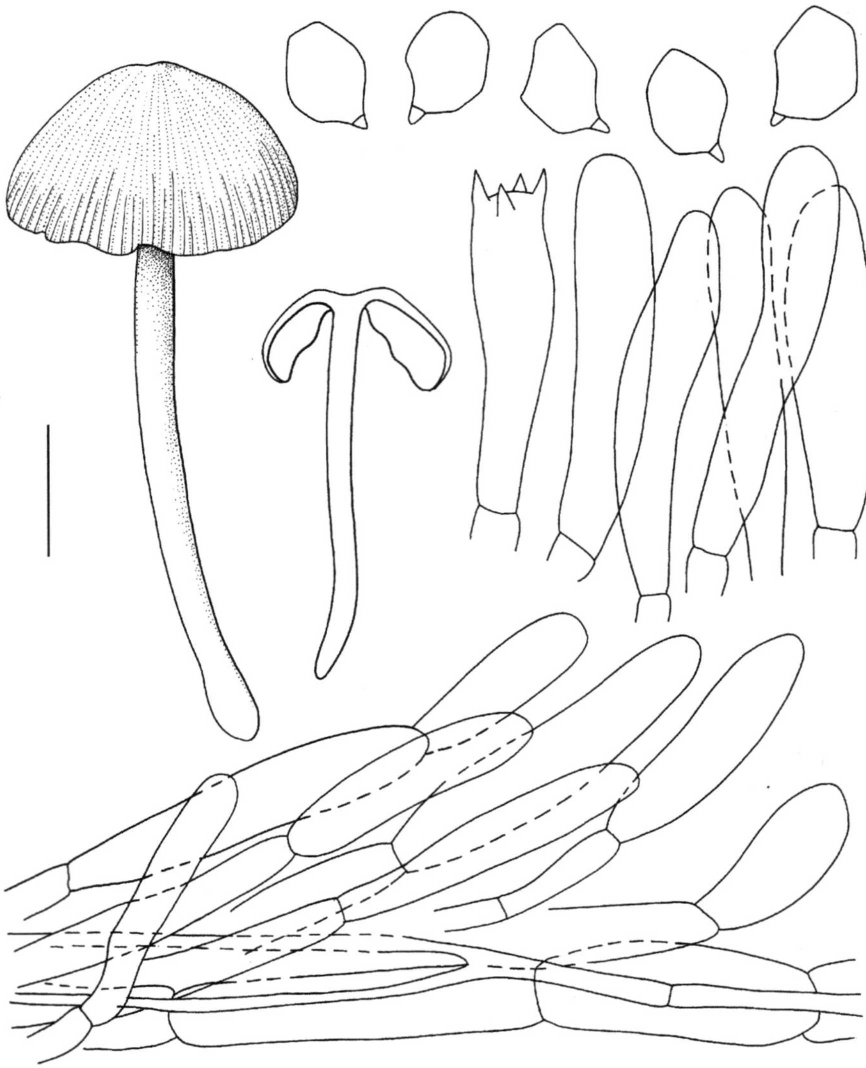

Fig. 3.59 *Entoloma aurantiolabes*. Habit, spores, basidia, and pileipellis. Bar = 10 mm or 10 μm

E, 18 Feb. 1999, G. Gates E 123; Geeveston, Donnellys Rd., 43° 07′ S, 146° 54′ E, 18 Oct. 1998, G. Gates E 81, E 83, E 85–E 87; —, 16 March 1999, G. Gates E 238; —, 3 April 1999, G. Gates E 355 & E 357; —, 12 June 1999, G. Gates E 596; —, 23 May 2000, G. Gates E 925, G. Gates E 929; —, 24 Oct. 2000, G. Gates E 1002; —, 14 Nov. 2000, G. Gates E 1010; —, 5 April 2001, G. Gates E 1077; —, 19 June 2001, G. Gates E 1253; —, 30 Oct. 2001, G. Gates E 1308; —, 30 July 2002, G. Gates E 1617; —, 10 April 2003, G. Gates E 1700; —, 2 May 2006, M.E. Noordeloos 2006051; Grass Tree Hill, 42° 47′ S, 147° 20′ E, 6 Oct. 2002, G. Gates E 1640;

Junee Caves State Reserve, 42° 44′ S, 146° 36′ E, 28 March 2009, M.E. Noordeloos 2009076 & 77 (L); Kermandie Falls, Lower Track, 43° 12′ S, 146° 52′ E, 16 March 1999, G. Gates E 230, G. Gates E 235; —, 11 May 1999, G. Gates E 491; —, 23 May 2000, G. Gates E 938a; —, 29 May 2001, G. Gates E 1198; —, 16 May 2002, G. Gates E 1527; —, 8 June 2002, G. Gates E 1565; —, 24 March 2009, M.E. Noordeloos 2009053 (L); Marriotts Falls, 42° 43′ S, 146° 39′ E, 10 June 1999, G. Gates E 583; Mt Arthur, 41° 16′ S, 147° 16′ E, 19 Jan. 2002, G. Gates E 1427; Mt Field National Park, Lady Barron Falls Track, 42° 41′ S, 146° 42′ E, 1 April 1999, G. Gates E 346; —, Lyrebird Walk, 42° 41′ S, 146° 40′ E, 9 May 1999, G. Gates E 478; —, River Walk, 42° 41′ S, 146° 42′ E, 4 May 2000, G. Gates E 890; —, 24 April 2001, G. Gates E 1125; —, Tall Trees Track, 4 Nov. 1999, G. Gates E 721; —, 13 Nov. 1999, G. Gates E 752; Mt Wellington, Fern Glade, 42° 55′ S, 147° 15′ E, 4 March 1999, G. Gates E 169; —, Myrtle Gully, 42° 54′ S, 147° 15′ E, 2 Sept. 1998, G. Gates E 54; —, 9 March 1999, G. Gates E 180; —, 17 Jan. 2002, G. Gates E 1411; —, Old Farm Rd, 13 April 1999, G. Gates E 404; North West Bay River, 42° 57′ S, 147° 12′ E, 7 June 2001, G. Gates E 1225; —, 14 Aug. 2001, G. Gates E 1273; —, 4 Oct. 2001, G. Gates E 1287; —, 17 Nov. 2001, G. Gates E 1320; —, 31 March 2003, G. Gates E 1665; —, 1 May 2003, G. Gates E 1755; —, 24 June 2003, G. Gates E 1887; —, 18 May 2004, G. Gates E 2000 (holotype, HO 543544; isotype, L); —, 21 April 2005, G. Gates E 2129; —, 25 Oct. 2005, G. Gates E 2209; Stump Track, Darcy Link Rd, 43° 25′ S, 146° 53′ E, 20 Aug. 2002, G. Gates E 1627; Waterfall Bay, 43° 04′ S, 147° 57′ E, 13 March 1999, G. Gates E 208, E 221, E 223 & E 224; Wielangta, 42° 42′ S, 147° 51′ E, 29 Jan. 1999, G. Gates E 120; —, 20 March 1999, G. Gates E 264; —, 29 April 1999, G. Gates E 450, G. Gates E 457; —, 2 Nov. 1999, G. Gates E 719; —, 29 March 2001, G. Gates E 1049; —, 8 Nov. 2001, G. Gates E 1311; —, 15 Jan. 2002, G. Gates E 1400.

This very difficult to identify *Cyanula* usually starts with a distinctly violet-brown pileus, particularly at centre, which fades with age to golden brown or paler, losing the violaceous tinge completely. The same applies to the stipe, which is a delicate blue-grey-violet in youth and fades to brown with age with a hint of blue sometimes persisting at the apex. Another remarkable feature is the orange mottling that occurs on the pileus, lamellae and stipe, particularly when bruised. Microscopically this species has rather irregularly shaped spores. Taxonomically it is placed near the group *E. longistriatum* (Peck) Noordel.

60. **Entoloma rufobasis** G.M. Gates & Noordel., Persoonia 19: 212. 2007.

Original diagnosis: Habitus collybioideus. Pileus 15–40 mm latus brunneus paulisper hygrophanus translucido-striatus radialiter fibrillosus centro squamulosus. Stipes 40–50 × 2–5 mm albidulus versus basim rufescens politus. Sporae 10–11 × 7.0–8.0 μm 6–7-angulatae. Cheilocystidia 40–60 × 8–14 μm cylindracea vel clavata. Pileipellis trichodermalis elementis 40–60 × 8–14 μm pigmentis intracellulosis. Fibulae desunt.

Holotype: Australia, Tasmania, Growling Swallet, 42° 41′ S, 146° 30′ E, 19 May 1999, G. Gates E 525 (HO 543560; isotype in L).

Plate 3.58 *Entoloma rufobasis* (Photo Michael Pilkington)

Etymology: rufus (Lat.) = reddish; basis (Lat.) = base, referring to the reddening base of the stipe.

Main characters: habit collybioid; pileus tan-brown, translucently striate, rather smooth; lamellae with brown edge; stipe very pale beige, polished, with reddening base (Plate 3.58).

Description:
Pileus 15–40 mm diam, convex to hemispheric, often truncate with slightly umbilicate centre, with involute to deflexed then straight margin, weakly hygrophanous, translucently striate when moist to 2/3 of radius, warm tan-brown (6C4–5, 6D4–5, 5C4–5, 5D4–5), uniformly coloured or with darker brown to grey-brown central spot, pallescent on drying, glabrous or radially fibrillose, except for the very finely squamulose centre. Lamellae broadly adnexed or adnate-emarginate, segmentiform to ventricose, moderately distant, up to 8 mm broad, pink or ochre pink, with brown, fimbriate edge, L = 28–36, l = 3–5. Stipe 40–50 × 2–5 mm, cylindrical, broadening towards base, very pale beige to almost white, glabrous, polished, base with white tomentum, turning orange-red or apricot pink with age or when bruised. Odour and taste not distinct.

Spores 10–11 × 7.0–8.0 μm, Q = 1.2–1.6, 6–7-angled in side-view. Basidia 28–35 × 5–11 μm, 4-spored, clampless. Lamella edge sterile. Cheilocystidia 40–60 × 8–14 μm, clavate with brown, intracellular pigment. Hymenophoral trama regular, made up of cylindrical hyphae, to 15 μm broad. Pileipellis a cutis of cylindrical hyphae, 4–16 μm in diameter, with scattered trichodermal tufts of clavate terminal elements, 40–60 ×

8–14 μm. Pigment golden brown, intracellular in pileipellis and upper pileitrama. Pileitrama regular to irregular, made up of cylindrical to inflated hyphae to 25 μm in diameter. Brilliant granules abundant in trama. Clamp-connections absent (Fig. 3.60).

Habitat & distribution: widespread in wet sclerophyll forests.

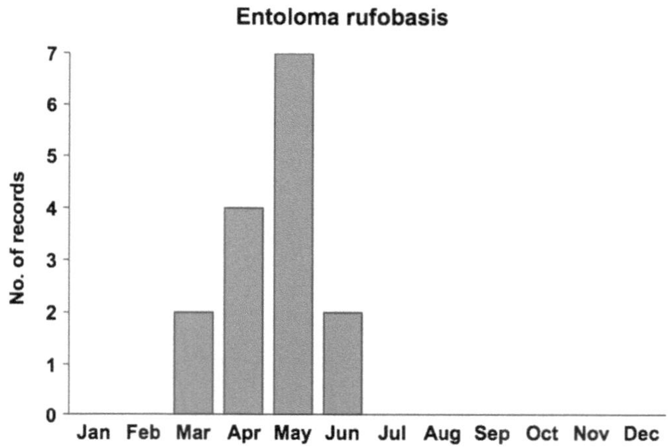

Collections examined. Australia, Tasmania, Douglas Apsley National Park, 41° 52′ S, 148° 11′ E, 2 May 2011, G. Gates (M.E. Noordeloos 2011024); Duckhole Lake Track, 43° 22′ S, 146° 53′ E, 6 June 2002, G. Gates E 1556; Growling Swallet, 42° 41′ S, 146° 30′ E, 19 May 1999, G. Gates E 525 (holotype, HO 543560; isotype, L); —, 11 May 2000, G. Gates E 901; —, 1 May 2001, G. Gates E 1147; —, 4 June 2002, G. Gates E 1552; —, 9 May 2011, M.E. Noordeloos 2011051 (L); Kermandie Falls, Upper Track, 43° 12′ S, 146° 52′ E, 30 May 2000, G. Gates E 951; Lake Skinner Track, 42° 57′ S, 146° 44′ E, 11 April 1999, G. Gates E 392; Old Hartz Mtn. Track, 43° 12′ S, 146° 52′ E, 19 March 2002, G. Gates E 1462.

This *Cyanula* is remarkable because of the vivid tan-coloured pileus, markedly contrasting with the pale stipe, the orange-red discolouration of the stipe base and the brown lamella edge. The translucently striate pileus places it in the vicinity of *E. longistriatum* (Peck) Noordel., from which it mainly differs in the more brightly coloured pileus and orange-red discolouration of the stipe. *Entoloma rubescentipes* E. Horak belongs to the group of *Entoloma phaeomarginatum* E. Horak (now including *E. rubescentipes* and some interpretations of *E. asprelloides* G. Stev. and *E. fuscum* (Cleland) E. Horak, according to Horak 2008), and clearly differs in the estriate, nonhygrophanous pileus, and fibrillose-striate stipe surface. Both species are reminiscent of the European species *Entoloma turci* (Bres.) M.M. Moser, which has an estriate, nonhygrophanous pileus, subpolished to striate stipe with a red staining base (Noordeloos 2004).

61. **Entoloma stramineopallescens** G.M. Gates & Noordel., Persoonia 19: 214. 2007.

Original diagnosis: Habitus collybioideus. Pileus 5–35 mm latus pallide stramineus pallescent hygrophanus translucido-striatus glabrus centro fibrillosus. Stipes 10–40 × 10–6.5 mm albidulus politus. Sporae 9.0–12 × 6.0–8.0 μm 5–7-angulatae.

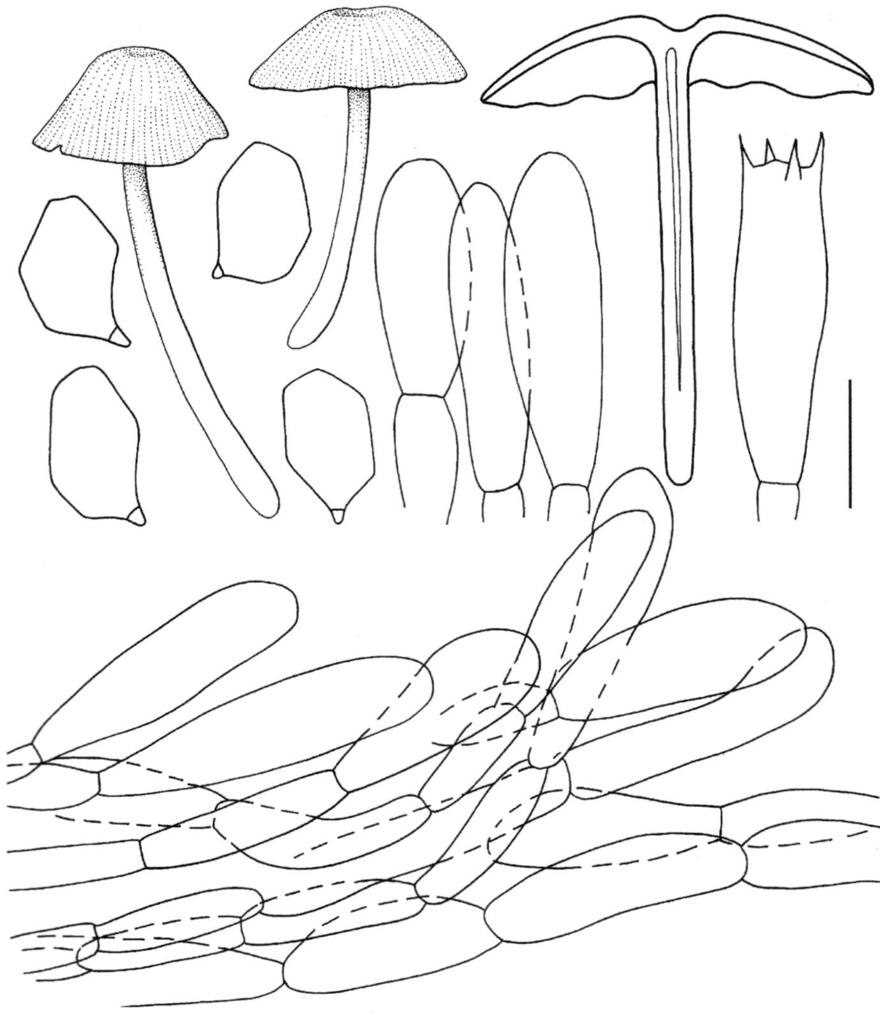

Fig. 3.60 *Entoloma rufobasis*. Habit, spores, basidia, cheilocystidia, and pileipellis. Bar = 10 mm or 10 μm

Cheilocystidia 20–40 × 5–12 μm cylindraceus vel clavatus. Pileipellis cutis elementis 8–12 μm latis in trichoderma transiens pigmentis intracellulosis. Fibulae desunt.

Holotype: Australia, Tasmania, Hobart, Waterworks Reserve, 42° 54′ S, 147° 17′ E, 21 Feb. 1999, G. Gates E 125 (HO 543545; isotype in L).

Etymology: stramineus (Lat.) = straw-coloured; pallescens (Lat.) = pallescent.

Main characters: basidiocarps very pale; pileus finely squamulose at centre; stipe polished; spores large; cheilocystidia present (Plate 3.59).

Plate 3.59 *Entoloma stramineopallescens* (Photo Machiel Noordeloos)

Description:
Pileus 5–35 mm diam, hemispheric to convex, truncate-campanulate, expanding to applanate, with slight central depression to distinctly umbilicate, with deflexed then straight margin, hygrophanous, when moist translucently striate at margin up to half the radius, rather pale yellow-brown, straw yellow, yellowish buff or buff-white (3A2), with a distinctly darker centre, pallescent on drying to off-white, fibrillose on disc, finely squamulose at centre, or entirely minutely squamulose. Lamellae adnate to adnate-emarginate, often with decurrent tooth, segmentiform, moderately distant, up to 4 mm broad, moderately thick, creamy buff with hint of pink, becoming deeper pink with age, with entire, concolorous edge. Stipe 10–40 × 10–6.5 mm, cylindrical, equal or with slightly broadened base, very pale yellow-buff to almost white, glabrous, polished, with white basal tomentum. Odour strongly spermatic or reminiscent of chlorine. Taste saliva inducing.

Spores 9.0–12 × 6.0–8.0 μm, Q = 1.2–1.5(–1.6), 5–7-angled with blunt, somewhat irregular angles. Basidia 20–40 × 8–10 μm, 4-spored, clampless. Lamella edge sterile. Cheilocystidia 20–40 × 5.0–12 μm, cylindrical to clavate, clampless. Hymenophoral trama regular, compact, made up of cylindrical elements to 20 μm in diameter, with abundant brilliant granules. Pileipellis a cutis of repent, cylindrical hyphae of 8–12 μm in diameter with scattered tufts of clavate terminal elements, which form

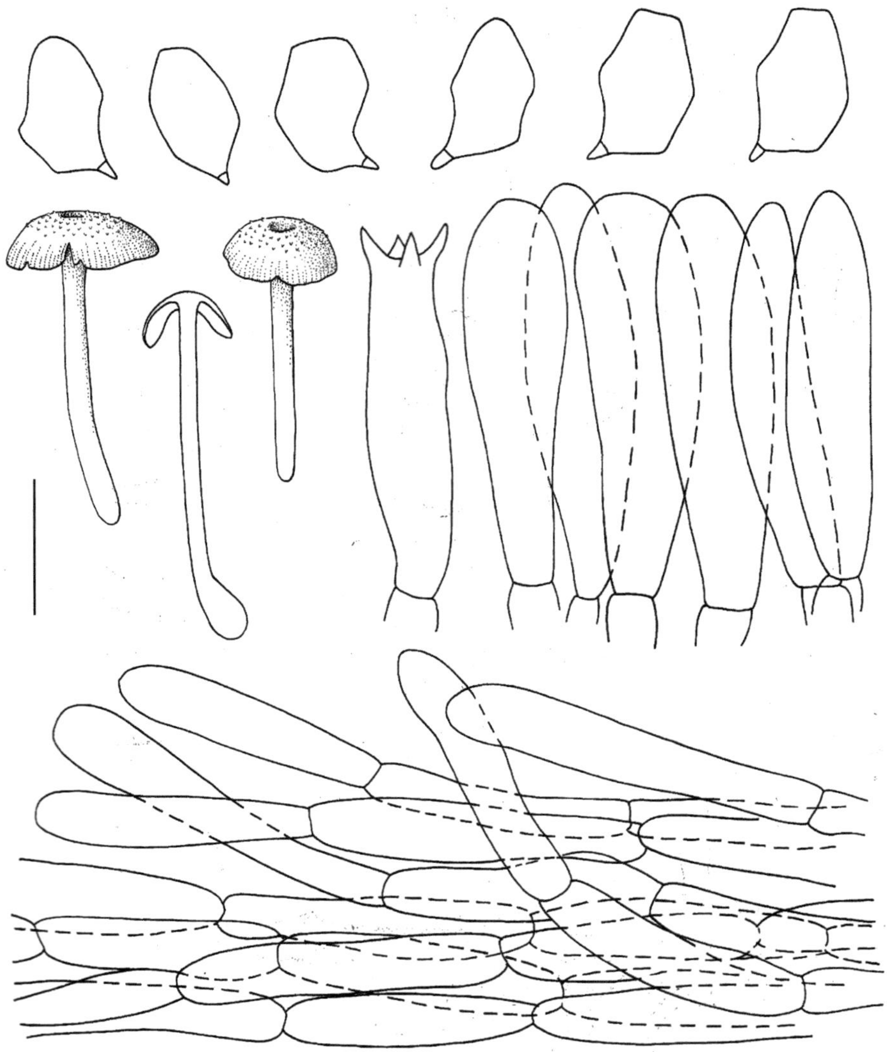

Fig. 3.61 *Entoloma stramineopallescens*. Habit, spores, basidia, cheilocystidia, and pileipellis. Bar = 10 mm or 10 μm

scales macroscopically at the centre of the pileus, 20–70 × 9.0–20 μm in diameter. Pigment pale brownish, intracellular. Clamp-connections absent (Fig. 3.61).

Habitat & distribution: common, widespread in wet eucalypt forest, often where *Pomaderris apetala* is present. Fruiting from late spring to late autumn or early winter (Group 2, in Chap. 1, see section "Fruiting Patterns of Species of Entolomataceae").

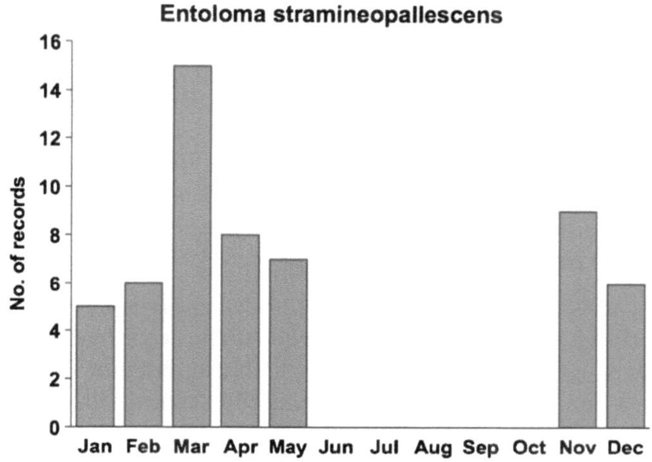

Collections examined. Australia, Tasmania, Chauncy Vale, 42° 37′ S, 147° 16′ E, 22 Nov. 2001, G. Gates E 1325, —, 29 Nov. 2001; G. Gates E 1349; Clarks Cliffs, 43° 06′ S, 147° 47′ E, 13 March 2004, G. Gates E 1952; Gowrie Park, O'Neills Creek, 41° 28′ S, 146° 13′ E, 2 May 2005, G. Gates E 2152; Hobart, Waterworks Reserve, 42° 54′ S, 147° 17′ E, 21 Feb. 1999, G. Gates E 125 (holotype, HO 543545; isotype, L); —, 16 Nov. 2000, G. Gates E 1013; Junee Caves State Reserve, 42° 44′ S, 146° 36′ E, 9 May 2011, M.E. Noordeloos 2011053 (L); Kermandie Falls, Lower Track, 43° 12′ S, 146° 52′ E, 9 Jan. 2005, G. Gates E 275; —, 24 March 2009, M.E. Noordeloos 2009060 & 2009061 (L); —, Upper Track, 30 March 1999, G. Gates E 314; —, 10 Jan. 2002, G. Gates E 1391; —, 15 April 2003, G. Gates E 1711; Lake St Clair National Park, Echo Point to Cynthia Bay, 42° 04′ S, 146° 09′ E, 22 April 2000, G. Gates E 864; —, 21 March 2009, M. E. Noordeloos 2009034; Mt Field National Park, Lady Barron Falls Track, 42° 41′ S, 146° 42′ E, 9 Dec. 1999, G. Gates E 780; —, River Walk, 42° 41′ S, 146° 42′ E, 4 May 2000, G. Gates E 891; —, Tall Trees Track, 13 Nov. 1999, G. Gates E 749; —, 31 March 2001, G. Gates E 1056; Mt Wellington, Myrtle Gully, 42° 54′ S, 147° 15′ E, 17 Jan. 2002, G. Gates E 1420; —, 26 April 2003, G. Gates E 1745; Wielangta, 42° 42′ S, 147° 51′ E, 23 Feb. 1999, G. Gates E 134 & E 140; —, 29 March 2001, G. Gates E 1052; —, 3 April 2003, G. Gates E 1669.

Entoloma stramineopallescens is a typical *Cyanula*, characterized by its rather pale colours, with distinctly darker, finely scaly centre of the pileus. This phenomenon occurs in many members of subgenus *Cyanula*, and was called a "calotte" (French) by Romagnesi (1937). *Entoloma stramineum* E. Horak from New Zealand is very similar, but has a wax yellow pileus without a darker central spot (Horak 2008).

Stirps ***Phaeomarginatum***

Basidiocarps brown to dark brown; lamella edge brown or concolorous with sides. Pigment exclusively brown, intracellular in pileipellis.

This stirps encompasses all dark brown species.

62. **Entoloma phaeomarginatum** E. Horak, *Beihefte Nova Hedwigia 43*: 70, 1973.

Syn. — *Leptonia fuscomarginata* Cleland, Trans. & Proc. Roy. Soc. S. Australia 57: 190. 1933; *Entoloma fuscomarginatum* (Cleland) E. Horak, Beih. Nova Hedwigia 65: 291 (1980) non *Entoloma fuscomarginatum* P.D. Orton 1960.

Misapplied Names. — *Entoloma asprelloides* G.Stev. [sensu Horak 1980, pro parte]. — *Entoloma fuscomarginatum* (Cleland) E. Horak [sensu Horak 1973, pro parte], nom. illeg. non P.D.Orton 1960; *Entoloma fuscum* (Cleland) E. Horak [sensu Horak 1973, pro parte; sensu Wölfel & Hausknecht 1999].

Original diagnosis: Pileo 10–30 mm lato, e convexo plano-depresso, subumbilicato, brunneo, fibrillis fuligineis obtecto, sicco, neque striato neque hygrophano. Lamellis adnato-adnexis, griseis dein roseis, acie fimbriata brunnea instructis. Stipite 10–25 × 1.5–3 mm, cylindraceo, brunneolo, fibrillis concoloribus instructo, sicco, fistuloso. Odore saporeque nullo. Sporis 8.5–11 × 6–7.5 μm, 5–6-angulatis. Basidiis 4-sporigeris. Cheilocystidiis fasciculatis (25–45 × 7–11 μm), pigmento brunneo et plasmatico impletis. Septis defibulatis. Ad detritum Dicksoniae squarrosae. Novazelandia.

Holotype (PDD 27052): 'Ulva Island, N.Z.; leg. Horak, 23 March 1969'.

Etymology: phaeos (Gr.) = brown; marginatus (Lat.) = margin (of the lamellae).

Selected literature: Grgurinovic, Larger fungi of South Australia: 376. 1997; Horak Beihefte Nova Hedwigia 43: 54, 70. 1973; Horak, Beihefte Nova Hedwigia 65: 222, 238. 1980; Horak, Agaricales of New Zealand 1: 268. 2008; Wölfel & Hausknecht, Österr. Z. Pilzk. 8: 130. 1999.

Main characters: habit collybioid; pileus dark brown to almost black, fibrillose; lamellae often with brown margin; stipe paler than pileus, polished, with pinkish orange blotch at base (Plate 3.60).

Description:
Pileus 20–50 mm diam, convex to plano-convex, umbilicate, with deflexed margin, not translucently striate when young and with maturity only at margin, blackish brown to sepia or burnt umber, radially fibrillose, rarely minutely villose at centre only, often more or less lubricous. Lamellae adnate-emarginate, segmentiform to ventricose, moderately distant, up to 5 mm broad, brown to grey-brown with pink tinge, with brown margin, lamellulae in two tiers. Stipe 30–40 × 3–5, cylindrical, grey-brown, paler at base, innately fibrillose, base with basal tomentum, which turns pinkish orange or apricot pink with age. Odour none. Taste bitter at first, soon disappearing.

Spores 10–13 × 7–8 μm, heterodiametric, 5–7-angled in side-view with pronounced angles. Basidia 28–40 × 8–11 μm, 4-spored, clampless. Lamella edge sterile. Cheilocystidia 20–50 × 6–14 μm, subcylindrical clavate with abundant, dark brown intracellular pigment. Hymenophoral trama regular, made up of cylindrical to inflated elements 70–180 × 5–22 μm. Pileipellis a cutis with transitions to a trichoderm, made up of inflated terminal elements, 30–75 × 8–20 μm with brown intracellular

Plate 3.60 *Entoloma phaeomarginatum* (Photo Michael Pilkington)

pigment. Pileitrama regular, made up of cylindrical to inflated hyphae, 6–24 μm wide. Brilliant granules abundant. Clamp-connections absent (Fig. 3.62).

Habitat & distribution: very common and widely distributed in wet and dry sclerophyll forests and rainforests. Fruiting throughout the year, with a fair number of records in spring and summer months (Group 5, in Chap. 1, see section "Fruiting Patterns of Species of Entolomataceae").

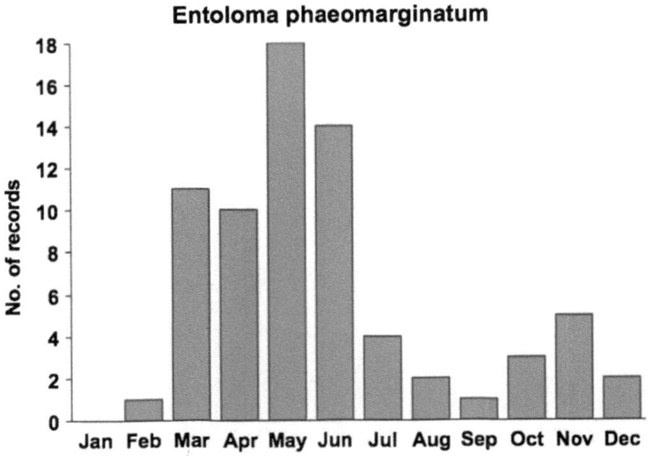

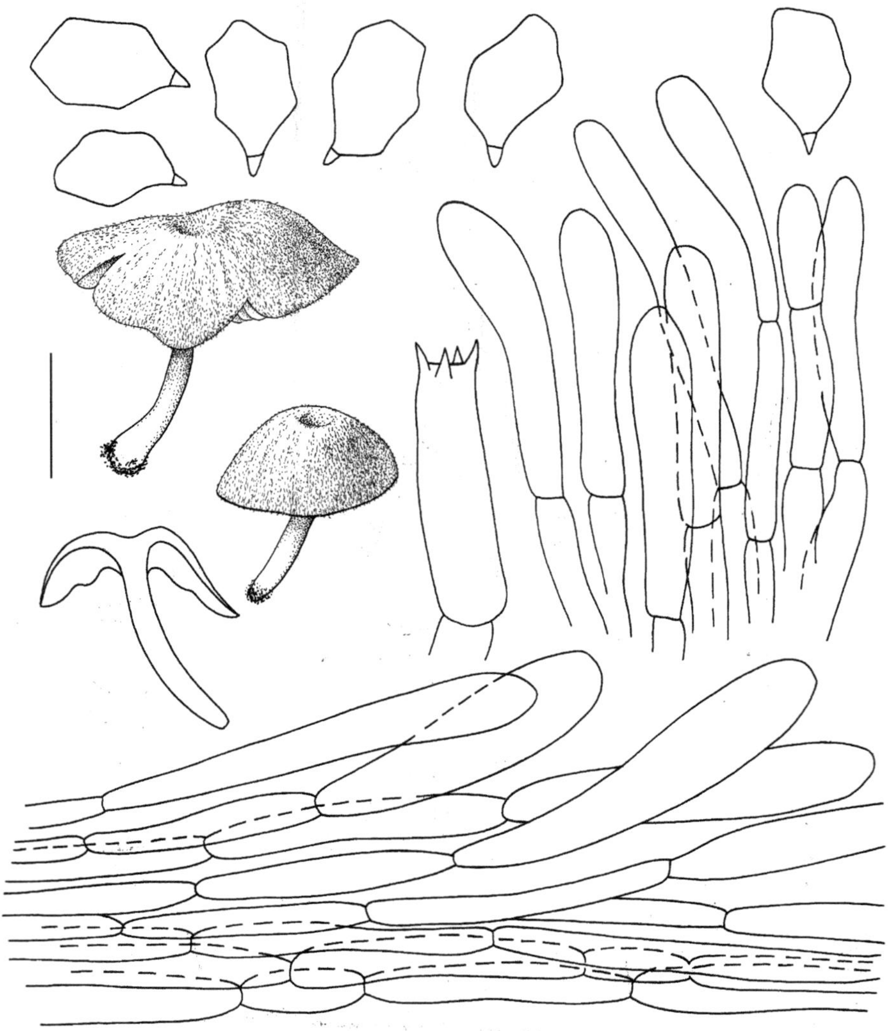

Fig. 3.62 *Entoloma phaeomarginatum*. Habit, spores, basidia, cheilocystidia, and pileipellis. Bar = 10 mm or 10 μm

Collections examined. Australia, South Australia, Encounter Bay, 26 May 1931, Cleland 13535 (holotype, K); — Tasmania, Adamsons Peak Track, 43° 20′ S, 146° 54′ E, 1 May 1999, G. Gates E 462; Arve River Picnic Area, 43° 10′ S, 146° 48.5′ E, 10 April 2001, G. Gates E 1090; Bruny Island, Mt Mangana, 43° 22′ S, 147° 17′ E, 1 June 1999, G. Gates E 553; —, 5 Feb. 2004, G. Gates E 1929; —, 27 May 2004, G. Gates E 2022 & E 2023; Cape Pillar Track, 43° 09′ S, 147° 56′ E, 29 May 1999, G. Gates E 548; Chauncy Vale, 42° 37′ S, 147° 16′ E, 13 Oct. 2001, G. Gates E 1240 & E 1290; —, 20 Nov. 2001, G. Gates E 1322; —, 10 June 2003, G. Gates E 1858; —, 28 June 2003, G. Gates E 1890; —, 22 June 2004, G. Gates E 2040; Collinsvale, Fire Trail, 42° 52′ S, 147° 09′ E, 15 April 1999, G. Gates E 419 & E 420; —, Myrtle Forest, 1 July 2003, G. Gates E 1892; Creepy Crawly, 42° 50′ S, 146° 22′ E, 15 April 2000, G. Gates E 841; Dial Range, Mt Duncan, 41° 12′ S, 146° 02′ E, 7 June 1999, G. Gates E 570; Duckhole Lake Track, 43° 22′ S, 146° 53′ E, 15 May 2003, G. Gates E 1812; Geeveston, Donnellys Rd, 43° 07′ S, 146° 54′ E, 13 Aug. 1998, G. Gates E 36, E 38 & E 41; —, 18 Oct. 1998, G. Gates E 80a & E 82 —, 12 June 1999, G. Gates E 591; Growling Swallet, 42° 41′ S, 146° 30′ E, 19 May 1999, G. Gates E 528; —, 4 June 2002, G. Gates E 1551; Kermandie Falls, Lower Track, 43° 12′ S, 146° 52′ E, 29 May 2001, G. Gates E 1193; —, Upper Track, 30 March 1999, G. Gates E 319; —, 6 May 1999, G. Gates E 472; Lake St. Clair National Park, Echo Point to Cynthia Bay, 42° 04′ S, 146° 09′ E, 22 April 2000, G. Gates E 851 & E 866; —, 26 March 2005, G. Gates E 2105; Mt Field National Park, Lady Barron Falls Track, 42° 41′ S, 146° 42′ E, 9 Nov. 1999, G. Gates E 740; —, Lyrebird Walk, 42° 41′ S, 146° 40′ E, 1 April 1999, G. Gates E 326; —, 2 May 2000, G. Gates E 887; —, 31 March 2001,G. Gates E 1060; —, River Walk, 42° 41′ S, 146° 42′ E, 9 Nov. 1999, G. Gates E 732; Mt Wellington, Fern Glade, 42° 55′ S, 147° 15′ E, 11 March 2004, G. Gates E 1940; —, Fingerpost Track, 4 March 1999, E 170; —, Middle Track, 4 May 1999, G. Gates E 466; —, Myrtle Gully, 42° 54′ S, 147° 15′ E, 9 March 1999, G. Gates E 181; —, 13 April 1999, G. Gates E 403 & E 409; —, 15 June 1999, G. Gates E 605 & E 610; —, Old Farm Rd., 8 July 1998, G. Gates E 25; —, 17 Oct. 1998, G. Gates E 77; —, 6 Sept. 1999, G. Gates E 700; —, Pipeline Track, 42° 55′ S, 147° 15′ E, 26 Dec. 2001, G. Gates E 1376; —, Silver Falls Service Track, 4 March 1999, G. Gates E 164; —, Woods Track, 25 March 1999, G. Gates E 282; North West Bay River, 42° 57′ S, 147° 12′ E, 14 Aug. 2001, G. Gates E 1275; —, 18 May 2004, G. Gates E 2005; Tahune, Huon Pine Walk, 43° 06′ S, 146° 44′ E, 27 May 2000, G. Gates E 946; —, 2 July 2002, G. Gates E 1595; —, 19 June 2003, G. Gates E 1882; Wielangta, 42° 42′ S, 147° 51′ E, 14 July 1998, G. Gates E 16; —, 20 March 1999, G. Gates E 262 & E 271; —, 29 April 1999, G. Gates E 452; —, 22 June 1999, G. Gates E 622. — New Zealand, Stewart Island, Ulva Island, Patterson Inlet, 23 March 1969, E. Horak (holotype of *E. fuscomarginatum* E. Horak, PDD 27052).

Originally, *Entoloma phaeomarginatum* was described from New Zealand (Horak 1973) as a new name for the illegitimate name *Entoloma fuscomarginatum* (Cleland) E. Horak (non *E. fuscomarginatum* P.D. Orton). The Tasmanian collections have slightly larger spores than in the description of Horak (2008). We think that our collections could well represent *Entoloma fuscomarginatum* Cleland non sensu Horak (1973). We studied the holotype in Kew, which had large spores (10–14×7–9 µm), a sterile lamella edge with brown pigmented cheilocystidia, brilliant granules in the trama and a pileipellis in the form of a transition between a cutis and a trichoderm of clavate terminal elements, 20–30×10–16 µm. This fits well with our collections of the species. *Entoloma asprelloides* G. Stev. is also very close (Horak 2008). Our species differs in the brown instead of blue lamella edge, the lack of blue in the pileus, and the lack of green tinges in the stipe and stipe cortex. *Entoloma phaeomarginatum* is very similar to the European *E. turci* (Bres.) M.M. Moser, which also has a reddish orange staining at the stipe base, and mainly differs by the distinctly fibrillose stipe. In the field this species is also reminiscent of *Entoloma pseudocoelestinum* Arnolds from Europe (Noordeloos 2004), from which it definitely differs by the brown coloured lamella edge. Contrary to Horak (2008) we do not accept the synonymy of *Entoloma rubescentipes* E. Horak. From a type study, we think that the latter may be a species in its own right not occurring in Tasmania.

63. **Entoloma persimile** E. Horak, Agaricales New Zealand 5: 271. 2008.

Original diagnosis: Pileus 15–30(–40) mm, primo convexus, dein applanatus vel subumbilicatus, griseobrunneus vel fuscus, conspicue hygrophanus, ad discum squamulis concoloribus distinctis instructus, marginem striatum versus fibrillis innatis obtectus. Lamellae late adnatae vel emarginatae, albidulae vel argillaceae dein griseorosaceae, ad aciem subbrunneum fimbriatae. Stipes 30–70×2–4 mm, cylindricus, pallide brunneus, glaber vel innate fibrillosus. Odor haud distinctus vel subfarinosus. Sapor amarus vel rancidus. Basidiosporae 8.5–10(–10.5)× 6–7.5 µm, distincte 5–6-angulatae. Basidia (26–)28–40×8–9(–10) µm, 4-spora, defibulata. Cheilocystidia (25–)30–85(–100)×9–18(–22) µm, cylindrica vel clavata, pigmento brunneo (KOH) plasmatico impleta. Pleurocystidia et caulocystidia nulla. Pileipellis ex hyphis cylindraceis trichodermium formantibus, cellulis terminalibus cylindricoclavatis (25–)35–80 (–140)×5–15(–20) µm, pigmento brunneo (KOH) plasmatico impletis. Fibulae nullae. Ad terram in silvis mixtis. Nova Zelandia.

Holotype: New Zealand, Lake Okataina, E-Track, 26 April 2000, E. & A. Horak (PDD 71246; isotype ZT 8560).

Etymology: per (Lat.) = strongly; similis (Lat.) = similar.

Main characters: habit collybioid; pileus very dark brown, translucently striate at margin; lamella edge brown; stipe polished; spore size variable in Tasmanian material.

63a. **Entoloma persimile** var. **persimile** (Plate 3.61)

Description:
Pileus 15–50 mm diam, hemispherical to convex then plano-convex with slightly to distinctly depressed centre, with deflexed margin, not hygrophanous, translucently striate at margin only, uniformly grey-brown (5F3) with darker centre, minutely

Plate 3.61 *Entoloma persimile* var. *persimile* (Photo Genevieve Gates)

squamulose at centre, distinctly innately radially fibrillose towards margin. Lamellae adnate-emarginate or with small decurrent tooth, segmentiform, moderately distant, up to 6 mm broad, pale grey-brown then pinkish brown with contrasting brown, entire edge, L=20–36, l=3–5(–7). Stipe 20–40×2–5 mm, cylindrical, slightly broadened at base, grey-brown, almost as dark as pileus, smooth, glabrous, polished, with white basal tomentum which often turns orange with age. Odour none. Taste bitter at first, then mild.

Spores 8.5–10.5×6–8 μm, Q=1.2–1.55, Qav =1.3–1.4, 5–6(–7)-angled in side-view with pronounced angles. Basidia 22–45×7–11 μm, 4-spored, clampless. Lamella edge sterile. Cheilocystidia 23–50×5–9 μm, cylindrical to clavate without pigment. Hymenophoral trama regular, made up of cylindrical to inflated elements 5–22 μm wide. Pileipellis a cutis with transitions to a trichoderm, made up of clavate elements, 30–55×5–9 μm. Pigment brown, intracellular in pileipellis. Pileitrama regular, made up of cylindrical to slightly inflated elements, 4–18 μm wide. Brilliant granules abundant. Stipitipellis a cutis of narrow, cylindrical, 2–5 μm wide hyphae. Caulocystidia absent. Clamp-connections absent (Fig. 3.63).

Habitat & distribution: widespread in wet eucalypt forest.

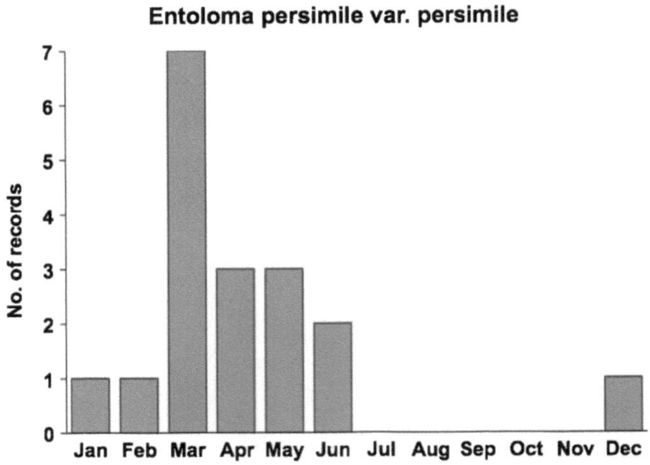

Collections examined. Australia, Tasmania, Arve River, Keoghs Walk, 43° 09′ S, 146° 47′ E, 10 April 2001, G. Gates E 1088; —, Zig-Zag Track, 43° 06′ S, 146° 45′ E, 23 May 1999, G. Gates E 531; Bruny Island, Mt Mangana, 43° 22′ S, 147° 17′ E, 30 April 2011, M.E. Noordeloos 2011023 (L); Collinsvale, Myrtle Forest, 42° 52′ S, 147° 09′ E, 18 March 1999, G. Gates E 247 & E 250; Geeveston, Donnellys Road, 43° 07′ S, 146° 54′ E, 5 April 2001, G. Gates E 1075; Kermandie Falls, Lower Track, 43° 12′ S, 146° 52′ E, 11 May 1999, G. Gates E 488; Mt Cripps, Snowy Mt Track, 41° 36′ S, 145° 46′ E, 23 March 2005, G. Gates E 2102; Mt Field National Park, Tall Trees Track, 42° 41′ S, 146° 52′ E, 4 Dec. 2001, G. Gates E 1367; Mt Wellington, Circle Track, 42° 55′ S, 147° 15′ E, 11 March 1999, G. Gates E 201; —, 25 March 1999, G. Gates E 293; —, Myrtle Gully, 42° 54′ S, 147° 15′ E, 3 May 2001, G. Gates E 1163; North West Bay River, 42° 57′ S, 147° 12′ E, 31 March 2003, G. Gates E 1664; Wielangta, 42° 42′ S, 147° 51′ E, 23 Feb. 1999, G. Gates E 143; —, 29 April 1999, G. Gates E 449. — New Zealand, Lake Okataina, E-Track, 26 April 2000, E. & A. Horak (holotype, PDD 71246).

The Tasmanian collections, part of the group nicknamed "brown bitter disappearing", resemble very much *Entoloma persimile* E. Horak with respect to the dark brown and translucently striate pileus, bitter taste, and small spores. The New Zealand collections, however, have a paler stipe surface and lack the distinct orange blotch at the base of the stipe, which is very distinct in all Tasmanian collections studied. Part of the Tasmanian collections of this group have distinctly larger spores, more within the range of *Entoloma phaeomarginatum* E. Horak, which differs by a non-translucent, fibrillose-tomentose pileus and a fibrillose stipe surface. For that reason these collections are described here as a large-spored variety.

63b. **Entoloma persimile** var. **macrosporum** Noordel. & G.M. Gates, var. nov.

MycoBank #564501.

Diagnosis: Differs from the typical variety by the distinctly larger spores, 10–12.5(–13) × 6.5–8 μm, Q = 1.3–1.6, Qav = 1.4–1.5, 6–7(–8), angled.

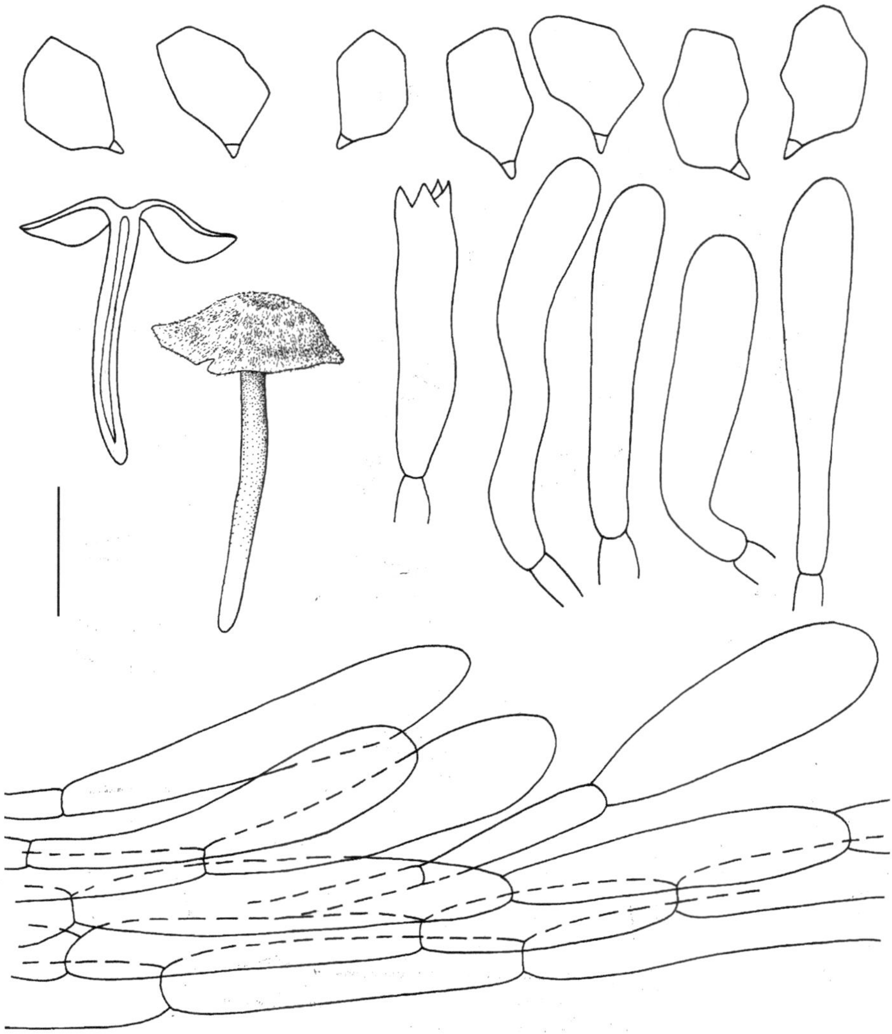

Fig. 3.63 *Entoloma persimile* var. *persimile*. Habit, spores, basidia, cheilocystidia, and pileipellis. Bar = 10 mm or 10 μm

Holotype: Australia, Tasmania, North West Bay River, 42° 57′ S, 147° 12′ E, 1 May 2003, G. Gates E 1758 (HO 564352).

Description:
Similar in all respects to the type variety, but with consistently larger spores. Spores 10–12.5(–13) × 6.5–8 μm, Q = 1.3–1.6, Qav = 1.4–1.5, 6–7(–8)-angled in side-view.

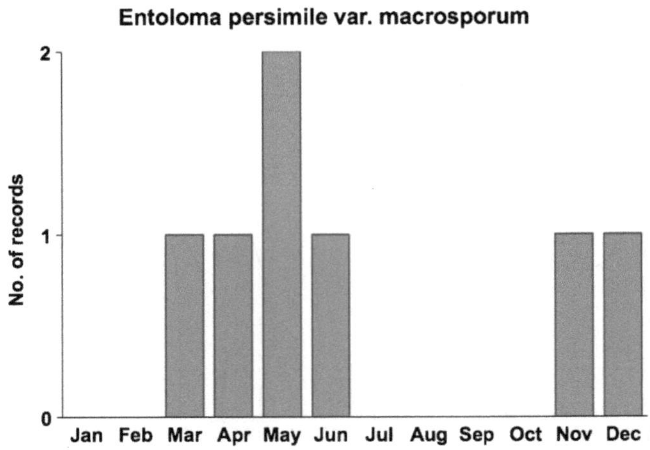

Collections examined. Australia, Tasmania, Collinsvale, Myrtle Forest, 42° 52′ S, 147° 09′ E, 18 March 1999, G. Gates E 256; —, 19 April 2001, G. Gates E 1104; —, 27 Dec. 2001, G. Gates E 1378; Geeveston, Donnellys Rd, 43° 07′ S, 146° 54′ E, 19 June 2001, G. Gates E 1256; Kermandie Falls, Lower Track, 43° 12′ S, 146° 52′ E, 11 May 1999, G. Gates E 493; Mt Field NP, Tall Trees Track, 42° 41′ S, 146° 42′ E, 1 Nov. 2011, G. Gates E 2307; North West Bay River, 42° 57′ S, 147° 12′ E, 1 May 2003, G. Gates E 1758 (holotype, HO 564352).

Entoloma persimile var. *macrosporum* is also similar to *E. asprelloides* G. Stev. from New Zealand, which differs by the blue-grey lamellae with blue, fimbriate lamella edge. *Entoloma phaeomarginatum* has a darker, non-translucently striate pileus that is not squamulose in any of the collections examined, and a fibrillose stipe surface.

Stirps ***Amarum***

Pileus dark brown to blackish brown, occasionally with hints of blue or violet, stipe with grey-blue tinges, taste often bitter.

Species in this stirps come close to those of stirps *Phaeomarginatum*, but have distinct blue or violaceous pigments in the pileus or stipe.

64. **Entoloma amarum** G.M. Gates & Noordel., spec. nov.

Entoloma fuscomarginatum (Cleland) E. Horak sensu Horak 1973 non 2008, nec *Entoloma fuscomarginatum* P.D. Orton 1960.

MycoBank #564502.

Diagnosis: Basidiocarps collybioid. Pileus 25–50 mm, convex, expanding, umbilicate, blackish brown with violaceous tinges, particularly near margin, translucently striate, minutely squamulose at centre, fibrillose at margin. Lamellae adnate, grey-brown with pink tinge with brown edge. Stipe 50–60 × 2–5 mm, greyish blue to purple-blue, glabrous, polished. Taste bitter. 9–12 × 6–8 μm, 5–7-angled. Cheilocystidia 40–60 × 8–14 μm, cylindrical to clavate with brown

Plate 3.62 *Entoloma amarum* (Photo Machiel Noordeloos)

intracellular pigment. Pileipellis a trichoderm of up to 30 μm wide inflated terminal elements with intracellular pigment. Brilliant granules present. Clamp-connections absent.

Holotype: Australia, Tasmania, Canoe Bay, 43° 09′ S, 147° 57′ E, 22 May 2004, G. Gates E 2012 (HO 561500; isotype in L).

Etymology: amarus (Lat.) = bitter, referring to the taste of the context.

Main characters: habit collybioid; pileus blackish brown with slight violaceous red tinge; stipe blackish blue; lamella edge brown; taste bitter; spores 9–12 um long (Plate 3.62).

Description:

Pileus 25–50 mm diam, campanulate-conical to convex then plano-convex with deflexed margin, with distinctly umbilicate centre, not hygrophanous or very slightly so, non translucently striate, brownish black with slight red-violet sheen (7F4–3, 8F4–3, 12F2–3), paler towards margin (closer to 7D3–4, 7E3–4, 12F3) or very dark brown without blue or violaceous tinges, entirely velvety in young specimens, then minutely tomentose or squamulose in central part, innately fibrillose toward margin. Lamellae L = ca. 40, l = 3–5, moderately distant, broadly emarginate to adnexed, ventricose, up to 5 mm broad, sordid grey-brown pink with brown, fimbriate edge. Stipe 50–60 × 2–5 mm, cylindrical, slightly broadened at base, grey-blue to purplish blue (14F1–3, 15F1–3), polished or fibrous with darker grey fibrils on purplish blue background, with white tomentum at base, developing a pink hue when bruised or with age. Context concolorous in surface, white in inner part, pink in base of stipe. Odour indistinct. Taste bitter at first then mild.

Spores 9–12×6–8 μm, Q=1.2–1.7, Qav=1.3–1.5, 5–7-angled with rather weak angles. Basidia 20–30×8–10 μm, 4-spored, clampless. Lamella edge sterile. Cheilocystidia 50–70×6–20 μm, clavate to sphaeropedunculate, with brown, intracellular pigment. Hymenophoral trama regular, made up of cylindrical hyphae, up to 20 μm wide. Brilliant granules abundant. Vascular hyphae present. Pileipellis a cutis with transitions to a trichoderm, made up of inflated terminal elements, up to 30 μm wide. Pigment pinkish brown, abundant, intracellular in pileipellis. Pileitrama regular, made up of cylindrical hyphae 8–12 μm wide. Clamp-connections absent (Fig. 3.64).

Habitat & distribution: common in wet eucalypt forests and rainforests. Fruiting throughout the year, with a fair number of records in spring and summer months (Group 5, in Chap. 1, see section "Fruiting Patterns of Species of Entolomataceae").

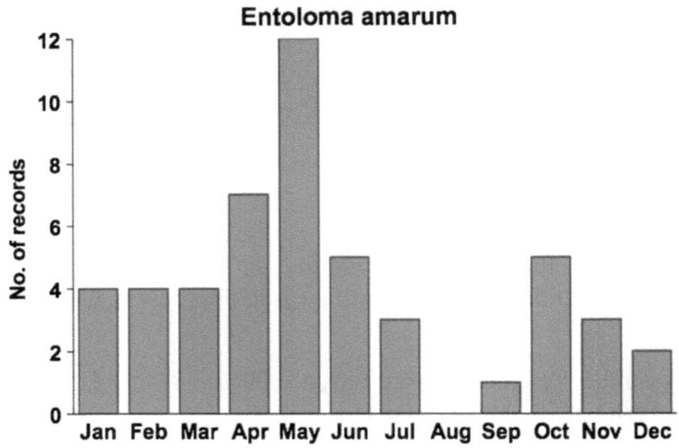

Collections examined. Australia, Tasmania, Arve Road, Keoghs Walk, 43° 09′ S, 146° 47′ E, 18 Oct. 2001, G. Gates E 1293; Bruny Island, Mt Mangana, 43° 22′ S, 147° 17′ E, 8 April 1999, G. Gates E 380; —, 31 March 2009, M.E. Noordeloos 2009089 (L); Canoe Bay, 43° 08′ S, 147° 57′ E, 22 May 2004, G. Gates E 2012 (holotype, HO 561500); Chauncy Vale, 42° 37′ S, 147° 16′ E, 25 Oct. 2001, G. Gates E 1301; —, 6 Oct. 2005, G. Gates E 2201; —, 28 June 2009, G. Gates E 2278; Collinsvale, Myrtle Forest, 42° 52′ S, 147° 09′ E, 18 March 1999, G. Gates E 257; Duckhole Lake Track, 43° 22′ S, 146° 53′ E, 8 April 2003, G. Gates E 1686; Geeveston, Donnellys Rd, 43° 07′ S, 146° 54′ E, 3 April 1999, G. Gates E 356; —, 23 May 2000, G. Gates E 926; —, 30 July 2002, G. Gates E 1616; —, 19 May 2005, G. Gates E 2169; Judbury Conservation Area, 24 June 2004, G. Gates E 2042; Kermandie Falls, Lower Track, 43° 12′ S, 146° 52′ E, 14 Nov. 2000, G. Gates E 1012; —, 5 Jan. 2002, G. Gates E 1387; —, 3 June 2003, G. Gates E 1853; —, 9 Jan. 2005, G. Gates E 2076; —, Upper Track, 12 June 1999, G. Gates E 602; Lake St Clair National Park, Echo Point to Cynthia Bay, 42° 04′ S, 146° 09′ E, 22 April 2000, G. Gates E 867 & E 868; Marriotts Falls, 42° 43′ S, 146° 39′ E, 20 Oct. 2005, G. Gates E 2207; Mt

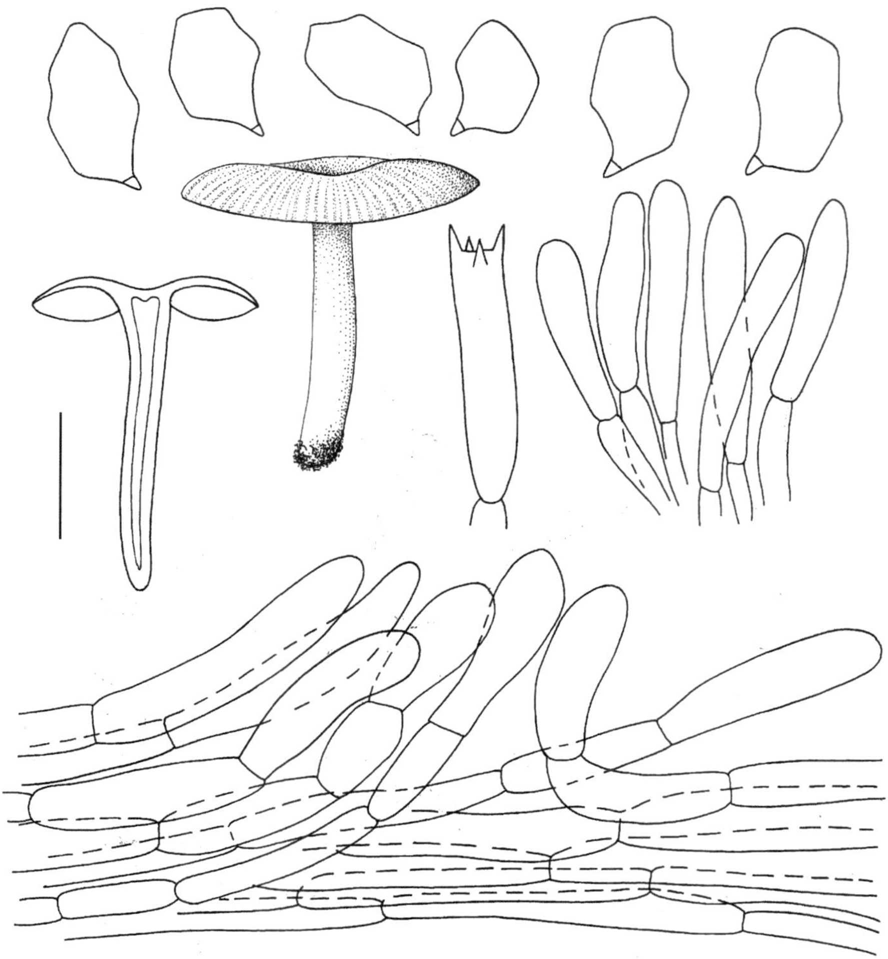

Fig. 3.64 *Entoloma amarum*. Habit, spores, basidia, cheilocystidia, and pileipellis. Bar = 10 mm or 10 μm

Arthur, 19 Jan. 2002, G. Gates E 1419; —, 9 Dec. 2005, G. Gates E 2223; Mt Field National Park, Lady Barron Falls Track, 42° 41′ S, 146° 42′ E, 1 April 1999, G. Gates E341; —, 23 Oct. 2001, G. Gates E 1298; —, Lyrebird Walk, 42° 41′ S, 146° 40′ E, 1 April 1999, G. Gates E333; —, Lake Dobson, 6 May 2001, G. Gates E 1171; —, Tall Trees Track, 42° 41′ S, 146° 42′ E, 9 Dec. 2004, G. Gates E 2067; Mt Wellington, Myrtle Gully, 42° 54′ S, 147° 15′ E, 9 March 1999, G. Gates E 190, E 191 & E 192; —, Reservoir Track, 42° 55′ S, 147° 15′ E, 19 Feb. 2002, G. Gates E 1438; —, Strickland Falls, 21 Nov. 2000, G. Gates E 1020; North West Bay River, 42° 57′ S, 147° 12′ E, 21 April 2005, G. Gates E 2136; Waterfall Bay, 43° 04′ S, 147° 57′ E, 13 March 1999, G. Gates E 212; Wielangta, 42° 42′ S, 147° 51′ E, 23 Feb. 1999, G. Gates E 144.

Plate 3.63 *Entoloma moabus* (Photo Machiel Noordeloos)

Entoloma phaeomarginatum E. Horak is similar, but lacks blue tinges in the basidiocarps, and has slenderer cheilocystidia. *E. asprelloides* G. Stev. (Horak 2008) is also very close. The main difference is found in the brown instead of blue lamella edge, the red-violet sheen in the pileus instead of bluish tones, and the lack of green tinges in the stipe and stipe cortex.

65. **Entoloma moabus** Noordel. & G.M. Gates, spec. nov.

MycoBank #564503.

Diagnosis: Pileus 15–35 mm, convex, expanding, applanate to umbilicate, not translucently striate, greyish black with brown tinge, entirely tomentose. Lamellae adnate, grey then greyish pink with concolorous edge. Stipe 25–35 × 3–5 mm, cylindrical, dark blue then grey, polished. Spores 8–10 × 6–7 μm, 5–6-angular. Lamella edge sterile. Cheilocystidia 20–50 × 6–12 μm, cylindrical-clavate. Pileipellis a trichoderm of clavate terminal elements, 50–110 × 15–22 μm, with brown intracellular pigment. Clamp-connections absent.

Holotype: Australia, Tasmania, Wielangta, 42° 42′ S, 147° 51′ E, 19 Feb. 2005, G. Gates E 2083 (HO 564353).

Etymology: moabus (aboriginal) = dark, referring to the dark pileus.

Main characters: habit collybioid; fruit body small, very dark, with concolorous lamella edge; taste mild; spores small (Plate 3.63).

Description:
Pileus 15–35 mm diam, convex then plano-convex with plane or depressed centre, with deflexed to straight margin, uplifted with age, not hygrophanous, not translucently striate, grey-black to grey with brown hue, tomentose all over, breaking up into small squamules with age. Lamellae adnate with small decurrent tooth, ventricose, up to 5 mm broad, pale grey then grey with pink tinge, with subentire, concolorous edge. Stipe 25–35×3–5 mm, cylindrical with swollen base, dark blue then becoming grey, with faint yellow tinge at base, with fine fibrils all over the length of the stipe. Odour none. Taste salty.

Spores 8–10×5.5–6 μm, Q=1.2–1.5, Qav=1.3, 5–6-angled. Basidia 20–35×8–11 μm, 4-spored, clampless. Lamella edge sterile. Cheilocystidia 20–50×6–12 μm, cylindrical to clavate, without pigmentation. Hymenophoral trama regular, made up of cylindrical 5–20 μm wide hyphae. Brilliant granules abundant. Pileipellis a cutis with transitions to a trichoderm, made up of clavate terminal elements 50–110×15–22 μm. Pileitrama regular, made up of cylindrical to slightly inflated hyphae 5–18 μm wide. Pigment intracellular, brown in pileipellis. Stipitipellis a cutis of narrow, cylindrical hyphae 4–8 μm wide. Caulocystidia absent. Clamp-connections absent (Fig. 3.65).

Habitat & distribution: in wet eucalypt forests. Uncommon.

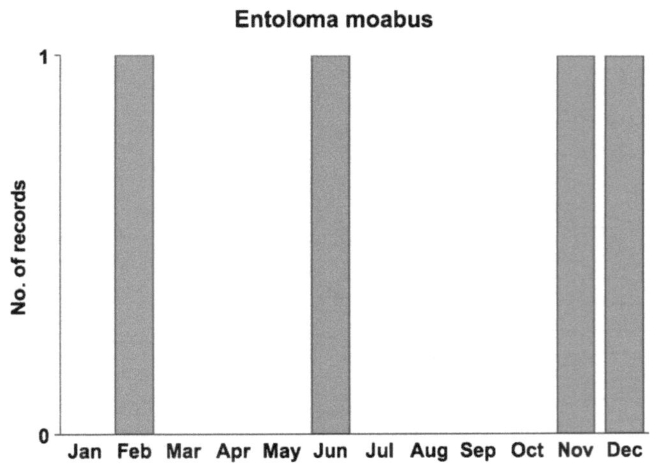

Collections examined. Australia, Tasmania, Mt Field National Park, Lady Barron Falls Track, 42° 41′ S, 146° 42′ E, 1 June 2010, M.E. Noordeloos 2010067; Mt Wellington, Middle Track, 42° 55′ S, 147° 15′ E, 18 Nov. 2000, G. Gates E 1019; —, 18 Dec. 2001, G. Gates E 1371; Wielangta, 42° 42′ S, 147° 51′ E, 19 Feb. 2005, G. Gates E 2083 (holotype, HO 564353).

Entoloma moabus keys out near *Entoloma amarum* from which it differs by the dark grey pileus, concolorous lamella edge, mild taste and much smaller spores.

66. **Entoloma obscureogracile** Noordel. & G.M. Gates, spec. nov.

MycoBank #564504.

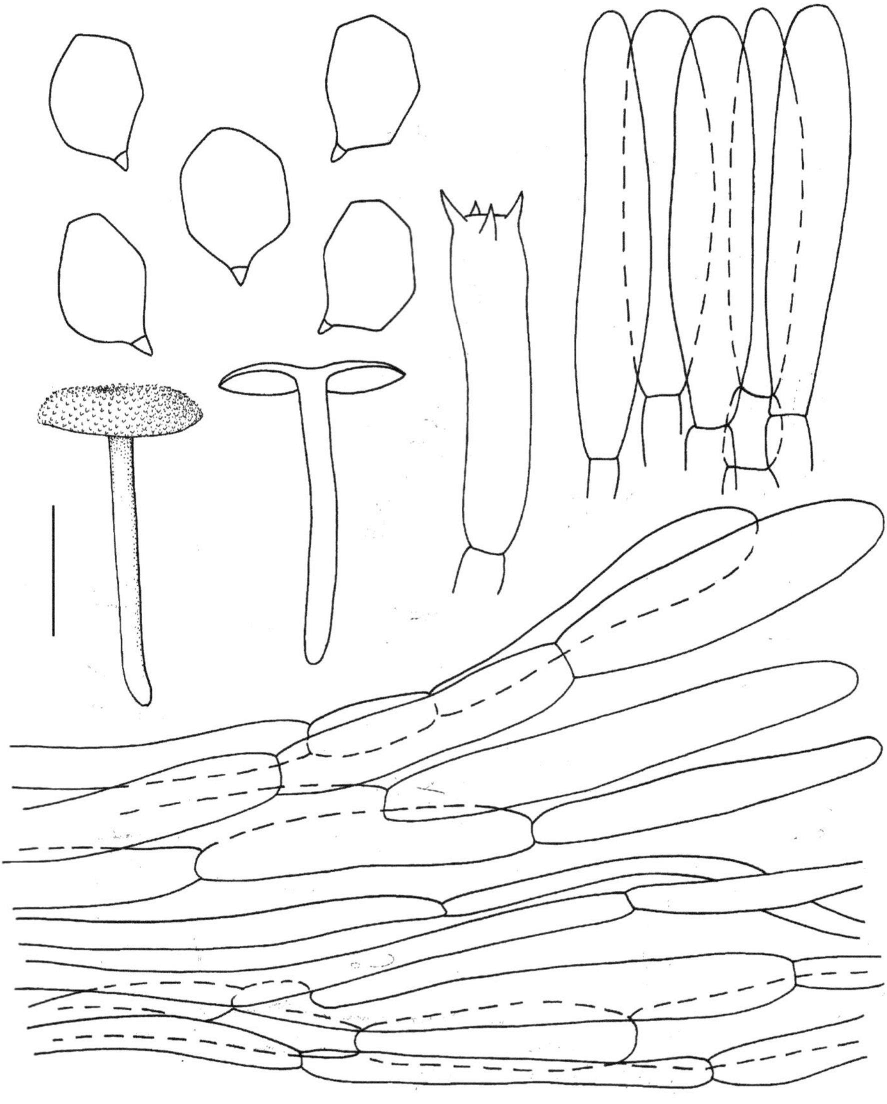

Fig. 3.65 *Entoloma moabus*. Habit, spores, basidia, cheilocystidia, and pileipellis. Bar = 10 mm or 10 μm

Diagnosis: Pileus up to 40 mm, truncate conical to convex, expanding, black to blackish violet, tomentose. Lamellae moderately distant, adnate-decurrent, sordid brownish pink with concolorous edge. Stipe concolorous with pileus, polished. Taste strong, astringent to bitter. Spores 10–12.5 × 7.5–9 μm, 6–7-angled. Cheilocystidia 22–45 × 5–11 μm, cylindrical to clavate. Pileipellis a trichoderm of 6–16 μm wide, cylindrical to clavate elements with dark brown, intracellular pigment. Clamp-connections absent.

Plate 3.64 *Entoloma obscureogracile* (Photo Machiel Noordeloos)

Holotype: Australia, Tasmania, Collinsvale, Myrtle Forest, 42° 52′ S, 147° 09′ E, 18 April 2003, G. Gates E 1717 (HO 564354, isotype in L).

Etymology: obscurus (Lat.) = dark; gracilis (Lat.) = slender.

Main characters: habit collybioid; pileus very dark blackish brown; stipe blue-grey to violaceous grey, polished; spores large; lamella edge concolorous (Plate 3.64).

Description:
Pileus 8–35 mm diam, conico-convex to convex, expanding with age to almost applanate, with deflexed margin, truncate to slightly depressed centre, not hygrophanous, not translucently striate, black to blackish violet, then brown-black, sometimes with burnt carmine tinge, tomentose all over, not squamulose. Lamellae adnate, often with decurrent tooth, ventricose, moderately distant, up to 2.5 mm broad, pale grey-blue or almost white at first, then sordid brownish pink, with entire, concolorous edge, L = 20–36, l = 3–5. Stipe 15–50 × 1.5–4 mm, cylindrical, sometimes broadened at base to 4 mm, delicate violet-grey, glabrous, polished, with white basal tomentum. Odour none. Taste mild to astringent to bitter, somewhat acrid.

Spores 10–12.5 × 7.5–9 µm, Q = 1.2–1.6, Qav = 1.4–1.5, 6–7-angled in side-view. Basidia 24–36 × 8–10 µm, 4-spored, clampless. Lamella edge sterile. Cheilocystidia 22–45 × 5–11 µm, cylindrical to clavate, hyaline. Hymenophoral trama regular, made up of cylindrical to slightly inflated elements, 5–22 µm wide. Brilliant granules

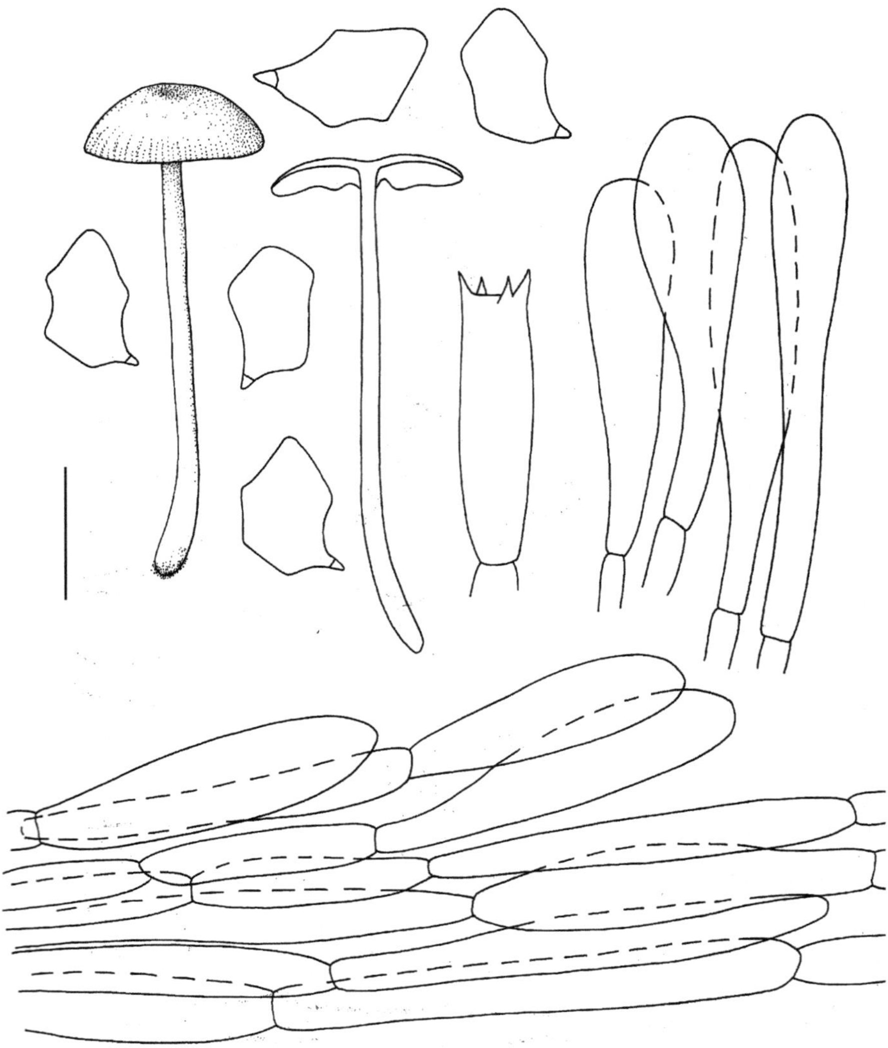

Fig. 3.66 *Entoloma obscureogracile*. Habit, spores, basidia, cheilocystidia, and pileipellis. Bar = 10 mm or 10 μm

present. Pileipellis a cutis with transitions to a trichoderm, made up of cylindrical to clavate elements, 30–70 × 6–16 μm. Pigment brown, intracellular. Stipitipellis a cutis of narrow, cylindrical hyphae, 3–15 μm wide. Caulocystidia absent. Clamp-connections absent (Fig. 3.66).

Habitat & distribution: widespread but uncommon in wet eucalypt forests.

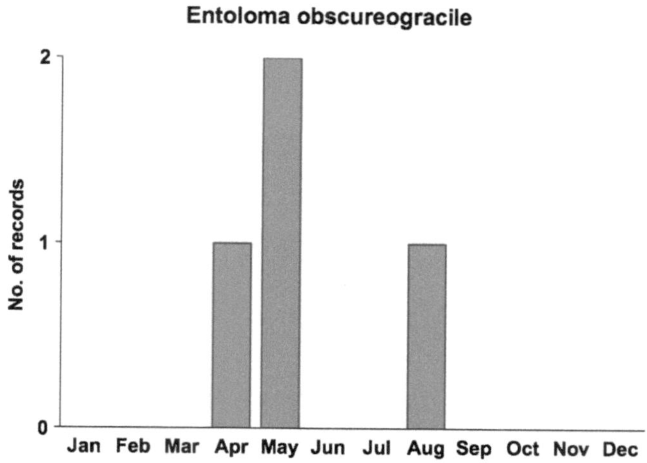

Collections examined. Australia, Tasmania, Collinsvale, Myrtle Forest, 42° 52′ S, 147° 09′ E, 18 April 2003, G. Gates E 1717 (holotype, HO 564354); Kermandie Falls, Upper Track, 43° 12′ S, 146° 52′ E, 30 May 2000, G. Gates E 949; Liffey Falls, 41° 42′ S, 146° 46′ E, 7 May 2011, M.E. Noordeloos 2011051 (L); Mt Field National Park, Tall Trees Track, 42° 41′ S, 146° 52′ E, 8 Aug. 2002, G. Gates E 1625.

Entoloma obscureogracile is a delicate species with a very dark pileus, large spores and concolorous lamella edge. It is similar to *E. amarum*, from which it differs by the mild taste and slightly paler stipe.

67. **Entoloma saponicum** G.M. Gates & Noordel., Cryptogamie Mycologie 30: 116. 2009.

Original diagnosis: Pileus 20–40 mm latus, convexus, centro depressus, haud hygrophanus, haud translucido-striatus, atro-brunneus vel griseus, fibrillosus, centro minute squamulosus. Lamellae adnatae, confertae, sordide brunneo-griseae, demum roseo-tinctae, acies lamellarum brunneo-tinctae. Stipes 30–60 × 2–6 mm, cylindraceus, griseo-brunneus, politus. Sporae 6–10 × 5–7 µm, 5–6-angulatae. Acies lamellarum sterilis. Cheilocystidia 30–60 × 4–12 µm contento brunneo. Pileipellis cutis vel trichoderma, ex elementis inflatis, usque ad 20 µm latis, constituta pigmentis intracellularibus. Granula lucentia abundantia. Fibulae desunt.

Holotype: Australia, Tasmania, Clarks Cliffs, 43° 06′ S, 147° 47′ E, 29 Dec. 2001, G. Gates E 1380 (HO 548313; isotype L).

Etymology: sapo (Lat.) = soap, the name saponicum reflecting its soapy, hot burning taste.

Main characters: habit collybioid, very dark brown or grey-brown species with umbilicate, smooth pileus except for a minutely squamulose centre; lamellae adnate-decurrent, lamella edge brown; taste burning, soapy (Plate 3.65).

Plate 3.65 *Entoloma saponicum* (Photo Michael Pilkington)

Description:

Pileus 20–40 mm diam, convex, then irregularly shaped with lobed margin, umbilicate, sometimes even almost infundibuliform, with involute to deflexed margin, not truly hygrophanous, not translucently striate except in old weathered specimens, very dark blackish brown to greyish brown or deep sepia brown, very minutely squamulose throughout or minutely squamulose at centre only, elsewhere innately radially fibrillose, dry. Lamellae adnate with pronounced decurrent tooth to distinctly decurrent, segmentiform, crowded, up to 6 mm broad, brown or grey-brown with pink tinge, edge blackish brown, more or less entire, L=25–40, l=3–5. Stipe 30–60×2–6 mm, cylindrical or compressed with groove, brittle, grey-brown, usually paler than pileus, smooth, glabrous, polished or with scattered fibrils. Context thin, concolorous. Odour none. Taste often soapy and burning.

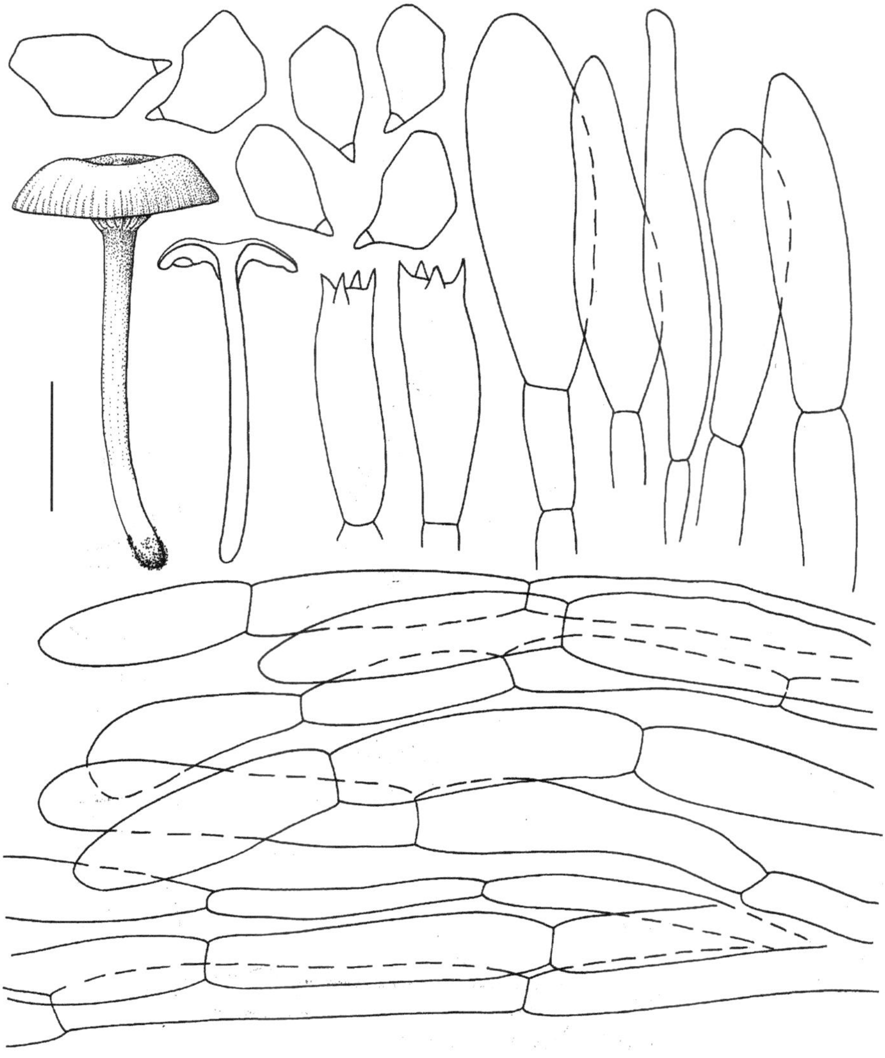

Fig. 3.67 *Entoloma saponicum*. Habit, spores, basidia, cheilocystidia, and pileipellis. Bar = 10 mm or 10 μm

Spores 6–10 × 5–7 μm, Q = 1.15–1.65(–2), 5–6-angled in side-view. Basidia 30–40 × 9–10 μm, 4-spored, clampless. Lamella edge sterile. Cheilocystidia 30–70 × 4–12 μm, narrowly to broadly clavate, fusiform or lageniform with rounded, rarely capitate apex, with dark grey or brown, intracellular pigment. Hymenophoral trama regular, made up of 4–20 μm wide, cylindrical hyphae. Pileipellis a cutis with transitions to a trichoderm at centre, consisting of cylindrical hyphae up to 12 μm wide with inflated terminal elements 20–70 × 4–18 μm. Pigment abundant, grey-brown, intracellular in pileipellis. Pileitrama regular, made up of inflated hyphae to 20 μm wide. Brilliant granules abundant. Clamp-connections absent (Fig. 3.67).

Habitat & distribution: terrestrial in small groups in forest litter in wet and dry sclerophyll forest and mixed forest; also found in a marsupial pasture. Fruiting from late spring to early winter (Group 2, in Chap. 1, see section "Fruiting Patterns of Species of Entolomataceae").

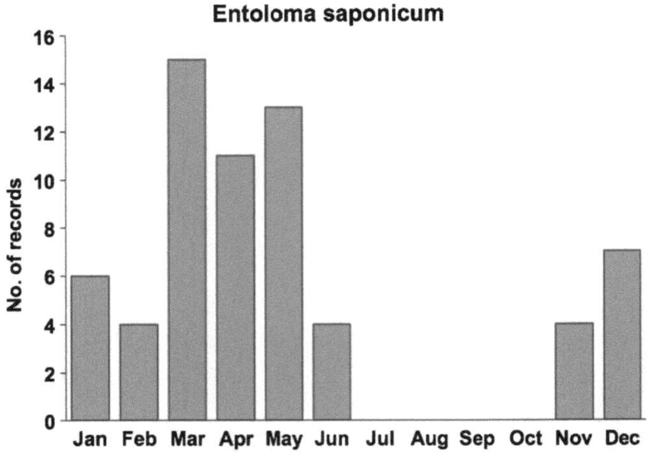

Collections examined. Australia, Tasmania, Cape Pillar Track, 43° 09′ S, 147° 56′ E, 6 March 1999, G. Gates E 174; Chauncy Vale, 42° 37′ S, 147° 16′ E, 22 Nov. 2001, G. Gates E 1329; —, 29 Nov. 2001, G. Gates E 1363; —, 10 June 2003, G. Gates E 1863; —, 11 June 2009, G. Gates E 2274; Clarks Cliffs, 43° 06′ S, 147° 47′ E, 29 Dec. 2001, G. Gates E 1380 (holotype, HO 548313; isotype, L); —, 5 March 2002, G. Gates E 1448 & E 1449; —, 4 May 2002, G. Gates E 1502; —, 13 March 2004, G. Gates E 1947 (HO 548314); —, 13 May 2005, G. Gates E 2167; Dismal Swamp, 40° 57′ S, 144° 51′ E, 5 May 2011, M.E. Noordeloos 2011038; Douglas Apsley National Park, 41° 52′ S, 148° 11′ E, 2 May 2011, M.E. Noordeloos 2011029; Florentine Rd, Lady Binney Track, 42° 43′ S, 146° 31′ E, 21 Dec. 2004, G. Gates E 2069; —, 17 Feb. 2005, G. Gates E 2080; Geeveston, Donnellys Road, 43° 07′ S, 146° 54′ E, 14 Nov. 2000, G. Gates E 1008; Junee Caves State Reserve, 42° 44′ S, 146° 36′ E, 28 March 2009, M.E. Noordeloos 2009065 & 2009066; Kermandie Falls, Lower Track, 43° 12′ S, 146° 52′ E, 16 March 1999, G. Gates E 234; —, 5 Jan. 2002, G. Gates E 1390; —, 5 April 2001, G. Gates E 1069; Lake Skinner Track, 42° 57′ S, 146° 44′ E, 11 April 1999, G. Gates E 384; Mt Arthur, 41° 16′ S, 147° 16′ E, 19 Jan. 2002, G. Gates E 1418; —, 9 Dec. 2005, G. Gates E 2222; Mt Field National Park, Lady Barron Falls, 42° 41′ S, 146° 42′ E, 1 April 1999, G. Gates E 343; —, 24 April 2001, G. Gates E 1123; Mt Wellington, Betts Vale Track, 42° 55′ S, 147° 15′ E, 15 Dec. 2001, G. Gates E 1348; —, Circle Track, 25 March 1999, G. Gates E 278 (HO 548315); —, Jacksons Track, 25 March 1999, G. Gates E 285; —, Myrtle Gully, 42° 54′ S, 147° 15′ E, 9 March 1999, G. Gates E 179 & E 187; —, 13 April 1999, G. Gates E 410; —, 3 May 2001, G. Gates E 1161; —, 27 Nov. 2001, G. Gates E 1345; —, 17 Jan. 2002, G. Gates E 1410; —, 8 Dec. 2005, G. Gates E 2220; North West Bay River, 42° 57′ S, 147° 12′ E, 7 June 2001, G. Gates E 1229; —, 17 Nov. 2001, G. Gates E 1315 &

E 1317; —, 1 Jan. 2002, G. Gates E 1384; —, 14 Feb. 2002, G. Gates E 1434; —, 1 May 2003, G. Gates E 1753; —, 9 March 2004, G. Gates E 1939; —, 18 May 2004, G. Gates E 2006; —, 21 April 2005, G. Gates E 2132; Taranna Rainforest Walk, 43° 04′ S, 147° 53′ E, 22 May 2004, G. Gates E 2010; Waterfall Bay, 43° 04′ S, 147° 57′ E, 13 March 1999, G. Gates E 214, E 218, E 222 & E 225; —, 22 Jan. 2002, G. Gates E 1421; —, 10 May 2003, G. Gates E 1789; Wielangta, Rainforest Loop Walk, 42° 42′ S, 147° 51′ E, 23 Feb. 1999, G. Gates E 145; —, 20 March 1999, G. Gates E 272; —, 5 May 2005, G. Gates E 2159.

Entoloma saponicum is a typical collybioid species with distinctly umbilicate pileus, segmentiform, adnate-emarginate lamellae often with very prominent decurrent tooth, and stiff, polished stipe. The colour of the fruit bodies may vary from grey, moderately dark brown to very dark grey-brown, and are uniformly coloured, or have a distinctly darker centre. Variants with a non-pigmented lamella edge also occur. *Entoloma saponicum* belongs, therefore, to a rather large group of similar species, which are not easy to distinguish. The Australian *Entoloma fuscum* (Cleland) E. Horak differs particularly by having a fertile lamella edge (Grgurinovic 1997; Horak 2008). From New Zealand, *Entoloma phaeomarginatum* E. Horak differs by the distinctly tomentose to squamulose pileus and fibrillose stipe, and *Entoloma asprelloides* G. Stev. has a blue lamella edge; in addition, both species have much larger spores (Horak 2008). Also from New Zealand, *Entoloma persimile* E. Horak differs by a paler, very hygrophanous, distinctly translucent-striate pileus with squamulose centre, while *Entoloma inventum* E. Horak differs particularly in having a very hygrophanous, deeply translucent-striate pileus with squamulose centre, and a strong rancid-farinaceous smell and taste. *Entoloma turci* (Bres.) M.M. Moser may represent a Northern Hemisphere counterpart of our species, with similar dark brown colours, differing by having an entirely tomentose pileus, and reddish bruising on the base of the stem with age, and larger spores. The record of *Entoloma turci* from Papua New Guinea by Horak (1980, pp. 226–227) is probably a misidentification since it describes a species with a fertile lamella edge. *Entoloma spermatiolens* E. Horak from Papua New Guinea differs in smell, a more pronouncedly squamulose pileus, and larger, differently shaped spores.

Stirps ***Formosum***

Pileus and stipe vivid yellow, yellow-brown or red-brown; stipe polished.

68. **Entoloma pyropus** Noordel. & G.M. Gates, spec. nov.

MycoBank #564505.

Diagnosis: Habit collybioid. Pileus 10–20 mm, convex, expanding, distinctly translucent-striate, fibrillose at margin, squamulose at centre, warm reddish yellow to orange. Lamellae distant, pinkish brown with reddish yellow edge. Stipe 30–40 × 1–2 mm, pale yellow, polished. Spores 8.5–11 × 5.5–7.5 μm, 5–7-angled. Cheilocystidia 22–45 × 7–12 μm, cylindrical to clavate, with pale brown intracellular pigment. Pileipellis trichoderm of inflated terminal elements with brown, intracellular pigment. Clamp-connections absent.

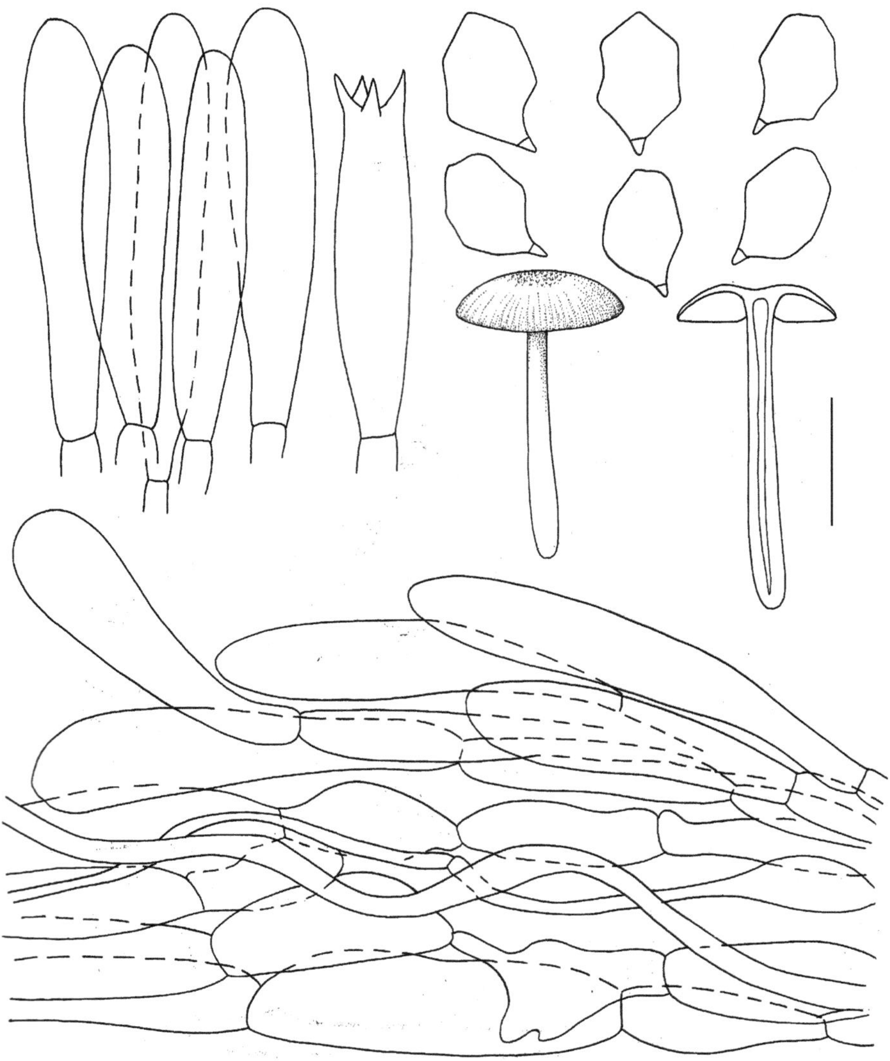

Fig. 3.68 *Entoloma pyropus*. Habit, spores, basidia, cheilocystidia, and pileipellis. Bar = 10 mm or 10 μm

Holotype: Australia, Tasmania, Cradle Mtn/Lake St Clair National Parks, Frog Flats, 41° 50′ S, 146° 01′ E, 1 May 2004, G. Gates E 1987 (HO 561502, isotype in L).

Etymology: pyropus (Lat.) = bronze, referring to the bronze-brown colour of the pileus.

Main characters: habit collybioid; pileus warm reddish yellow to orange, translucent-striate; lamella edge concolorous; stipe rather pale yellow; spores small (Fig. 3.68).

Description:
Pileus 10–20 mm diam, convex to plano-convex with slightly depressed centre, with deflexed to straight margin, hygrophanous, deeply translucently striate, vivid red-yellow, bronze or golden brown, darker at centre, rarely with touch of olive green, minutely squamulose at centre, radially fibrillose on margin. Lamellae adnate, ventricose, moderately distant, up to 4 mm broad, pinkish brown with reddish yellow edge. Stipe 30–40 × 1–2 mm, cylindrical, pale yellow (straw yellow) or yellow-grey, base sometimes stronger yellow, pruinose at apex, downwards smooth, glabrous, polished, base with white tomentum. Odour mild. Taste somewhat farinaceous.

Spores 8.5–11 × 5.5–7.5 μm, average 9.1 × 6.5 μm, Q = 1.2–1.7, Qav = 1.5, regular to irregular 5–7-angled in side-view. Basidia 20–50 × 7–12 μm, 4-spored, clampless. Lamella edge sterile. Cheilocystidia 22–45 × 7–12 μm, cylindrical to clavate, with pale brown intracellular pigment. Hymenophoral trama regular, made up of cylindrical hyphae 6–19 μm wide. Brilliant granules present. Pileipellis a cutis with transition to a trichoderm at centre, made up of repent, cylindrical to inflated hyphae, 8–16 um wide with clavate terminal elements, 60–120 × 14–22 μm. Pigment brown, intracellular. Pileitrama regular, made up of long, cylindrical to slightly inflated elements, up to 22 μm wide. Stipitipellis a cutis of narrow, cylindrical hyphae, 5–12 μm wide. Caulocystidia absent. Clamp-connections absent.

Habitat & distribution: widespread in wet eucalypt forests, but recorded from only five localities.

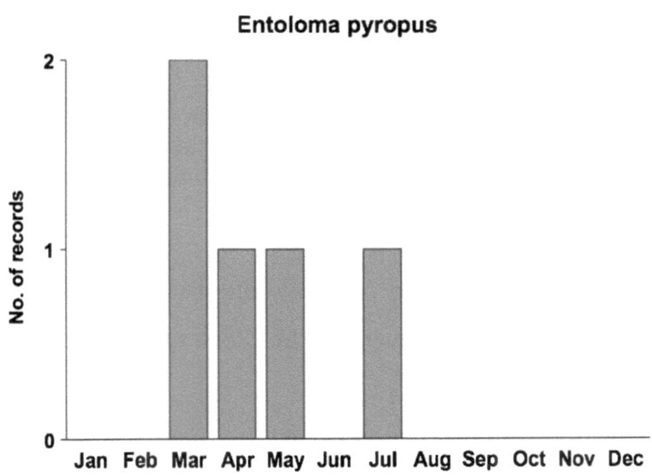

Collections examined. Australia, Tasmania, Cradle Mtn/Lake St Clair National Parks, Frog Flats, 41° 50′ S, 146° 01′ E, 1 May 2004, G. Gates E 1987 (HO 561502, holotype); Duckhole Lake Track, 43° 22′ S, 146° 53′ E, 13 July 2002, G. Gates E 1599; Hellyer Gorge, 41° 16′ S, 145° 37′ E, 26 April 2002, G. Gates E 1491; Mt Wellington, Circle Track, 42° 55′ S, 147° 15′ E, 11 March 1999, G. Gates E 199, E204, E205 & E207; Wielangta, 42° 42′ S, 147° 51′ E, 29 March 2001, G. Gates E 1050.

Entoloma pyropus is distinguished from *E. rufobasis* mainly by the more vividly coloured pileus and stipe, and the lack of an orange blotch at the stipe base. It resembles also *Entoloma formosum* (Fr.) Noordel. from Europe and North America. An aberrant collection with slightly larger spores and concolorous lamella edge is described below.

Description:
Pileus 12–26 mm diam, convex with uplifted marginal zone, with slightly depressed centre, with deflexed margin, not hygrophanous, translucently striate at margin, broad bronze brown, reddish brown, with minute darker squamules, denser at centre, spreading towards margin. Lamellae adnate, ventricose, fairly distant, pale pink with irregular, concolorous edge. Stipe 26–35 × 1–3 mm, cylindrical, pale brown (no trace of grey), pruinose at apex, downwards polished with hardly any basal mycelium. Odour somewhat aromatic. Taste nutty.

Spores 10–12 × 7–9.5 µm, Q = 1.1–1.6, Qav = 1.4, 6–7(–8)-angled in side-view. Basidia 4-spored, clampless. Lamella edge sterile. Cheilocystidia 22–45 × 7–13 µm, cylindrical to clavate, without pigment. Hymenophoral trama regular, made up of inflated to cylindrical elements, 5–12 µm wide. Pileipellis a cutis with transitions to a trichoderm, made up of clavate terminal elements, 20–85 × 5–19 µm wide. Pigment pale brown, intracellular. Pileitrama regular, made up of cylindrical hyphae 5–18 µm wide. Brilliant granules abundant. Stipitipellis a cutis of narrow, cylindrical hyphae, 3–8 µm wide. Caulocystidia absent. Clamp-connections absent.

Habitat & distribution: in wet eucalypt forest. Uncommon.

Collection examined: Australia, Tasmania, Kermandie Falls, Lower Track, 43° 12′ S, 146° 52′ E, 16 March 1999, G. Gates E 233.

This collection is similar to *E. pyropus* with respect to the colour of the fruit body, but the lamella edges are concolorous with the sides, and the spores are larger. For the time being it is not given a formal taxonomic status, needing more collections to study the variability of the species.

Stirps ***Indistinctum***

Pileus medium brown; stipe grey or yellow with slate blue tinge, polished

69. **Entoloma indistinctum** Noordel. & G.M. Gates, spec. nov.

MycoBank #564506.

Diagnosis: Pileus 15–30 mm, convex, depressed, translucently striate, medium brown to light grey-brown with darker, minutely squamulose centre. Lamellae distant, adnate-emarginate, pale pink with concolorous edge. Stipe 20–60 × 1–5 mm, pale grey, polished. Spores 8–10 × 6–8 µm, 5–7-angled. Cheilocystidia 30–60 × 4–10 µm, cylindrical to clavate. Pileipellis a cutis to trichoderm up to 20 µm wide, inflated elements with intracellular pigment. Clamp-connections absent.

Holotype: Australia, Tasmania, Collinsvale, Myrtle Forest, 42° 52′ S, 147° 09′ E, 3 May 2005, G. Gates E 2146 (HO 561503, isotype in L).

Plate 3.66 *Entoloma indistinctum* var. *indistinctum* (Photos Genevieve Gates)

Etymology: indistinctum (Lat.) = indistinct.

Key to the varieties:

1. Spores 8–10(–11) × 6–7.5(–8) µm	69a. var. ***indistinctum***
1. Spores 9–12(–14) × 7–9 µm	69b. var. ***macrosporum***

69a. **Entoloma indistinctum** var. **indistinctum** (Plate 3.66)

Main characters: habit collybioid; pileus moderately dark brown, yellow-brown or grey-brown, moderately to deeply translucently striate; stipe pale grey; lamellae and lamella edge concolorous; spores 8–10(–11) × 6–7.5(–8) µm.

Description:
Pileus 15–30 mm diam, convex with depressed centre, slightly hygrophanous, when moist translucently striate at margin or up to centre, medium brown to light grey-brown with dark reddish brown or grey-brown centre, minutely squamulose at centre, fibrillose, and sometimes sulcate on margin. Lamellae adnate-emarginate

with decurrent tooth, ventricose, fairly distant, up to 5 mm broad, pale pink with concolorous edge. Stipe 20–60 × 1–5 mm, cylindrical, pale grey, glabrous, polished, base with white tomentum. Odour and taste indistinct.

Spores 8–10 × 6–8 μm, Q = 1.2–1.7, Qav = 1.3–1.4, heterodiametric, 5–7-angled in side-view. Basidia 20–32 × 7–15 μm, 4-spored, clampless. Lamella edge sterile (serrulatum-type). Cheilocystidia 30–60 × 4–10 μm, cylindrical to clavate without pigment. Hymenophoral trama regular, made up of cylindrical elements, 5–20 μm wide. Pileipellis a cutis with transitions to a trichoderm, made up of cylindrical to clavate terminal elements, 40–90 × 8–22 μm. Pigment brown, intracellular in pileipellis. Pileitrama regular, made up of cylindrical, 9–20 μm wide elements. Brilliant granules present. Stipitipellis a cutis of cylindrical hyphae, 3–9 μm wide. Caulocystidia absent. Clamp-connections absent (Fig. 3.69).

Habitat & distribution: widespread but uncommon in wet eucalypt forest.

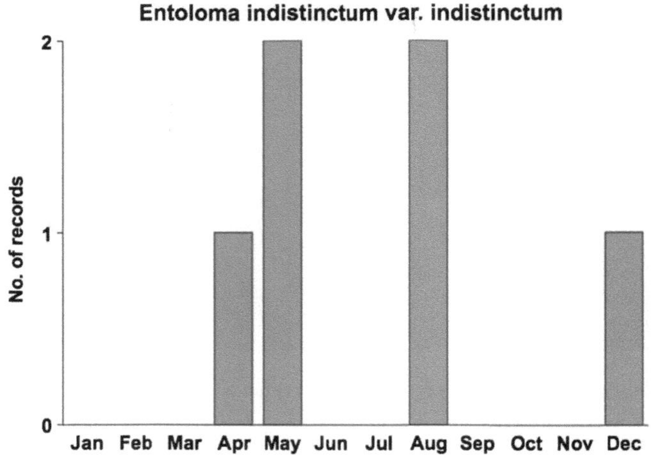

Collections examined. Australia, Tasmania, Arve Rd, Keoghs Walk, 43° 09′ S, 146° 47′ E, 22 Dec. 2001, G. Gates E 1373; Collinsvale, Myrtle Forest, 42° 52′ S, 147° 09′ E, 3 May 2005, G. Gates E 2146 (holotype, HO 561503); Growling Swallet, 42° 41′ S, 146° 30′ E, 14 May 2001, G. Gates E 1146; Mt Field National Park, Lyrebird Walk, 42° 41′ S, 146° 40′ E, 24 April 2001, G. Gates E 1114; —, Tall Trees Track, 8 Aug. 2002, G. Gates E 1625; Ridgeway, 42° 56′ S, 147° 18′ E, 15 Dec. 2001, G. Gates E 1369; Warra LTER site, 43° 06′ S, 146° 41′ E, 23 Aug. 2005, G. Gates E 2187.

69b. **Entoloma indistinctum** var. **macrosporum** Noordel. & G.M. Gates, var. nov. MycoBank #564508.

Diagnosis: Differs from the type variety by larger spores, 9–12(–14) × 7–9 μm.

Holotype: Australia, Tasmania, Mt. Field National Park, Lady Barron Falls, 42° 41′ S, 146° 42′ E, 1 April 1999, G. Gates E 342 (HO 564355, isotype in L).

Etymology: macro = large, referring to the large spores.

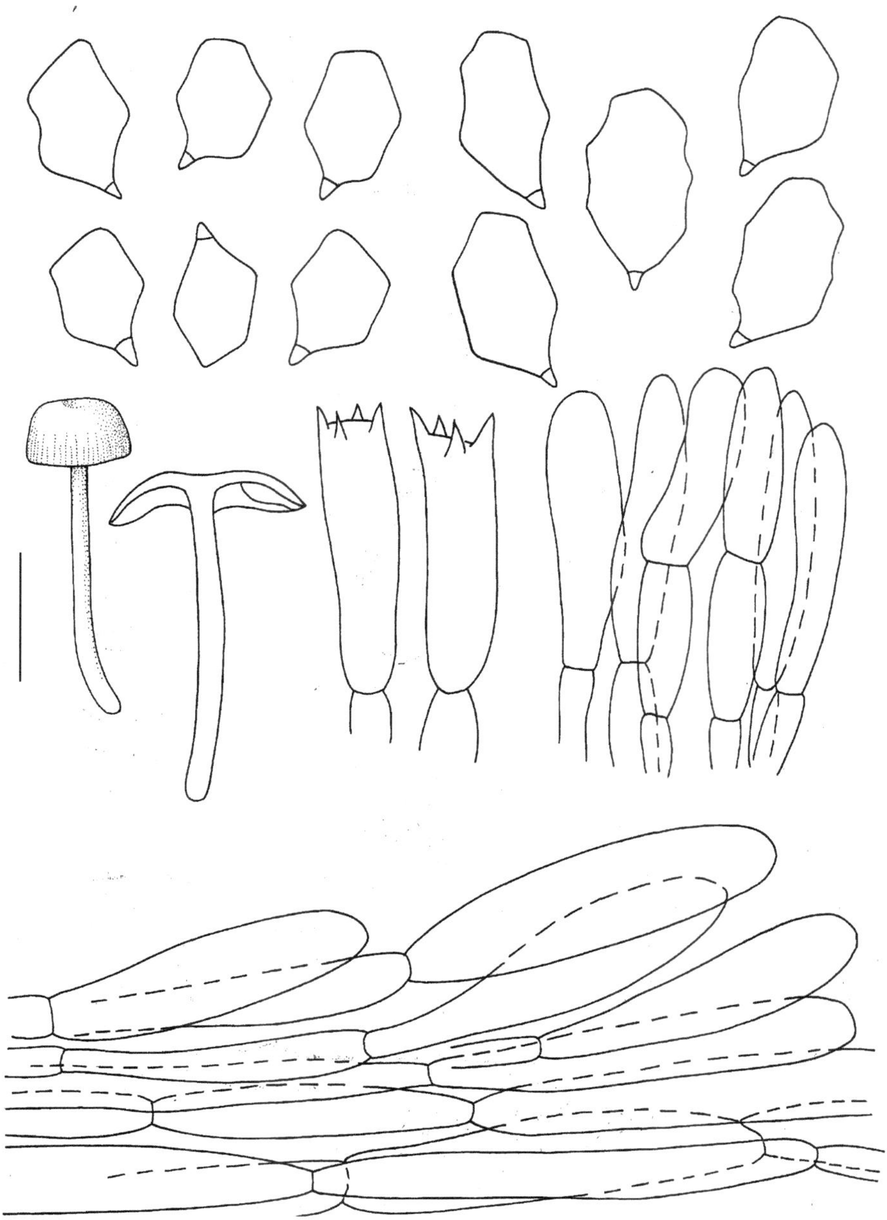

Fig. 3.69 *Entoloma indistinctum*. Habit, spores, basidia, cheilocystidia, and pileipellis. Bar = 10 mm or 10 μm. The drawings of habit, basidia, cheilocystidia and pileipellis apply to both varieties. However, the 6 spores in the upper left-hand corner apply to var. *indistinctum* and the 5 spores in the upper right-hand corner apply to var. *macrosporum*

Characters as for the type variety, apart from the larger spores.

Habitat & distribution: wet eucalypt forests.

Entoloma indistinctum var. macrosporum

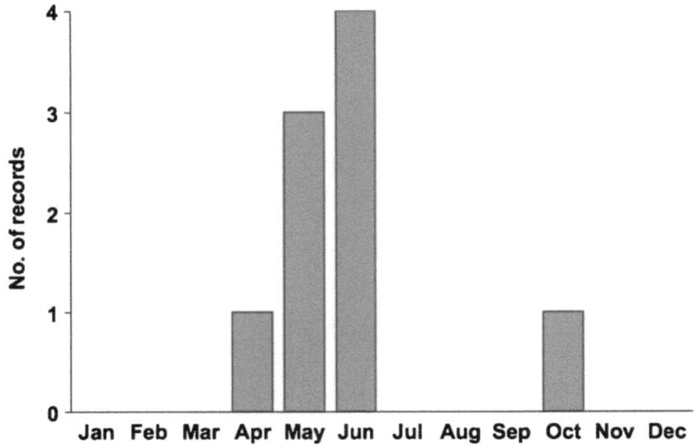

Collections examined. Australia, Tasmania, Duckhole Lake Track, 43° 22′ S, 146° 53′ E, 22 May 2008, G. Gates E 2269; Growling Swallet, 42° 41′ S, 146° 30′ E, 1 May 2001, G. Gates E 1139; Mt Field National Park, Lady Barron Falls, 42° 41′ S, 146° 42′ E, 1 April 1999, G. Gates E 342 (holotype, HO 564355); —, 1 June 2010, F. Karstedt 1550; —, Tyenna River Walk, 23 Oct. 2001, G. Gates E 1299; Mt Wellington, Lower Sawmill Track, 12 June 2001, G. Gates E 1242; North West Bay River, 42° 57′ S, 147° 12′ E, 7 June 2001, G. Gates E 1224; Wielangta, Sandspit Forest Reserve, Rainforest Walk, 42° 42′ S, 147° 51′ E, 14 June 2001, G. Gates E 1249.

Under the name *E. indistinctum*, we have included a fairly large number of collections with dull yellow-ochre, pinkish brown or grey-brown pileus, a rather pale grey stipe, and pink lamellae with concolorous edge. It is a duller coloured species than *E. pyropus*, which has a more vivid orangy yellow-brown pileus and a stipe without grey. Within this complex, spore size and shape vary considerably, and further research may well demonstrate that it is complex of several cryptic species. Two varieties are distinguished at present, the typical variety with spores 8–10 × 6–7.5 μm, and variety *macrosporum* with considerably larger spores, ranging from 9 to 12(–14) × 7–9 μm. *Entoloma austrosarcitulum* is very similar, differing by the brown lamella edge, and the very pale stipe with a hint of blue-grey.

70. **Entoloma austrosarcitulum** Noordel. & G.M. Gates, spec. nov.

MycoBank #564564.

Diagnosis: Habit collybioid. Pileus 20–55 mm, convex, expanding, depressed, translucently striate, medium brown with darker centre. Lamellae adnexed, greyish orange with pale brown edge. Stipe yellowish with faint blue tinge. Spores 9–11 × 7–9 μm. Cheilocystidia 35–45 × 9–12 μm, clavate-cylindrical with golden brown intracellular pigment. Pileipellis a cutis to trichoderm of inflated elements, up to 20 μm wide with brown intracellular pigment. Clamp-connections absent.

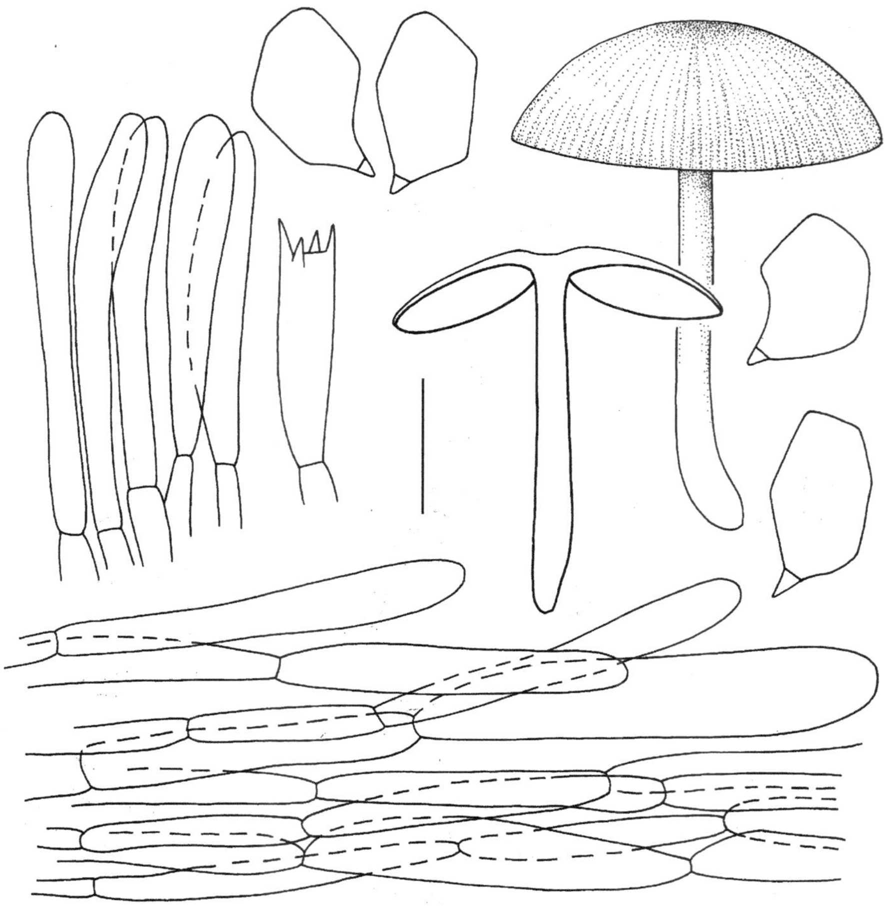

Fig. 3.70 *Entoloma austrosarcitulum*. Habit, spores, basidia, cheilocystidia, and pileipellis. Bar = 10 mm or 10 μm

Holotype: Growling Swallet, 22 April 2004, M.E. Noordeloos 2004074 (HO 564356).

Etymology: austro = southern, being a southern counterpart of *Entoloma longistriatum* var. *sarcitulum* (P.D. Orton) Noordel.

Main characters: habit collybioid; pileus medium brown; lamellae greyish orange with brown edge; stipe yellowish with blue hue (Fig. 3.70).

Description:
Pileus 20–55 mm diam, convex with deflexed margin then expanding with straight margin, with slight central depression, not distinctly hygrophanous, translucently striate up to two-thirds of the radius, medium dark brown to grey-brown with darker centre, innately fibrillose at margin, finely squamulose at centre. Lamellae broadly adnexed,

ventricose, fairly crowded, up to 7 mm broad, thin, greyish orange (6B4), with pale brown edge, lamellulae in two tiers. Stipe 20–55 × 3–4 mm, cylindrical, very pale yellowish white, but often also with a slight bluish grey hue, glabrous, polished, with white basal tomentum. Odour not distinct. Taste bitter becoming mild.

Spores 9–11 × 7–9 μm, Q = 1.25–1.7, regularly 5–7-angled. Basidia 20–40 × 6–9 μm, 4-spored, clampless. Lamella edge sterile. Cheilocystidia clavate-cylindrical, 35–45 × 9–12 μm with golden brown, intracellular pigment. Hymenophoral trama regular, made up of cylindrical hyphae, up to 20 μm wide, with very abundant brilliant granules. Pileipellis a cutis with transitions to a trichoderm, made up of inflated hyphae, 8–12 μm wide with clavate terminal elements, up to 20 μm wide, with brown intracellular pigment. Pileitrama regular, made up of inflated hyphae, up to 20 μm wide with abundant brilliant granules. Stipitipellis a cutis of cylindrical hyphae, up to 10 μm wide. Caulocystidia absent. Clamp-connections absent.

Habitat & distribution: known from only two locations, both in wet eucalypt forests.

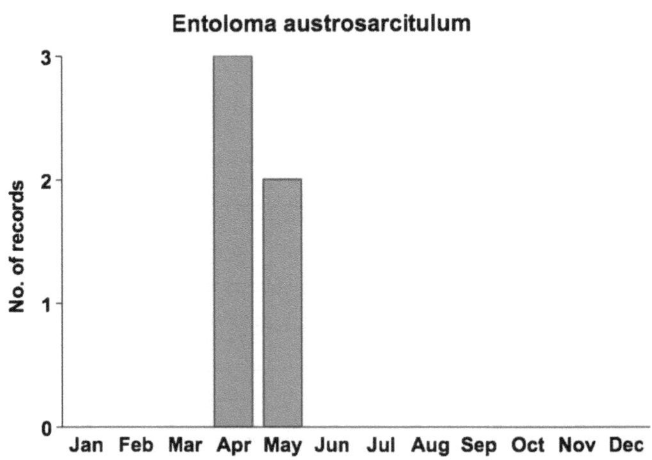

Collections examined. Australia, Tasmania, Growling Swallet, 42° 41′ S, 146° 30′ E, 11 May 2000, G. Gates E 901; —, 1 May 2001, G. Gates E 1147; Mt Field National Park, 42° 41′ S, 146° 42′ E, 1 April 1999, G. Gates E 345; —, 17 April 2004, M. E. Noordeloos 2004043; —, 22 April 2004, M.E. Noordeloos 2004074 (holotype, HO 564356).

Entoloma austrosarcitulum resembles very much *E. longistriatum* (Peck) Noordel. (syn. *E. sarcitulum*), a widespread temperate species of the Northern Hemisphere (Noordeloos 2004), particularly on account of the vivid brown pileus, pallid yellow stipe with a faint hue of blue, and coloured lamella edge. *Entoloma persimile* is much darker brown, and lacks the faint bluish tinge in the stipe.

Stirps ***Melanocephalum***

Pileus very dark blackish-blue; lamellae frequently tinged blue, especially in youth, with or without coloured lamella edge; stipe concolorous with pileus or paler, polished to subfibrillose; pigment in pileipellis often blue and brown, often in different layers.

71. **Entoloma melanocephalum** G. Stev., Kew Bulletin 16: 231. 1962.

Original diagnosis: Pileus 2.5–3.5 cm diam, fuscus, aliquando atro-venetotinctus, late convexus, interdum breviter umbilicatus, subtiliter fibrillosus vel squamulis parvis fibrillosis ornatus; caro lazulino-ardisiaca. Lamellae adnexae vel adnatae, subconfertae, griseo-lazulinae, aciebus obscurioribus, demum sporis obscure roseis maculatae. Stipes 4.5–5.5 cm × 3–5 mm, atrolazulinus vel lineus, basi albidus, e farcto cavus. Sporae 7×10.5–12 μm, irregulariter angulatae, in cumulo sordide roseae. Acies lamellarum sterilis, ex hyphis subgelatinosus.

Holotype: New Zealand, Wellington, Keith George Park, 30 May 1949, G. Stevenson 597.

Etymology: melanos (Gr.)=black; cephalon (Gr.)=head, referring to the almost black pileus.

Main characters: habit collybioid; basidiocarps very dark blue-black; lamella edge blackish-serrulate; spores 8–12 μm long (Plate 3.67).

Description:
Pileus 20–70 mm diam, hemispherical to convex, expanding to convex or plano-convex, sometimes truncated, often distinctly umbilicate, with deflexed then straight margin, not hygrophanous, not or indistinctly translucent striate in aged specimens, uniformly blue-black or violet-black (16F2–3), entirely finely fibrillose-tomentose, breaking up into small concentrically arranged squamules with age, slightly radially grooved on back of lamellae with age. Lamellae adnate-emarginate, segmentiform to subventricose, moderately distant, dark blue then blue or violaceous blue with pinkish grey tinge, with or without a darker blackish violet or blackish blue edge, L=20–40, l=3–5. Stipe 20–60×(1–)2–6 mm, cylindrical or compressed with groove, dark blue concolorous with pileus, apex with blackish dots, downwards glabrous, polished, with white basal tomentum, which occasionally turns orange when handled or with age. Odour indistinct. Taste indistinctly to distinctly farinaceous-rancid, not bitter.

Spores 8–12×7–9 μm, 5–7-angled in side-view. Basidia 19–34×7–10 μm, 4-spored, clampless. Lamella edge sterile with densely clustered cheilocystidia, 20–40×7–16 μm with or without blackish intracellular pigment. Pileipellis a cutis with transitions to a trichoderm or a true trichoderm, made up of clusters of inflated terminal elements 20–70×6–20 μm; suprapellis of relatively slender, cylindrical hyphae 4–12 μm wide, gradually passing into pileitrama. Pigment in suprapellis purple-brown and blackish blue, in subpellis mainly blue-granular. Brilliant granules abundant in trama. Stipitipellis dense clusters of cylindrico-clavate terminal elements ('caulocystidia'), similar to the trichodermal tufts in pileipellis, with blue and brown intracellular pigment at apex. Clamp-connections absent (Fig. 3.71).

Habitat & distribution: widespread in wet eucalypt forests.

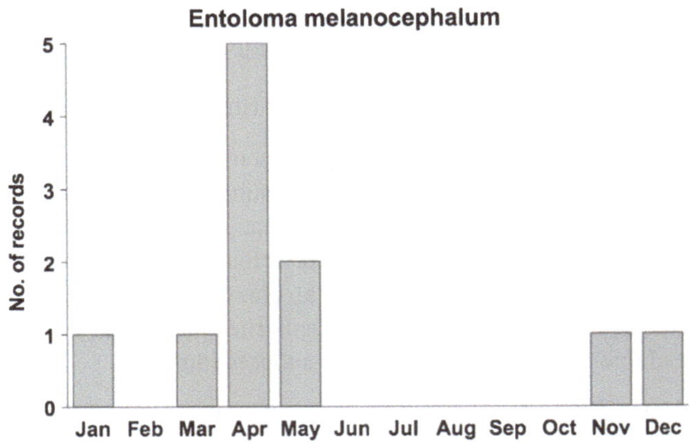

Plate 3.67 *Entoloma melanocephalum* (Photo Machiel Noordeloos)

Collections examined. Australia, Tasmania, Bruny Island, Mt Mangana, 43° 22′ S, 147° 17′ E, 26 May 2001, G. Gates E 1147; —, 31 March 2009, M.E. Noordeloos 2009088; Duckhole Lake Track, 43° 22′ S, 146° 53′ E, 16 April 2005, G. Gates E 2111; Ellendale, 42° 37′ S, 146° 43′ E, 5 April 1999, G. Gates E 359; Growling Swallet, 42° 41′ S, 146° 30′ E, 22 April 2004, M.E. Noordeloos 2004073; —, 21 April 2007, G. Gates E 2252; Mt Field National Park, Tall Trees Track, 42° 41′ S, 146° 42′ E, 13 Nov. 1999, G. Gates E 746; North West Bay River, 42° 57′ S, 147° 12′ E, 18 April 2007, G. Gates 2249; Styx Valley, Christmas Tree Grove, 42° 50′ S,

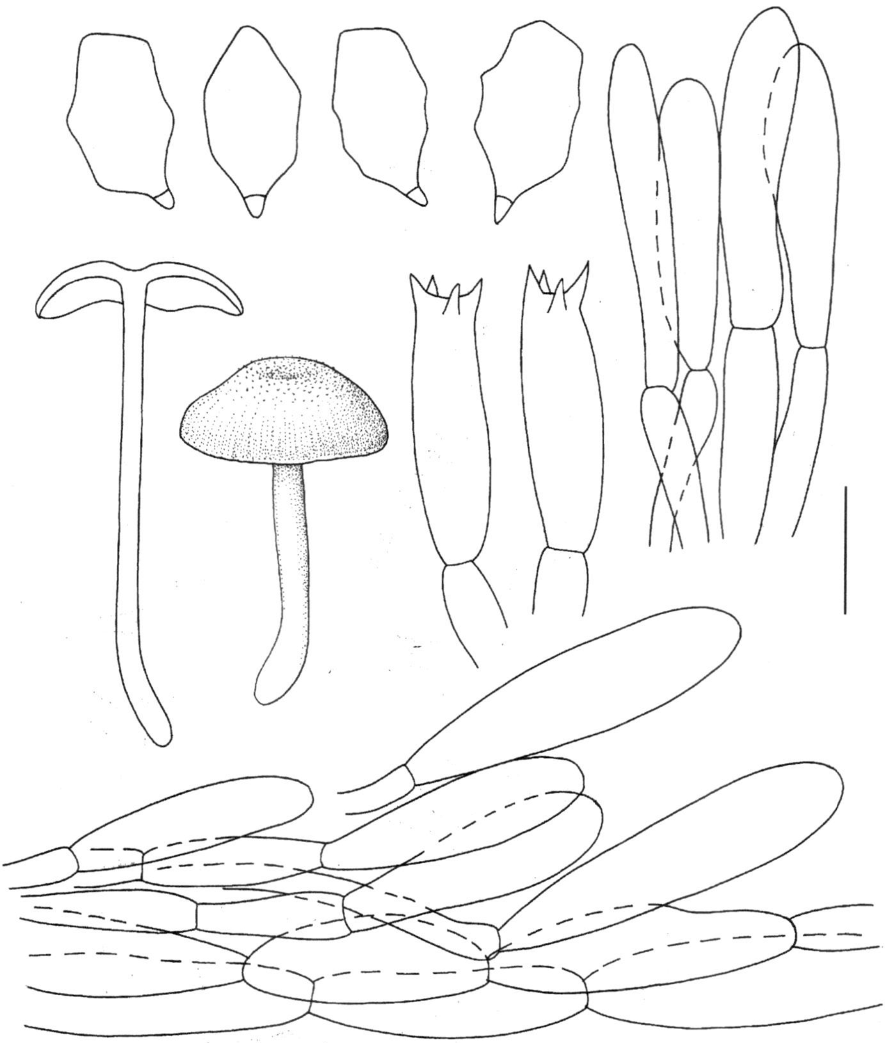

Fig. 3.71 *Entoloma melanocephalum.* Habit, spores, basidia, cheilocystidia, and pileipellis. Bar = 10 mm or 10 μm

146° 41′ E, 13 May 2010, M. E. Noordeloos 2010016. — New Zealand, Wellington, Keith George Park, 30 May 1949, G. Stevenson 597 (holotype, K).

An interesting observation on fresh material (M.E. Noordeloos 2010016) is that in the terminal elements of the suprapellis, some specimens have brown-purple, and others deep blue, intracellular pigment. *Entoloma melanocephalum* was originally described by Stevenson (1962) as a *Leptonia* with very dark blackish brown pileus, occasionally with dark indigo-blue tinges, lamellae with blue tinge and coloured edge, and dark blue, polished stipe. Her description and colour plate leave no doubt that it is very similar to the European *E. chalybaeum* (Pers.) Zerova. Our collections from Tasmania fit well with the original concept of Stevenson (1962), which is supported by a study of the holotype.

Horak (1973) gave a redescription of *E. melanocephalum* that deviates from the original by the distinct greenish-blue or olive tinges in the stipe, not mentioned by Stevenson. In the same paper, he described *Entoloma decolorans* E. Horak, with very similar characters, however, lacking green or olivaceous tinges in the stipe and with a concolorous lamella edge, and also *E. rubromarginatum* E. Horak with a reddish lamella edge. Later, Horak (2008) reduced both species to synonymy under *E. melanocephalum*, apparently accepting a great deal of variability for that species. Interestingly, the description of the emended concept of *Entoloma melanocephalum* (Horak 2008) does not contain the character states thought to be distinctive for the synonymised taxa (concolorous or reddish lamella edge respectively). Green or olivaceous tinges have, however, never been observed in our collections. According to Horak (2008), *Entoloma melanocephalum* has a wide geographical range, including New Zealand, Papua New Guinea and S.E. Asia. *Entoloma discrepans* Noordel. & G.M. Gates is similar, differing by the generally more slender basidiocarps, concolorous or reddish brown edge, and smaller spores.

72. **Entoloma discrepans** Noordel. & G.M. Gates, spec. nov.

MycoBank #564509.

Diagnosis: Habit collybioid. Pileus entirely tomentose-squamulose, very dark blackish blue; stipe very dark blackish blue; lamellae blue-grey with pink tinge, with or without brown edge; spores 8–10×6–7 μm, 5–7-angled. Cheilocystidia 20–50×10–20 μm, clavate or cylindrical, with or without brown intracellular pigment. Pileipellis a trichoderm of inflated, up to 25 μ wide terminal elements with blue and brown intracellular pigment. Clamp-connections absent.

Holotype: Australia, Tasmania, Warra LTER site, 43° 06′ S, 146° 41′ E, 27 April 2004, G. Gates E 1979 (HO 564357; isotype in L).

Etymology: discrepans (Lat.) = confusing (Plate 3.68).

Description:
Pileus 10–40 mm diam, truncate-campanulate, hemispherical to convex, with deflexed then straight margin, slightly radially grooved along back of lamellae with age, not hygrophanous, not translucently striate or slightly so in old specimens, very dark blackish blue, entirely finely tomentose then minutely squamulose, dull. Lamellae broadly adnate with decurrent tooth, slate blue-grey, paler than pileus, finally with pink tinge, with more or less entire, concolorous or reddish-brown edge, L = 25–50, l = 3–5. Stipe 30–50×2–5 mm, cylindrical or compressed with groove, dark blue, concolorous with pileus, apex with scattered bluish flocks, downwards more or less glabrous, dull, polished, with white basal tomentum. Odour indistinct. Taste very acidulous or mild.

Spores 8–10(–11)×6–7 μm, Q = 1.3–1.5, Qav = 1.4, heterodiametric, 5–7-angled in side-view with pronounced angles. Basidia 20–34×7–10 μm, 4-spored, clampless. Lamella edge sterile or fertile, not of the serrulatum-type. Cheilocystidia

Plate 3.68 *Entoloma discrepans* (Photos Michael Pilkington (*top*), Machiel Noordeloos (*bottom*))

20–50 × 10–20 µm, clavate or cylindrical, with or without brown intracellular pigment. Hymenophoral trama regular, made up of cylindrical to inflated elements 30–90 × 5–20 µm. Brilliant granules present. Pileipellis a trichoderm of densely clustered, clavate terminal elements, 40–80 × 10–25 µm wide, subpellis made up of radially arranged, relatively slender, cylindrical elements 30–90 × 5–10 µm wide. Pigment of two types: (1) brown, diffuse, intracellular in terminal elements of suprapellis, and (2) blue, granular and also deposited in fine grains on inner side of hyphal walls in subpellis and in upper part of pileitrama. Pileitrama regular, made up of cylindrical to inflated hyphae, 8–20 µm wide, mixed with much narrower, cylindrical connecting hyphae, 3–6 µm wide. Stipitipellis a cutis of narrow, cylindrical

Fig. 3.72 *Entoloma discrepans*. Habit, spores, basidia, cheilocystidia, and pileipellis. Bar = 10 mm or 10 μm

hyphae, 4–10 μm wide with pale grey-blue intracellular pigment. Caulocystidia present as clusters of clavate terminal elements at apex of stipe only, 20–70 × 6–20 μm, with pale brown, intracellular pigment. Clamp-connections absent (Fig. 3.72).

Habitat & distribution: widespread in dry and wet sclerophyll forest. Fruiting throughout the year, with a fair number of records in spring and summer months (Group 5, in Chap. 1, see section "Fruiting Patterns of Species of Entolomataceae").

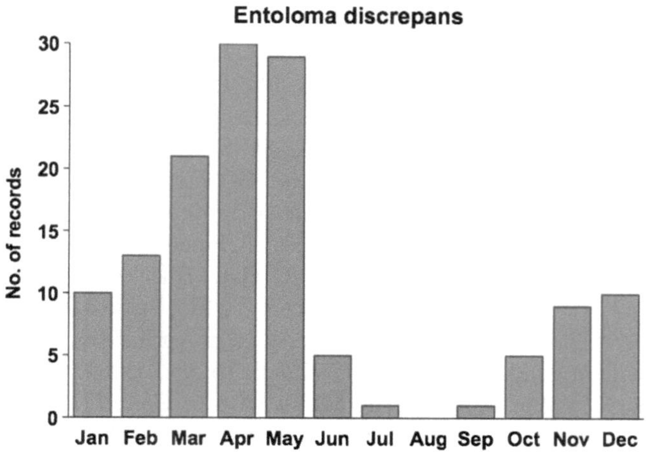

Collections examined. Australia, Tasmania, Adamsons Peak Track, 43° 20′ S, 146° 54′ E, 1 May 1999, G. Gates E 461; Arve Road, Zig-Zag Track, 43° 06′ S, 146° 45′ E, 23 Nov. 1999, G. Gates E 762; Bermuda Road, 43° 04′ S, 146° 54′ E, 17 April 2003, G. Gates E 1715; Bruny Island, Mavista Falls Picnic Area, 8 April 2001, G. Gates E 1083; —, 28 March 2002, G. Gates E 1464; —, 5 Feb. 2004, G. Gates E 1928; —, Mt Mangana, 43° 22′ S, 147° 17′ E, 12 April 2003, G. Gates E 1702; Cape Pillar Track, 43° 09′ S, 147° 56′ E, 13 Nov. 2001, G. Gates E 1312; Chauncy Vale, 13 Oct. 2001, G. Gates E 1289; —, 29 Nov. 2001, G. Gates E 1357; Clarks Cliffs, 43° 06′ S, 147° 47′ E, 13 March 2004, G. Gates E 1948; Coles Bay, Hazards Beach Track, 42° 10′ S, 148° 17′ E, 1 Dec. 2001, G. Gates E 1364; Collinsvale, Fire Trail, 42° 52′ S, 147° 09′ E, 15 April 1999, G. Gates E 416; —, Myrtle Forest, 18 March 1999, G. Gates E 252; Duckhole Lake Track, 43° 22′ S, 146° 53′ E, 8 April 2003, G. Gates E 1677; —, 4 Jan. 2005, G. Gates E 2073; —, 29 May 2010, M.E. Noordeloos 2010067; Florentine Valley, Lady Binney Track, 42° 43′ S, 146° 31′ E, 27 Dec. 2002, G. Gates E 1649; Geeveston, Donnellys Rd, 43° 07′ S, 146° 54′ E, 18 Oct. 1998, G. Gates E 78; —, 12 June 1999, G. Gates E 590; —, 23 May 2000, G. Gates E 928, 25 May 2000, G. Gates E 944; —, 24 Oct. 2000, G. Gates E 1001; —, 14 Nov. 2000, G. Gates E 1011; —, 5 April 2001, G. Gates E 1076; Growling Swallet, 42° 41′ S, 146° 30′ E, 27 March 2001, G. Gates E 1043; —, 13 May 2003, G. Gates E 1798; —, 28 March 2009, M.E Noordeloos 2009084; Halls Falls, 41° 15′ S, 148° 02′ E, 21 Oct. 2004, G. Gates E 2061; Kermandie Falls, Lower Track, 43° 12′ S, 146° 52′ E, 8 June 2002, G. Gates E 1560; —, 10 April 2003, G. Gates E 1696; —, 24 March 2009, M.E. Noordeloos 2009052 (L); —, Upper Track, 30 March 1999, G. Gates E 318; —, 11 April 2000, G. Gates E 838; —, 30 May 2000, G. Gates E 952; —, 27 Jan. 2004, G. Gates E 1924; Lake Skinner Track, 42° 57′ S, 146° 44′ E, 11 April 1999, G. Gates E 394; Lake St Clair National Park, Echo Point to Cynthia Bay, 42° 04′ S, 146° 09′ E, 22 April 2000, G. Gates E 855; —, 19 April 2003, G. Gates E 1727; —, 21 March 2009, M.E. Noordeloos 2009039 (L); Mt Field NP, 42° 41′ S, 146° 42′ E, 9 May 1999, G. Gates E 486; —, 31 May 2003, G. Gates E 1845; —, Lady Barron Falls Track, 1 April 1999, G. Gates E 347; —, 7

Dec. 2004, G. Gates E 2066; —, River Walk, 31 March 2001, G. Gates E 1065; —, 23 Oct. 2001, G. Gates E 1297; —, Tall Trees Track, 4 Nov. 1999, G. Gates E 730; Mt Wellington, Betts Vale Track, 42° 55′ S, 147°15′ E, 15 Dec. 2001, G. Gates E 1370; —, Circle Track, 25 March 1999, G. Gates E 294; —, Fern Glade, 27 Feb. 1999, G. Gates E 153; —, Myrtle Gully, 42° 54′ S, 147° 15′ E, 13 April 1999, G. Gates E 405, E 407 & E 408; —, 9 Nov. 2000, G. Gates E 1006; —, 17 Jan. 2002, G. Gates E 1406; —, 26 April 2003, G. Gates E 1740; —, Pillinger Drive, 42° 55′ S, 147° 15′ E, 19 Feb. 2002, G. Gates E 1439; —, Pipeline Track, 21 July 2004, G. Gates E 2048; —, Silver Falls, 27 Feb. 1999, G. Gates E 154; North West Bay River, 42° 57′ S, 147° 12′ E, 21 April 2005, G. Gates E 2135; Scottsdale, Forester Road, 41° 06′ S, 147° 37′ E, 9 June 2004, G. Gates s.n.; 29 April 2006, M.E. Noordeloos s.n.; Skemps Fern Gully, 41° 18′ S, 147° 22′ E, 16 April 2001, G. Gates E 1097; Timbs Track, 42° 44′ S, 146° 25′ E, 27 March 2004, G. Gates E 1961; Warra LTER site, 43° 06′ S, 146° 41′ E, 27 April 2004, G. Gates E 1979 (HO 564357, holotype); Wielangta, 42° 42′ S, 147° 51′ E, 23 Feb. 1999, G. Gates E 135; —, 9 March 2000, G. Gates E 800; —, 15 Jan. 2002, G. Gates E 1403.

Entoloma discrepans is close to *Entoloma melanocephalum* G. Stev. from New Zealand and Tasmania, from which it mainly differs by the purer blue colours, the smaller spores, and red-brown lamella edge. Forms occur with a concolorous lamella edge (e.g. Noordeloos 2009052), which are otherwise exactly the same. *Entoloma moongum* Grgur. from Australia has a dark brown or purplish brown pileus without an obvious blue tinge, distinctly larger spores, and lacks cheilocystidia. Also, the surface of the pileus is described as finely fibrillose with a pileipellis consisting of narrow hyphae. *Entoloma chalybaeum* (Pers.) Zerova from Europe is similar, but differs by the distinctly larger spores and polished stipe. In North America several similar species have been described. *Leptonia chalybaea* var. *squamulosipes* Largent resembles our fungus in having small spores but the stipe is described as minutely squamulose and longitudinally striate. Furthermore, the lamella edge is heterogeneous and not pigmented (Largent 1994). *Leptonia nigra* Murrill has larger spores, and lacks cystidia (Largent 1977); *Leptonia pigmentosipes* (Largent) Largent has a blue punctate stipe, but larger spores and a fertile lamella edge (Largent 1994). *Entoloma calliviolaceum* Manim. & Noordel. from Kerala State, India, is more violaceous in colour, and has much larger spores (Manimohan et al. 2006).

An aberrant collection with smaller spores was found once, with the following characters:

Description:

Pileus 12–30 mm diam, convex then applanate, not depressed, sometimes with very weak umbo, with sulcate margin, not hygrophanous, not translucently striate, black, glabrous, becoming rimose with age. Lamellae adnate with small decurrent tooth, segmentiform, up to 5 mm broad, whitish then pale grey with greyish violet edge. Stipe 17–47 × 2–3.5 mm, cylindrical, dark violet at first then blue-grey, glabrous, polished, with white basal mycelium. Odour indistinct. Taste fresh.

Spores 7.5–8 × 5.5–6 um, Q = 1.3, heterodiametric, 5–6-angled in side-view. Lamella edge serrulatum-type with cylindrical to clavate cheilocystidia with blue intracellular pigment. Pileipellis a cutis of cylindrical hyphae with clavate terminal elements, up to 24 μm wide with brown, intracellular pigment. Pileitrama regular, made up of cylindrical to slightly inflated elements. Brilliant granules rare to abundant. Clamp-connections not observed.

Habitat & distribution: found once only, in dry sclerophyll forest.

Collection examined. Australia, Tasmania, Chauncy Vale, 42° 37′ S, 147° 16′ E, 29 Sept. 2005, G. Gates E 2199.

This is a remarkable bluish taxon, similar to the typical forms of the species, but with less blue in the pileus, and with smoother, smaller spores, and a differently coloured lamella edge. More material is needed to establish its taxonomic status.

73. **Entoloma purpureofuscum** Noordel. & G.M. Gates, spec. nov.

MycoBank #564510.

Diagnosis: Habit collybioid. Pileus 15–30 mm diam, conical, convex, expanding, dark brown to black, with violaceous tinge, translucently striate at margin, centre squamulose. Lamellae adnate-emarginate, pink with subtle purple coloured edge. Stipe 20–40×2–3 mm, grey-blue fading to pale purple-brown, polished. Spores 8–11×6–8 μm, 5–7-angled. Cheilocystidia 29–45×7–11 μm, cylindrical to clavate with pale brown-purple intracellular pigment. Pileipellis cutis to trichoderm of inflated terminal elements, up to 18 μm wide with brown-purple intracellular pigment. Clamp-connections absent.

Holotype: Australia, Tasmania, Kermandie Falls, 43° 12′ S, 146° 52′ E, Lower Track, 24 March 2009, M.E. Noordeloos 2009048 (HO 564358; isotype, L).

Etymology: purpureus (Lat.) = purple; fuscus (Lat.) = dark brown, referring to the purple-brown pileus.

Main characters: habit collybioid; pileus dark purple-brown, minutely squamulose at centre; lamellae pink with purple-brown or concolorous edge; stipe blue-grey (Plate 3.69).

Description:

Pileus 15–30 mm diam, conical then convex to plano-convex with low umbo, or more or less truncate to slightly depressed but not umbilicate, with deflexed then straight margin, not hygrophanous, translucently striate up to half the radius, very dark purple-brown, paler brown towards margin, distinctly minutely squamulose with very small pointed squamules in the central part, radially fibrillose towards margin. Lamellae adnate-emarginate, ventricose, moderately crowded, up to 4 mm broad, pale then dark pink with subserrulate, concolorous or pale purple-brown edge. Stipe 20–40×2–3 mm, cylindrical, pale blue-grey fading soon to grey or brown-purple, much paler than pileus, smooth, glabrous, polished, with white basal tomentum. Odour none. Taste not noted.

Spores 8–10(–11)×6–8 μm, 5–7-angled. Basidia 22–40×7–10 μm, 4-spored. Lamella edge sterile of serrulatum-type. Cheilocystidia in dense tufts arising from a strand of hyphae along the lamella edge 29–45×7–11 μm, cylindrical to clavate with pale brown-purple intracellular pigment. Hymenophoral trama regular, made up of cylindrical to inflated hyphae 5–20 μm wide. Brilliant granules present.

Plate 3.69 *Entoloma purpureofuscum* (Photo Machiel Noordeloos)

Pileipellis a cutis at margin, trichodermal at centre, made up of cylindrical to inflated hyphae with cylindrical to clavate terminal elements 40–115×9–18 µm. Pigment brown-purple, intracellular, diffuse and granular in pileipellis. Clamp-connections absent (Fig. 3.73).

Habitat & distribution: occasional, in wet sclerophyll forest.

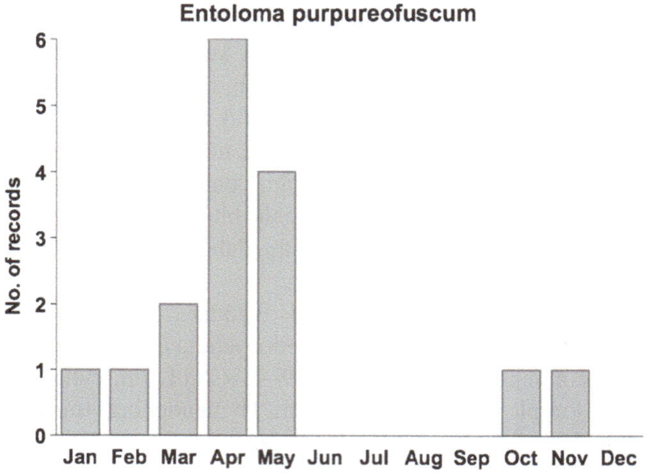

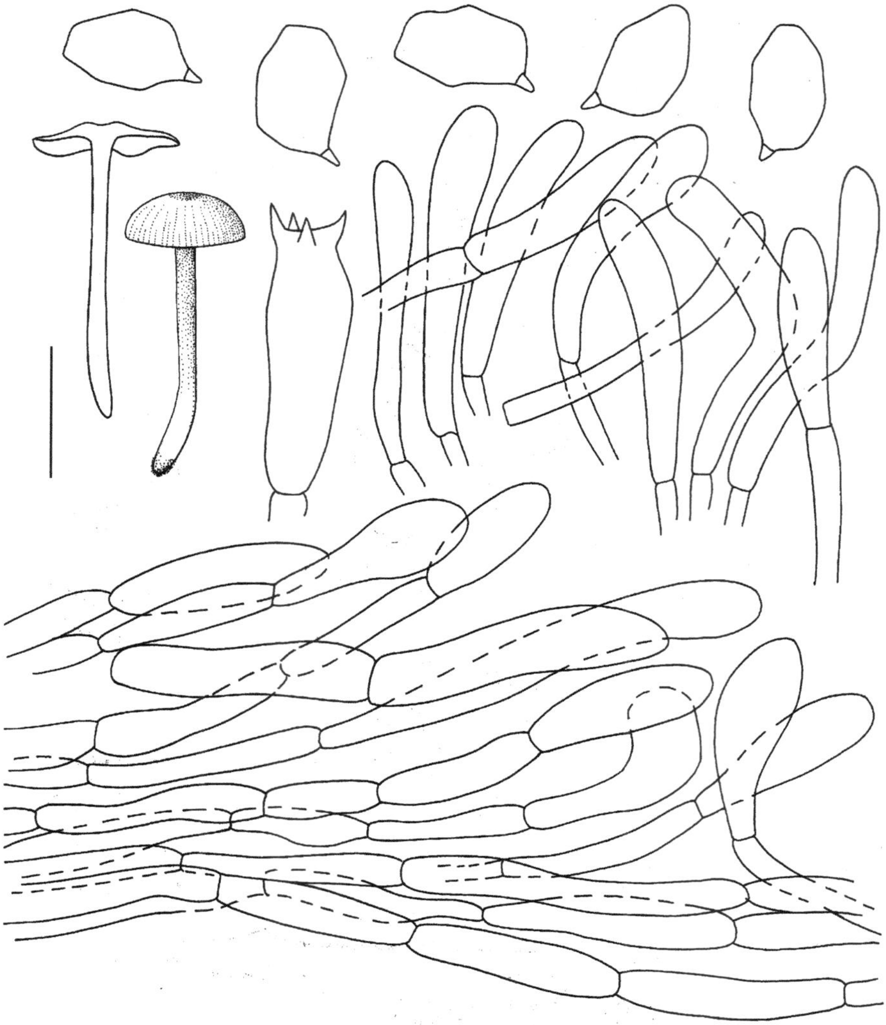

Fig. 3.73 *Entoloma purpureofuscum*. Habit, spores, basidia, cheilocystidia, and pileipellis. Bar = 10 mm or 10 μm

Collections examined. Australia, Tasmania, Black Glen, Lachlan River, 42° 52′ S, 147° 02′ E, 27 Oct. 2005, G. Gates E 2210; Collinsvale, Myrtle Forest, 42° 52′ S, 147° 09′ E, 10 Nov. 2005, G. Gates E 2213; —, 11 May 2010, F Karstedt 1368 & 1381; Duckhole Lake Track, 43° 22′ S, 146° 53′ E, 24 Jan. 2004, G. Gates E 1923; Kermandie Falls, Lower Track, 43° 12′ S, 146° 52′ E, 24 March 2009, M.E. Noordeloos 2009048 (HO 564358, holotype) & 2009049; Mt Wellington, Myrtle Gully, 42° 54′ S, 147° 15′ E, 11 Feb. 1999, G. Gates E 1636; North West Bay River, 42° 57′ S, 147° 12′ E, 1 May 2003, G. Gates E 1754; —, 18 April 2007, G. Gates E 2249; —, 22 April 2008, G. Gates E 2266; —, 2 April 2011, G. Gates E 2290.

Entoloma purpureofuscum is very close to *Entoloma duplocoloratum* E. Horak from New Zealand, which differs in colour: *E. duplocoloratum* has a fuscous cap with argillaceous pink margin, and a pale brown lamella edge. *Entoloma sassafras* G.M. Gates & Noordel. differs by the blue tinges in the lamellae, which also has a blue edge, and the brown to yellow-brown pileus. *Entoloma melanophthalmum* G.M. Gates & Noordel. has much larger spores, and distinct blue tinges in the pileus, at least when young. *Entoloma pseudocoelestinum* Arnolds from Europe has similar colours, and small spores.

74. **Entoloma duplocoloratum** E. Horak, Agaricales New Zealand 1, vol. 5: 185. 2008.

Original diagnosis: Pileus 15–20(–30) mm, primo ex ovato convexus dein applanatus, fuscus vel fuliginosus, in centro minute squamulosus, fibrillosus marginem striatum versus, siccus, fragilis. Lamellae subliberae vel adnexae, isabelinae dein rosaceae, ad aciem concolores. Stipes 30–45 × 1–1.5 mm, cylindricus, elongatus, fragilis, pallide azureus dein azureogriseus, glaber. Odor saporque nulli vel subfarinacei. Basidiosporae 8.5–10.5 × 6–7 µm, (5–)6-angulatae. Cheilocystidia 25–60 × 10–16 µm, cylindrica vel clavata, hyalina. Pleurocystidia et caulocystidia nulla. Pileipellis ex hyphis cylindraceis cutem vel trichodermium formantibus, cellulae terminales 7–12 µm diam, pigmento coeruleobrunneo impletae. Fibulae nullae. Ad terram in silvis leptospermarum. Nova Zelandia.

Holotype. New Zealand, Warkworth, The Dome, 5 June 1981, E. Horak 886 (PDD 74722, isotype ZT 886).

Etymology: duplus (Lat.) twofold; coloratus (Lat.) coloured.

Main characters: dark red-brown collybioid *Cyanula* with striate pileus; lamellae pale brownish pink with or without pigmented edge; stipe grey-blue, polished; spores small; pileipellis composed of relatively narrow hyphae (Plate 3.70).

Description:
Pileus 7–23 mm diam, truncate-conical to hemispherical then convex to plano-convex, with deflexed margin, not hygrophanous, broadly translucently striate at least up to half the radius, red-brown to bronze brown with much darker centre or central spot, entirely tomentose to subsquamulose at first, then becoming fibrillose-rimose in marginal zone, remaining minutely squamulose at centre. Lamellae adnate, seceding, ventricose, moderately crowded, up to 5 mm broad, pale brown-pink, sometimes greyish pink with irregular, concolorous or brown edge. Stipe 10–40 × 2–4 mm, cylindrical, grey-blue, smooth, polished, with white basal tomentum. Context concolorous with surface, cartilaginous. Odour none or spermatic. Taste saliva inducing.

Spores 8–10 × 6.5–8(–9) µm, Q = 1.2–1.6, Qav = 1.3–1.4, heterodiametric, 5–7-angled in side-view with pronounced angles. Basidia 20–40 × 8–12 µm, 4-spored, clampless. Lamella edge sterile. Cheilocystidia 20–45 × 5–16 µm, cylindrical to clavate with or without brown, intracellular pigment. Hymenophoral trama regular, made up of cylindrical hyphae 3–20 µm wide. Pileipellis a cutis with transitions to a trichoderm, made up of cylindrical 6–14 µm wide hyphae with subcylindrical or

Plate 3.70 *Entoloma duplocoloratum* (Photo Machiel Noordeloos)

narrowly clavate terminal elements 6–12(–15) µm wide. Pigment brown and blue, intracellular in pileipellis. Brilliant granules abundant. Stipitipellis a cutis of narrow, 4–12 µm wide, cylindrical hyphae. Clamp-connections absent (Fig. 3.74).

Habitat & distribution: terrestrial in groups in forest litter of wet sclerophyll forest.

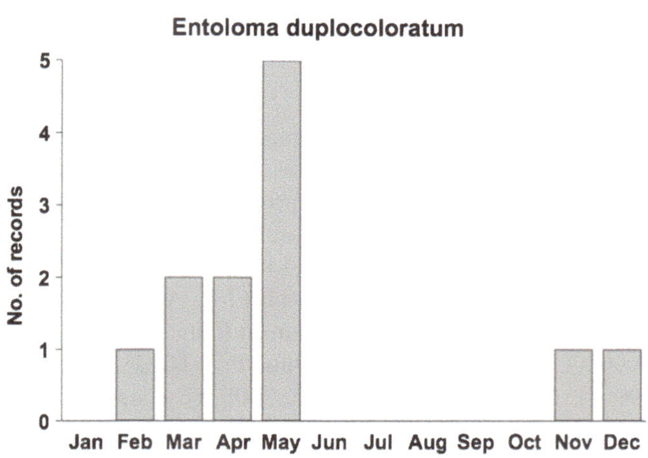

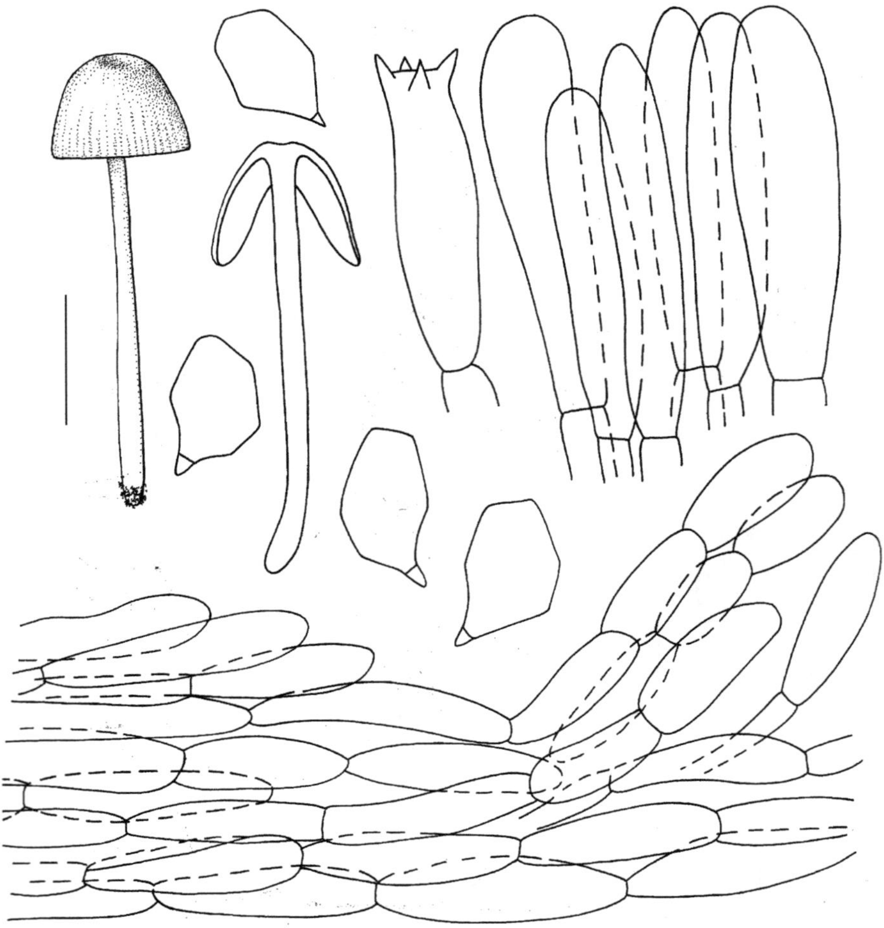

Fig. 3.74 *Entoloma duplocoloratum*. Habit, spores, basidia, cheilocystidia, and pileipellis. Bar = 10 mm or 10 μm

Collections examined. Australia, Tasmania, Bruny Island, Labillardiere Estate, 43° 27′ S, 147° 11′ E, 15 May 2010, M.E. Noordeloos 2010036; Collinsvale, Myrtle Forest, 42° 52′ S, 147° 09′ E, 28 April 2005, E. Gates 2142; Junee Caves State Reserve, 42° 44′ S, 146° 36′ E, 28 March 2009, M.E. Noordeloos 2009069; Kermandie Falls, Lower Track, 43° 12′ S, 146° 52′ E, 11 May 1999, G. Gates E 489; Lake Skinner Track, 42° 57′ S, 146° 44′ E, 11 April 1999, G. Gates E 359 & E 389; Mt Field National Park, Lady Barron Falls Track, 42° 41′ S, 146° 42′ E, 9 Nov. 1999, G. Gates E 73; Mt Wellington, Fern Glade, 42° 55′ S, 147°15′ E, 27 Feb. 1999, G. Gates E 156; —, 1 Dec. 2005, G. Gates E 2219; Wielangta, Long Walk, 42° 42′ S, 147° 51′ E, 5 May 2005, G. Gates E 2160. — New Zealand, Warkworth, The Dome, 5 June 1981, E. Horak 886 (PDD 74722, holotype).

Entoloma duplocoloratum has a rather wide distribution in Tasmania. It resembles *E. purpureofuscum*, which differs by the purple tinges in the pileus and lamella edge. *Entoloma asprellopsis* has a darker pileus and much larger spores. Our collections fit well with the type collection from New Zealand. *Entoloma duplocoloratum* is similar to the European complex of *E. poliopus* (Romagn.) Noordel. and the North American group of *E. parvum* (Peck) Hesler (Largent 1977, 1994; Noordeloos 2004).

Stirps ***Asprellum***

Pileus distinctly translucently striate, usually squamulose at centre only; stipe blue or blue-grey, polished.

75. **Entoloma albidocoeruleum** G.M. Gates & Noordel., Persoonia 19: 216. 2007.

Original diagnosis: Habitus collybioideus. Pileus 15–47 mm latus pallidus hygrophanus translucido-striatus leviter fibrillosus subglabrus centro minute squamulosus. Stipes 35–70×3–7 mm coeruleus politus. Sporae 8.0–10×6.0–8.0 μm 5–7-angulatae. Cheilocystidia 20–45×7–15 μm clavatus vel vesiculosus. Pileipellis cutis in trichoderma transiens pigmentis intracellulosis. Fibulae desunt.

Holotype: Australia, Tasmania, Clarks Cliffs, 43° 06–07′ S, 147° 47–48′ E, 13 March 2004, G. Gates E 1950 (HO 543548; isotype in L).

Etymology: albidus (Lat.) = whitish; (Lat.) coeruleum = sky blue, referring to the colours of the basidiocarp.

Main characters: habit collybioid; pileus very pale brown or whitish, strongly contrasting with the sky blue polished stipe; spores small; cheilocystidia present (Plate 3.71).

Description:

Pileus 15–50(–65) mm diam, conico-convex or campanulate, then convex or expanded, often more or less truncate with slightly depressed centre, sometimes with an umbo-like elevation, with deflexed then straight margin, pallid beige to very pale brown, sometimes almost white with ephemeral violet margin, often with slightly darker central spot, translucently striate in marginal zone when moist, hygrophanous, distinctly pallescent on drying to off-white or silky buff, glabrous or innately fibrillose, except for the minutely scaly disc. Lamellae adnate or with small decurrent tooth, subventricose, moderately crowded, up to 7 mm broad, thin, very pale pink, almost white, with entire concolorous edge, lamellulae in two tiers. Stipe 35–70×3–7 mm, cylindrical or laterally compressed and longitudinally grooved, with slightly inflated base, slender, brittle, pale sky blue, becoming grey-blue and finally grey with age, an orange splotch developing at base with age, glabrous, polished. Odour and taste slightly to distinctly farinaceous.

Spores 8.0–10×6.0–8.0 μm, Q = 1.2–1.5, 5–7-angled in side-view. Basidia 19–34×7–11 μm, 4-spored, clampless. Lamella edge sterile. Cheilocystidia 22–45×7–15 μm, clavate to cylindrical. Hymenophoral trama regular, made up of cylindrical elements, 40–120×5–12 μm. Brilliant granules present. Pileipellis a

Plate 3.71 *Entoloma albidocoeruleum* (Photo Machiel Noordeloos)

cutis with transitions to a trichoderm, made up of clavate terminal elements, 30–70 × 7–22 µm. Pigment pallid brown, intracellular. Pileitrama regular, made up of cylindrical elements, 40–140 × 5–22 µm. Stipitipellis a simple cutis of narrow, cylindrical hypae, 5–12 µm wide. Caulocystidia absent. Clamp-connections absent (Fig. 3.75).

Habitat & distribution: common in litter on floor of wet eucalypt forest. Main fruiting occurring in autumn (March–May) with most of the remaining records occurring in June (Group 3, in Chap. 1, see section "Fruiting Patterns of Species of Entolomataceae").

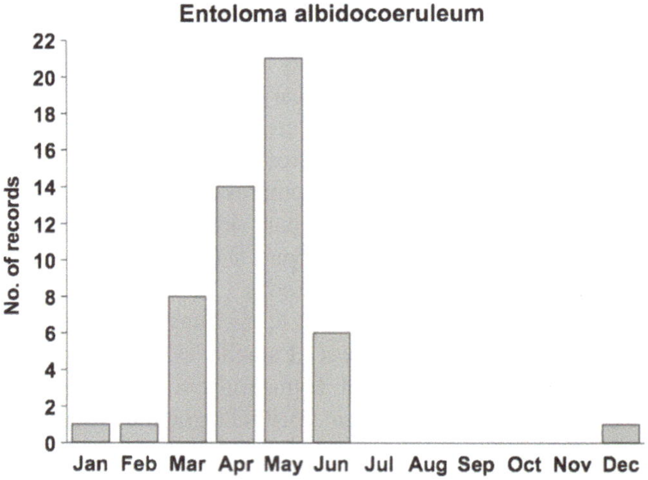

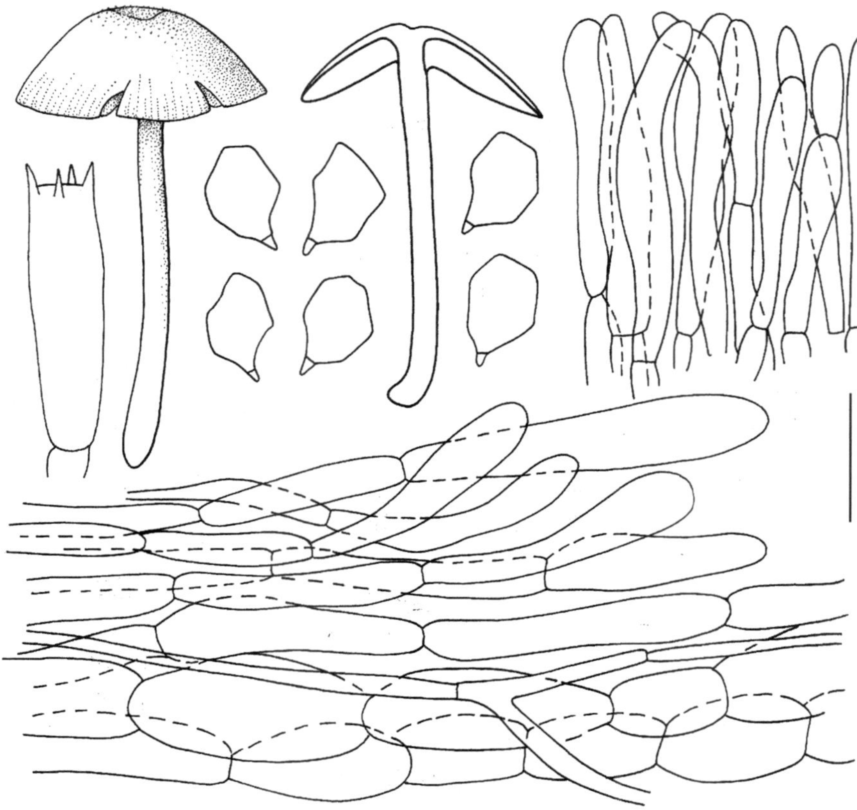

Fig. 3.75 *Entoloma albidocoeruleum*. Habit, spores, basidia, cheilocystidia, and pileipellis. Bar = 10 mm or 10 μm

Collections examined. Australia, Tasmania, Bruny Island, Mt Mangana, 43° 22′ S, 147° 17′ E, 7 May 2000, G. Gates E 898; —, 7 April 2001, G. Gates E 1084; Clarks Cliffs, 43° 06′ S, 147° 47′ E, 13 March 2004, G. Gates E 1950 (holotype, HO 543548; isotype, L); Duckhole Lake Track, 43° 22′ S, 146° 53′ E, 9 March 2002, G. Gates E 1451; —, 30 March 2002, G. Gates E 1465; —, 12 Feb. 2004, G. Gates E 1932; —, 20 April 2004, M.E. Noordeloos 2004058; —, 8 May 2004, G. Gates E 1993; —, 20 Dec. 2005, G. Gates E 2225; —, 29 May 2010, M.E. Noordeloos 2010066; Geeveston, Donnellys Road, 43° 07′ S, 146° 54′ E, 17 June 1999, G. Gates E 618; Junee Caves State Reserve, 42° 44′ S, 146° 36′ E, 9 May 2011, M.E. Noordeloos 2011055 (L); Keoghs Walk, 43° 08′ S, 146° 46′ E, 10 April 2001, G. Gates E 1086; Kermandie Falls, Lower Track, 43° 12′ S, 146° 52′ E, 16 March 1999, G. Gates E 229; —, 11 May 1999, G. Gates E 496; —, 23 May 2000, G. Gates E 938; —, 5 April 2001, G. Gates E 1074; —, 16 May 2002, G. Gates E 1509; —, 8 June 2002, G. Gates E 1564; —, 10 April 2003, G. Gates E 1692; —, 20 April 2005, G. Gates E 2124; —, 24 March 2009, M.E. Noordeloos 2009057 (L); Marriotts Falls, 42° 43′ S, 146° 39′ E, 9 April 2000, G. Gates E 835; —, 24 May 2003, G.

Gates E 1821; Reuben Falls, 43° 00′ S, 146° 40′ E, 15 May 1999, G. Gates E 508; Wielangta Forest Reserve, 42° 42′ S, 147° 51′ E, 9 March 2000, G. Gates E 798.

Entoloma albidocoeruleum can be placed in stirps *Asprellum* of subgenus *Cyanula*. *Entoloma consanguineum* E. Horak from New Zealand is similar in having a very pale pileus and pale blue to grey-blue stipe, but has slightly smaller spores, and a fertile lamella edge without cheilocystidia (Horak 2008). The small spores and pale colours, as well as the often slightly decurrent lamellae are reminiscent also of *Entoloma lividocyanulum* Noordel. from Europe, which has a slightly different, very pale yellow-brown pileus and a fertile lamella edge without cystidia (Noordeloos 1992, 2004).

76. **Entoloma asprellopsis** G.M. Gates & Noordel., Cryptogamie Mycologie 30: 118. 2009.

Original diagnosis: Pileus 21–25 mm latus, hemisphaericus vel convexus, centro depressus, haud hygrophanus, haud translucido-striatus, brunneus vel obscure brunneus, fibrillosus, centro minute squamulosus. Lamellae adnatae, moderate distantes, sordide coeruleo-griseae, demum roseo-tinctae. Stipes 40–60 × 1–3 mm, cylindraceus, coeruleo-griseus, politus. Sporae 9–12 × 6–8 µm, 5–6-angulatae. Acies lamellarum sterilis. Cheilocystidia 28–45 × 8–15 µm contento brunneo. Pileipellis cutis vel trichoderma, ex elementis inflatis, usque ad 19 µm latis, constituta pigmentis intracellularibus. Granula lucentia abundantia. Fibulae desunt.

Holotype: Australia, Tasmania, North West Bay River, 42° 57′ S × 147° 12′ E, 18 May 2004, G. Gates E 2004 (HO 548318; isotype L).

Etymology: opsis (Gr.) = similar, referring to the similarity with *E. asprellum* and *E. asprelloides*.

76a. **typical form with brown coloured lamella edge**

Main characters: habit collybioid; pileus medium to dark brown, centre squamulose, margin fibrillose; lamellae grey-blue with brown margin; stipe blue to blue-grey, polished (Plate 3.72a).

Description:
Pileus 12–25 mm diam, hemispherical to convex with depressed centre, expanding to plano-convex or applanate, with deflexed margin in youth, not distinctly hygrophanous, translucently striate at margin up to half the radius, medium to dark brown, darker at centre, minutely squamulose at centre, fibrillose at margin, not really pallescent on drying, but increasingly shiny-fibrillose. Lamellae adnate, seceding, moderately distant, up to 4 mm broad, subventricose, sordid blue-grey, tinged pink, with brown edge. Stipe 40–60 × 1–3 mm, cylindrical, slender, blue-grey, glabrous, polished, with white basal tomentum. Odour somewhat spermatic. Taste saliva inducing.

Spores 9–12 × 6–8 µm, Q = 1.2–1.5, Qav = 1.35, heterodiametric, 5–6-angled in side-view. Basidia 25–35 × 9–11 µm, 4-spored, clampless. Lamella edge sterile with dense clusters of cylindrico-clavate cheilocystidia, 28–45 × 8–15 µm with brown, intracellular pigment. Hymenophoral trama regular, made up of inflated cylindrical elements 5–18 µm wide with brown intracellular pigment. Pileipellis a cutis

Plate 3.72a *Entoloma asprellopsis* (Photos Michael Pilkington)

of radially arranged, cylindrical to inflated hyphae 6–12 μm wide with scattered trichodermal tufts of clavate terminal elements 20–70 × 8–19 μm. Pigment brown, intracellular in pileipellis and upper pileitrama. Brilliant granules abundant. Vascular hyphae present. Stipitipellis a cutis of cylindrical hyphae 7–13 μm wide with pale blue intracellular pigment. Clamp-connections absent (Fig. 3.76).

Habitat & distribution: terrestrial in small groups in forest litter in wet sclerophyll forest with abundant broad-leaved species such as *Bedfordia salicina* DC and *Pomaderris apetala* Labill. Fruiting throughout the year, with a fair number of records in spring and summer months (Group 5, in Chap. 1, see section "Fruiting Patterns of Species of Entolomataceae").

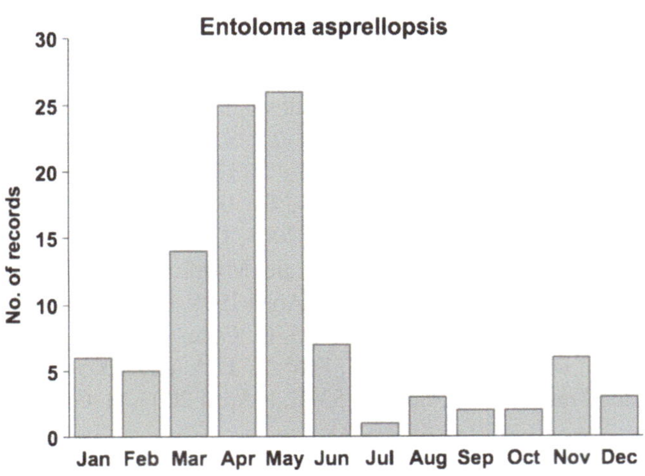

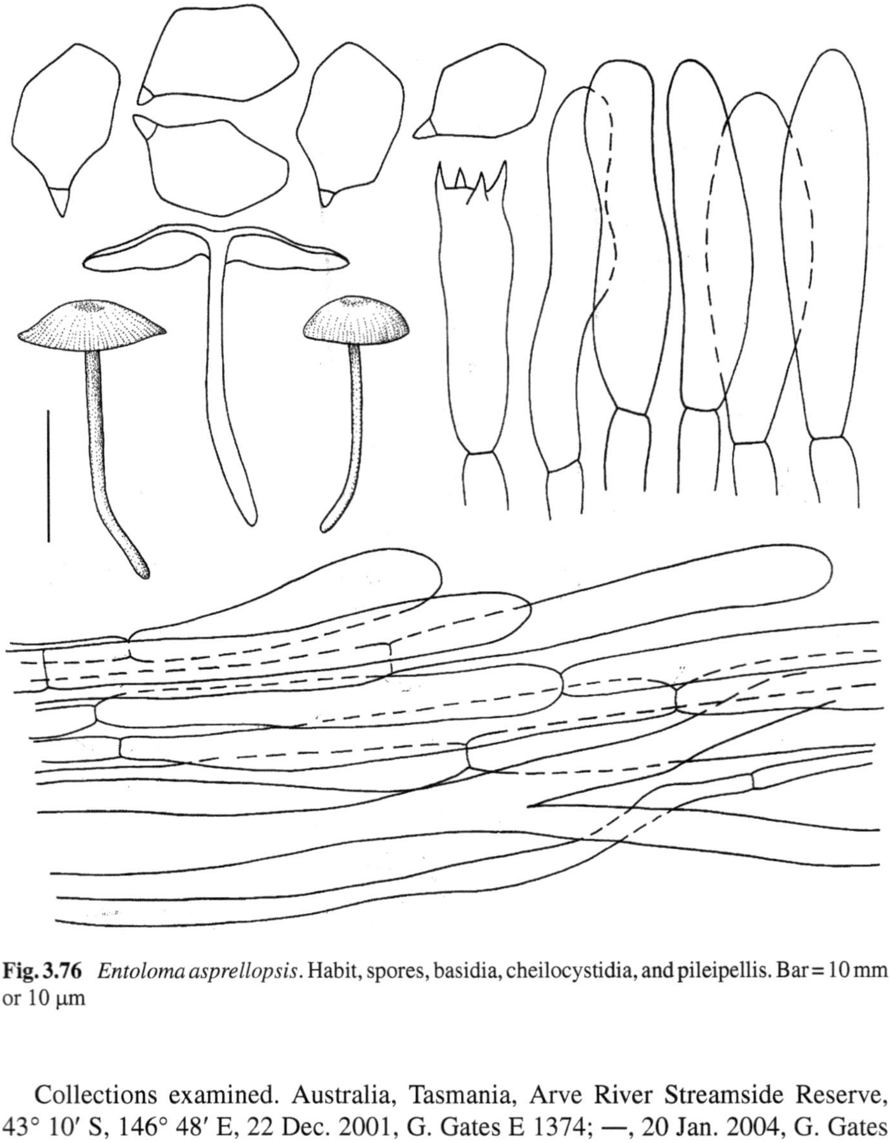

Fig. 3.76 *Entoloma asprellopsis*. Habit, spores, basidia, cheilocystidia, and pileipellis. Bar = 10 mm or 10 μm

Collections examined. Australia, Tasmania, Arve River Streamside Reserve, 43° 10′ S, 146° 48′ E, 22 Dec. 2001, G. Gates E 1374; —, 20 Jan. 2004, G. Gates E 1920; Bruny Island, Mt Mangana, 43° 22′ S, 147° 17′ E, 31 March 2009, M.E. Noordeloos 2009090; Chauncy Vale, 42° 37′ S, 147° 16′ E, 29 Nov. 2001, G. Gates E 1355 & E 1356; Clarks Cliffs, 43° 06′ S, 147° 47′ E, 4 May 2002, G. Gates E 1495 & E 1499; —, 21 April 2003, G. Gates E 1735; —, 13 March 2004, G. Gates E 1941 & E 1946; Collinsvale, Myrtle Forest, 42° 52′ S, 147° 09′ E, 18 March 1999, G. Gates E 244; —, 15 April 1999, G. Gates E 429; —, 6 April 2004, G. Gates E 1966; —, 28 April 2005, G. Gates E 2141; —, 15 Jan. 2011, G. Gates E 2285; Dora Falls, 42° 58′ S, 146° 55′ E, 5 June 2005, G. Gates E 2178; Duckhole Lake Track, 43° 22′ S, 146° 53′ E, 8 April 2003, G. Gates E 1681; —, 8

May 2004, G. Gates E 1990; —, 29 May 2010, M.E. Noordeloos 2010065; Florentine Valley, Pagoda Hut Site, 42° 30′ S, 146° 27′ E, 29 April 2003, G. Gates E 1748; Geeveston, Donnellys Road, 43° 07′ S, 146° 54′ E, 23 May 2000, G. Gates E 927; Junee Caves State Reserve, 42° 44′ S, 146° 36′ E, 28 March 2009, M.E. Noordeloos 2009072; —, 9 May 2011, M.E. Noordeloos 2011052 (L); Kermandie Falls, Upper Track, 43° 12′ S, 146° 52′ E, 30 March 1999, G. Gates E 312; —, 6 May 1999, G. Gates E 471; —, 12 June 1999, G. Gates E 600; —, 16 Nov. 1999, G. Gates E 758; —, 30 May 2000, G. Gates E 948; —, 26 April 2001, G. Gates E 1129; —, 15 April 2003, G. Gates E 1712; Lake St Clair National Park, Echo Point to Cynthia Bay, 42° 04′ S, 146° 09′ E, 22 April 2000, G. Gates E 869; —, 24 March 2001, G. Gates E 1037; —, 20 April 2002, G. Gates E 1483; —, 21 March 2009, M.E. Noordeloos 2009040 (L); Mt Field National Park, 42° 41′ S, 146° 42′ E, 4 April 1999, G. Gates E 720; —, 9 May 1999, G. Gates E 485; —, 31 March 2001, G. Gates E 1055; —, 31 May 2003, G. Gates E 1849; —, Lady Barron Falls, 9 Nov. 1999, G. Gates E 737; —, Tall Trees Track, 8 Aug. 2002, G. Gates E 1624; Mt Wellington, Lower Sawmill Track, 42° 55′ S, 147° 15′ E, 8 Aug. 1998, G. Gates E 33; —, Myrtle Gully, 42° 54′ S, 147° 15′ E, 9 March 1999, G. Gates E 173 & E 188; —, 13 April 1999, G. Gates E 411; —, 12 Aug. 1999, G. Gates E 676; —, 31 July 2001, G. Gates E 1269; —, 27 Nov. 2001, G. Gates E 1343 & E 1344; —, 17 Jan. 2002, G. Gates E 1405, G. Gates E 1407; —, 12 Sept. 2002, G. Gates E 1635; —, 24 Oct. 2002, G. Gates E 1645; —, 26 April 2003, G. Gates E 1743; —, 8 Dec. 2005, G. Gates E 2221; —, Pillinger Drive, 42° 55′ S, 147° 15′ E, 18 Nov. 2000, G. Gates E 1018; —, Silver Falls; 4 March 1999, G. Gates E 166 & E 168; —, 27 March 2003, G. Gates E 1657; North West Bay River, 42° 57′ S, 147° 12′ E, 14 Feb. 2002, G. Gates E 1432; —, 1 May 2003, G. Gates E 1761; —, 18 May 2004, G. Gates E 2004 (holotype, HO 548318); —, 21 April 2005, G. Gates E 2126; Taranna Forest Reserve, 43° 04′ S, 147° 53′ E, 10 May 2003, G. Gates E 1788; Wedge Forest Reserve, 42° 50′ S, 146° 16′ E, 15 April 2000, G. Gates E 840; —, 21 April 2001, G. Gates E 1127; Wielangta Forest Reserve, 42° 42′ S, 147° 51′ E, 3 April 2003, G. Gates E 1673; —, 19 Feb. 2005, G. Gates E 2082.

76b. **form with concolorous lamella edge** (Plate 3.72b)

Collections examined. Australia, Tasmania, Growling Swallet, 42° 41′ S, 146° 30′ E, 1 May 2001, G. Gates E 1155; Junee Caves State Reserve, 42° 44′ S, 146° 36′ E, 15 April 2000, G. Gates E 849; Lake St Clair National Park, Echo Point to Cynthia Bay, 42° 04′ S, 146° 09′ E, 24 March 2001, G. Gates E 1039; Mt Field National Park, River Walk, 42° 41′ S, 146° 42′ E, 31 March 2001, G. Gates E 1055; Mt Wellington, Myrtle Gully, 42° 54′ S, 147° 15′ E, 9 March 1999, G. Gates E 173; —, 27 Nov. 2001, G. Gates E 1344; —, 17 Jan. 2002, G. Gates E 1405; —, 24 Oct. 2002, G. Gates E 1645.

Entoloma asprelloides G. Stev. from New Zealand differs by having a more or less equally coloured dark brown to fuliginous pileus and stipe, and lamellae with a blue, serrulatum-type lamella edge (Horak 2008). *Entoloma asprellopsis* can have a brown or concolorous lamellae edge, and in this way it resembles *Entoloma*

Plate 3.72b *Entoloma asprellopsis* forma (Photo Michael Pilkington)

poliopus (Romagn.) Noordel. from Europe (Noordeloos 2004), which, however, never has blue tinges in the lamellae. *Leptonia gracilipes* Peck from North America is also similar, and differs mainly by the concolorous lamella edge (Largent 1977, 1994).

77. **Entoloma melanophthalmum** G.M. Gates & Noordel., Cryptogamie Mycologie 30: 125. 2009.

Original diagnosis: Pileus 13–30 mm latus, convexus vel plano-convexus, umbilicatus, haud hygrophanus, haud translucido-striatus, obscure brunneus vel atrobrunneus, versus marginem violaceus, tomentosus, centro minute squamulosus. Lamellae adnatae, distantes, pallidae demum roseae. Stipes 30–45 × 2–5 mm, cylindraceus, coeruleo-griseus, politus. Sporae 10–15(–17) × 8–11 µm, 5–7-angulatae. Acies lamellarum sterilis. Cheilocystidia 24–60 × 6–15 µm, clavata. Pileipellis cutis vel trichoderma, ex elementis inflatis, usque ad 20 µm latis, constituta pigmentis intracellularibus. Granula lucentia abundantia. Fibulae absentes.

Holotype: Australia, Tasmania, Mt Field NP, 42°41′S, 146°42′E, 31 May 2003, G. Gates E 1842 (HO 548326; isotype L).

Etymology: melanos (Gr.) = black; ophthalmus (Gr.) = eye, referring to the black "eye" at centre of pileus.

Main characters: habit collybioid, pileus a rather variable shade of brown with distinct, darker "eye", with distinct lilac-violet hue at least at margin, sometimes over whole pileus; stipe grey-blue, polished; spores large, heterodiametric with a large oil drop; cheilocystidia present (Plate 3.73).

Description:
Pileus 10–40 mm diam, convex with deflexed margin, with depressed centre, not hygrophanous, not translucently striate or striate at margin, entirely dark greyish

Plate 3.73 *Entoloma melanophthalmum* (Photo Machiel Noordeloos)

black at first, soon sordid yellow-brown to brown, frequently retaining a distinct lilac-violet tinge all over or at least at margin, while expanding retaining the dark colour at centre as a distinctly marked velutinous or squamulose disc, the rest yellowish or golden brown (5D6), with a distinct lilac-violet hue in the marginal zone, radially fibrillose-squamulose. Lamellae adnate, segmentiform, moderately distant, up to 5 mm broad, pallid then flesh-coloured pink with entire, concolorous edge. Stipe 30–45 × 2–5 mm, cylindrical, grey-blue, fading to grey with age, glabrous, polished, with white basal tomentum. Odour slightly of iodine or spermatic. Taste none.

Spores 10–15(−17) × 8–11 μm, Q = 1.25–1.9, Qav = 1.5, heterodiametric, 5–7-angled in side-view, with large oil drop. Basidia 22–40 × 8–12 μm, 4-spored. Lamella edge sterile. Cheilocystidia 24–60 × 6–15 μm, narrowly to broadly clavate. Pileipellis a cutis with transition to a trichoderm, with inflated terminal elements up to 20 μm wide. Pigment pinkish brown, intracellular in upper layer of pileipellis. Pileitrama regular, made up of inflated elements to 22 μm wide, mixed with narrow, cylindrical hyphae 4–11 μm wide. Brilliant granules abundant in pileitrama. Stipitipellis a cutis of narrow, cylindrical 4–14 μm wide hyphae. Caulocystidia absent. Clamp-connections absent (Fig. 3.77).

Habitat & distribution: terrestrial in groups in forest litter in wet sclerophyll forest. Fruiting spread from late spring to early winter (Group 2, in Chap. 1, see section “Fruiting Patterns of Species of Entolomataceae”).

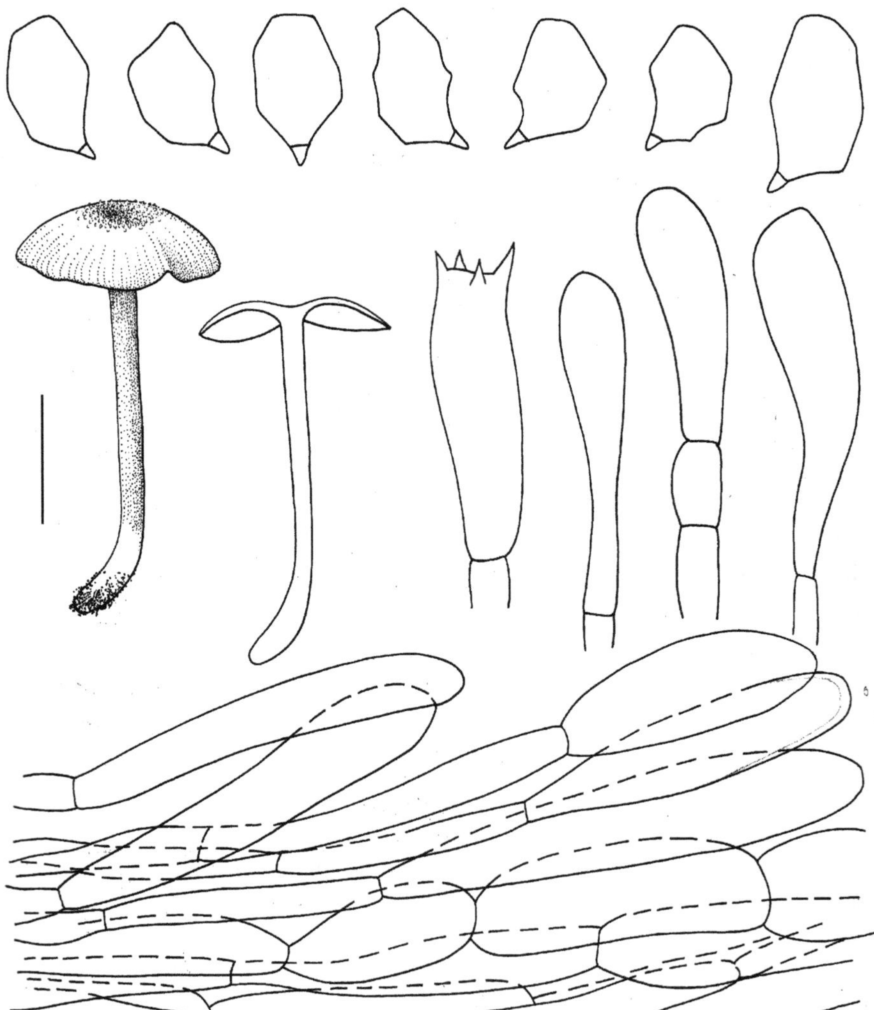

Fig. 3.77 *Entoloma melanophthalmum*. Habit, spores, basidia, cheilocystidia, and pileipellis. Bar = 10 mm or 10 μm

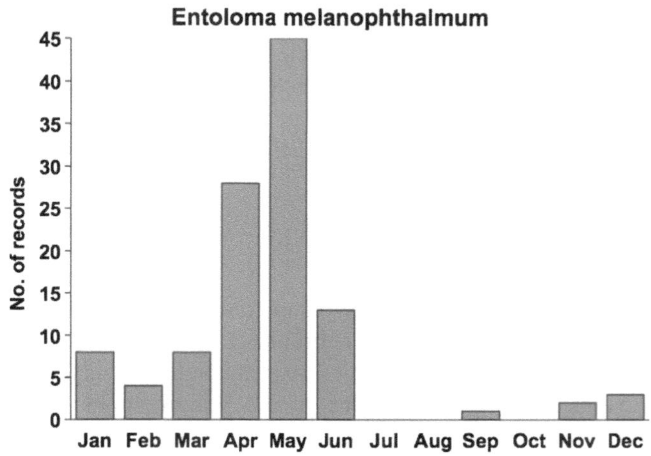

Collections examined. Australia, Tasmania, Arve Rd, Keoghs Walk, 43° 09′ S, 146° 47′ E, 10 April 2001, G. Gates E 1087; Collinsvale, Myrtle Forest, 42° 52′ S, 147° 09′ E, 7 Jan. 1999, G. Gates E 114; Ellendale, 42° 37′ S, 146° 43′ E, 5 April 1999, G. Gates E 359; Evercreech Forest Reserve, 41° 24′ S, 147° 58′ E, 16 June 2003, G. Gates E 1875; Geeveston, Donnellys Rd, 43° 07′ S, 146° 54′ E, 10 April 2003, G. Gates E 1698; Growling Swallet, 42° 41′ S, 146° 30′ E, 13 May 2003, G. Gates E 1795; Hellyer Gorge, 41° 16′ S, 145° 37′ E, 26 April 2002, G. Gates E 1494; Junee Caves State Reserve, 42° 44′ S, 146° 36′ E, 9 May 2011, M.E. Noordeloos 2011054 (L); Kermandie Falls, Lower Track, 43° 12′ S, 146° 52′ E, 10 April 2003, G. Gates E 1693; —, 22 Dec. 2005, G. Gates E 2228; —, Upper Track, 30 May 2000, G. Gates E 950 & E 955; —, 26 April 2001, G. Gates E 1128; —, 15 April 2003, G. Gates E 1713; Lake St Clair National Park, Echo Point to Cynthia Bay, 42° 04′ S, 146° 09′ E, 21 March 2009, M. E. Noordeloos 2009037; Marriotts Falls, 42° 43′ S, 146° 39′ E, 2 June 2001, G. Gates E 1211; —, 24 May 2003, G. Gates E 1817 & E 1818; —, 29 May 2003, G. Gates E 1831; Mt Field National Park, 42° 41′ S, 146° 42′ E, 31 May 2003, G. Gates E 1842 (holotype, HO 548326); —, Lady Barron Falls, 28 May 2002, G. Gates E 1539; —, 7 May 2005, G. Gates E 2162, E 2162a & E 2163; —, 22 Nov. 2005, G. Gates E 2214; —, Tall Trees Track, 6 May 2001, G. Gates E 1173; Mt Wellington, Betts Vale Track, 42° 55′ S, 147° 15′ E, 26 May 2004, G. Gates E 2017; North West Bay River, 42° 57′ S, 147° 12′ E, 14 Feb. 2002, G. Gates E 1429 & E 1431; —, 31 March 2003, G. Gates E 1667a; —, 18 May 2004, G. Gates E 2001 & E 2003; —, 21 April 2005, G. Gates E 2127; Notley Gorge, 41° 21′ S, 146° 55′ E, 13 Sept. 2003. G. Gates E 1911; Tahune, Huon Pine Walk, 43° 06′ S, 146° 44′ E, 17 June 2000, G. Gates E 970; Tasman Peninsula, Canoe Bay, 43° 08′ S, 147° 57′ E, 22 May 2004, G. Gates E 2011; Warra LTER site, 43° 06′ S, 146° 41′ E, 15 April 2004, G. Gates E 1970; —, 3 Feb. 2004, G. Gates E 1926; —, 27 April 2004, G. Gates E 1977; —, 11 May 2004, G. Gates E 1997; —, 25 May 2004, G. Gates E 2015; —, 20 April 2006, G. Gates E 2238; Wielangta Forest Reserve, 42° 42′ S, 147° 51′ E, 29 April 1999, G. Gates E 453; —, 3 April 2003, G. Gates E 1668.

Entoloma melanophthalmum is distinctive because of the violaceous margin to the brownish pileus contrasting with the grey-blue stipe, and the persistent dark spot at the centre of the pileus. No species known from New Zealand (Horak 2008) resembles the present species. It can be classified among the species of stirps *Poliopus* (Noordeloos 1992), but differs from all known species by its colours.

78. **Entoloma tenuicystidiatum** G.M. Gates & Noordel., Cryptogamie Mycologie 30: 127. 2009.

Original diagnosis: Pileus 9–14 mm latus, plano-convexus, umbilicatus, hygrophanus, translucido-striatus, pallide brunneus vel brunneus, fibrillosus, centro minute squamulosus. Lamellae adnatae, moderate distantes, pallide brunneo-roseae. Stipes 35–46 × 1–2.5 mm, cylindraceus, coeruleo-griseus, politus. Sporae 10–15 × 8–10 μm, 6–7-angulatae. Acies lamellarum sterilis. Cheilocystidia 40–110 × 5–9 μm, filiformia vel anguste clavata. Pileipellis cutis vel trichoderma, ex elementis inflatis, usque ad 10 μm latis, constituta pigmentis intracellularibus. Granula lucentia absentia. Fibulae absentes.

Holotype: Australia, Tasmania, Warra LTER Site, at roadside, 43° 06′ S, 146° 41′ E, 26 April 2005, G. Gates E 2140 (HO 548329; isotype L).

Etymology: tenuis (Lat.) = thin, referring to the slender cheilocystidia.

Main characters: small collybioid species; pileus pale brown with darker centre; cheilocystidia long and slender; spores large (Plate 3.74).

Description:
Pileus 9–14 mm diam, plano-convex with slight central depression, becoming applanate, pale brown with darker grey-brown central disc, hygrophanous, translucently striate, becoming sulcate along the backs of the lamellae, velutinous on disc, glabrous elsewhere. Lamellae adnate, seceding, ventricose, moderately crowded, up to 3 mm broad, moderately thick, flesh pink with faint brown edge. Stipe 35–46 × 1–2.5 mm, cylindrical, fragile, hyaline, white or pale brown, glabrous, polished, with white basal mycelium. Odour not distinct. Taste not distinct.

Spores 10–15 × 8–10 μm, heterodiametric, 6–7-angled in side-view with pronounced angles. Basidia 24–30 × 11–13 μm, 4-spored, clampless. Lamella edge sterile. Cheilocystidia 40–110 × 5–9 μm, cylindrical-filiform to slenderly clavate, with brown, intracellular pigment. Hymenophoral trama regular, consisting of cylindrical hyphae up to 10 μm wide. Pileipellis a cutis of long narrow hyphae, up to 6 μm wide with scattered trichodermal tufts of narrowly clavate terminal elements to 10 μm wide. Pileitrama regular, composed of cylindrical hyphae to 20 μm wide. Pigment golden brown, intracellular in pileipellis. Stipitipellis a cutis of narrow hyphae to 8 μm wide. Clamp-connections absent (Fig. 3.78).

Habitat & distribution: terrestrial in groups in somewhat disturbed habitats, like roadsides and marsupial pasture, surrounded by wet sclerophyll forest containing, for example, *Pomaderris apetala* Labill. and *Gahnia grandis* (Labill.) S.T. Blake.

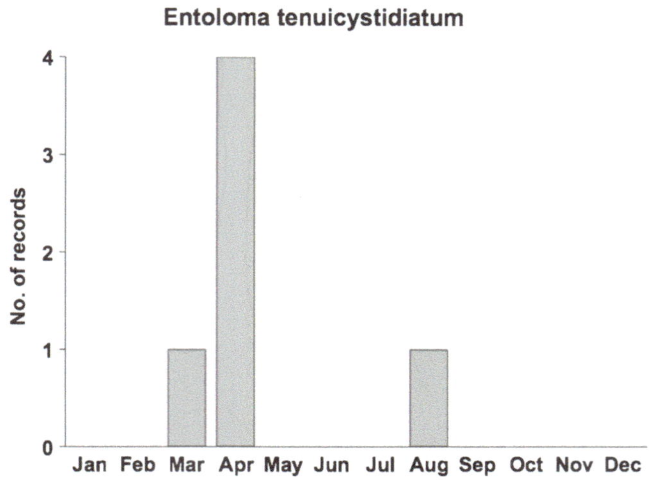

Plate 3.74 *Entoloma tenuicystidiatum* (Photos Machiel Noordeloos)

Collections examined. Australia, Tasmania, Duckhole Lake Track, 43° 22′ S, 146° 53′ E, 16 April 2005, G. Gates E 2112; Junee Cave State Reserve, 42° 44′ S, 146° 36′ E, 22 March 2001, G. Gates E 1033; Warra LTER site, at roadside, 43° 06′ S, 146° 41′ E, 15 April 2004, G. Gates E 1969 & M.E. Noordeloos 2004090; —, 27 April 2004, G. Gates E 1974; —, 26 April 2005, G. Gates E 2140 (holotype, HO 548329).

Entoloma tenuicystidiatum keys out in the rather variable group of *Entoloma longistriatum* (Peck) Noordel. but is distinct because of the overall pale brown colour of the fruit body and brown-edged lamellae with abundant, very long, slender

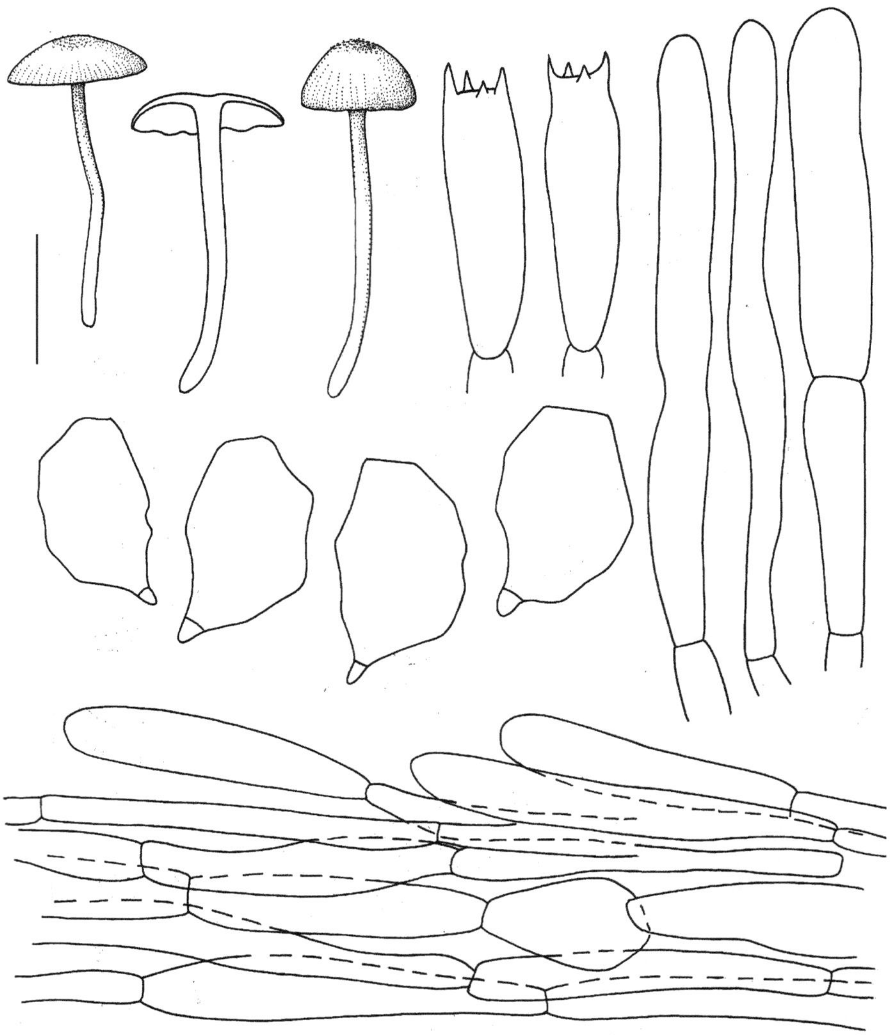

Fig. 3.78 *Entoloma tenuicystidiatum*. Habit, spores, basidia, cheilocystidia, and pileipellis. Bar = 10 mm or 10 μm

cheilocystidia. It is not unlike *Entoloma mutabilipes* Noordel. & Liiv from Europe. However, the latter species has stronger colours, blue tinges in the stipe, as well as smaller spores and cheilocystidia. *Entoloma gnophodes* (Berk. & Broome) E. Horak, a poorly understood species from Sri Lanka, is somewhat similar in habit and colour, but has differently shaped spores and broader cheilocystidia (up to 20 μm wide). The only species from New Zealand that resemble *E. tenuicystidiatum* are *E. corneum* E. Horak and *E. perplexum* E. Horak, both of which, however, have a hymenidermal pileipellis and more or less capitate cheilocystidia (Horak 2008).

Plate 3.75 *Entoloma griseosquamulosum* (Photo Genevieve Gates)

79. **Entoloma griseosquamulosum** G.M. Gates & Noordel., Cryptogamie Mycologie 30: 128. 2009.

Original diagnosis: Pileus 10–15 mm latus, convexus demum plano-convexus, umbilicatus, haud hygrophanus, haud translucido-striatus, obscure griseo-brunneus vel ater, fibrillosus, centro minute squamulosus. Lamellae adnatae, confertae, pallide griseo-roseae. Stipes 20–35 × 1.5–2 mm, cylindraceus, griseo-violaceus, politus. Sporae 9–12 × 7–9 μm, 5–7-angulatae. Acies lamellarum sterilis. Cheilocystidia 45–60 × 10–16 μm, clavata. Pileipellis cutis vel trichoderma, ex elementis inflatis, usque ad 15 μm latis, constituta pigmentis intracellularibus. Granula lucentia abundantia. Fibulae absentes.

Holotype*:* Australia, Tasmania, Myrtle Bank, Skemps, Fern Gully, 41° 18′ S, 147° 22′ E, 20 March 2004, G. Gates E 1956 (HO 548336, isotype L).

Etymology: griseus (Lat.) = grey; squamulosus (Lat.) = squamulose, referring to the grey, squamulose surface of the pileus (Plate 3.75).

Main characters: small collybioid species with grey, squamulose cap and grey-violet stipe.

Description:
Pileus 10–15 mm diam, convex, expanding to plano-convex, umbilicate, with deflexed margin, not hygrophanous, translucently striate at margin or not, dark grey to

grey-black (2E1), squamulose in central part, fibrillose on margin, dry. Lamellae adnate, segmentiform, crowded, up to 3 mm broad, pallid grey then with pink tinge, with entire, concolorous edge. Stipe 20–32 × 1.5–2 mm, cylindrical, slender, grey-violet, glabrous, with white basal tomentum. Odour spermatic. Taste saliva inducing.

Spores 9–12 × 7–9 µm, Q = 1.25–1.9, Qav = 1.55, heterodiametric, 5–7-angled. Basidia 4-spored, clampless. Lamella edge sterile. Cheilocystidia 45–60 × 10–16 µm, clavate. Pileipellis a cutis with transitions to a trichoderm at centre, made up of cylindrical hyphae 8–10 µm wide, with clavate terminal elements to 15 µm wide. Pigment of pileipellis violet-brown, in intracellular clumps. Pileitrama regular, made up of cylindrical to inflated hyphae to 20 µm wide, with clumps of violet-brown, intracellular pigment. Brilliant granules abundant. Stipitipellis a cutis of narrow, cylindrical hyphae 4–12 µm wide. Clamp-connections absent (Fig. 3.79).
Habitat & distribution: terrestrial in groups in forest litter of wet sclerophyll forest.

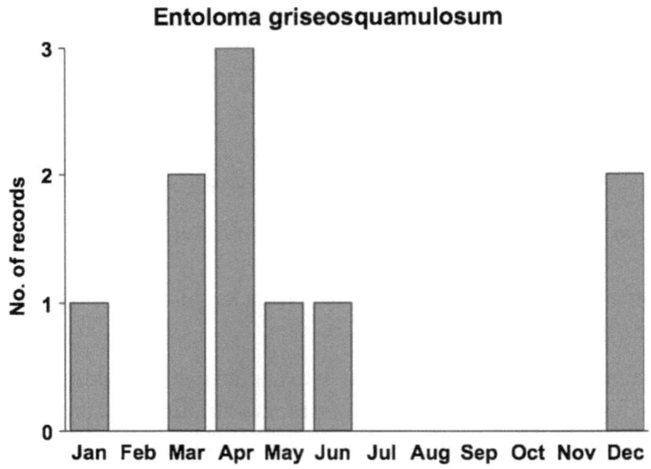

Collections examined. Australia, Tasmania, Collinsvale, Myrtle Forest, 42° 52′ S, 147° 09′ E, 27 Dec. 2001, G. Gates E 1377; Duckhole Lake Track, 43° 22′ S, 146° 53′ E, 15 Jan. 2005, G. Gates E 2077; Kermandie Falls, Lower Track, 43° 12′ S, 146° 52′ E, 10 April 2003, G. Gates E 1697; Mt Field National Park, Lady Barron Falls, 42° 41′ S, 146° 42′ E, 9 May 2002, G. Gates E 1507; —, Tall Trees Track, 4 Dec. 2001, G. Gates E 1366; Myrtle Bank, Skemps, Fern Gully, 41° 18′ S, 147° 22′ E, 20 March 2004, G. Gates E 1956 (holotype, HO 548336); Warra LTER Site, 43° 06′ S, 146° 41′ E, 16 March 2004, G. Gates E 1953; —, 15 April 2004, G. Gates E 1968 (HO 548333); Wielangta Forest Reserve, 42° 42′ S, 147° 51′ E, 29 April 1999, G. Gates E 456.

Entoloma griseosquamulosum can be placed in subgenus *Cyanula*, stirps *Asprellum* (Noordeloos 1992) on account of the grey pileus and grey-violet, polished stipe. It keys out near *Entoloma poliopus* (Romagn.) Noordel. (Noordeloos 2004), which differs, however, by the brown, often distinctly translucently striate pileus and blue tinges in the stipe. No species described by Horak (2008) from New Zealand matches *E. griseosquamulosum*.

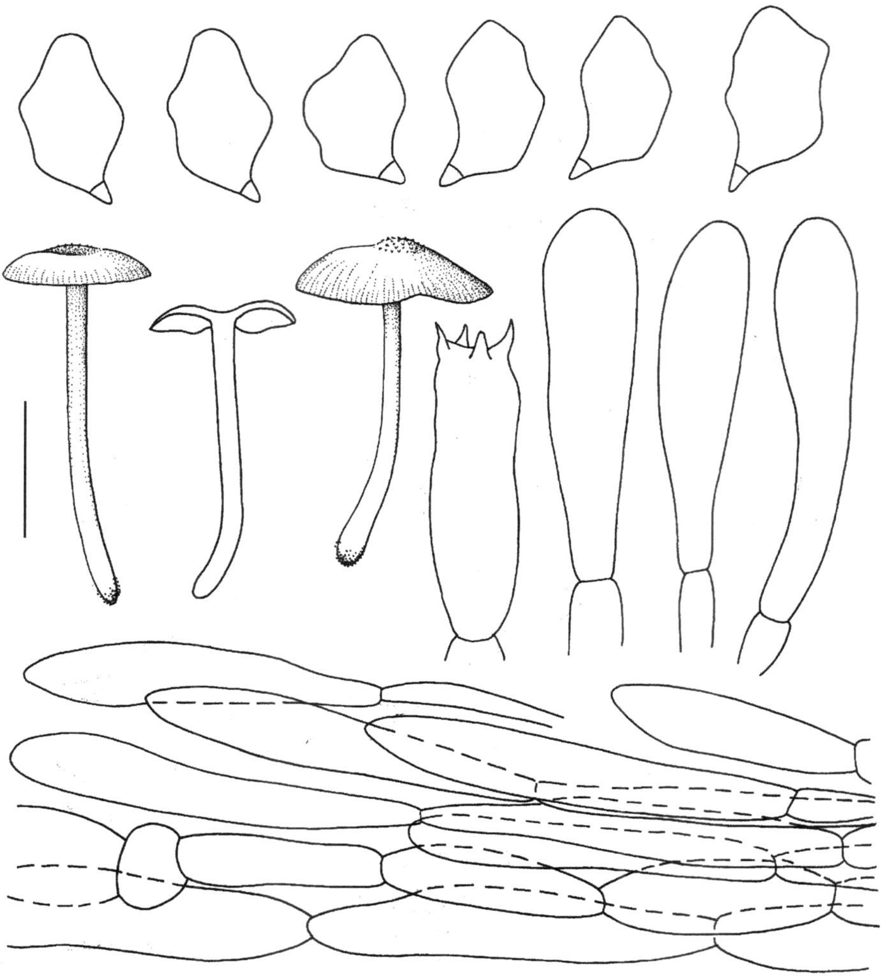

Fig. 3.79 *Entoloma griseosquamulosum*. Habit, spores, basidia, cheilocystidia, and pileipellis. Bar = 10 mm or 10 μm

Stirps ***Incanum***

Basidiocarps with green-yellow or olivaceous-yellow tinges.

80. **Entoloma rodwayi** (Massee) E. Horak, Beih. Nova Hedwigia 65: 303. 1980.

Leptonia rodwayii Massee, Bull. Misc. Inf. Kew: 124. 1898. — *Entoloma chloroxanthum* G. Stev., Kew Bulletin 16: 236. 1962. — *Entoloma citreostipitatum* G. Stev., Kew Bulletin 16: 231. 1962. — *Entoloma pteridicola* G. Stev., Kew Bulletin 16: 235. 1962.

Original diagnosis: Pileo membranaceo, convexo-explanato, profunde umbilicato, margine subundulato, estriato, obscure caesio-virenti, minute squamuloso, sicci-

Plate 3.76 *Entoloma rodwayi* (Photos Machiel Noordeloos (*left*), Michael Pilkington (*right*))

tate pallescenti, 2–3 cm lato; stipite fistuloso, flexuoso, fibrilloso, viridi-olivaceo v. luteo-virenti; lamellis adnatis, postice sinuato-uncinatis, latis, subdistantibus, pallidis; sporis valde irregularibus, roseis, 8–10×7.

Holotype. Australia, Tasmania, Hobart, 18 June 1893, Rodway 47 (K).

Etymology: named after the collector Leonard Rodway.

Main characters: habit collybioid; fruit body with pronounced green-yellow colours, becoming green-blue upon drying or damage; lamella edge fertile, lacking cheilocystidia; odour indistinct (Plate 3.76).

Description:
Pileus 10–57 mm diam, conico-campanulate to convex, often truncate, depressed to umbilicate at centre, with deflexed margin, expanding with age to plano-convex with umbilicate centre, slightly but distinctly hygrophanous, deeply translucently striate, when moist dark green to greenish brown with darker centre, then green to yellow-brown with darker, almost black centre, minutely squamulose at centre, fibrillose towards margin, often radially sulcate at margin when old. Lamellae adnate-emarginate or with short decurrent tooth, moderately distant, pale green then sordid brownish pink with eroded, concolorous edge, up to 7 mm broad. Stipe 30–50×1–2 mm, cylindrical, slightly broadened towards base, entirely greenish to greenish yellow, glabrous, polished, turning greenish blue when bruised, with white basal tomentum. Odour indistinct. Taste slightly farinaceous-rancid. Dries greenish blue.

Spores 10–13×7–8 μm, Q=1.2–1.6, Qav=1.4, 6–7-angled in side-view. Basidia 28–40×9–12 μm, 4-spored, clampless. Lamella edge fertile. Cystidia absent. Pileipellis a cutis with transitions to a trichoderm, made up of cylindrical to clavate terminal elements, 35–90×9–23 μm. Pigment yellow-green, intracellular in pileipellis and stipitipellis. Stipitipellis a cutis of narrow, cylindrical hyphae, 4–11 μm wide.

Fig. 3.80 *Entoloma rodwayi*. Habit, spores, basidia, and pileipellis. Bar = 10 mm or 10 μm

Brilliant granules present in trama of lamellae and pileus. Stipitipellis a cutis of narrow, cylindrical hyphae, 3–8 μm wide, with green intracellular pigment. Clamp-connections absent (Fig. 3.80).

Habitat & distribution: a common species in eucalypt forests and rainforests in Tasmania. Fruiting throughout the year, with a fair number of records in spring and summer months (Group 5, in Chap. 1, see section "Fruiting Patterns of Species of Entolomataceae").

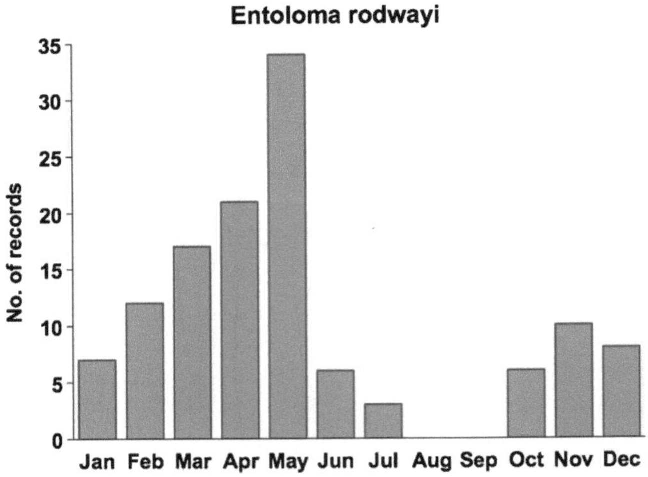

Collections examined. Australia, Tasmania, Hobart, 18 June 1893, Rodway 47 (holotype, K); Clarks Cliffs, 43° 06′ S, 147° 47′ E, 13 March 2004, G. Gates E 1949; Duckhole Lake Track, 43° 22′ S, 146° 53′ E, 8 April 2003, G. Gates E 1679; Geeveston, Donnellys Rd, 43° 07′ S, 146° 54′ E, 17 June 1999, G. Gates E 617; —, 30 Oct. 2001, G. Gates E 1305; Kermandie Falls, Lower Track, 43° 12′ S, 146° 52′ E, 24 March 2009, M.E. Noordeloos 2009059 (L); Lake St Clair National Park, Echo Point to Cynthia Bay, 42° 04′ S, 146° 09′ E, 21 March 2009, M.E. Noordeloos 2009035 (L); North West Bay River, 42° 57′ S, 147° 12′ E, 1 May 2003, G. Gates E 1752; Styx Valley, Big Tree, 42° 49′ S, 146° 38′ E, 19 March 2009, M. E. Noordeloos 2009022 (L). Many additional observations not conserved. — New Zealand, Wellington, Wellington Botanic Gardens, 26 May 1949, G. Stevenson 591 (holotype of *E. pteridicola*); —, 2 June 1949, G. Stevenson 615 (holotype of *E. citreostipitatum*); Nelson, Fringed Hill, 3 June 1956, G. Stevenson 1116 (holotype of *E. chloroxanthum*).

Stevenson (1962) described three very similar species from New Zealand, viz. *E. chloroxanthum, E. citreostipitatum,* and *E. pteridicola*, all characterized by predominantly greenish colours. Horak (2008) studied the types, and concluded they were all identical, and should be named *E. chloroxanthum. Entoloma rodwayi* is very similar to *E. chloroxanthum*, and the latter can be considered synonymous. This has been confirmed by our own study of Stevenson's type material in Kew Herbarium. *Entoloma rodwayi* is also very similar to the European *E. incanum* (Fr.) Hesler, which however, differs strikingly in having a very strong, unpleasant smell, like mice, and this species also shows a much more intense colour change to blue-green when bruised (Noordeloos 2004, Vesterholt 2002).

Section **Griseorubida** (Romagn.) Noordel. in Persoonia 11: 147. 1981.

Rhodophyllus sect. *Griseorubida* Romagn. in Bull. mens. Soc. linn. Lyon 43: 328. 1974.

Habit collybioid; pileus fibrillose to squamulose; spores often rather large and irregularly shaped; cheilocystidia present, large protruding; clamp-connections present.

Type species: *Entoloma griseorubidum* (Kühner) Noordel.

81. **Entoloma cystidiosum** G.M. Gates & Noordel., Cryptogamie Mycologie 30: 134. 2009.

Original diagnosis: Pileus 20–40 mm latus, conico-convexus, demum plano-convexus, leviter umbilicatus vel umbonatus, paulisper hygrophanus, leviter translucido-striatus, brunneus, fibrillosus, centro minute squamulosus. Lamellae adnatae, moderate distantes, roseobrunneae. Stipes 20–40×2–5 mm, cylindraceus, pallide cremeus vel brunneus, politus. Sporae (9–)10–12×(7–)8–10 µm, 5–7-angulatae. Acies lamellarum heterogeneae (cheilocystidia et basidia adsunt). Cheilocystidia 40–90×8–15 µm, clavata vel utriformia. Pileipellis cutis ex hyphis cylindraceis, 8–15 µm latis, constituta pigmentis intracellularibus, vel minute incrustatis. Granula lucentia absentia. Fibulae presentes.

Holotype: Australia, Tasmania, Wielangta Forest Reserve, 42° 42′ S, 147° 51′ E, 14 July 1998, G. Gates E 14 (HO 548347; isotype L).

Etymology: cystidiosum (Lat.) = with cystidia, referring to the abundant cheilocystidia.

Main characters: habit collybioid; pileus variably shaped, umbonate or umbilicate, brown, more or less glabrous; stipe contrastingly pale brown or light yellow-brown, glabrous; spores irregularly shaped, sometimes almost quadrate, or irregularly 5–7-angled; cystidia large, lageniform, abundant on gill edge (Plate 3.77).

Description:
Pileus 20–40 mm diam, conico-convex, convex or plano-convex, centre flat, slightly umbilicate or umbonate, with deflexed then straight margin, brown, slightly translucently striate, sometimes hygrophanous, becoming micaceous-lustrous pale buff-brown, radially fibrillose, at centre sometimes minutely squamulose. Lamellae adnate with pronounced decurrent tooth, moderately distant, up to 3 mm broad, brown with pink tinge, with entire, concolorous edge. Stipe 20–40×2–5 mm, cylindrical, sometimes slightly broadened at base, pallid brownish beige, pale yellow-brown, paler than pileus, polished. Odour not distinct. Taste not distinct.

Spores (9–)10–12×(7–)8–10 µm, isodiametric to heterodiametric, usually irregularly 5–7-angled, but some 4-angled, pseudo-cuboid spores are found. Basidia 30–38×10–12 µm, 4-spored, clampless. Lamella edge heterogeneous. Cheilocystidia 40–90×8–15 µm, fairly abundant, variably shaped from clavate to utriform, lageniform or sphaeropedunculate with rounded, frequently subcapitate apex. Pileipellis a differentiated cutis of cylindrical to inflated hyphae 8–15 µm wide, with brown to olivaceous brown intracellular pigment and sometimes fine encrustations. Pileitrama regular, made up of medium-sized, inflated elements. Clamp-connections absent (Fig. 3.81).

Habitat & distribution: terrestrial, widespread in wet sclerophyll forest and other wet habitats. Fruiting mainly in autumn (March-May) and winter (June-August) and only rarely in the other months (Group 4, in Chap. 1, see section "Fruiting Patterns of Species of Entolomataceae").

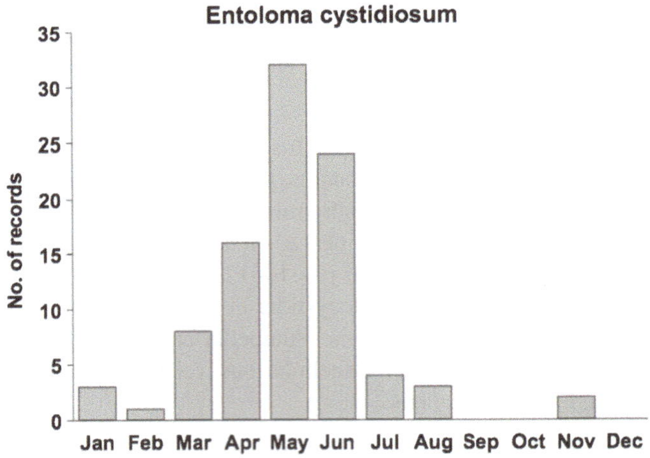

Plate 3.77 *Entoloma cystidiosum* (Photos Michael Pilkington)

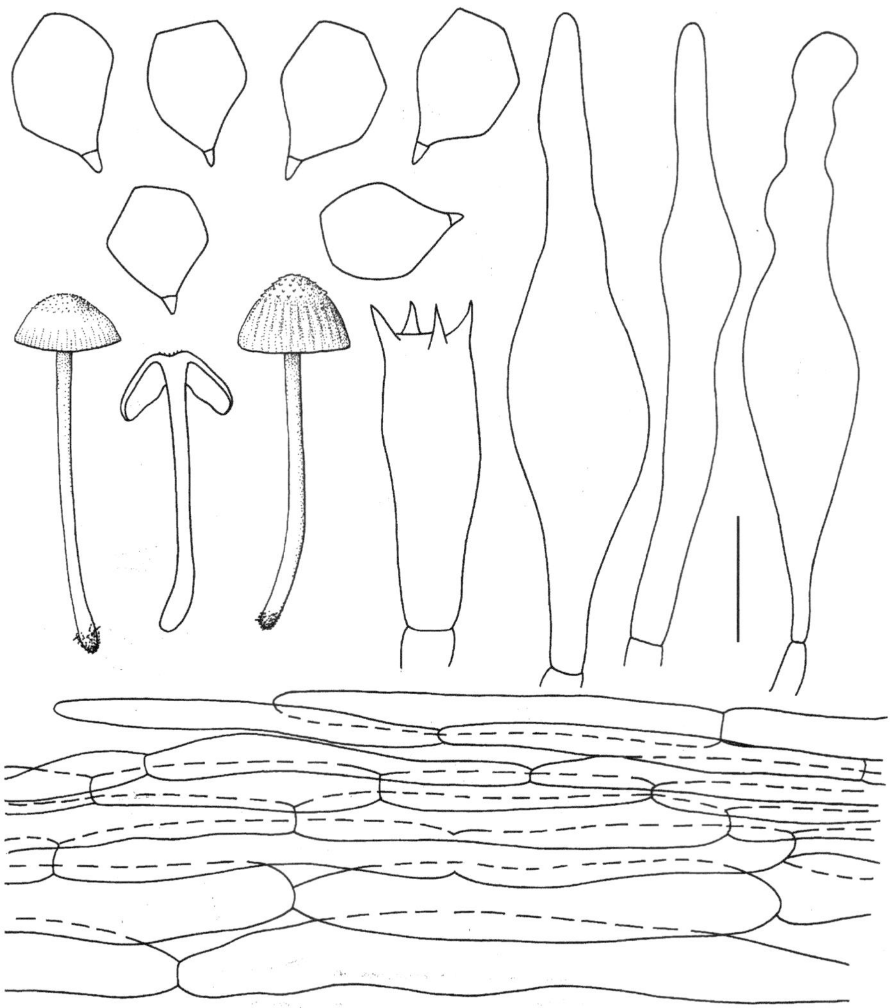

Fig. 3.81 *Entoloma cystidiosum*. Habit, spores, basidia, cheilocystidia, and pileipellis. Bar = 10 mm or 10 μm

Collections examined. Australia, Tasmania, Arthur River, West Point Reserve, 40° 56′ S, 144° 37′ E, 15 June 2002, G. Gates E 1578; Arve Road, 43° 08′ S, 146° 46′ E, 29 Nov. 2005, G. Gates E 2216; Bruny Island, Mavista Falls Picnic Area, 43° 23′ S, 147° 19′ E, 17 May 2010, F. Karstedt 1426; —, Mt Mangana, 43° 22′ S, 147° 17′ E, 8 April 1999, G. Gates E 378, E 379 & E 382; —, 1 June 1999, G. Gates E 556, E 560 & E 562; —, 7 May 2000, G. Gates E 896; —, 10 June 2000, G. Gates E 966; —, 26 May 2001, G. Gates E 1184 & E 1185; —, 27 May 2003, G. Gates E 1829; Chauncy Vale, 42° 37′ S, 147° 16′ E, 29 Nov. 2001, G. Gates E 1353; —, 1 Aug. 2002, G. Gates E 1619; —, 28 June 2009, G. Gates E 2279; Clarks Cliffs,

43° 06′ S, 147° 47′ E, 4 May 2002, G. Gates E 1505; Collinsvale, Myrtle Forest, 42° 52′ S, 147° 09′ E, 19 April 2001, G. Gates E 1105; —, 18 April 2003, G. Gates E 1722; Cradle Mt National Park, Pine Forest Moor, 41° 45′ S, 145° 58′ E, 30 April 2004, G. Gates E 1982; Duckhole Lake Track, 43° 22′ S, 146° 53′ E, 15 May 2003, G. Gates E 1807; —, 16 April 2011, G. Gates E 2291; Growling Swallet, 42° 41′ S, 146° 30′ E, 1 May 2001, G. Gates E 1142 & E 1143; Judbury Conservation Area, 43° 00′ S, 147° 00′ E, 3 April 1999, G. Gates E 353; Junee Caves State Reserve, 42° 44′ S, 146° 36′ E, 22 March 2001, G. Gates E 1036; —, 28 March 2009, G. Gates E 2271; Kermandie Falls, Lower Track, 43° 12′ S, 146° 52′ E, 11 May 1999, G. Gates E 500; —, 23 May 2000, G. Gates E 934; —, 29 May 2001, G. Gates E 1192 & E 1199; —, 16 May 2002, G. Gates E 1513; —, 19 May 2010, M.E. Noordeloos 2010049; —, Upper Track, 30 March 1999, G. Gates E 311; —, 6 May 1999, G. Gates E 470; —, 11 May 1999, G. Gates E 498; —, 3 July 2001, G. Gates E 1257; —, 10 Jan. 2002, G. Gates E 1394; Lake Skinner Track, 42° 57′ S, 146° 44′ E, 11 April 1999, G. Gates E 390; Little Florentine River, F5 Road, 42° 44′ S, 146° 25′ E, 26 May 1999, G. Gates E 539; Marriotts Falls, 42° 43′ S, 146° 39′ E, 2 June 2001, G. Gates E 1212 & E 1215; Mt Field National Park, Lady Barron Falls, 42° 41′ S, 146° 42′ E, 24 April 2001, G. Gates E 1119; Mt Barrow, 41° 22′ S, 147° 26′ E, 20 March 2004, G. Gates E 1957; Mt Wellington, Jacksons Bend, 42° 55′ S, 147° 15′ E, 25 March 1999, G. Gates E 286; —, Myrtle Gully, 42° 54′ S, 147° 15′ E, 9 March 1999, G. Gates E 184; —, 3 May 2001, G. Gates E 1164a; —, 21 Feb. 2002, G. Gates E 1441; —, O'Grady's Falls Track, 42° 55′ S, 147° 15′ E, 17 May 2001, G. Gates E 1175; North West Bay River, 42° 57′ S, 147° 12′ E, 7 June 2001, G. Gates E 1228; —, 1 May 2003, G. Gates E 1760; —, 18 May 2004, G. Gates E 2007; —, 21 April 2005, G. Gates E 2134; Pelverata Falls, 43° 04′ S, 147° 08′ E, 22 May 1999, G. Gates E 529; Ralphs Falls, 41° 18′ S, 147° 51′ E, 15 June 2003, G. Gates E 1872; Reuben Falls, 43° 00′ S, 146° 40′ E, 15 May 1999, G. Gates E 516; —, 17 June 2000, G. Gates E 969; Snug Falls, 16 July 2002, G. Gates E 1604; Tahune, 43° 06′ S, 146° 44′ E, 23 May 1999, G. Gates E 532; Warra LTER site, 43° 06′ S, 146° 41′ E, 12 June 2003, G. Gates E 1866; —, 25 May 2004, G. Gates E 2014; —, 7 June 2005, G. Gates E 2181; Wielangta Forest Reserve, 42° 42′ S, 147° 51′ E, 14 July 1998, G. Gates E 14 (holotype, HO 548347); —, 14 June 2001, G. Gates E 1246 & E 1247; —, 15 Jan. 2002, G. Gates E 1395; —, 20 June 2002, G. Gates E 1585.

Entoloma cystidiosum is very close to *E. deprensum* E. Horak from New Zealand, from which the former mainly differs by its larger fruit bodies, with paler, yellow-brown colours, predominantly intracellular pigment in the pileipellis, and larger spores. *Entoloma farinolens* E. Horak from New Zealand is a clamped species with heavily encrusted hyphae in the pileipellis. *Entoloma indutoides* (P.D. Orton) Noordel. from Europe has a similar microscopy, but differs in its clamped hyphae, lack of encrusting pigment, presence of pleurocystidia and it differs macroscopically in colour. *E. cystidiosum* is best classified in section *Griseorubida* Noordel. (Noordeloos 2004).

Subgenus ***Claudopus*** (Gill.) Noordel.

Claudopus Gillet, Hymen. Fr.: 426. 1876; *Rhodophyllus* subgenus *Claudopus* (Gill.) Romagn. in Beih. Nova Hedwigia 59: 41. 1978; *Entoloma* subgenus *Claudopus* (Gill.) Noordel. in Persoonia 11: 147. 1981.

Habit pleurotoid or omphalioid; pileus usually depressed, umbilicate or infundibuliform; stipe central, eccentric, lateral or lacking; lamellae usually arcuate-decurrent, sometimes adnate with decurrent tooth or rarely adnate-emarginate; pigment encrusting, sometimes in addition also intracellular, absent in white taxa.

Lectotype (Konrad 1934): *Agaricus byssisedus* Pers.: Fr.

Section ***Claudopus*** (Gill.) Noordel.

Entoloma sect. *Claudopus* (Gill.) Noordel., Fungi Europaei 5: 604. 1992.

Habit pleurotoid; pileus regularly or irregularly circular, flabelliform or spathuliform; stipe very reduced, rarely central in young specimens only, then eccentric, lateral or lacking, then basidiocarps sessile.

Type species: *Entoloma byssisedum* (Pers.: Fr.) Donk.

82. **Entoloma pitereka** Noordel. & G.M. Gates, spec. nov.

MycoBank #564511.

Diagnosis: Basidiocarps tiny, pleurotoid; pileus 1–15 mm across, white, tomentose; lamellae subdistant, thick, pink; stipe reduced, white, fibrillose; odour spermatic; spores 8–12×6–8 μm, 5–6-angled; cheilocystidia and terminal elements of pileipellis capitate or subcapitate; clamp-connections present at base of basidia.

Holotype: Australia, Tasmania, Wielangta, 42° 42′ S, 147° 51′ E, 19 Feb. 2005, G. Gates E 2081 (HO 564359; isotype in L).

Etymology: pitereka = aboriginal for white, referring to the white basidiocarps.

Main characters: habit pleurotoid; fruit body very small, white (Plate 3.78).

Description:

Pileus 1–15 mm across, orbicular at first then reniform, convex to applanate in side view, with slightly inrolled margin, white, not hygrophanous, not translucently striate, covered with aeriferous fibrils or tomentose. Lamellae adnate to free, subdistant, ventricose, thick, white with concolorous edge. Stipe eccentric to lateral, up to 2.5 mm long and 0.5 mm wide, fragile, white with aeriferous fibrils. Odour spermatic. Taste strong, nutty. Basal rhizomorphs very prominent, forming a subiculum.

Spores 8–12×6–8 μm, 5–6-angled in side-view. Basidia 28–34×9–11 μm, 4-spored, clamped. Lamella edge fertile or heterogeneous. Cheilocystidia sometimes found, 30–60×4–8 μm, cylindrical with subcapitate apex. Hymenophoral trama regular, made up of cylindrical hyphae, 4–10 μm wide. Pileipellis an entangled trichoderm of interwoven cylindrical hyphae, 3–9 μm wide with cylindrical to narrowly clavate terminal endings with rounded, sometimes slightly swollen apex. Pigment none, but hyphae of lower layers in pileipellis sometimes with granular content. Pileitrama regular, like hymenophoral trama. Clamp-connections observed only at base of basidia (Fig. 3.82).

Habitat & distribution: in troops on underside of rotten wood in wet eucalypt forests and rainforests.

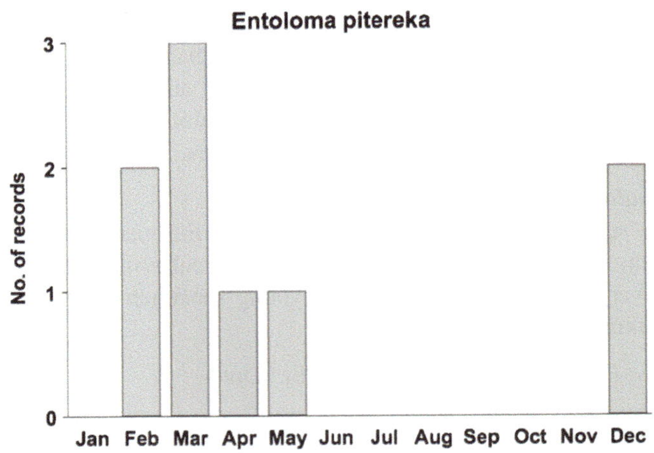

Plate 3.78 *Entoloma pitereka* (Photos Machiel Noordeloos (*top*), Michael Pilkington (*bottom*))

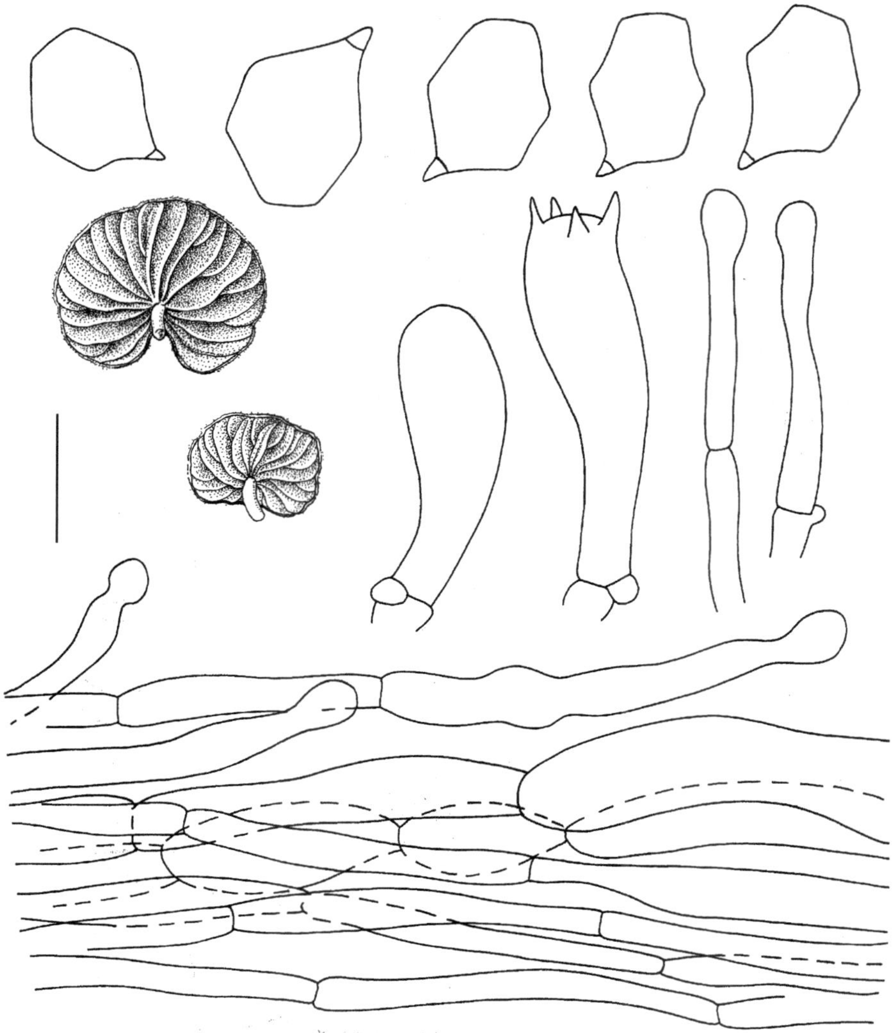

Fig. 3.82 *Entoloma pitereka*. Habit, spores, basidia, cheilocystidia, and pileipellis. Bar = 10 mm or 10 μm

Collections examined. Australia, Tasmania, Florentine Rd, Lady Binney Track, 42° 43′ S, 146° 31′ E, 27 Dec. 2002, G. Gates E 1650; —, 8 March 2003, G. Gates E 1652; Geeveston, Donnellys Rd, 43° 07′ S, 146° 54′ E, 19 Dec. 1999, G. Gates E 783; Growling Swallet, 42° 41′ S, 146° 30′ E, 6 April 2000, G. Gates E 826; Kermandie Falls, Upper Track, 43° 12′ S, 146° 52′ E, 23 Feb. 2000, G. Gates E 789; Tahune, Stringybark Loop, 43° 06′ S, 146° 44′ E, 4 March 2000, G. Gates E 794; Wielangta, Rainforest Walk, 42° 42′ S, 147° 51′ E, 9 March 2000, G. Gates E 801; —, 19 Feb. 2005, G. Gates E 2081 (holotype, HO 564359).

Entoloma pitereka belongs to the group of closely related taxa around *E. byssisedum*, with small, pleurotoid basidiocarps, resembling a *Crepidotus* species in the field. Formerly they were accommodated in the genus *Claudopus*. There are a few pure white species in the literature, which differ mainly by microscopic characters. The European *Entoloma exiguum* Esteve-Rav. & M. de la Cruz seems to be closest to our species, differing slightly in spore morphology and pileipellis structure (Noordeloos 2004); *Entoloma albotomentosum* Noordel. & Hauskn. from Europe has much larger spores.

Section ***Undati*** (Romagn.) Noordel.

Rhodophyllus sect. *Undati* Romagn. in Bull. mens. Soc. linn. Lyon 43: 327. 1974. — *Entoloma* sect. *Undati* (Romagn.) Noordel., Fungi Europaei 5: 613. 1992.

Habit omphalioid; pileus convex with depressed to deeply infundibuliform centre; stipe well-developed, central; pigment coarsely encrusting, sometimes also intracellular.

Type species: *Entoloma undatum* (Fr.) M.M. Moser.

83. **Entoloma austrorhodocalyx** G.M. Gates & Noordel., Persoonia 17: 221. 2007.

Original diagnosis: Habitus omphalinoideus. Pileus 3–15 mm latis stipesque pallide brunneus. Stipes 20–30 × 1–2 mm tenuis fragilis. Sporae 7.8–9.5 × 7.0–9.5 μm 6–8-angulatae. Cheilocystidia desunt. Pileipellis cutis hyphis elementis septatis clavates usque ad 15 μm latis. Fibulae numerosae.

Holotype: Australia, Tasmania, Duckhole Lake Track, 43° 22′ S, 146° 53′ E, 20 April 2004, M.E. Noordeloos 2004047 (HO 543557; isotype in L).

Etymology: austro (Lat.) = southern; rhodocalyx = resembling *E. rhodocalyx*.

Main characters: basidiocarps small, omphalioid, pale grey or grey-brown; spores isodiametric; cheilocystidia lacking; clamp-connections present (Plate 3.79).

Description:
Pileus 3–15 mm diam, convex to plano-convex with umbilicate centre and deflexed crenate margin, pale brown-grey, sometimes with a pinkish hue (4–5C3, 4B2, 4A2) with darker centre (4D3), dry, very translucently striate when moist, hygrophanous, clearly pallescent, innately radially fibrillose, becoming micaceous to lustrous grey-buff on drying. Lamellae broadly decurrent, triangular, arcuate, very distant, pale greyish pink with hyaline, sometimes slightly darker edge, L = 16–20, l = 1–3. Stipe 20–30 × 1–2 mm, very slender, fragile, cylindrical, often with distinctly swollen base, pale brown to almost white at apex, darker towards base (grey-brown like centre of pileus), slightly pruinose at apex, otherwise glabrous, polished or with a

Plate 3.79 *Entoloma austrorhodocalyx* (Photo Machiel Noordeloos)

few longitudinal fibrils but never silvery striate, with white basal tomentum. Odour and taste strongly spermatic.

Spores 7.8–9.5 × 7.0–9.5 μm, on average 7.8–8.9 × 6.7–8.4 μm, Q = 1.0–1.2, isodiametric or subisodiametric, 6–8-angled in side-view with irregular, rather blunt angles. Basidia 18–25 × 6–9 μm, 4-spored, clamped. Lamella edge fertile. Cystidia absent. Pileipellis a differentiated cutis of septate, cylindrical hairs with slightly inflated clavate terminal elements, to 15 μm in diameter. Pigment yellow-brown, minutely encrusting and parietal in the hyphae of pileipellis. Pileitrama regular, made up of long, fusiform elements, to 200 × 5–18 μm. Stipitipellis a cutis of narrow, cylindrical hyphae 4–9 μm in diameter, finely encrusted. Clamp-connections present (Fig. 3.83).

Habitat & distribution: widespread, but not common, in wet sclerophyll forests. Almost all of the fruiting occurs in autumn (March-May) and winter (June-August), rarely in other months (Group 4, in Chap. 1, see section "Fruiting Patterns of Species of Entolomataceae").

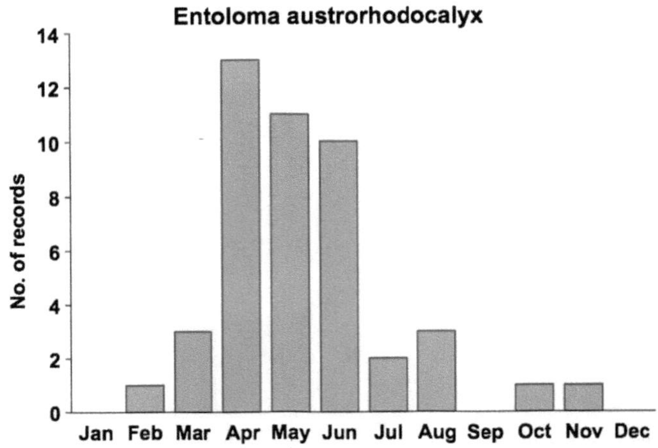

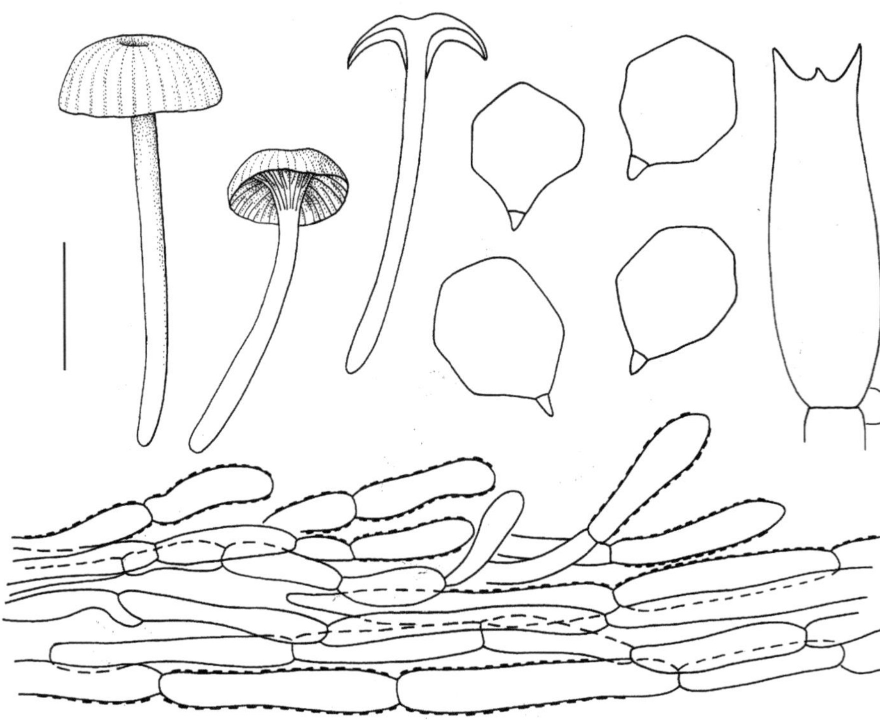

Fig. 3.83 *Entoloma austrorhodocalyx*. Habit, spores, basidia, cheilocystidia, and pileipellis. Bar = 10 mm or 10 μm

Collections examined. Australia, Tasmania, Bermuda Road, 43° 04′ S, 146° 54′ E, 25 July 2002, G. Gates E 1613; Bruny Island, Mt Mangana, 43° 22′ S, 147° 17′ E, 12 April 2003, G. Gates E 1703; Creepy Crawly, 42° 50′ S, 146° 22′ E, 15 April 2000, G. Gates E 844; Duckhole Lake Track, 43° 22′ S, 146° 53′ E, 8 April 2003,

G. Gates E 1683; —, 15 May 2003, G. Gates E 1810; —, 20 April 2004, M.E. Noordeloos 2004047 (holotype, HO 543557) & 2004054; Marriotts Falls, 42° 43′ S, 146° 39′ E, 29 May 2003, G. Gates E 1832; Mt Field NP, 42° 41′ S, 146° 42′ E, 31 May 2003, G. Gates E 1840; —, Lady Barron Falls, 31 March 2001, G. Gates E 1053; —, 24 April 2001, G. Gates E 1118; —, Lyrebird Track, 3 May 2003, G. Gates E 1767; Mt Wellington, Lower Sawmill Track, 42° 55′ S, 147° 15′ E, 12 June 2001, G. Gates E 1241; North West Bay River, 42° 57′ S, 147° 12′ E, 7 June 2001, G. Gates E 1230; Tahune, Huon Pine Walk, 43° 06′ S, 146° 44′ E, 2 July 2002, G. Gates E 1594; Timbs Track, 42° 44′ S, 146° 25′ E, 29 April 2000, G. Gates E 876; —, 13 May 2000, G. Gates E 876a; —, 21 April 2001, G. Gates E 1106; —, 23 May 2002, G. Gates E 1535; Warra LTER site, 43° 06′ S, 146° 39′ E, several observations between March 2005 and April 2007.

This small species fits well in the current concept of section *Undati*, and keys out close to the European species *E. rhodocalyx* (Lasch) M.M. Moser from which it differs mainly by the absence of cheilocystidia. *Entoloma rancidulum* E. Horak from New Zealand may be close, but differs in having much larger spores, and a strong rancid-farinaceous taste.

84. **Entoloma lillirium** Noordel. & G.M. Gates, spec. nov.

MycoBank #564512.

Diagnosis: basidiocarps small, omphalioid, very dark greyish brown all over, pileus squamulose; lamella decurrent; spores 7–8(–8.5) × 5.5–7 μm, isodiametric; pileipellis a trichoderm with cystidioid terminal elements, up to 50 μm wide with coarse encrusting pigment; clamp-connections absent.

Holotype: Australia, Tasmania, Taroona, Truganini Track, 42° 56′ S, 147° 21′ E, 8 June 2005, G. Gates E 2176a (HO 564360, isotype in L).

Etymology: lilliria (aboriginal) = shade or shadow (Plate 3.80).

Description:
Pileus 6–10 mm diam, plano-convex with shallow central depression, with straight entire margin, dry, not translucently striate, not hygrophanous, dark grey-brown (5D3), darker at centre, velutinous at first then breaking up into small squamules entirely to margin. Lamellae arcuate-decurrent, narrow, crowded to moderately distant, up to 1.5 mm broad, grey-brown, with concolorous edge. Stipe 15–30 × 1–1.5 mm, cylindrical, brown, dry, glabrous, with white basal tomentum. Odour and taste indistinct.

Spores 7–8(–8.5) × 5.5–7 μm, Q = 1.15–1.3, Qav = 1.2, very thin-walled, weakly angled with (5–)6–7(–8) angles. Basidia 20–33 × 6–9 μm, 4-spored, clampless. Lamella edge fertile. Cystidia absent. Hymenophoral trama regular, made up of cylindrical hyphae, with brown, heavily encrusted walls. Pileipellis a cutis with transitions to a trichoderm, made up of very regularly shaped, inflated, septate hyphae with cystidioid terminal elements, up to 50 μm in diameter. Pigment brown, intracellular and encrusting the hyphae of pileipellis and pileitrama. Clamp-connections absent (Fig. 3.84).

Plate 3.80 *Entoloma lillirium* (Photo Genevieve Gates)

Habitat & distribution: rare, known from only one location, a wet eucalypt gully at the back of suburbia.

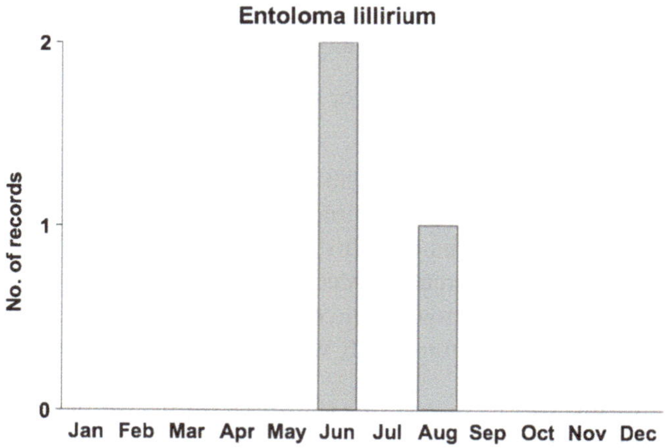

Collections studied: Australia, Tasmania, Taroona, Truganini Track, 42° 56′ S, 147° 21′ E, 2 June 2005, G. Gates E 2176a (holotype, HO 564360); —, 8 June 2005, G. Gates E 2176b; —, 8 Aug. 2005, G. Gates s.n. — New Zealand, Otago Prov. Southland, L.D. near Denton River, Te Anan Downs, Wallace Co. leg. E. & A. Horak (holotype of *E. crinitum,* PDD 27022);

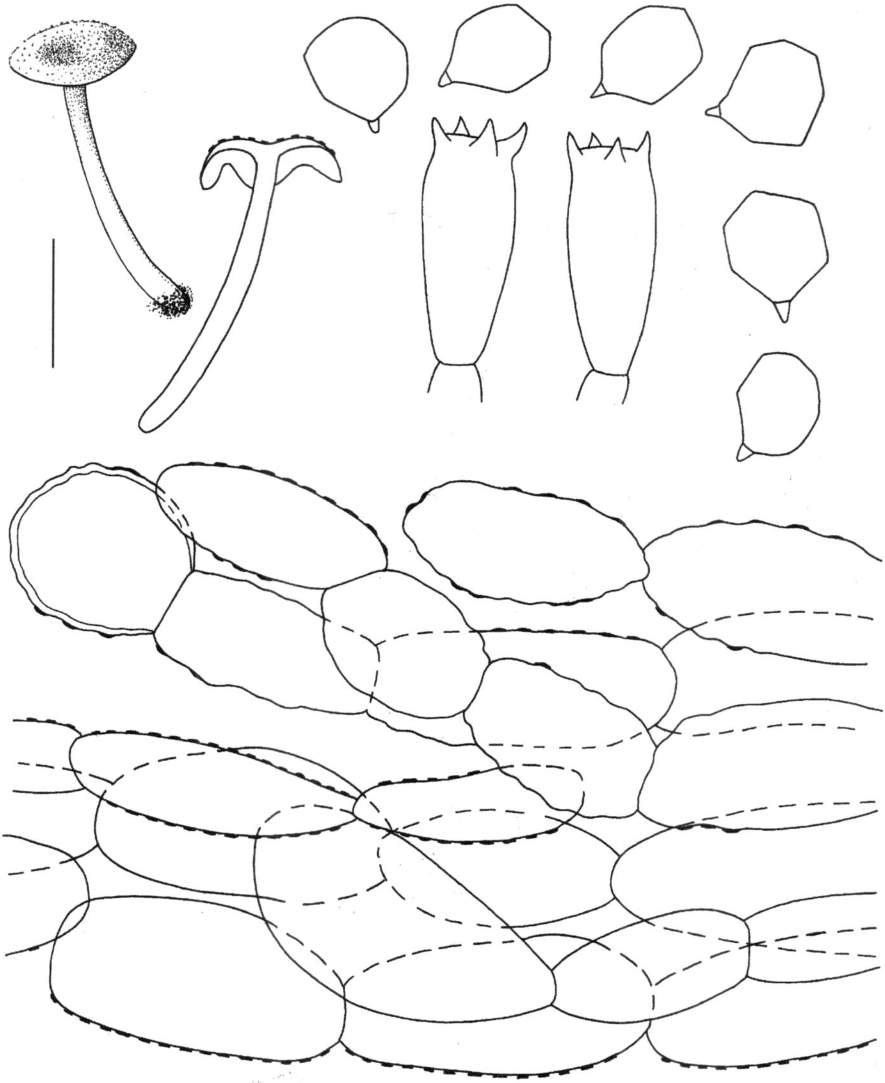

Fig. 3.84 *Entoloma lillirium*. Habit, spores, basidia, and pileipellis. Bar = 10 mm or 10 μm

At first we thought this represented *Entoloma crinitum* E. Horak from New Zealand, on account of the striking similarities with respect to the dark omphalioid basidiocarps and isodiametric spores. However, the structure of the pileipellis differs too much to confirm the conspecificity. *E. lillirium* has a trichoderm of very inflated, large, clavate to globose elements, whereas *Entoloma crinitum* has a simple cutis with transitions to a trichoderm of subcylindrical hyphae. Both species, and the related *E. percrinitum*, fit well in the current concept of section *Undati* by virtue of

Plate 3.81 *Entoloma percrinitum* (Photos Machiel Noordeloos)

the small, isodiametric spores, and heavily encrusting pigment (Noordeloos 2004). *Entoloma percrinitum*, as described below, differs, in particular, by the very different structure of the pileipellis and larger stature, but is in other respects very similar.

85. **Entoloma percrinitum** G.M. Gates & Noordel., Persoonia 17: 222. 2007.

Original diagnosis: Habitus omphalinoideus. Pileus 10–35 mm latus obscure griseobrunneus toto aerifere-villoso. Lamellae confertae obscure griseae. Stipes 15–37 × 1–3 mm obscure griseobrunneus glabrus politus. Sporae 7.0–8.5 × 6.0–7.5 µm 5–7-angulatae. Cheilocystidia desunt. Pileipellis cutis vel trichoderma elementis inflatis 5–12 µm latis minute incrustatis. Fibulae numerosae.

Holotype: Australia, Tasmania, Chauncy Vale, 42° 37′ S, 147° 16′ E, 28 June 2003, G. Gates E 1888 (HO 543547; isotype in L).

Etymology: resembling *E. crinitum* E. Horak.

Main characters: habit omphalioid, pileus dark coloured almost black, covered in concentric aeriferous fibrils especially when fresh, not hygrophanous, not striate; spores small, isodiametric; cystidia absent; clamp-connections present (Plate 3.81).

Description:
Pileus 10–35 mm diam, convex with or without a small, acute umbo in central depression, with deflexed margin, uniformly dark grey-brown (5D1–5F2, 6D1–6F2; 10YR 3–2/1; 7.5YR 3/2) to black, not translucently striate, dry, slightly hygrophanous,

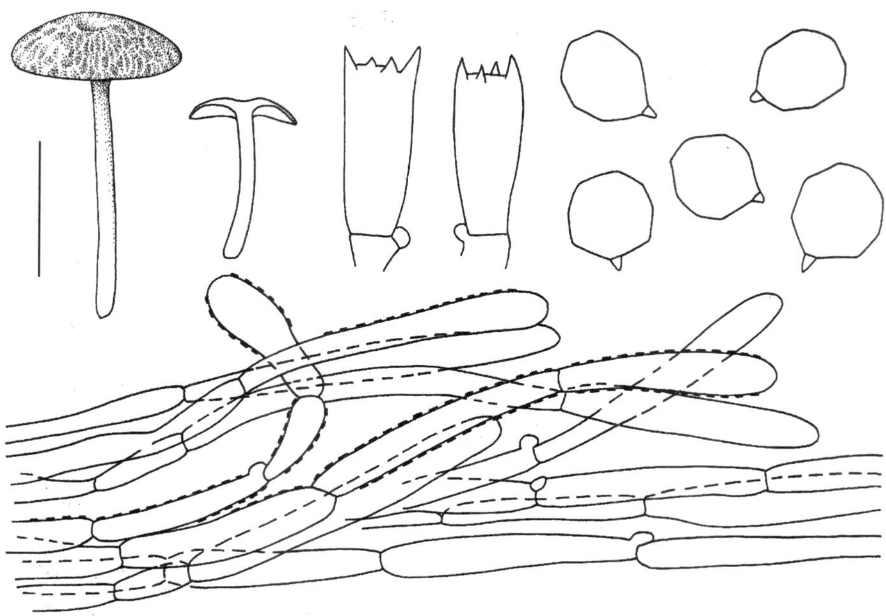

Fig. 3.85 *Entoloma percrinitum*. Habit, spores, basidia, and pileipellis. Bar = 10 mm or 10 μm

pallescent on drying along concentric zones to pale grey-brown, entirely subrugulose to hoary-aeriferously fibrillose, not squamulose. Lamellae adnate to subdecurrent, segmentiform or subarcuate, moderately crowded, 5–9 mm broad, thin or thickish, moderately dark to dark grey with concolorous or somewhat hyaline edge, L = 30–40, l = 1–5. Stipe 15–40 × 1–3 mm, cylindrical, dark grey-brown, more or less concolorous with pileus, glabrous, polished with white basal tomentum. Odour of iodine. Taste none.

Spores 7.0–8.5 × 6.0–7.5 μm, average 7.5 × 6.3 μm, Q = 1.0–1.2, isodiametric to subisodiametric, 5–7-angled in side-view with relatively thin walls. Basidia 12–30 × 6–9 μm, 4-spored, clamped. Lamella edge fertile. Cheilocystidia absent. Hymenophoral trama regular, made up of cylindrical to inflated elements, 50–95(–120) × 7–25 μm with brown, coarsely encrusted walls. Pileipellis a differentiated cutis with transitions to a trichoderm, made up of fascicles of septate, cylindrical hyphae, 5–12 μm in diameter, with cylindrical to narrowly clavate terminal elements. Pigment brown, parietal to coarsely encrusting. Pileitrama regular, made up of cylindrical to inflated elements, 40–100 × 8–17 μm with brown encrusted walls. Stipitipellis a cutis of cylindrical hyphae, 4–10 μm in diameter with encrusted walls. Caulocystidia absent. Clamp-connections present in hymenium and covering layers of pileus and stipe (Fig. 3.85).

Habitat & distribution: widespread, but uncommon, known only from a few locations in Tasmania.

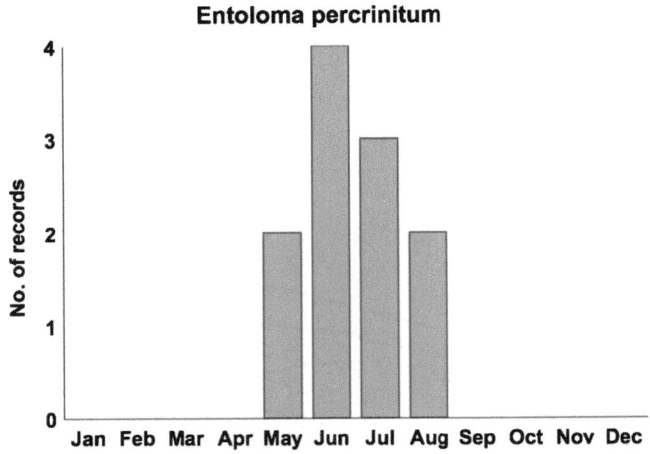

Collections examined. Australia, Tasmania, Birralee, Black Sugarloaf, 41° 24′ S, 146° 48′ E, 5 June 2006, G. Gates E 2241; Blue Tier, 41° 11′ S, 148° 00′ E, 1 June 2004, G. Gates E 2027; Chauncy Vale, 42° 37′ S, 147° 16′ E, 1 Aug. 2002, G. Gates E 1618; —, 28 June 2003, G. Gates E 1888 (holotype, HO 543547; isotype, L); —, 29 July 2003, G. Gates E 1898; —, 22 June 2004, G. Gates E 2037; Vale of Belvoir, 41° 33′ S, 145° 54′ E, 25 May 2010, M.E. Noordeloos 2010059 & 2010060.

Entoloma percrinitum keys out in section *Undati* on account of the more or less omphalioid habit, isodiametric spores, and differentiated pileipellis of septate, heavily encrusted hyphae. In the field it can be taken for a rather dark *Cyanula*, but the surface of the pileus with its rugulose-hoary subzonate covering make it easy to recognize. *Entoloma crinitum* E. Horak from New Zealand is similar, but is a much smaller species and has a brown pileal surface with rather pronounced squamules that are very different microscopically.

Subgenus ***Pouzarella*** (Mazzer) Noordel.

Entoloma subgenus *Pouzarella* (Mazzer) Noordel. in Persoonia 12: 196. 1984.

Syn.: Pouzarella Mazzer in bibltca. mycol. 46: 69. 1976. *Rhodophyllus* subgenus *Inopilus* Romagn. in Bull. mens. Soc. linn. Lyon. 43: 329. 1974 non Romagn. 1978 (=subgenus *Inocephalus* Noordel.).

Excluded: *Entoloma* subgenus *Pouzaromyces* sensu Noordel. in Persoonia 11: 142. 1981; sensu Moser in Kl. KryptogFl. 2(b/2), 4. Aufl.: 191. 1978; *Rhodophyllus* subgenus *Pouzaromyces* sensu Romagn. in Beih. Nova Hedwigia 59: 50. 1978; *Pouzaromyces* Pilàt in Acta Mus. nat. Prag. (B)9: 60. 1953 sensu Pilàt; Horak in Beih. Nova Hedwigia 65: 37. 1978.

Habit mycenoid with distinctly fibrillose-hairy to fibrillose-squamulose pileus, reminiscent of a species of *Inocybe*, not hygrophanous, not translucently striate; lamellae often very dark grey-brown, stipe usually with encrusted hairs, base with radiating hairs; pileipellis a cutis with transitions to a trichoderm of septate hairs with encrusted pigment; clamp-connections absent; cheilocystidia, if present, cylindrical to clavate, often encrusted.

Plate 3.82 *Entoloma farinosum* (Photos Michael Pilkington (*left*), Genevieve Gates (*right*))

Holotype: *Entoloma nodosporum* (G.F. Atk.) Noordel.

Section ***Dysthales*** (Mazzer) Noordel.

Pouzarella sect. *Dysthales* Mazzer in Bibltca mycol. 46: 92. 1976; *Entoloma* sect. *Dysthales* (Mazzer) Noordel. in Beih. Nova Hedwigia 91: 66. 1987.

Pigment exclusively encrusting; cheilocystidia, if present, cylindrical to (broadly) clavate.

Holotype: *Entoloma nodosporum* (G.F. Atk.) Noordel.

86. **Entoloma farinosum** (Largent & Skye Moore) Noordel. & G.M. Gates, comb. nov.

Pouzarella farinosa Largent & Skye Moore in Mycotaxon 117: 455. 2011.

Original diagnosis: Habitu Entolomate dysthali (subg. Pouzarellae) similis, sed sapore farinaceo, squamulis pilei aurantiaco-griseis, cheilocystidiis latioribus, 21.1–51.1 × 16.7–30.9 μm differt.

Holotype: Australia, New South Wales, Cook Region, Central Hunter District, Barrington Tops National Park, Pool of Reflections Track, 32° 08′ 15.8″ S, 151° 30′ 38.2″ E, 727.6 m, 22 April 2010, DL Largent 9934 (holotype DAR).

Main characters: pileus tomentose with a thick covering of pale, ochraceous tipped hairs matted together to appear grey in mass; spores 12–20 × 9–10 μm (Plate 3.82).

Description:
Pileus 5–17 mm diam, conical with blunt apex to convex, with straight or reflexed margin, not hygrophanous, not translucently striate, dark grey-brown, tomentose all over with a dense mat of ochraceous tipped fibrils which give a greyish appearance overall. Lamellae adnate, subventricose, distant, up to 2 mm broad, thickish, dark grey-brown with white, fimbriate edge, L = 20–25, l = 1–3. Stipe 10–80 × 1–2 mm,

cylindrical, equal, brown, dry, covered with a dense matted tomentum of ochraceous hairs which are white at the apex of stipe. Odour none.

Spores 12–20×9–10 μm, Q=1.5–2.0, Qav=1.6–1.7, many-angled, nodulose, thick-walled. Basidia 38–78×14–20 μm, 4-, rarely also 2-spored, clampless; so-called abortive basidia with brown, granular content frequent in hymenium. Lamella edge sterile. Cheilocystidia 25–90×10–35 μm, septate with clavate to broadly clavate or obovoid terminal elements, with thick, encrusted walls. Hymenophoral and pileitrama regular, made up of 6–20 μm wide, heavily encrusted hyphae. Pileipellis a cutis with transitions to a trichoderm, made up of long, thick-walled, septate, 6–20 μm wide, heavily encrusted hyphae. Stipitipellis similar to pileipellis. Clamp-connections absent (Fig. 3.86).

Habitat & distribution: wet eucalypt forests, widespread but not common.

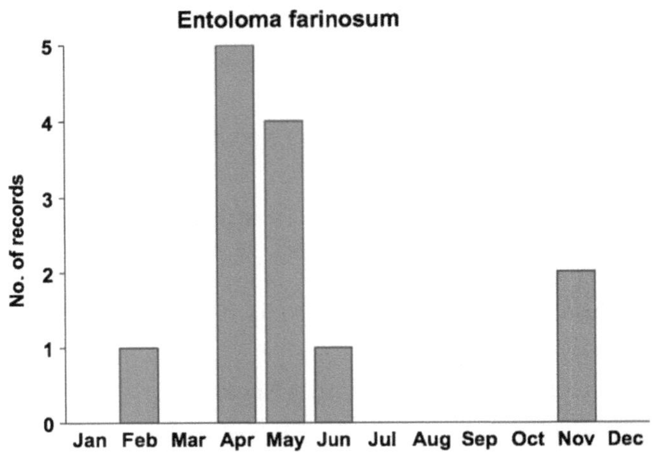

Collections examined. Australia, Tasmania, Kermandie Falls, Lower Track, 43° 12′ S, 146° 52′ E, 5 April 2001, G. Gates E 1057; 29 May 2001, G. Gates E 1190; —, 31 May 2001, G. Gates E 1190a; —, 20 April 2005, G. Gates E 2123; Mt Field National Park, River Walk, 42° 41′ S, 146° 42′ E, 9 Nov. 1999, G. Gates E 733; North West Bay River, 42° 57′ S, 147° 12′ E, 22 April 2008, G. Gates E 2265; —, 14 Feb. 2002, G. Gates E 1430.

Largent et al. (2011) recently described ten new species of *Pouzarella* from Queensland and New South Wales. We had no difficulty recognizing our species as *Pouzarella farinosa,* because of the very similar macroscopic and microscopic characters. The only question mark is the obvious lack of a distinct farinaceous taste in our specimens.

Pouzaromyces lasius (Berk. & Broome) Pegler, known from Ceylon and Papua New Guinea (Horak 1980), is macroscopically similar with ochraceous brown hairs on the pileus and stipe, but has considerably shorter spores (10–13.5 μm), and lacks cheilocystidia. *Pouzarella minuta* (E. Horak) E. Horak from New Zealand, has similar large spores, but has much smaller basidiocarps, and lacks true cheilocystidia. Our species is also similar to *Entoloma dysthales* (Peck) Sacc. from which it mainly differs by the distinct ochraceous grey sheen in the pileus and stipe.

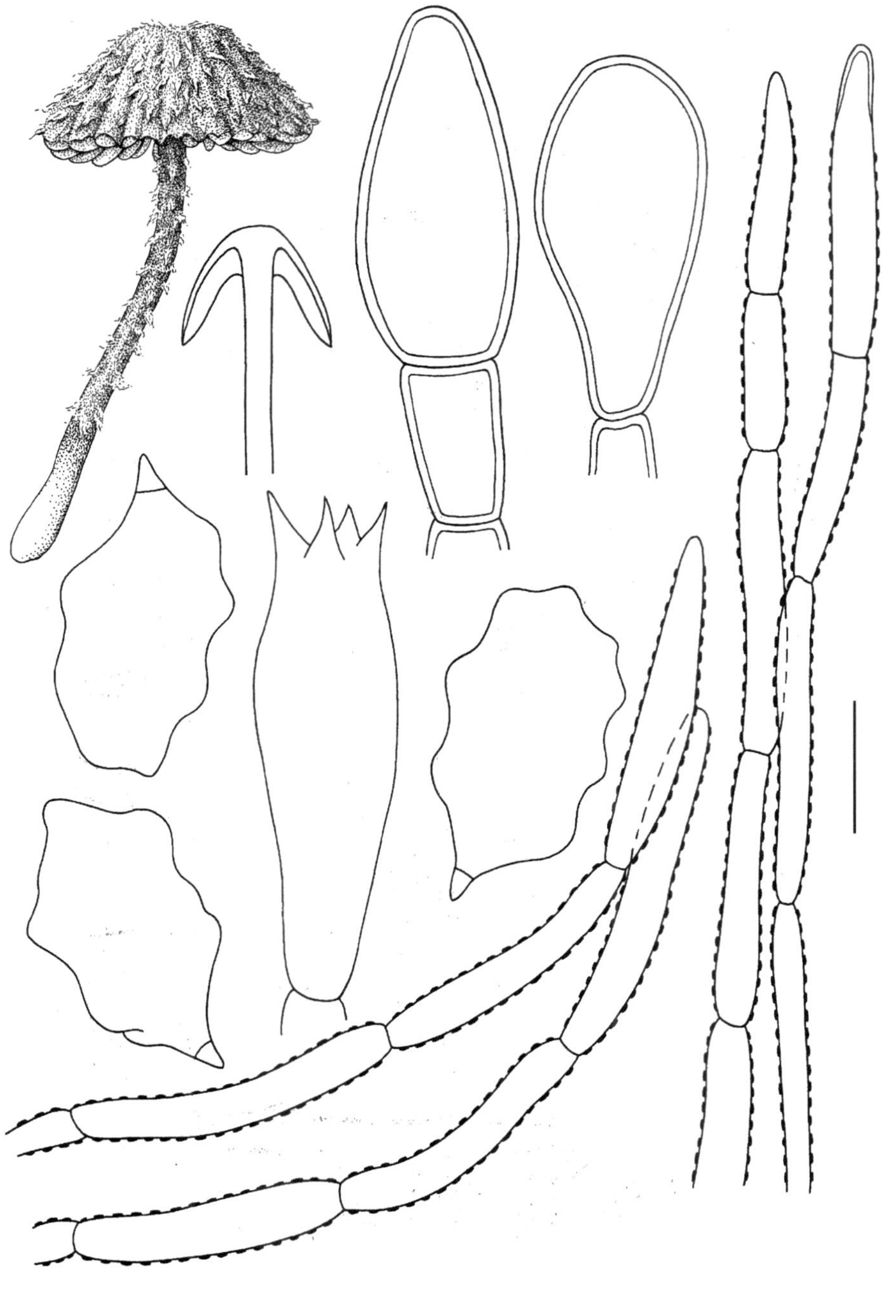

Fig. 3.86 *Entoloma farinosum*. Habit, spores, basidia, hairs on stipe (bottom), and hairs on pileus (right). Bar = 10 mm or 10 µm. The thick-walled structures at the top centre of the drawing are cheilocystidia

Taxa of Unknown Taxonomic Position

87. **Entoloma camarophyllus** G.M. Gates & Noordel., Persoonia 19: 219. 2007.

Original diagnosis: Habitus omphalinoideus. Pileus 10–35 mm latus fuscobrunneus paulisper hygrophanus translucido-striatus glaber sed centro leviter fibrilloso-ruguloso. Lamellae distantes crassae. Stipes 25–40×2–4 mm pileo concolorus cartilagineus glaber politus. Odore saporeque rancido-farinaceis. Sporae 8.5–16(–16.5)×7.5–11.5 µm valde 4–9-angulatae. Cheilocystidia sparsa irregulariter clavata vel lageniformia. Pileipellis cutis compactis hyphis cylindraceis 1.5–7 µm latis. Pigmentis intracellulosis vel minute incrustatis. Fibulae desunt.

Holotype: Australia, Tasmania, Growling Swallet, 42° 41′ S, 146° 30′ E, 6 April 2000, G. Gates E 469 (HO 543554; isotype in L).

Etymology: referring to the habit which resembles a *Camarophyllus* species.

Main characters: habit omphalioid, brownish or greyish blonde; lamellae decurrent, very distant, often forming a pseudocollarium around the stipe; spores large and complex; clamp-connections absent (Plate 3.83).

Description:
Pileus 10–35 mm diam, convex to plano-convex with inflexed then deflexed margin, with slightly to broadly umbilicate centre to infundibuliform, slightly hygrophanous, translucently striate at margin only or up to 1/3 of radius when moist, uniformly rather dark red-brown, grey-brown or light grey-brown when young, then with paler marginal zone, remaining rather dark brown at centre (10YR 3/2–3, 10YR 4/3, 7.5 YR 3/2, margin 10YR 6/3–4, 10YR 7/4, outermost margin 10YR 8/4), becoming lighter grey-buff on drying, somewhat greasy when fresh, finely innately fibrillose under a hand lens, very slightly fibrillose-subrugulose at centre. Lamellae deeply decurrent, arcuate, very distant (reminiscent of a *Camarophyllus*), sometimes forming a pseudocollarium, pallid brown or pallid pink, then tinged darker pink (10YR 8/4), with entire, thickened, sometimes hyaline edge, waxy, thick, L = ca. 20, l = 0–3. Stipe 25–40×2–4 mm, cylindrical, often curved or flexuous, often distinctly broadened at base to bulbous, pale brown or pale greyish blonde, much paler than pileus (10YR 8–7/4), entirely glabrous, dull. Context hyaline, whitish, somewhat cartilaginous, inner parts of pileus and stipe paler, fibrous. Odour indistinct. Taste none or sometimes like cucumber.

Spores 8.5–16(–16.5)×7.5–11.5 µm, on average 11.8–13.2×9.0–9.8 µm, Q = 1.15–1.9, Qav = 1.4, very irregularly shaped, with 4–9 rather pronounced angles in side-view, sometimes appearing almost cruciform or cuboid with a large, protruding apiculus. Basidia 21–40×6–14 µm, 4- rarely 2-spored, clampless. Lamella edge sterile or with scattered cheilocystidia. Cheilocystidia irregularly shaped clavate, lageniform or moniliform, often with slightly thickened, refringent wall in upper part. Hymenophoral trama irregular, made up of short, cylindrical or inflated elements, with parietal and minutely encrusted pigment in places. Pileipellis a compact cutis of cylindrical elements, 1.5–7 µm in diameter, sometimes with some tufts of ascending,

Plate 3.83 *Entoloma camarophyllus* (Photo Michael Pilkington)

coralloid or clavate terminal elements, to 7 μm in diameter, gradually passing into pileitrama. Pileitrama subregular, made up of short, inflated elements, 22–55 (–70) × 8–15 μm. Brilliant granules and oily droplets abundant when observed fresh in water. Pigment difficult to locate, intracellular and parietal, or with few encrustations. Stipitipellis a cutis of cylindrical hyphae 2–9 μm in diameter. Caulocystidia absent. Clamp-connections absent (Fig. 3.87).

Habitat & distribution: widespread in wet forests, and also found on button-grass plains at high altitudes. Main fruiting in April and May, with almost all of the remaining records occurring in March and June (Group 3, in Chap. 1, see section "Fruiting Patterns of Species of Entolomataceae").

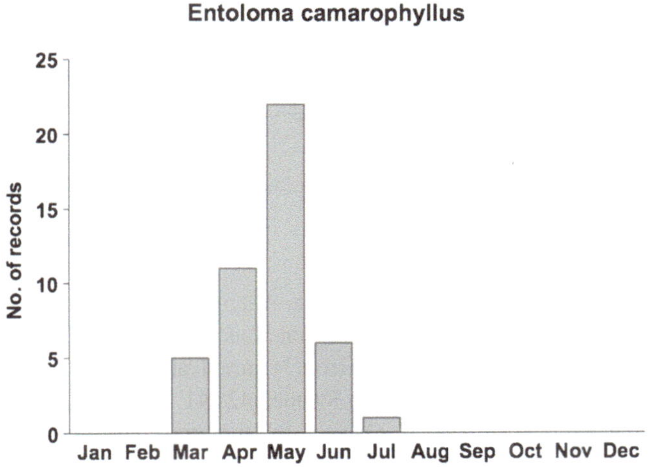

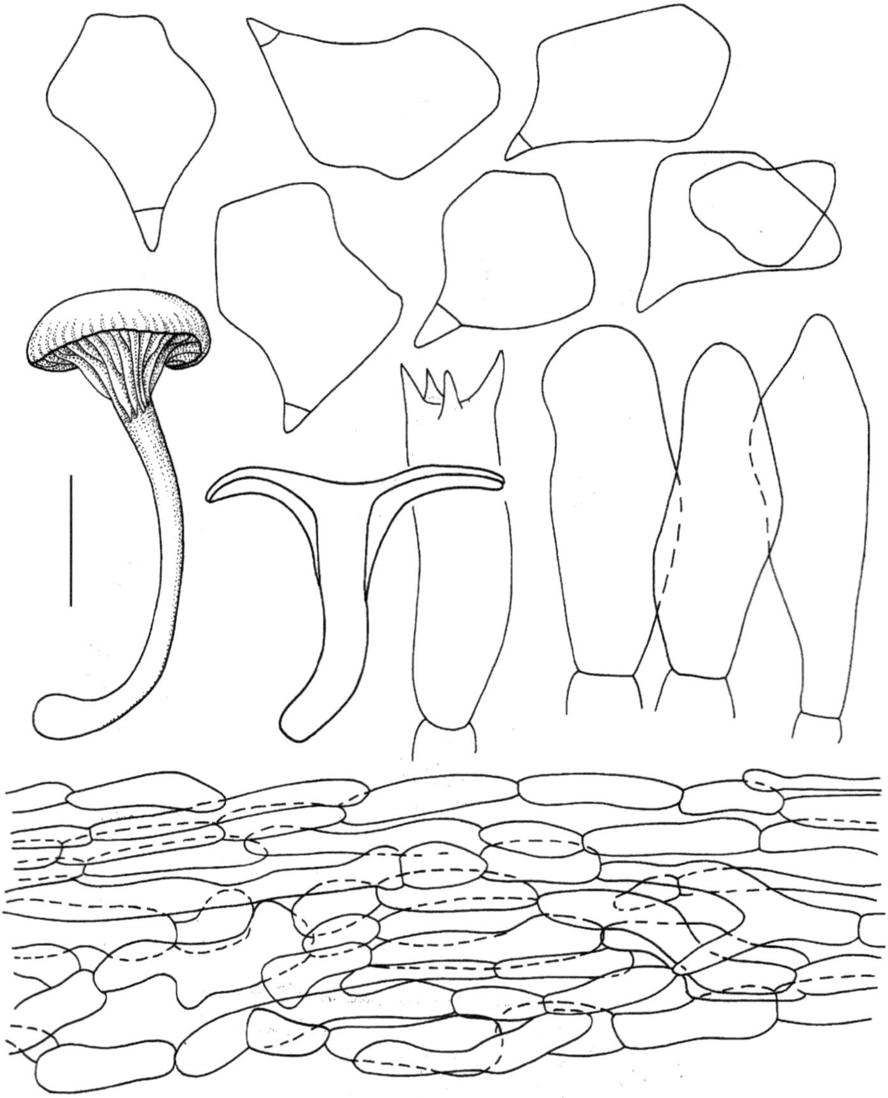

Fig. 3.87 *Entoloma camarophyllus*. Habit, spores, basidia, cheilocystidia, and pileipellis. Bar = 10 mm or 10 μm

Collections examined. Australia, Tasmania, Adamsons Peak Track, 43° 20′ S, 146° 54′ E, 31 May 2001, G. Gates E 1201; Duckhole Lake Track, 43° 22′ S, 146° 53′ E, 15 May 2003, G. Gates E 1813; Franklin River Nature Trail, 42° 13′ S, 146° 01′ E, 22 March 2005, G. Gates E 2095; Growling Swallet, 42° 41′ S, 146° 30′ E, 19 May 1999, G. Gates E 519; —, 6 April 2000, G. Gates E 469 (holotype, HO 543554); —, 11 May 2000, G. Gates E 907; —, 11 April 2002, G. Gates E 1472; —, 4 June 2002, G. Gates

Plate 3.84 *Entoloma choanomorphum* (Photos Michael Pilkington (*left*), Genevieve Gates (*right*))

E 1555; —, 13 May 2003, G. Gates E 1800; —, 22 April 2004, M.E. Noordeloos 2004062; Kermandie Falls, Lower Track, 43° 12′ S, 146° 52′ E, 11 May 1999, G. Gates E 497; Milkshake Hills, 41° 06′ S, 145° 10′ E, 14 June 2002, G. Gates E 1576; North West Bay River, 42° 57′ S, 147° 12′ E, 31 March 2003, G. Gates E 1666; Reuben Falls, 43° 00′ S, 146° 40′ E, 15 May 1999, G. Gates E 511; Styx Valley, Christmas Tree Grove, 42° 50′ S, 146° 41′ E, 26 April 2011, M. E. Noordeloos 2011002; Warra LTER site, Coupe WR008J, 43° 06′ S, 146° 41′ E, 27 April 2004, G. Gates E 1980. Wedge Forest Reserve, 42° 50′ S, 146° 16′ E, 6 May 2003, G. Gates E 1773.

The spores of this remarkable species, which mimics a *Camarophyllus* or *Camarophyllopsis*, are very large, very pronouncedly angled and with a conspicuously protruding, pointed apiculus. The species is difficult to accommodate in the current classification. Preliminary results of phylogenetic studies confirm its isolated status.

88. **Entoloma choanomorphum** G.M. Gates & Noordel., Persoonia 19: 223. 2007.

Original diagnosis: Habitus omphalinoideus. Pileus 15–40 mm latus obscure fuligineus leviter hygrophanus leviter translucido-striatus glabrus. Stipes 30–55 × 1.5–5 mm pileo pallidior glabrus politus. Odore valde nauseosus. Sporae 10–14 × 8.0–11 µm 6–7-angulatae. Cheilocystidia 40–100 × 6–12 × 2–5 µm lageniformia vel fusiformia. Pileipellis cutis hyphis cylindraceis 6–12 µm latis pigmentis intracellulsis. Fibulae desunt.

Holotype: Australia, Tasmania, Wielangta, 42° 42′ S, 147° 51′ E, 5 May 2005, G. Gates E 2158 (HO 543546; isotype in L).

Etymology: choano (Lat.) = funnel; morphum (Lat.) = shaped.

Main characters: habit omphalioid, dark brown; odour strong, like burnt rubber or sulphur; cheilocystidia large (Plate 3.84).

Description:
Pileus 15–40 mm diam, convex with depressed centre, margin deflexed, becoming straight, translucently striate at margin, dark brown (5F3, 6F4–6) or a paler brown (5D6) with darker centre, slightly hygrophanous, glabrous, radially fibrillose, often becoming radially split with age. Lamellae decurrent, arcuate to segmentiform, moderately distant to distant, up to 6 mm broad, thickish, whitish then pinkish brown with slightly irregular, concolorous edge, lamellulae in two tiers. Stipe 30–55 × 1.5–5 mm, cylindrical or attenuated downwards, pale brown (blonde, horn), glabrous or with slight white bloom and white basal tomentum. Odour strong, sulphurous or like burnt rubber. Taste burning.

Spores 10–14 × 8.0–11 μm, Q = 1.2–1.5, irregularly 6–7-angled in side-view. Basidia 24–46 × 7–12 μm, 2–4-spored, clampless. Lamella edge heterogeneous. Cheilocystidia scattered to abundant, mixed among basidia, 40–100 × 6–12 × 2–5 μm, lageniform to fusiform, sometimes with broadly swollen base and long, gradually tapering or moniliform neck, thin-walled. Hymenophoral trama regular, made up of cylindrical to subfusiform elements, to 150 × 4–18 μm. Brilliant granules not observed. Pileipellis a cutis of cylindrical to inflated hyphae, 6–12 μm in diameter, with subclavate to clavate terminal repent or slightly ascending elements, 30–60 × 8–15 μm. Pigment dark brown, intracellular in pileipellis and upper pileitrama. Pileitrama regular, made up of cylindrical to inflated elements, 40–170 × 5–20 μm. Clamp-connections absent (Fig. 3.88).

Habitat & distribution: widespread, but not common, in wet sclerophyll forests.

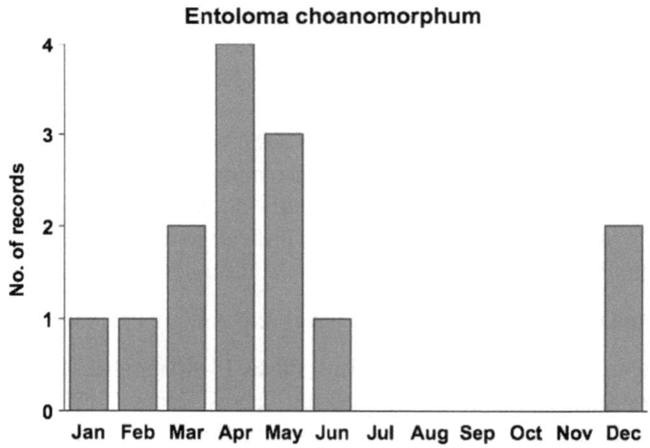

Collections examined. Australia, Tasmania, Duckhole Lake Track, 43° 22′ S, 146° 53′ E, 9 March 2002, G. Gates E 1450; —, 8 April 2003, G. Gates E 1642; —, 4 Jan. 2005, G. Gates E 2072; —, 20 Dec. 2005, G. Gates E 2226; —, 14 Feb. 2006, G. Gates E 2237; North West Bay River, 42° 57′ S, 147° 12′ E, 1 May 2003, G. Gates E 1759; —, 21 April 2005, G. Gates E 2128; Waterfall Bay, 43° 04′ S, 147° 57′ E, 13 March 1999, G. Gates E 217; Wielangta, 42° 42′ S, 147° 51′ E, 5 May 2005, G. Gates E 2158 (holotype, HO 543546).

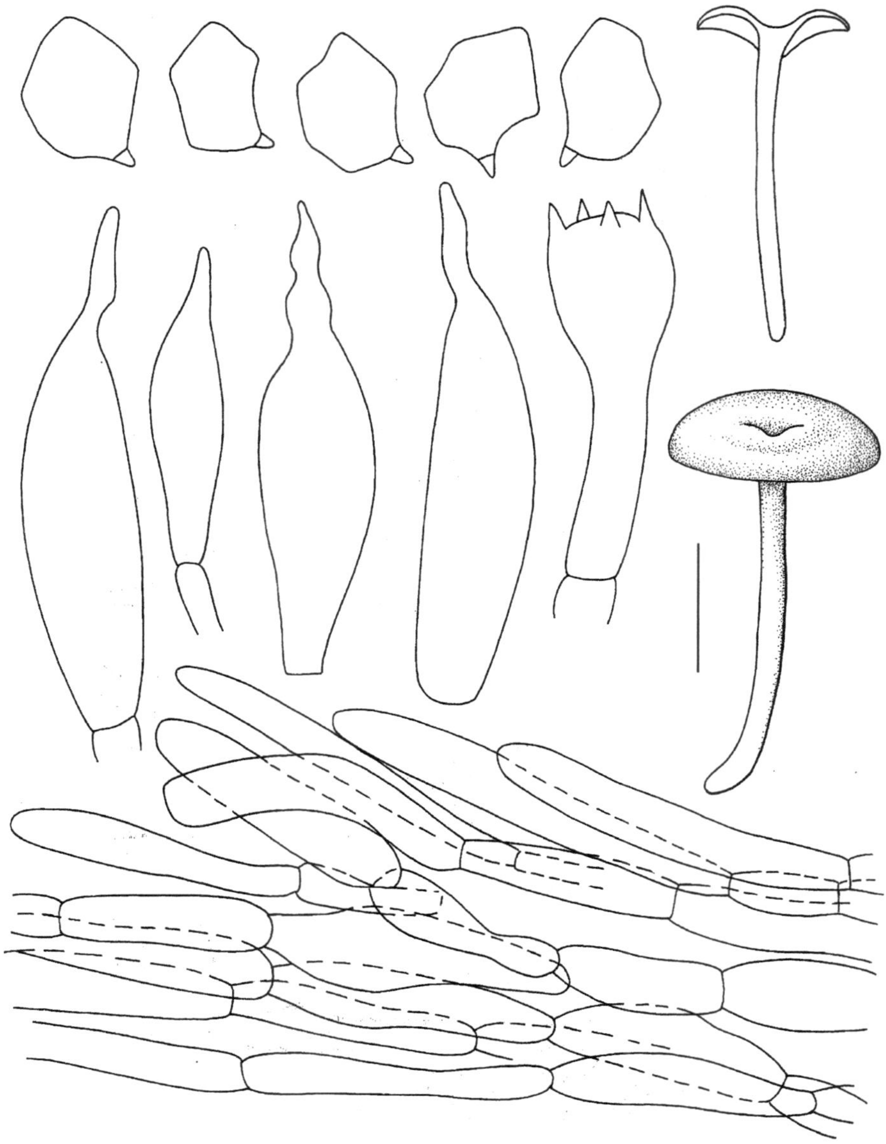

Fig. 3.88 *Entoloma choanomorphum*. Habit, spores, basidia, cheilocystidia, and pileipellis. Bar = 10 mm or 10 μm

Entoloma choanomorphum is a distinct species with its omphalioid habit and large cheilocystidia. It shows some resemblance to *E. disputatum* E. Horak from Chile, which differs, however, in its paler colours, more simple shaped spores, and acidulous smell. *Entoloma nausiosme* Noordel. from Europe and *E. odoriferum* Hesler from North America have similar dark coloured omphalioid basidiocarps

Plate 3.85 *Entoloma uliginicola* (Photos Genevieve Gates)

and a strong smell, but differ in the shape of the cystidia and in their abundant clamp-connections.

89. **Entoloma uliginicola** E. Horak, Beih. Nova Hedwigia 65: 247. 1980 (as uligincola).

Original diagnosis: Pileus –20 mm, convexus dein planus, saepe centro depressus aetate, albus, griseo-brunneo-azureus tinctu, conspicue fibrillosus. Lamellae adnato-emarginatae. Stipes –60/–2 mm, cylindricus, pallide azureus, mycelio albostrigosus. Caro farinacea. Sporae 9–12 × 5.5–6 µm. In locis paludosis. Nova Zelandia.

Holotype: New Zealand, Nelson, W. of Tophouse, 4 March 1968, E. Horak (PDD 27146).

Etymology: uliginosus (Lat.) = marsh; cola (Lat.) = inhabitant.

Selected literature: Horak, Agaricales of New Zealand 1, Figs. 54, 112.3, Plate 4E. 2008.

Main characters: habit collybioid; pileus with blue-lilac hue, aeriferously fibrillose; stipe concolorous; spores many-angled, nodulose (Plate 3.85).

Description:
Pileus 5–20 mm diam, truncate-convex to plano-convex with slightly depressed centre, convex with deflexed, more or less crenate margin, slightly hygrophanous, not translucently striate, greyish lilac (12B2–C2, 11B2–C2, centre 11E2, 17–18D4), sometimes with brown hues, pallescent on drying to a frosty grey-lilac, finely aeriferous-fibrillose somewhat micaceous, sometimes subzonate. Lamellae adnate to broadly adnate with small decurrent tooth, subventricose,

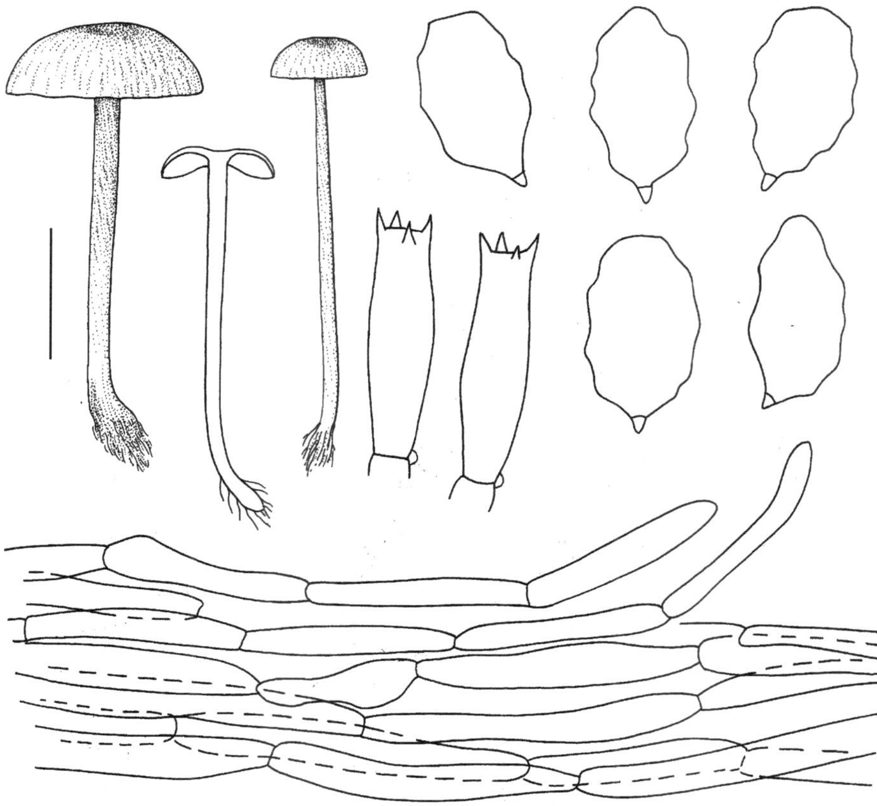

Fig. 3.89 *Entoloma uliginicola*. Habit, spores, basidia, and pileipellis. Bar = 10 mm or 10 μm

crowded, up to 3 mm broad, thin, yellowish pink or flesh pink (6A2, 10YR 7/6) with concolorous, entire or pruinose edge, L = ca. 30, l = 1–6. Stipe 30–45 × 1–2 mm, cylindrical, slender, brittle, lilac or greyish lilac becoming grey with age, finely fibrillose or with a whitish bloom, with strigose hairs at base. Odour and taste spermatic.

Spores 10.5–12 × 7.0–8.0 μm, mostly 10.7–11.3 × 6.9–7.4 μm, heterodiametric, many-angled nodulose in outline, thin-walled. Basidia 22–30 × 7–10 μm, 4-spored, clamped. Lamella edge fertile. Cystidia absent. Hymenophoral trama regular, made up of inflated elements, 40–100 × 6–15 μm. Pileipellis a differentiated cutis with transitions to a trichoderm, made up of septate hyphae, 5–12 μm wide, with terminal elements often irregularly shaped, cystidiform, 20–50(–70) × 8–10 μm. Pigment abundant, bluish, intracellular, rather diffuse, often more or less granular. Brilliant granules absent. Stipitipellis a trichoderm of septate terminal hyphae ("hairs"), 5–11 μm wide, often irregularly shaped with intracellular pigment. Abundant clamp-connections observed in hymenium (Fig. 3.89).

Habitat & distribution: widespread in litter of wet sclerophyll and mixed forest. Uncommon.

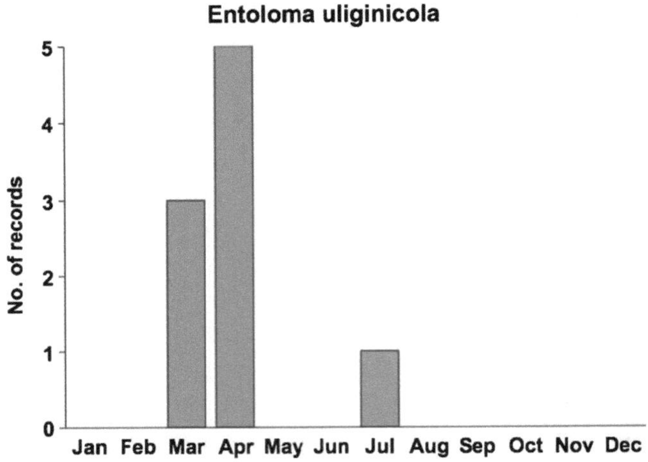

Collections examined. Australia, Tasmania, Bermuda Road, 43° 04′ S, 146° 54′ E, 17 April 2003, G. Gates E 1716; Birralee, Black Sugarloaf, 41° 24′ S, 146° 48′ E, 18 July 2001, G. Gates E 1655; Florentine River Valley, Pagoda Hut, 42° 30′ S, 146° 27′ E, 29 April 2003, G. Gates E 1746; Growling Swallet, 42° 41′ S, 146° 30′ E, 22 April 2004, M.E. Noordeloos 2004070; Huon Pine Walk, 43° 06′ S, 146° 44′ E, 31 March 2005, G. Gates E 2108; Mt Field National Park, Lyrebird Track, 42° 41′ S, 146° 40′ E, 31 March 2001, G. Gates E 1066. — New Zealand, Raku Sadda, coll. E. & A. Horak, 20 April 2004 (PDD 75853).

Entoloma uliginicola is a very remarkable species with its heterodiametric, nodulose spores, and strigose hairs at the base of stipe, that are reminiscent of many species in subgenus *Pouzarella*. The nodulose spores also are similar to those of *Pouzarella*, but they have thinner walls. Horak (2008) compared the spores with those of *Rhodocybe*. The morphological characters of *E. uliginicola* place it in a rather isolated position within the genus *Entoloma*.

Entoloma subgenus ***Richoniella*** (Costantin & L.M. Dufour) Noordel., comb. & stat. nov.

MycoBank #562983.

Basionym: *Richoniella* Costantin & L.M. Dufour, Nouvelle Flore des Champignons, Ed. 5: 203. 1916.

Basidiocarps secotioid, more or less globose, whitish, growing submerged in soil.

Type species: *Richoniella leptoniispora* Costantin & L.M. Dufour.

Plate 3.86 *Entoloma gasteromycetoides* (Photo Michael Pilkington)

90. **Entoloma gasteromycetoides** Noordel. & Co-David
Richoniella pumila G. Cunn. N. Z. Jl Sci. Technol., ser. B22: 62. 1940.

Selected literature: Cunningham, New Zealand Journal Science Technology, B22: 62. 1940; Horak, Agaricales of New Zealand 1: 93–96. 2008.

Main characters: basidiocarps secotioid, more or less globose, whitish, growing submerged in soil (Plate 3.86).

Basidiocarps globose to irregularly globose, 10–30 mm diam, white, becoming sordid whitish cream or with greyish tinges with age, entirely covered with silky fibrils, dull, bursting open when mature. Gleba irregularly labyrinthiform, with small chambers, pure pink. Columella poorly developed or practically absent, pink. Stipe none or very rudimentary, attached to mycelium with whitish rhizomorphs. Smell and taste farinaceous.

Spores 8–12 × 7.5–11 µm, cuboid or 5-angled in side-view with thickened walls. Basidia 24–34 × 7–11 µm, 4-spored, clampless. Hymenial cystidia absent. Pileipellis a loose cutis of interwoven, cylindrical to slightly inflated hyphae, 4–11 µm wide, without visible pigment. Clamp-connections absent (Fig. 3.90).

Habitat & distribution: widespread in wet sclerophyll forest; uncommon.

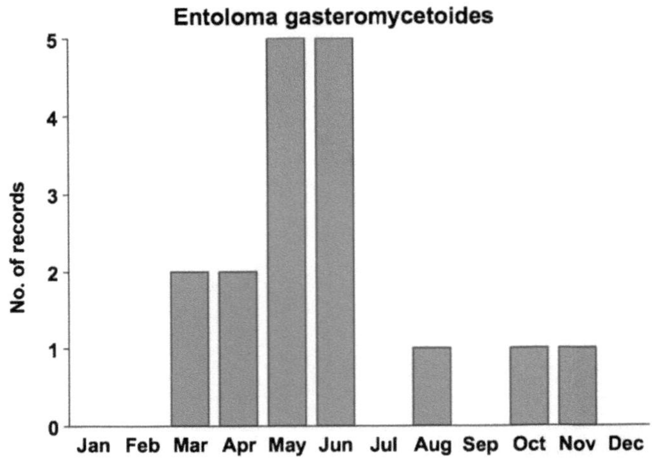

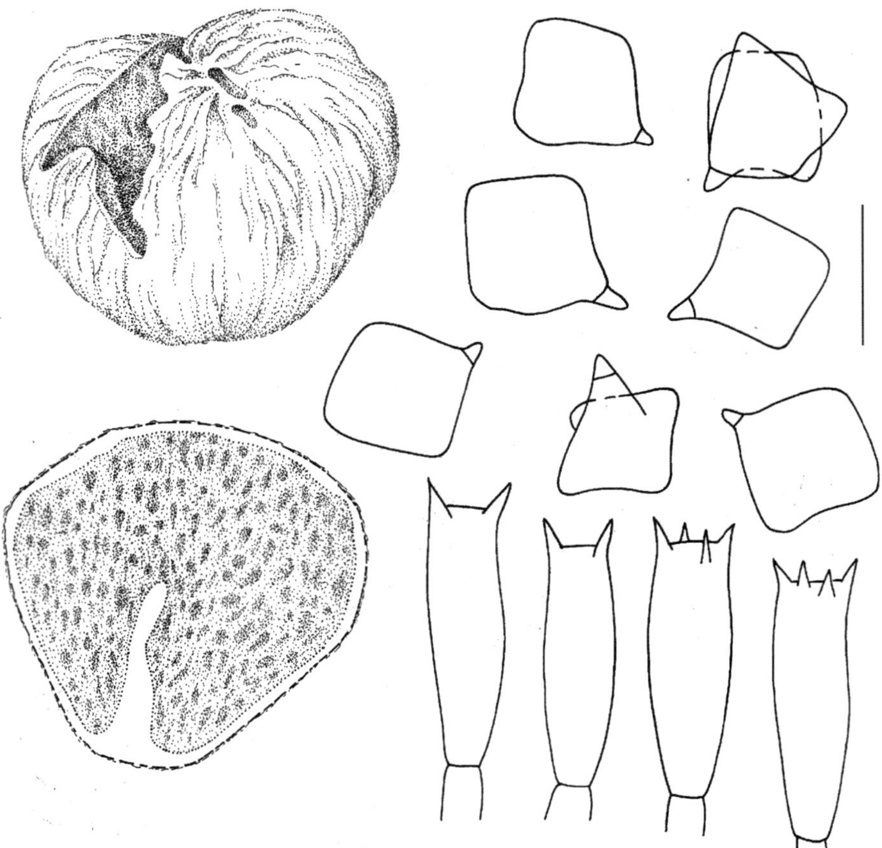

Fig. 3.90 *Entoloma gasteromycetoides*. Habit, spores, and basidia. Bar = 10 mm or 10 μm

Collections examined. Australia, Tasmania, Collinsvale, Myrtle Forest, 42° 52′ S, 147° 09′ E, 19 April 2001, G. Gates E 1102; Evercreech Forest Reserve, 41° 24′ S, 147° 58′ E, 7 June 2004, G. Gates E 2031; Geeveston, Donnellys Rd, 43° 07′ S, 146° 54′ E, 14 Nov. 2000, G. Gates E 1009; —, 19 June 2001, G. Gates E 1255; —, 30 Oct. 2001, G. Gates E 1307; Kermandie Falls, Lower Track, 43° 12′ S, 146° 52′ E, 31 May 2001, G. Gates E 1200; —, Upper Track, 30 March 1999, G. Gates E 321; Marriotts Falls, 42° 43′ S, 146° 39′ E, 13 May 2000, G. Gates E 909; Mt Cripps, Philrod Track, 41° 36′ S, 145° 46′ E, 24 March 2005, G. Gates E 2099.

Entoloma gasteromycetoides is recorded from New Zealand and Tasmania. Horak (2008) also described a 2-spored form from New Zealand, which so far has not been recorded from Tasmania. *Entoloma gasteromycetoides* represents one of the few known sequestrate *Entoloma* species. Formerly it was accommodated in the genus *Richoniella* Costantin & L.M. Dufour, but a molecular-phylogenetic study (Co-David et al. 2009) demonstrated that *Richoniella pumila* is firmly nested within a large clade, containing most of the agaricoid *Entoloma* taxa. Therefore, *Richoniella* does not deserve the rank of genus. That study equally demonstrated that *Rhodogaster* E. Horak is not the missing link between *Richoniella* and *Entoloma*, as suggested by Horak (2008), but represents another, unrelated secotioid lineage within the genus *Entoloma*.

References

Agerer, R., & Waller, K. (1993). Mycorrhizae of *Entoloma saepium*: parasitism or symbiosis? *Mycorrhiza, 3*, 145–154.

Baroni, T. J., & Matheny, P. B. (2011). A re-evaluation of gasteroid and cyphelloid species of Entolomataceae from eastern North America. *Harvard Papers in Botany, 16*(2), 293–310.

Baroni, T. J., Hofstetter, V., Largent, D. L., & Vilgalys, R. (2011). *Entocybe* is proposed as a new genus in the Entolomataceae (Agaricomycetes, Basidiomycota) based on morphological and molecular evidence. *North American Fungi, 6*, 1–19.

Co-David, D., Langeveld, D., & Noordeloos, M. E. (2009). Molecular phylogeny and spore evolution of *Entolomataceae*. *Persoonia, 23*, 147–176.

Donk, M. A. (1961). *The generic names proposed for the Agaricaceae* (Beihefte zur Nova Hedwigia, Vol. 2, pp. 1–320). Weinheim: Cramer.

Gates, G. M., & Noordeloos, M. E. (2007). Preliminary studies in the genus Entoloma in Tasmania – I. *Persoonia, 19*, 157–226.

Gates, G. M., Horton, B. M., & Noordeloos, M. E. (2009). A new *Entoloma* (Basidiomycetes, Agaricales) from Tasmania. *Mycotaxon, 107*, 175–179.

Grgurinovic, C. (1997). *Larger fungi of South Australia.* Adelaide: The Botanic Gardens of Adelaide and State Herbarium and the Flora and Fauna of South Australia Handbooks Committee. 725 pp.

Horak, E. (1973). Fungi Agaricini Novaezelandiae. I. *Entoloma* (Fr.) and related genera. *Beihefte Nova Hedwigia, 43*, 1–86. Cramer, Vaduz.

Horak, E. (1976). On cuboid-spored species of *Entoloma* (Agaricales). *Sydowia, 28*, 171–236.

Horak, E. (1977). Additions to 'On cuboid spored species of *Entoloma*'. *Sydowia, 29*, 289–299.

Horak, E. (1978). *Entoloma in South America. I. Sydowia, 30*, 40–111.

Horak, E. (1980). *Entoloma* (Agaricales) in Indomalaya and Australasia. *Beihefte Nova Hedwigia, 65*, 1–352. Cramer, Vaduz.

Horak, E. (1982). *Entoloma in South America. II. Sydowia, 35*, 75–99.

Horak, E. (2008). *Agaricales of New Zealand 1: Pluteaceae – Entolomataceae* (The Fungi of New Zealand, Vol. 5, Fungal Diversity Research Series, no. 19). Hong Kong: Fungal Diversity Press.

Largent, D. L. (1974). New or interesting species of *Claudopus* and *Entoloma* from the Pacific Coast. *Madrono, 22*, 363–373.

Largent, D. L. (1977). *The genus Leptonia on the Pacific Coast of the United States: Including a study of the North American types* (Bibliotheca mycologica 55). Vaduz: J. Cramer. 286 pp.

Largent, D. L. (1994). *Entolomatoid fungi of the Western United States and Alaska*. Eureka: Mad River Press. 516 pp.

Largent, D. L., Bergemann, S. E., Cummings, G. A., Ryan, K. L., Abell-Davis, S. E., & Moore, S. (2011). *Pouzarella* (Agaricales, Entolomataceae) species from New South Wales (Barrington Top National Park) and Northeastern Queensland (Australia). *Mycotaxon, 117*, 435–483.

Manimohan, P., Joseph, A. V., & Leelavathy, K. M. (1995). The genus *Entoloma* (Agaricales) in Kerala State, India. *Mycological Research, 99*, 1083–1097.

Manimohan, P., Noordeloos, M. E., & Dhanya, A. M. (2006). Studies on the genus *Entoloma* (Basidiomycetes, Agaricales) in Kerala State, India. *Persoonia, 19*, 45–93.

Morgado-Neves, L., Geml, J., Noordeloos, M. E., Co-David, D. L. V., & Lamoureux, Y. (2012 in prep.). Phylogenetic and taxonomic studies in selected groups of *Entoloma* species. *Persoonia 29*.

Noordeloos, M. E. (1980). *Entoloma* subgenus *Nolanea* in the Netherlands and adjacent regions with a reconnaissance of its remaining taxa in Europe. *Persoonia, 10*, 427–534.

Noordeloos, M. E. (1988). *Entolomataceae*. In C. Bas, T. W. Kuyper, M. E. Noordeloos, & E. C. Vellinga (Eds.), *Flora agaricina neerlandica* (Vol. 1, pp. 77–182). Rotterdam: Balkema.

Noordeloos, M. E. (1992). *Entoloma* s.l. In *Fungi Europaei* (Vol. 5). Alassio: Ed. Candusso.

Noordeloos, M. E. (2004). *Entoloma* s.l. In *Fungi Europaei* (Vol. 5a). Alassio: Ed. Candusso. 618 pp.

Pennycook, S. R. (2004). *Bibliographic checklist of agarics, boletes, and related fungi recorded from New Zealand* (Fungi of New Zealand, Vol. 1. Chapter 5/Fungal Diversity Research Series 14, pp. 165–362).

Romagnesi, H. (1937). Essai d'une sectionnement du genre Rhodophyllus. *Bulletin de la Société mycologique de France, 53*, 319–338.

Romagnesi, H. (1974). Essai d'une classification des Rhodophylles. *Bulletin mensuel de la Société Linnéenne de Lyon, 43*, 325–332.

Romagnesi, H., & Gilles, G. (1979). Les Rhodophylles des forêts côtières du Gabon et de Côte d' Ivoire. *Beihefte zur Nova Hedwigia, 59*, 1–649. Cramer, Vaduz.

Segedin, B. P., & Pennycook, S. R. (2001). A nomenclatural checklist of agarics, boletes, and related secotioid and gasteromycetous fungi recorded from New Zealand. *New Zealand Journal of Botany, 39*(2), 285–348.

Singer, R. (1951). The Agaricales in modern taxonomy. *Lilloa*, 22, 1–832.

Stevenson, G. (1962). The Agaricales of New Zealand: III. *Kew Bulletin, 16*(2), 227–237.

Vesterholt, J. (2002). *Contribution to the knowledge of species of Entoloma subgenus Leptonia* (Fungi non delineati 21). Alassio: Ed. Candusso.

Chapter 4
Genus *Clitopilus*

Abstract This chapter deals with the genus *Clitopilus* in Tasmania. This small genus is presented in the emended concept of Co-David et al. (Persoonia 23: 147–176, 2009), and includes also the former genus *Rhodocybe*. This revision is based on an earlier paper by Baroni and Gates (Aust Sys Bot 19: 343–358, 2006), expanded with more data, and with the addition of one species hitherto unknown to the Tasmanian mycoflora (*C. conchatus*). A generic description of the genus *Clitopilus* is followed by a dichotomous key to all known species of this genus in Tasmania. Each species is presented with full synonymy, reference to existing literature, followed by detailed macroscopic and microscopic descriptions, data on habitat, distribution, and phenology, usually followed by notes on taxonomic position, similar species, and other relevant facts. Line drawings are made for each species, containing a habit drawing, as well as diagnostic microscopic features, such as spores, basidia, cystidia, and pileipellis structures. For almost every species one or more coloured photographs are also included. In all, ten species are accepted. Two species, viz. *C.* cf. *prunulus*, and *C. hobsonii*, are morphologically very similar to European collections. Molecular studies are needed to reveal whether the Tasmanian collections represent cryptic species or not. Of the remaining eight species, most are so far known only from Tasmania or mainland Australia, except for *C. conchatus* and *C. fuligineus*, both of which were described originally from New Zealand. Full references to the cited literature are given.

GENUS CLITOPILUS (Fr. ex Rabenh.) P. Kumm., Führ. Pilzk.: 23. 1871 emend. Co-David & Noordel.

Agaricus tribus *Clitopilus* Fr., Epicrisis: 148. 1838. — *Hexajuga* Fayod, Ann. Soc. Bot. VII.9: 389. 1889. — *Octojuga* Fayod, Ann. Soc. Bot. VII.9: 390. 1889. — *Orcella* Earle, Bull. New York Bot. Gard. 5: 430. 1909. — *Rhodocybe* Maire, Bull. Soc. mycol. France 40: 298. 1925. — *Clitopilopsis* Maire, Publ. Inst. Bot. 3, 4: 82. 1937. — *Hirneola* Velen., Nov. Myc.: 73. 1939. — *Pluteispora* Maire, Bull. Soc. mycol. France 50: xxvii. 1935 (1934). — *Rhodophana* Kühner ex Métrod, Rev. Mycol. 17: 69. 1952 (invalid).

M.E. Noordeloos and G.M. Gates, *The Entolomataceae of Tasmania*,
Fungal Diversity Research Series 22, DOI 10.1007/978-94-007-4679-4_4,

Main characters: basidiomata agaricoid; spore print pink or brownish pink; spores either provided with longitudinal ribs, appearing angular in polar view, or with bumps, which may be more or less evenly distributed or arranged in lines, never with true facets.

Type species: *Clitopilus prunulus* (Scop.) P. Kumm.

Clitopilus is treated here in the emended concept, including *Rhodocybe* Maire (Co-David et al. 2009). They demonstrated that in the /rhodocybe-clitopilus clade, *Rhodocybe* is paraphyletic and *Clitopilus* is, with significant support, well-nested within *Rhodocybe*. Previous studies had already suggested this relationship (Moncalvo et al. 2002; Matheny et al. 2006). Some *Rhodocybe* spores have bumps and ridges arranged linearly along the spore length (Baroni 1981), presenting a spore shape that approaches clitopiloid spores. Kühner (1980) had already united *Clitopilus* with *Rhodocybe* in one genus on account of the characters these groups share: predominantly clitocyboid habit, decurrent lamellae, and resemblance in basal structure of the ornamentation of the spores. As a consequence, the two genera were merged into one.

The genus *Rhodocybe* in Tasmania has been treated earlier (Baroni and Gates 2006). The descriptions and illustrations in this book are adapted from that publication, with additional observations and records.

Keys to the Species

1. Spores with longitudinal ribs; angular in polar view	**Key 1. (*Clitopilus*)**
1. Spores pustulate with undulating outline; angular in polar view	**Key 2. (*Rhodocybe*)**

Key 1. Species with longitudinally ribbed spores

1. Habit clitocyboid; pileus 40–90 mm diam.	**1. C.** cf. **prunulus**
1. Habit pleurotoid; pileus 3–10 mm across	**2. C. hobsonii**

Key 2. Species with pustulate spores with undulating outline

1. Habit crepidotoid (pleurotoid), with very reduced stipe	**10. C. conchatus**
1. Habit clitocyboid or tricholomatoid, with well-developed, central, rarely eccentric stipe	2.
2. Hymenial pseudocystidia with bright coloured content present	3.
2. Hymenial pseudocystidia absent	5.
3. Pileus up to 20 mm diam., overall colour pale brown or pinkish; taste very bitter	**5. C. acerbus**
3. Pileus larger and darker coloured; taste bitter or mild	4.
4. Spores 7–12.5 × 4–5.6 μm, amygdaliform in side-view with distinct apical papilla; small brownish fruit body	**3. C. tasmanicus**
4. Spores 5.6–8.8 × 4.5–6.4 μm, ellipsoid in side-view	**4. C. fuligineus**
5. Pileus either pallid, pinkish, pinkish brown, or orange-brown to red-brown (brick-coloured)	6.

5. Pileus grey or greyish brown	7.
6. Pileus orange-brown or red-brown (brick)	**6. C. lateritius**
6. Pileus pale, pinkish brown or flesh pink	**7. C. pseudopiperitus**
7. Pileus distinctly reticulate with concentric cracks and lines	**9. C. reticulatus**
7. Pileus glabrous	**8. C. pallidogriseus**

Species Descriptions

Section *Clitopilus*

1. **Clitopilus** cf. **prunulus** (Scop.) P. Kumm. (Plate 4.1)

Description:
Pileus 40–90 mm diam., convex when young, expanding to concave or infundibuliform, with involute margin, becoming irregularly shaped with age with undulating marginal zone. Not hygrophanous, not translucently striate, uniformly pale grey, sometimes with a slight brown tinge at centre (10YR 7/3–4), adpressed-tomentose all over. Lamellae arcuate-decurrent, grey-pink with entire, concolorous or more or less hyaline edge, crowded. Stipe 30–60 × 10–15 mm (apex), usually central, rarely eccentric, tapering towards base, white or with a greyish brown tinge like pileus, tomentose. Context rather thick in pileus, firm, white. Odour very strongly farinaceous-rancid. Taste farinaceous-rancid (Fig. 4.1).

Spores (8–)9–11 × 4.5–6 mm, Q = 1.7–2.1, Qav = 1.9, slenderly fusiform occasionally amygdaliform, thin-walled, distinctly ribbed lengthwise with 5–8 longitudinal ribs, angular when seen from above. Basidia 20–30 × 4–8 µm, 4-spored. Lamella edge fertile or with scattered subcylindrical cheilocystidia, 20–40 × 4–11 mm. Pileipellis a cutis of densely packed narrow cylindrical, 4–8 µm wide hyphae with dark brown coloured walls, and scattered fine encrustations. Clamp-connections absent.

Habitat & distribution: occasional in wet sclerophyll forest.

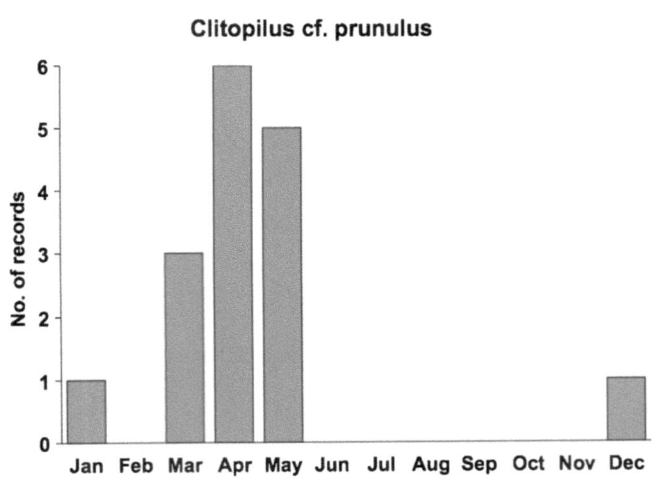

Plate 4.1 *Clitopilus* cf. *prunulus* (Photos Machiel Noordeloos (*left*), Michael Pilkington (*right*))

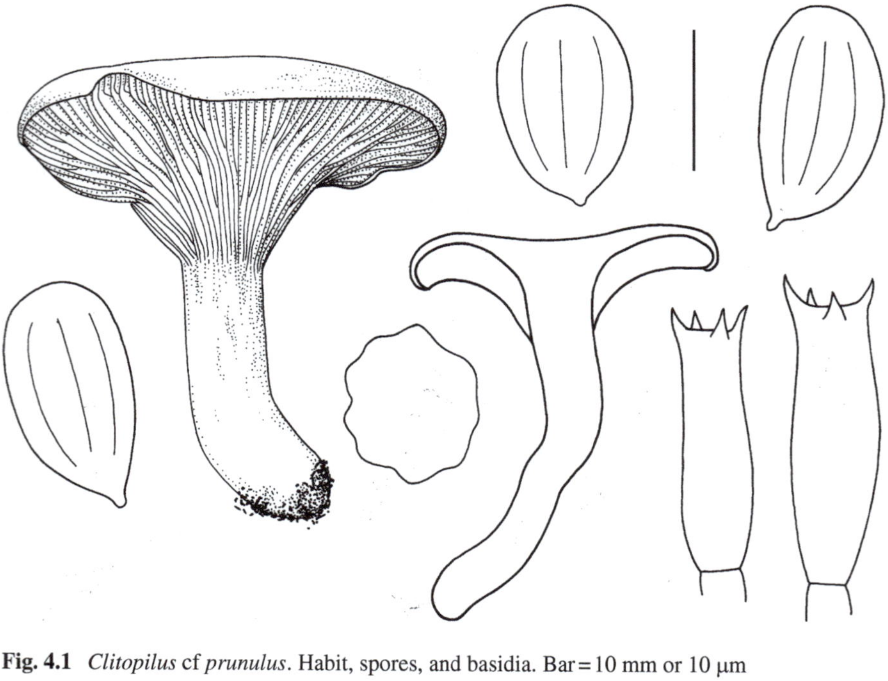

Fig. 4.1 *Clitopilus* cf *prunulus*. Habit, spores, and basidia. Bar = 10 mm or 10 μm

Collections examined. Australia, Tasmania, Kermandie Falls, Lower Track, 43° 12′ S, 146° 52′ E, 16 March 1999, G. Gates E 226; —, 23 May 2000, G. Gates E 936; —, 5 April 2001, G. Gates E 1072; —, 16 May 2002, G. Gates E 1508; —, 10 April 2003, G. Gates E 1694; —, 24 March 2009, M.E. Noordeloos 2009062 (L); —, Upper Track, 26 April 2001, G. Gates E 1131; Reuben Falls, 43° 00′ S, 146° 40′ E, 15 May 1999, G. Gates E 507; Tahune, hanging bridges walk, 43° 06′ S, 146° 43′ E, 14 March 2009, M.E. Noordeloos 2009001 (L).

Clitopilus prunulus and related species are widespread over the Northern Hemisphere (Noordeloos 1988, 1993, 2008; Hausknecht and Noordeloos 1999). A number of species have been described in the literature; however, species limits are often difficult to define due to the lack of good morphological characters. Recent studies confirm that the current concept of the morphospecies *C. prunulus* is polyphyletic (Hartley et al. 2009; Vizzini et al. 2011). *C. cystidiatus* Hauskn. & Noordel. and *C. chrischonensis* Vizzini et al. differ by the presence of cheilocystidia, but judging from the phylogeny presented in the last mentioned paper, the occurrence of cheilocystidia may well be of limited diagnostic value to distinguish species in this group. *C. amygdaliformis* Zhu L. Yang from Japan (Yang 2007) is distinguished by a farinaceous smell, the absence of cystidia, and broadly amygdaliform to limoniform spores with 5–6 ribs; *C. chalybescens* T.J. Baroni & Desjardin (Baroni et al. 2001) has much smaller (5.5–7.5 × 3.6–4.8 µm), ellipsoid basidiospores, no cystidia and a mealy-smelling context, often turning greyish brown, whilst *C. amarus* A. de Haan has an umbonate pileus, bitter taste, no cystidia, and much smaller [(5.1)5.5–6.5(7) × (3.4)4–4.5(5.4) µm], amygdaliform to rhomboid basidiospores (de Haan 1998). Further molecular work is needed in this group to establish sound species limits. For the time being we describe the Tasmanian collections here as *C.* cf. *prunulus*.

2. **Clitopilus hobsonii** (Berk.) P.D. Orton, Trans. Br. mycol. Soc. 43(2): 174. 1960.

Agaricus hobsonii Berk., Outl. Brit. Fung. (London): 138. 1860; — *Pleurotus hobsonii* (Berk.) Sacc., Syll. fung. 5: 382. 1887. — *Octojuga pleurotelloides* Kühner, Botaniste 17: 158. 1926; — *Clitopilus pleurotelloides* (Kühner) Joss., Bull. mens. Soc. linn. Lyon 10: 14. 1941; — *Pleurotus pleurotelloides* (Kühner) Trotter, Syll. 26: 795. 1972. — *Octojuga fayodii* Konrad & Maubl., Icon. Select. Fung. 6: 234. 1934.

Holotype: Britain, Apethorpe, Sept. 1859, leg. J.C. Hobson (K).

Original diagnosis: pileus membranaceous, reniform or dimidiate, stemless, pale grey, minutely downy; gills rather distant, pallid. On Larch-stumps. Apethorpe. Sept. 1859. Pileus 1–4 lines across; margin involute.

Etymology: named after Lieut. Julian C. Hobson.

Main characters: tiny, white, pleurotoid species with longitudinally ribbed spores (Fig. 4.2).

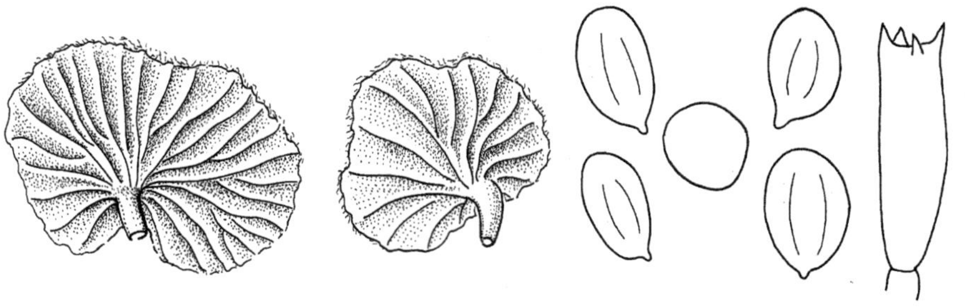

Fig. 4.2 *Clitopilus hobsonii*. Habit, spores, and basidia. Bar = 10 mm or10 μm

Description:
Habit pleurotoid. Pileus 3–10 mm across, flabelliform with undulating margin, white, tomentose. Lamellae adnexed, distant, unequal, sometimes forked, narrow, up to 0.5 mm broad, white, with concolorous edge. Stipe very reduced, lateral or virtually absent.

Spores 7–9 × 4–5(–6) μm, with 6–8 longitudinal ribs, ellipsoid to amygdaliform in side-view, angular in profile view, thin-walled, slightly cyanophilous. Basidia 15–22 × 6–9 μm, clampless. Lamella edge fertile. Cystidia absent. Hymenophoral and pileitrama made up of long, cylindrical to slightly inflated elements 6–19 μm. Pileipellis an entangled trichoderm of narrow, cylindrical, 6–8 μm wide hyphae. Pigment absent. Clamp-connections absent.

Habitat & distribution: widespread, on underside of wood in wet eucalypt forests and a suburban garden.

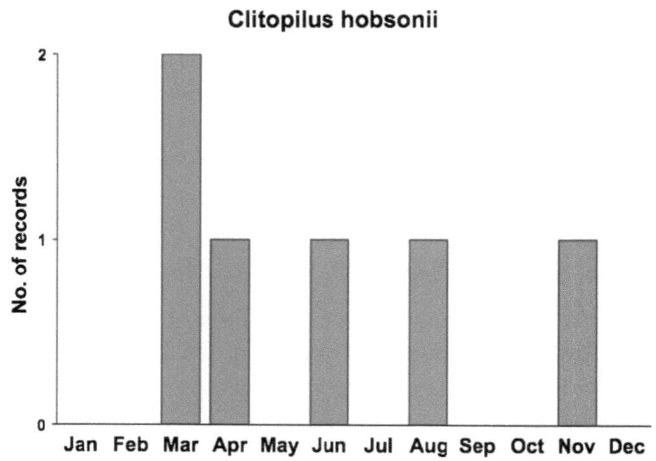

Collections examined. Australia, Tasmania, Mt Cripps Karst Area, Philrod Track, 41° 36′ S, 145° 46′ E, 24 March 2005, G. Gates E 2098; Notley Gorge, 41° 21′ S, 146° 55′ E, 25 April 2003, G. Gates E 1736; Taranna Forest Walk, 43° 04′ S, 147°

53′ E, 15 March 2005, G. Gates E 2092; Taroona, 9 Winmarleigh Avenue, 12 June 2011, G. Gates E 2300.

The Tasmanian collections belong to the collective species *Clitopilus hobsonii*. Similar collections are recorded from various places world-wide (Horak 2008; Noordeloos 1988, 1993, 2008). The lack of good separating morphological characters in this group of pleurotoid species warrants a molecular approach. For the time being we accept our collections as belonging within the limits of *C. hobsonii*.

Section ***Rhodocybe*** (R. Maire) Contu, in Bollettino AMER 77–78: 29. 2009.

Rhodocybe sect. *Rhodocybe* R. Maire in Bull. Soc. mycol. France 40: 298. 1924.

Habit collybioid, tricholomatoid, mycenoid or omphalioid; hymenial pseudocystidia present.

Type species: *Clitopilus caelatus* (Fr.) Vila & Contu.

3. **Clitopilus tasmanicus** (T.J. Baroni & G.M. Gates) Noordel. & Co-David, Persoonia 23: 164. 2009.

Rhodocybe tasmanica T.J. Baroni & G.M. Gates, Austr. Syst. Bot. 19: 350. 2006.

MycoBank #509951.

Original diagnosis: Pileus juvenilis brunneo-senatus, aetate multo pallidiore flavescens, 7–11 mm latus, primum convexus vel conice convexus, siccus, ut videtur glaber sed implicatus vel pannosus. Lamellae griseolo-flavae, decurrentes, modice arctae, lamellulis in gradum unicum dispositis. Stipes plusminusve pileo concolor vel aetate griseolo-bubalinus, ad apicem 3 mm latus, ad basem 2 mm latus, 20–22 mm longus, versus basem contractus, teres, siccus, primum uniformiter pruinosus, tantum maturitate supra basem pruinosus. Odor pungens, speciebus Agarici commercialibus similis. Sapor amarissimus. Basidiosporae 4.7–7.5 × 3.8–5.6 μm, amygdaliformes vel fusiformes, ad polum visae angulares, multae rostro apicali elongato armatae, valde vel modice undulato-pustulatae. Basidia 4-sterigmata, clavata vel anguste clavata, 10–30 % parietibus incrassatis praedita (0.8–1.6 μm). Cheilocystidia atque pleurocystidia ut pote pseudocystidia, in forma, amplitudine, coloure similia, pigmentis atro-ochraceis aeque implena, fusiformia, ventricosa, ventricoso-rostrata, rostro saepe longo, contracto vel toruloso vel undulato-cylindrico, interdum ramoso, 36–80×4.8–8.8 μm. Pileipellis hyalina, repens, e hyphis cylindricis, 1.6–6.4 μm diametro, cellulas ultimas sparsas, inflatas, versiformes efformans, e contextu non bene distincta praeter prope pellem hyphis contextualis pallide brunneis atque in hyphis angustissimis pigmentas encrustantes leniter brunneas formantibus. Fibulae nullae.

Holotype: G. Gates E 666 (HO 536115; isotype, CORT).

Main characters: small, collybioid species with decurrent lamellae; pileus brown, becoming grey-yellow with age; spores amygdaliform with distinct snout (Fig. 4.3).

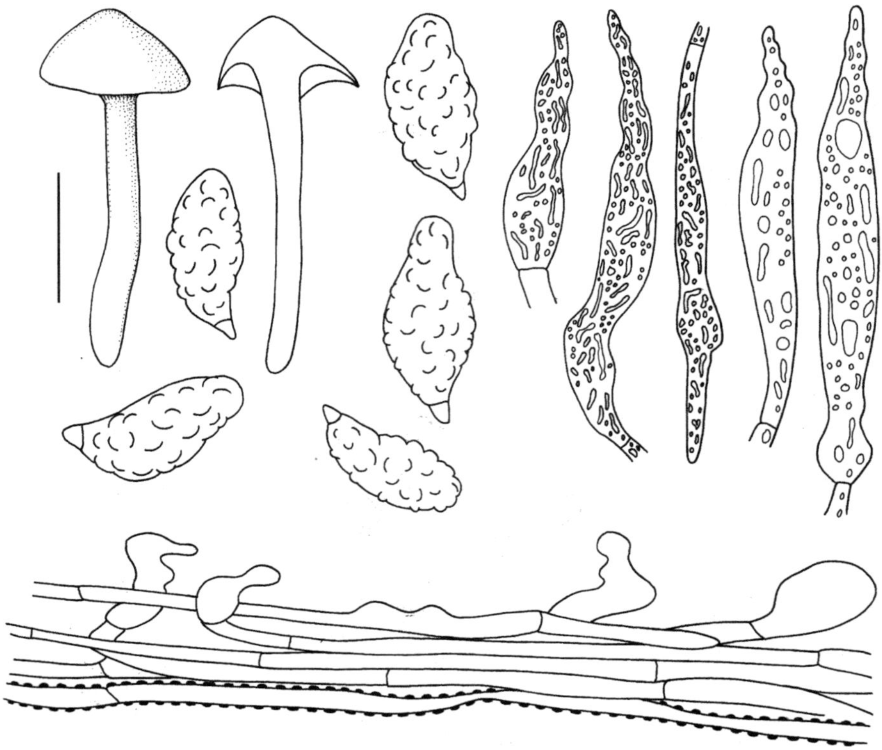

Fig. 4.3 *Clitopilus tasmanicus*. Habit, spores, pseudocystidia, and pileipellis. Bar = 10 mm or 10 μm

Description:
Pileus 7–11 mm diam., conico-convex to convex with incurved margin, not hygrophanous, not translucently striate, brown (6F5), becoming much paler brown with age (4B5), glabrous, matt, subfelted under a hand lens. Lamellae decurrent, arcuate, moderately crowded, up to 2 mm broad, pale yellowy brown (4B4), lamellulae in one tier. Stipe 20–22 × 3 (apex) 2 (base) mm, tapering, stuffed, concolorous with pileus at first then light greyish brown, pruinose all over when young, later on only in basal part. Odour pungent, mushroomy. Taste very bitter.

Spores 7–12.5 × 4–5.5 μm, Q = 1.45–2.4, Qav = 1.9, amygdaliform in profile view, fusiform in face view, angular in polar view, many spores with an elongated apical 'snout', moderately or very undulate-pustulate. Basidia 25–36(−46) × 6–9 μm, (2–)4-spored, clavate or narrowly clavate, 10–30 % with thickened walls (0.8–1.6 μm). Cheilo- and pleurocystidia 36–80 × 5–9 μm, as pseudocystidia, similar in shape, size and colour, fusiform, ventricose, ventricose-rostrate, rostrum often long, tapered or torulose or undulate cylindrical, occasionally branched, filled evenly with dark ochre pigments. Hymenophoral trama regular, made up of 3–12 μm wide cylindrical or inflated hyphae, with pale orange or yellow brownish walls.

Gloeoplerous hyphae 3–12 μm wide, scattered through pileitrama, filled with ochre or golden brown contents, connected to the pseudocystidia. Pileipellis a cutis of hyaline, cylindrical hyphae, 1.5–6.5 μm wide, producing scattered inflated, versiform terminal elements, not well differentiated from pileitrama. Pileitrama radially arranged or interwoven, made up of cylindrical to inflated, 3–8 μm wide hyphae with pale brown walls. Pigment parietal and encrusting in pileipellis. Stipitipellis a cutis of 3–7 μm wide, hyaline, cylindrical hyphae, some encrusted with brown pigment, with distinctive vesiculose terminal or intercalary cells in trama. Clamp-connections absent.

Habitat & distribution: in wet sclerophyll forests. Rare.

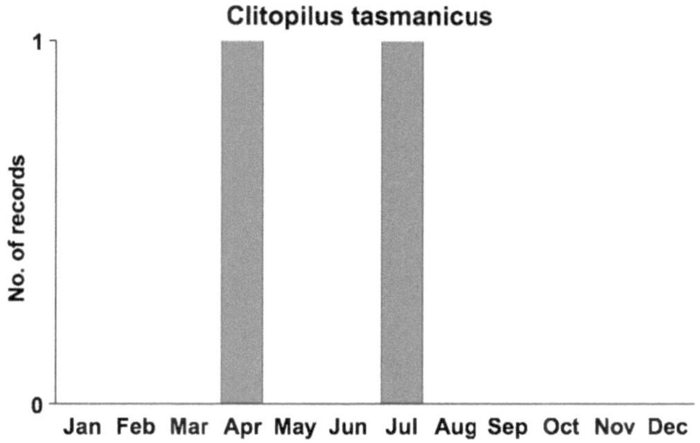

Collections examined. Australia, Tasmania, Lilydale Falls, 41° 14′ S, 147° 13′ E, 25 July 1999, G. Gates E 666 (holotype, HO 536115); Mt Wellington, Lenah Valley Track, 42° 55′ S, 147° 15′ E, 30 April 2001, S. McMullan-Fisher SMF1088.

Clitopilus tasmanicus is an unusual member of section *Rhodocybe* due to the sombre dark brown colours that change to very pale creamy yellow colours with age, the distinctly amygdaliform basidiospores that often produce an obvious 'snout' apically, the inflated erect end cells of the pileipellis and the vesiculose cells in the stipe trama. *Clitopilus caelatus* has similar amygdaliform basidiospores, but these are shorter than those of *C. tasmanicus*. The colour of the basidiomata for *C. caelatus* is grey or greyish brown and not becoming paler as in *C. tasmanicus*, and the pileus surface of *C. caelatus* is often wrinkled, areolate or reticulate-cracked with age. In addition, *C. caelatus* has a two layered pileipellis with a hyaline layer overlaying a brownish, heavily encrusted subpellis. The hyaline suprapellis layer is due to the delicate pubescence of the young pileus surface of *C. caelatus*, characters not found in *C. tasmanicus*. However, the spore shapes of these two taxa do indicate there may be some relationship. *Clitopilus dingleyae* (E. Horak) Noordel. & Co-David from New Zealand differs by having an ochre pileus with umbilicate to subinfundibuliform shape, and a tomentose surface, and quite small spores, which readily distinguish it from *C. tasmanicus*. *Clitopilus fuligineus* (E. Horak) Noordel. & Co-David, also originally described from New Zealand, differs by having a velutinous black-brown pileus, and small, ellipsoid spores.

Plate 4.2 *Clitopilus fuligineus* (Photos Genevieve Gates)

4. **Clitopilus fuligineus** (E. Horak) Noordel. & Co-David, Persoonia 23: 161. 2009.

Rhodocybe fuliginea E. Horak, New Zealand J. Botany 17:280. 1979
MycoBank #509884.

Original diagnosis:
Pileo –20 mm. umbilicato-depresso, fuligineo, minute velutino. Lamellis adnato-decurrentibus, griseo-brunneis vel porphyreis. Stipite –20×–2 mm, cylindrico, pileo concolori, glabro, centrico vel sublaterali. Odore sub-farinaceo. Sporis 6–7.5×4.5–6 μm, ovatis, rugulosis vel subangulatis. Pseudocystidiis –50×–7 μm, acuto-fusoideis. Hyphis defibulatis. Ad truncum putridum Cyatheae medullaris. Novazelandia. Holotype: New Zealand, Nelson, Golden Bay, N. of Pakawau, 20 m., on dead trunk of *Cyathea medullaris*, 15 May 1968, E. Horak (PDD 27142) (Plate 4.2)

Description:
Pileus 4–40 mm diam., convex, shallowly depressed at centre, with incurved then straight, crenate or finally lacerate margin, not hygrophanous, not translucently striate, pale grey, greyish brown with patches of greyish buff, greyish with ochraceous hues, becoming very dark greyish brown or brownish black with age, sometimes with dark golden brown hues, dry, glabrous, appearing suede-like or matted, opaque.

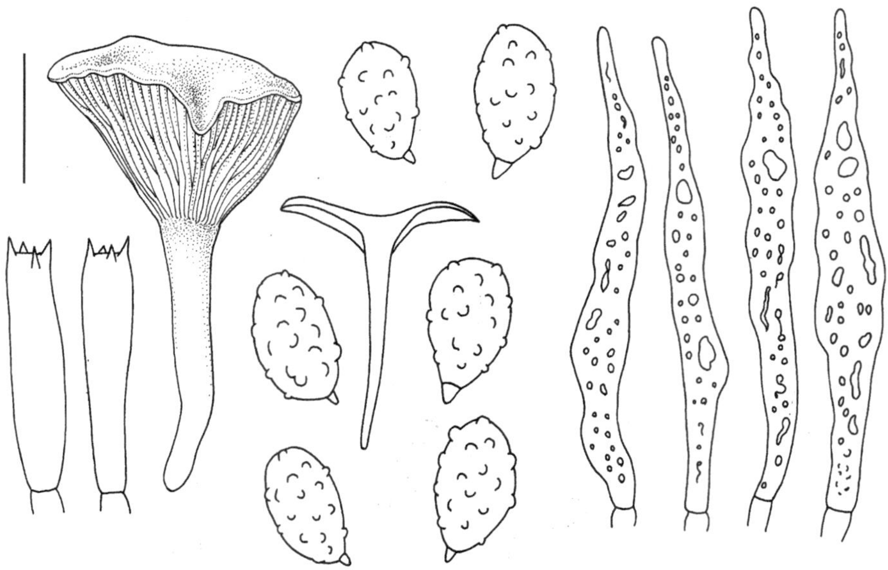

Fig. 4.4 *Clitopilus fuligineus*. Habit, spores, basidia, and pseudocystidia. Bar = 10 mm or 10 μm

Lamellae adnate with decurrent tooth, subdecurrent or decurrent, arcuate, crowded, 1–2(−6) mm broad, thick, light brown or greyish brown, becoming brownish orange with age (5C6), with slightly darker, entire edge, lamellulae in one layer. Stipe 15–30 × 1–4 mm, equal or tapering towards base, fistulose, light brown or greyish brown concolorous with pileus, whitely pruinose all over at first, glabrescent, dry, with white basal tomentum. Context pale greyish or very dark greyish brown, thick. Odour and taste none or farinaceous or bitter.

Spores 5.5–9 × 4.5–6.5 μm, Q = 1.1–1.5, Qav = 1.35, subglobose (or broadly elliptical in profile and face view), angular in polar view (8–10 facets), obviously undulate-pustulate, walls evenly cyanophilic, inamyloid. Basidia 20–30 × 6–8 μm, clavate, 4-spored. Cheilo- and pleuropseudocystidia similar in size, shape and colour, 30–70 × 5–8 μm, ventricose, ventricose-rostrate or fusiform, filled with dark ochre red granular pigments. Hymenophoral regular, made up of parallel, cylindrical hyphae, 2.5–4 μm wide, yellow-brown, with excretory hyphae of similar diameters scattered throughout trama, filled with ochre red or golden brown contents, connected to the pseudocystidia. Pileipellis a unilayered cutis of 2–7 μm wide, cylindrical hyphae with dark brown encrusted walls. Pileitrama regular, made up of hyaline or very pale brown, radially arranged or somewhat interwoven, cylindrical or slightly inflated hyphae, 2.4–8 μm in diameter. Clamp-connections absent (Fig. 4.4).

Habitat & distribution: on soil or decaying plant litter in dry sclerophyll and wet sclerophyll forests.

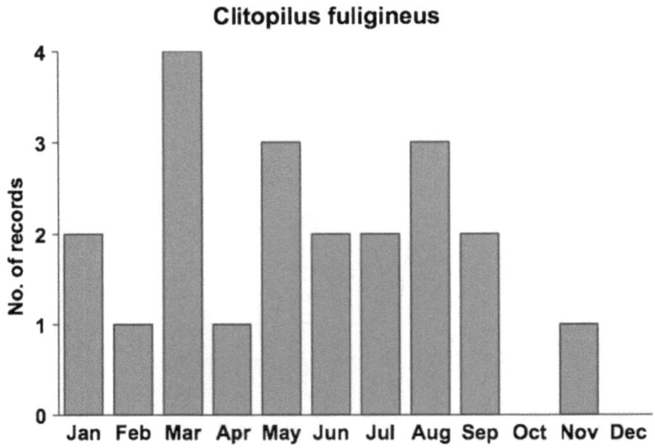

Collections examined. Australia, Tasmania, Bellerive, Waverley Flora Park, 42° 52′ S, 147° 23′ E, 25 May 1999, G. Gates E 537; Bruny Island, Mt Mangana, 43° 22′ S, 147° 17′ E, 9 Sept. 1999, G. Gates E 696; Growling Swallet, 42° 41′ S, 146° 30′ E, 19 May 1999, G. Gates E 524; Hospital Creek, 42° 45′ S, 147° 49′ E, 9 Aug. 2003, G. Gates E 1902; Lake St Clair National Park, Echo Point to Cynthia Bay, 42° 04′ S, 146° 09′ E, 22 April 2000, G. Gates E 863; Mt Field National Park, 42° 41′ S, 146° 42′ E, 6 July 1999, G. Gates E 655; —, 2 Sept. 1999, G. Gates E 688; Mt Wellington, Myrtle Gully, 42° 54′ S, 147° 15′ E, 13 May 1999, G. Gates E 502; North West Bay River, 42° 57′ S, 147° 12′ E, 14 Feb.2002, G. Gates E 1428; —, 4 Aug. 2005, G. Gates E 2185; Pelverata Falls, 43° 04′ S, 147° 08′ E, 1 July 1999, G. Gates E 645.

Clitopilus fuligineus is a rather widespread species in Tasmania and the basidiomata can be quite variable in colour. The pileus in very young specimens is a pale greyish colour, that soon becomes darker greyish brown, and often with some ochraceous hues developing as well. With age the colours become quite dark, fuliginous, and thus the character that prompted the specific epithet.

5. **Clitopilus acerbus** Noordel. & Co-David, Persoonia 23: 160. 2009.

Rhodocybe amara T.J. Baroni & G.M. Gates, Austr. Syst. Bot. 19: 352. 2006; non *Clitopilus amarus* de Haan 1998.

MycoBank #509853.

Original diagnosis: Pileus bubalinus disco fusciore pallide brunneo, hygrophanus, aetate inter centrum usque ad marginem decolorate bubalinus, 19 mm latus, convexus loco centrali depresso vel paene truncato-campanulatus, siccus, glaber. Lamellae pallide subroseolae, decurrentes, arctae, lamellulis in gradum unicum dispositis. Stipes bubalinus, ad apicem 3.5 mm latus, ad basem 2.5 mm latus, 40 mm longus, plusminusve aequus vel deorsum subcontractae, teres, laeves, siccus, elasticus vel flexiliter lentus, intus albus farctus. Odor non distinctus. Sapor amarissimus. Basidiosporae 6.3–8.8 (−11.3) × 4–5.6 μm, amygdaliformes vel ellipticae, ad polum visae angulares (faciebus 8–10), omnino undulato-pustulatae. Basidia 4-sterigmata, anguste clavata, interdum prope apicem constricta. Cheilocystidia atque pleurocystidia ut pote pseudocystidia, in amplitudine, forma, coloureque similia, pigmentis fuscis rufidulo-ochraceis implena, ventricosa, ventricoso-rostrata, saepe ad basem

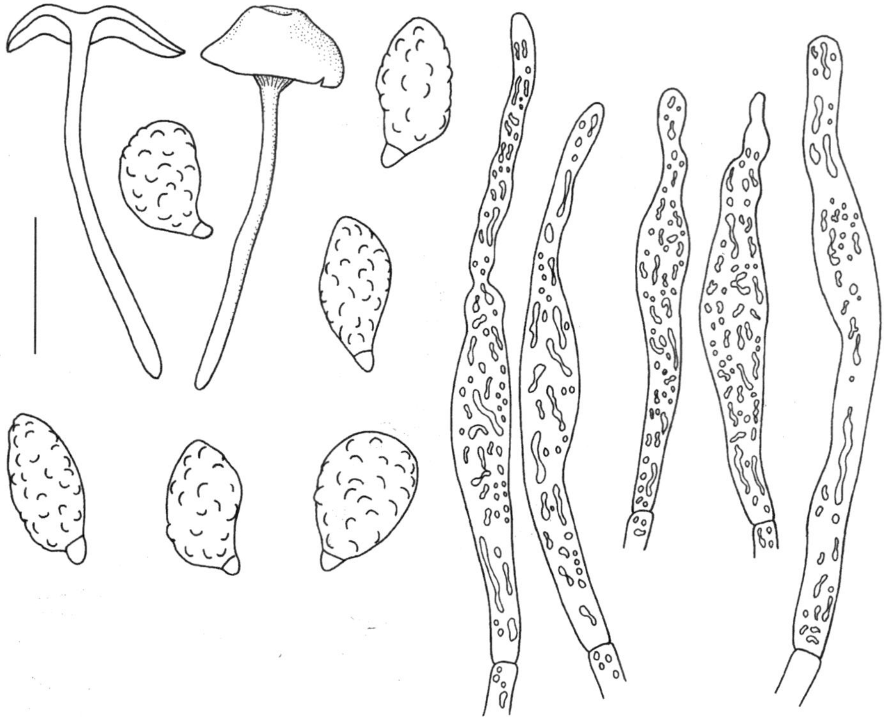

Fig. 4.5 *Clitopilus acerbus*. Habit, spores, and pseudocystidia. Bar = 10 mm or 10 μm

rostri septata, rostro saepe undulato, 36–70 × 4.8–8 μm, in praeparationibus 3% KOH crystallis rhomboideis sparsis supra basidia, basidiola et cystidia. Pileipellis 2-stratosa, stratis ambobus repentibus, suprapelle in strato hyalino tenuissisimo subpellem ex hyphis pallido-brunneis superimponente, in stratis ambabus hyphis cylindricis vel subinflatis, 2.4–7.2 μm diametro, in strato pallido-brunneo hyphis contextuque leniter tenuiterque incrustatis. Fibulae nullae.

Holotype: Australia, Tasmania, Lake St Clair National Park, Echo Point to Cynthia Bay, 42° 04′ S, 146° 09′ E, 22 April 2000, G. Gates E 875 (HO 536112; isotype, CORT) (Fig. 4.5).

Description:
Pileus 19 mm diam., convex with a central depression or somewhat truncate-campanulate, with entire, plane or slightly revolute margin with age, hygrophanous, more or less buff with flesh pink hues, darker at centre and pale brownish, lighter buff brown towards margin, becoming a faded buff from centre to margin on drying, glabrous. Lamellae decurrent, arcuate, close, to 2.5 mm broad, lamellulae in one tier. Stipe 40 × 3.5 (apex) × 2.5 mm (base) broad at base, equal or tapered slightly downwards, terete, smooth, dry, rubbery or wiry tough, buff, stuffed white inner. Odour not distinctive. Taste extremely bitter.

Spores 6.5–9.0 (–11.5) × 4–5.5 μm, Q = 1.35–2.1 (–2.4), Qav = 1.67 ± 0.18, amygdaliform or elliptical in profile and face views, angular in polar view (8–10 facets), undulate-pustulate in all views. Basidia 25–35×(5–)7–8 μm, 4-sterigmate, narrowly clavate, some constricted near apex. Cheilo- and pleurocystidia as pseudocystidia, similar in size, shape and colour, filled with dark reddish ochre pigment, ventricose, ventricose-rostrate, often septate at base of rostrum, rostrum often undulate, 36–70×5–8 μm, with scattered rhomboidal crystals lying over basidia, basidioles and cystidia in 3 % KOH mounts. Hymenophoral trama hyaline or pale brownish, parallel, cylindrical or some inflated, 2.4–8.8 μm in diameter, with repository hyphae of similar diameters scattered through trama, filled with ochre or golden brown contents, connected to the pseudocystidia. Pileipellis two-layered cutis, suprapellis a very thin hyaline layer overlying a subpellis of pale brown hyphae, hyphae in both layers cylindrical or slightly inflated, 2.4–7.2 μm in diameter, hyphae in pale brown layer and context finely and faintly encrusted. Pileitrama hardly differentiated from subpellis, radially arranged, cylindrical or inflated, 2–10 μm in diameter, brownish encrusted. Stipitipellis a hyaline cutis of cylindrical hyphae, 2–4 μm in diameter, with some scattered encrustations. Clamp-connections absent.

Habitat & distribution: very rare, only known from the type locality; terrestrial in a rainforest with *Nothofagus cunninghamii* and *Eucalyptus delegatensis* as dominants, an understory of *Anopterus glandulosus*, *Atherosperma moschatum*, *Phyllocladus aspleniifolius*, *Leptospermum lanigerum*, as well as *Gaultheria hispida, Coprosma quadrifida* and *Dicksonia antarctica.*

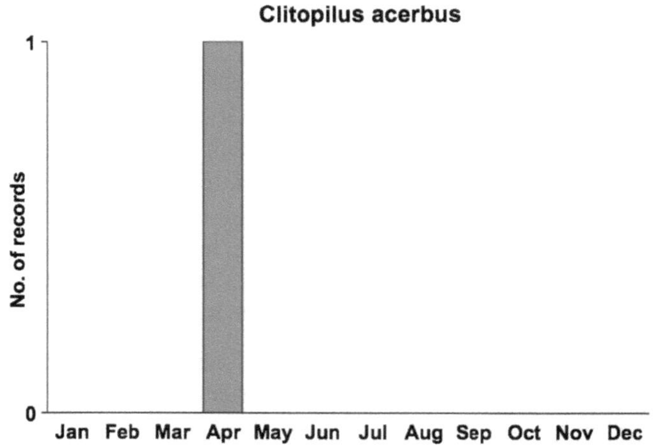

Collection examined. Australia, Tasmania, Lake St Clair National Park, Echo Point to Cynthia Bay, 42° 04′ S, 146° 09′ E, 22 April 2000, G. Gates E 875 (holotype, HO 536112).

Although *Clitopilus acerbus* is based on very limited material, it is fairly characteristic of members of section *Rhodocybe*, with its small, pale brownish fruit body and extremely bitter taste, and crystals in the hymenium and trama. *C. caelatus* is similar, but a darker coloured species, without or with a slightly bitter taste.

Section ***Rufobrunnei*** (T.J. Baroni) Contu, in Bollettino AMER 77–78: 29. 2009.
Rhodocybe sect. *Rufobrunnea* T.J. Baroni, Beih. Nova Hedwigia 67: 74. 1981.
Basidiocarps centrally stipitate; usually with brown colours; lamellae adnate to adnexed or decurrent; hymenial pseudocystidia absent.

Type species: *Clitopilus roseiavellaneus* (Murrill) Murrill.

6. **Clitopilus lateritius** (T.J. Baroni & G.M. Gates) Noordel. & Co-David, Persoonia 23: 162. 2009.
Rhodocybe lateritia T.J. Baroni & G.M. Gates, Austr. Syst. Bot. 19: 346. 2006.
MycoBank #509907.

Original diagnosis: Pileus rubro-senatus vel rubello-brunneus, primum hygrophanus deinde pallido-bubalinus vel bubalino-roseus, saepe maculis pallescentibus pruinoso-pellucidis aquosis, 21–85 mm latus, primo convexus cito plano-convexus, deinde margine elevatus, interdum late atque non profunde depressus, siccus, in statu madido sublubricus vel gelatinosus, in superficie implicatus vel pannosus, opacus. Lamellae juveniles pallide roseolae, deinde carneo-roseolae vel brunneolo-roseolae, decurrentes vel subdecurrentes, interdum primum adnatae, congestae, lamellulis in gradibus duobus dispositis. Stipes exalbescens vel subbubalinus, omnino pruina alba atque tomento albo basali, ad apicem 5–13 mm latus, ad basem 5–20 mm latus, 16–55 mm longus, aequus vel deorsum contractus vel ad basem protuberans subclavatusque, typice robustus sed interdum flexuosus, teres, siccus, glaber, intus albus farctus. Odor aromaticus, in statu sicco mucidus vel odore condimenti, interdum in statu naturali odore aromatico carens. Sapor non distinctus. Basidiosporae 5.6–11×4.5–7.2 μm, ellipticae vel subamygdaliformes, ad polum visae angulares (faciebus 8–11–13), leniter vel modice undulato-pustulatae, parietibus aeque cyanophilicis. Basidia 4-sterigmata. Cheilocystidia distanter sparsa, saepe paene procurrentia, nonnulla inclusa, maximum partem cylindrica, saepe ex base ramoso orientia, nonnulla interdum septata, flexuosa vel subcapitata vel angustate fusiformia, hyalinae, 36–66×3.2–6 μm. Pleurocystidia nulla. Pileipellis stratum pallide flavo-brunneum ex hyphis repentibus, laxe intertextis, cylindricis vel subinflatis, 4–10.5 μm diametro, crustulis carentibus, nonnullis hyphis atque cellulis ultimis pellicula tenui gelatinoidea, reagentibus quum in tinctura Congo Red dicta tinctis, cellulis ultimis (pileocystidiis) distinctis, sparsis, versiformibus praeditis, cellulis ultimis clavatis, fusiformibus, late lecythiformibus vel ventricoso-rostratis, nonnullis collo septato, 25–65×4–22.7 μm. Fibula nulla.

Holotype: Australia, Tasmania, Hobart, Waterworks Reserve, 42° 54′ S, 147° 17′ E, 27 June 2002, G. Gates E 1589 (HO 536113; isotype, CORT).

Etymology: lateritius (Lat.) = brick-coloured.

Main characters: robust species with brick red pileus; stipe pale (Plate 4.3).

Description:
Pileus 40–120 mm diam., convex to plano-convex when young with involute margin, becoming irregularly plano-convex to plano-concave with age with subdepressed

Plate 4.3 *Clitopilus lateritius* (Photo Machiel Noordeloos)

centre, and lobed, undulating marginal zone, hygrophanous, not translucently striate, more or less uniformly brick red, frequently with darker spots (water spots), which sometimes are arranged in neat concentric circles; pallescent on drying to ochraceous red, glabrous, dry to subviscid when moist, dull. Lamellae adnate-emarginate, segmentiform to subventricose, up to 7 mm broad, white then dirty pink to pinkish brown, with coarsely eroded, concolorous edge, with age often with darker rusty pink spots, crowded. Stipe relatively short, 20–40 × 12–25 mm (middle) × 25–35 (base), obclavate when young, then cylindrical in upper part, with very enlarged bulbous base, white when young, when old with dirty yellowish pink spots, subfelted. Context concolorous with surface in cortex of pileus, elsewhere white, firm. Odour pleasant, slightly fruity. Taste mild.

Spores 5.5–11 × 4.5–7.5 μm, Q = 1.2–1.7, Qav = 1.3–1.4, ellipsoid to amygdaliform in profile, ellipsoid in side-view, weakly to moderately undulate-pustulate, cyanophilic. Basidia 25–40 × 7–10 μm, 4-spored, clampless. Lamella edge heterogeneous. Cheilocystidia scattered, 30–70(–80) × 3–7 μm, cylindrical-flexuous or fusiform to lageniform, with rounded to subcapitate apex, very thin walled. Pleurocystidia absent. Hymenophoral trama regular, made up of cylindrical to slightly inflated hyphae 4–11 μm wide with some pale brown, diffuse, intracellular

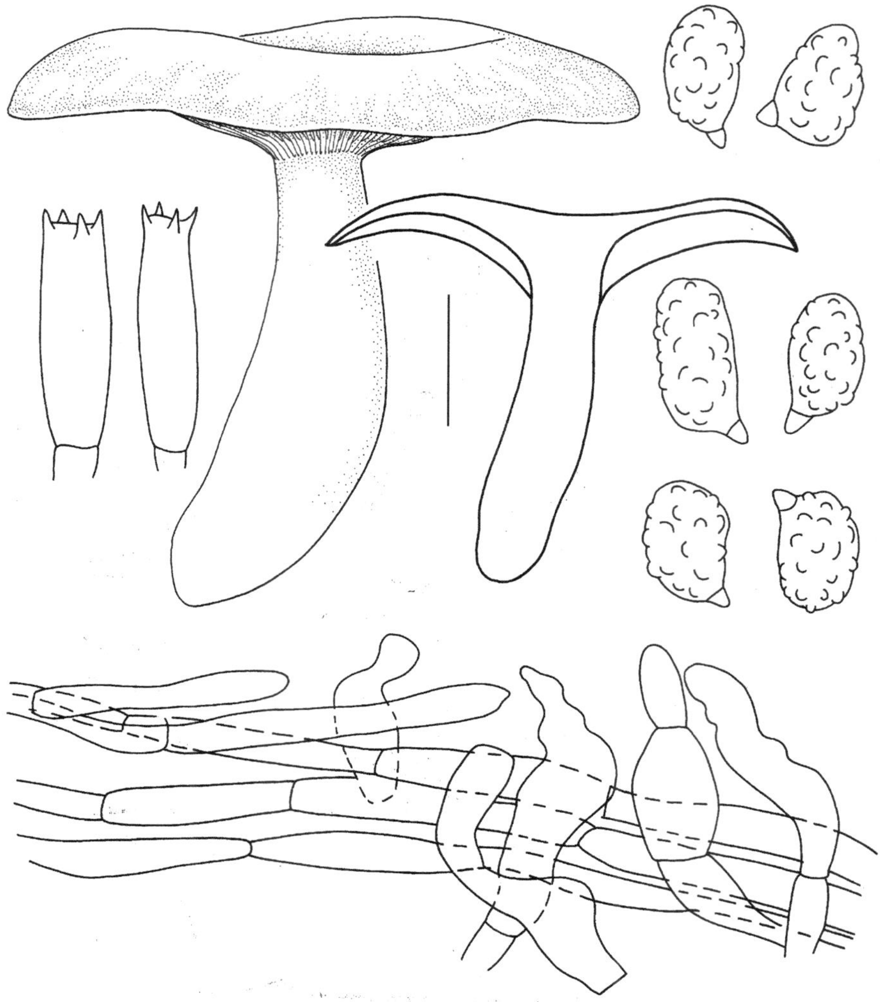

Fig. 4.6 *Clitopilus lateritius*. Habit, spores, basidia, and pileipellis. Bar = 10 mm or 10 μm

pigment. Pileipellis a cutis of cylindrical to slightly inflated hyphae 4–12 μm, sometimes with minutely encrusted walls, at intervals with ascending versiform, cystidioid terminal elements 20–70 × 6–20(–25) μm, with pale brown intracellular pigment. Pileitrama compact, more or less regular, of narrow, cylindrical 3–15 μm wide hyphae. Stipitipellis a cutis of narrow, cylindrical hyphae 3–10 μm wide, with occasional cystidioid terminal elements, as in pileipellis. Clamp-connections absent (Fig. 4.6).

Habitat & distribution: terrestrial, single or in small groups in wet and dry sclerophyll forests and wet sclerophyll gullies.

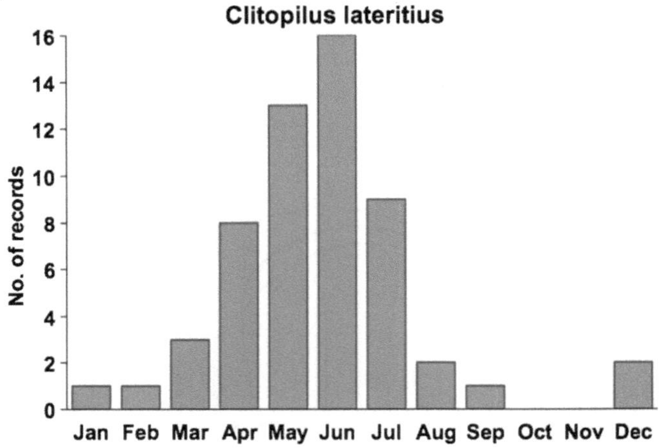

Collections examined. Australia, Tasmania, Collinsvale, Myrtle Forest, 42° 52′ S, 147° 09′ E, 11 May 2010, M. Noordeloos 2010003; Fortescue Bay, 43° 09′ S, 147° 57′ E, 1 June 2002, G. Gates E 1547; —, 22 May 2004, G. Gates E 2013; Gowrie Pk, O'Neills Creek, 41° 28′ S, 146° 13′ E, 2 May 2005, G. Gates E 2155; Hobart, Waterworks Reserve, 42° 54′ S, 147° 17′ E, 18 July 1999, G. Gates E 663; —, 27 June 2002, G. Gates E 1589 (holotype, HO 536113); —, 29 March 2003, G. Gates E 1661; Junee Cave State Reserve, 42° 44′ S, 146° 36′ E, 24 May 2003, G. Gates E 1824; —, 24 Feb. 2004, G. Gates E 1934; —, 9 May 2011, M.E. Noordeloos 2011056 (L); Lilydale Falls, 41° 14′ S, 147° 13′ E, 25 July 1999, G. Gates E 664; Maydena (Old Toilet Block), 42° 46′ S, 146° 36′ E, 11 July 2002, G. Gates E 1596; Mt Field National Park, Falls Loop, 42° 41′ S, 146° 42′ E, 9 Dec. 1999, G. Gates E 775; —, 29 June 2000, G. Gates E 978; —, 8 Aug. 2002, G. Gates E 1622; Mt Wellington, Fern Glade, 42° 55′ S, 147° 15′ E, 17 April 2005, G. Gates E 2114; —, Jacksons Track, 14 Aug. 2000, G. Gates E 996; —, 17 May 2001, G. Gates E 1176; —, Myrtle Gully, 42° 54′ S, 147° 15′ E, 9 March 1999, G. Gates E 196; —, 13 April 1999, G. Gates E 414; —, 3 May 2001, G. Gates E 1157; —, 17 Jan. 2002, G. Gates E 1408; Risdon Brook Reservoir, 42° 47′ S, 147° 20′ E, 1 July 2003, G. Gates E 1893; Sandy Bay, University of Tasmania, 42° 54′ S, 147° 19′ E, 16 June 1999, G. Gates E 613; —, 9 July 1999, G. Gates E 660; —, 3 June 2001, G. Gates E 1218; —, 9 July 2001, G. Gates E 1258; —, 4 June 2004, G. Gates E 2036; —, 3 June 2009, G. Gates E 2272; Taroona, Truganini Track, 42° 56′ S, 147° 21′ E, 23 July 2000, G. Gates E 990; —, 8 June 2005, G. Gates E 2180.

Clitopilus lateritius is similar to *Clitopilus geminus* (Fr.) Noordel. & Co-David with respect to the robust, brick-coloured basidiocarps. The spores, however, are larger, and, contrary to *C. geminus*, our species has distinctly differentiated terminal elements in the pileipellis ("pileocystidia"). The close relation is confirmed by molecular-phylogenetic studies where both taxa form a well supported monophyletic clade (Co-David et al. 2009).

7. **Clitopilus pseudopiperitus** (T.J. Baroni & G.M. Gates) Noordel. & Co-David, Persoonia 23: 163. 2009.

Rhodocybe pseudopiperita T.J. Baroni & G.M. Gates, Austr. Syst. Bot. 19: 345. 2006.

MycoBank #509940.

Original diagnosis: Pileus bubalino-roseolus colouribus ochraceis paululum tinctus, saepe supra centrum infuscatus brunnescensque, in areolis primum hygrophanus deinde pallide subroseo-bubalinus, (15–)24–47 mm latus, convexus, saepe leniter umbonatus, siccus vel interdum in statu madido lubricus, ut videtur glaber sed sub lente × 10 implicatus vel pannosus, laevis, interdum in superficie leniter rugulosus, opacus. Lamellae bubalino-roseolae, plerumque decurrentes sed nonnullae adnatae vel subdecurrentes, arctae vel coarctatae lamellulis in gradibus duobus dispositis. Stipes roseolo-bubalinus, intus roseolo-bubalinus vel arctus farctus, tomento albo basali atque rhizomorphis albis praeditus, 20–30 mm longus, ad apicem 2.5–4 mm latus, ad basem 2–6 mm latus, plusminusve aequus, teres, siccus, glaber vel sub lente × 10 leniter pruinosus, laevis vel subasper. Odor graminum nuper sectorum vel non distinctus. Sapor graminum viventium vel non distinctus. Basidiosporae 5.6–8 × 4–5.6 μm, ellipticae vel subamygdaliformes vel amygdaliformes, ad polum visae angulares (faciebus 7–10), omnino distincte undulato-pustulatae vel obscure vel in sporis maioribus vel maturis sublaeves, parietibus aeque cyanophilicis. Basidia 4-sterigmata. Cystidia hymenialia nulla. Pileipellis hyalina vel pallide straminea, repens, plerumque cylindrica, interdum subinflata, 3.2–7.2(−10.5) μm diametro, hyphis angustissimis crustulis leniter pallide flavo-brunneis (obscuris, interdum difficiliter inventis), nonnullis cellulis ultimis anguste clavatis vel anguste ventricosis hyalinis vel pigmentis plasmaticis sublutescentibus, 24–38 × 4–9.7 μm. Fibulae nullae.

Holotype: Australia, Tasmania, Mt Field, 42° 41′ S, 146° 42′ E, 9 June 2001, G. Gates E 1231 (HO 536114; isotype, CORT).

Etymology: pseudo (Lat.) = false, not genuine, referring to its likeness to *Rhodocybe piperita* (G. Stev.) E. Horak.

Main characters: slender, pinkish species; spores small, dimorphic; taste non-bitter (Plate 4.4).

Description:

Pileus 15–50 mm diam., convex then irregularly plano-convex to plano-concave with umbo or flattened to subdepressed centre with weak umbo, with involute margin, not hygrophanous, not translucently striate, pale pinkish brown, occasionally with ochre hue, often with some darker water-spots, lubricous when moist, dry, matt, glabrous or subfelted under a hand lens. Lamellae arcuate-decurrent, segmentiform, up to 4 mm broad, pink with concolorous, subserrulate edge, very crowded with two tiers of lamellulae. Stipe 20–30 × 2.5–6 mm, cylindrical, sometimes

Plate 4.4 *Clitopilus pseudopiperitus* (Photo Genevieve Gates)

compressed with groove or with slightly broadened base, straight or flexuous, yellowish pink to pinkish brown then concolorous with pileus, subglabrous, or with whitish fibrils, especially in lower part, base whitely tomentose. Context tough, pinkish white. Odour indistinct or like mown grass. Taste mild.

Spores dimorphic, type 1 5.5–6.5 × 4–5.5 μm, Q = 1.2–1.6, Qav = 1.4, ellipsoid to amygdaliform with weakly undulate-pustulate outline, appearing angular in polar view, cyanophilic, type 2 7.0–9.0 × 5.0–5.5 μm, more or less ellipsoid and appearing almost smooth. Basidia 22–40 × 5–9 μm, 4-spored, clavate, clampless. Lamella edge fertile. Cheilocystidia absent. Hymenophoral trama regular, made up of narrow, cylindrical 4–10 μm wide hyphae. Pileipellis a cutis of narrow, cylindrical to inflated hyphae 3–10 μm wide; some terminal elements inflated, clavate, 20–40 × 4–10 μm, with brown intracellular pigment. Pileitrama regular, compact, made up of narrow, cylindrical to slightly inflated hyphae 5–20 μm wide. Clamp-connections absent (Fig. 4.7).

Habitat & distribution: rather common; single or in small groups in forest litter of wet sclerophyll gullies or wet sclerophyll forests. Fruiting throughout the year.

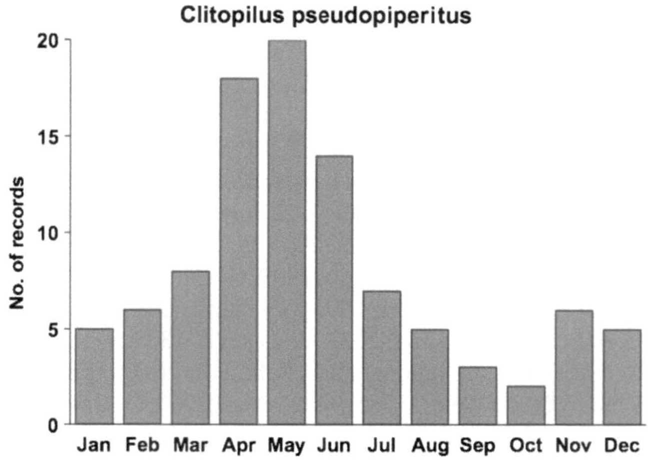

Collections examined. Australia, Tasmania, Arve River Streamside Reserve, 43° 10′ S, 146° 48.5′ E, 10 April 2001, G. Gates E 1092; Bermuda Road, 43° 04′ S, 146° 54′ E, 17 April 2003, G. Gates E 1720; Bruny Island, Mavista Falls Picnic Area, 43° 23′ S, 147° 19′ E, 8 April 2001, G. Gates E 1082; Cape Pillar Track, 43° 09′ S, 147° 56′ E, 29 May 1999, G. Gates E 551; Clarks Cliffs, 43° 06′ S, 147° 47′ E, 5 March

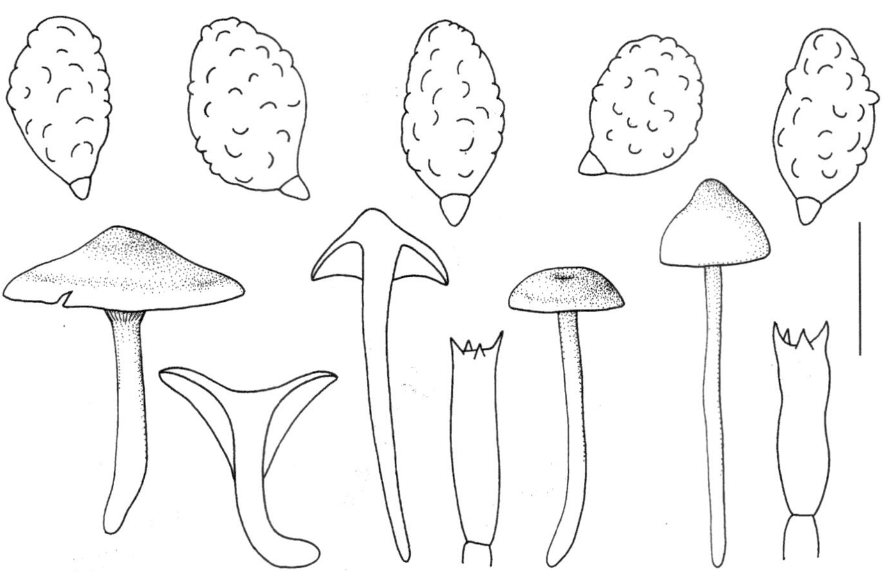

Fig. 4.7 *Clitopilus pseudopiperitus*. Habit, spores, and basidia. Bar = 10 mm or 10 μm

2002, G. Gates E 1447; Collinsvale, Myrtle Forest, 42° 52′ S, 147° 09′ E, 18 March 1999, G. Gates E 260; —, 15 April 1999, G. Gates E 431 & E 432; —, 11 May 2010, G. Gates (M.E. Noordeloos 2010032); Duckhole Lake Track, 43° 22′ S, 146° 53′ E, 8 May 2005, G. Gates E 2165; Ellendale, 42° 37′ S, 146° 43′ E, 5 April 1999, G. Gates E 361; Evercreech Forest Reserve, 41° 24′ S, 147° 58′ E, 16 June 2003, G. Gates E 1877; Growling Swallet, 42° 41′ S, 146° 30′ E, 5 Aug. 1999, G. Gates E 672; —, 4 June 2002, G. Gates E 1550; —, 22 July 2003, G. Gates E 1896; —, 21 Dec. 2004, G. Gates E 2071; —, 24 Sept. 2005, G. Gates E 2196; Kermandie Falls, Lower Track, 43° 12′ S, 146° 52′ E, 8 July 1999, G. Gates E 659; —, 23 May 2000, G. Gates E 937 & E 941; —, 5 April 2001, G. Gates E 1089; —, 29 May 2001, G. Gates E 1195 & E 1196; —, 5 Jan. 2002, G. Gates E 1388; —, 20 April 2005, G. Gates E 2120; Lake St Clair National Park, Echo Point to Cynthia Bay, 42° 04′ S, 146° 09′ E, 6 Jan. 2000, G. Gates E 785; —, 22 April 2000, G. Gates E 865; Lilydale Falls, 41° 14′ S, 147° 13′ E, 25 July 1999, G. Gates E 665 & E 668; Marriotts Falls, 42° 43′ S, 146° 39′ E, 10 June 1999, G. Gates E 585; —, 27 Nov. 1999, G. Gates E 765; Mt Field National Park, 42° 41′ S, 146° 42′ E, 9 May 1999, G. Gates E 477; —, 29 June 1999, G. Gates E 643 & E 644; —, 6 July 1999, G. Gates E 653; —, 9 June 2001, G. Gates E 1231 (holotype, HO 536114) & E 1232; —, Lady Barron Falls, 1 April 1999, G. Gates E 351; —, 9 Nov. 1999, G. Gates E 735; —, 9 Dec. 1999, G. Gates E 774, E 775 & E 776a; —, 19 July 2001, G. Gates E 1260; —, 15 Aug. 2002, G. Gates E 1628; —, Lyrebird Track, 42° 41′ S, 146° 40′ E, 16 March 2002, G. Gates E 1458; —, River Walk, 42° 41′ S, 146° 42′ E, 13 Nov. 1999, G. Gates E 751; —, 24 April 2001, G. Gates E 1121; —, 9 Feb. 2002, G. Gates E 1425 & E 1426; —, Russell Falls, 11 Sept. 2001, G. Gates E 1280; —, Tall Trees Track, 16 March 2002, G. Gates E 1460; —, 8 Aug. 2002, G. Gates E 1622; Mt Wellington, Betts Vale Track, 42° 55′ S, 147° 15′ E, 19 Feb. 2002, G. Gates E 1437; —, 2 Aug. 2003, G. Gates E 1900; —, Myrtle Gully, 42° 54′ S, 147° 15′ E, 13 April 1999, G. Gates E 413 & E 414; —, 13 May 1999, G. Gates E 501; —, 15 June 1999, G. Gates E 606; —, 18 May 2000, G. Gates E 922; —, 3 May 2001, G. Gates E 1159; —, 27 Nov. 2001, G. Gates E 1342; —, 17 Jan. 2002, G. Gates E 1408 & E 1409; —, 21 Feb. 2002, G. Gates E 1442 & E 1443; —, 26 April 2003, G. Gates E 1744; —, Silver Falls, 42° 55′ S, 147° 15′ E, 5 May 2001, G. Gates E 1238; Oldina Forest Reserve, Noel Jago Track, 41° 00′ S, 145° 40′ E, 1 May 2005, G. Gates E 2151; Reuben Falls, 43° 00′ S, 146° 40′ E, 15 May 1999, G. Gates E 506; Ulverstone, Paton Park, 41° 10′ S, 146° 06′ E, 27 March 1999, G. Gates E 306; Wielangta, 42° 42′ S, 147° 51′ E, 20 March 1999, G. Gates E 273; —, 22 June 1999, G. Gates E 620.

Clitopilus pseudopiperitus resembles the New Zealand *C. piperitus* (G. Stev.) Noordel. & Co-David, but differs by the lack of a bitter taste, smaller spores, and the presence of inflated terminal elements (pileocystidia) in the pileipellis.

C. pseudopiperitus shows an interesting dimorphic basidiospore morphology with most of the spores being distinctly undulate-pustulate and smaller (5.6–6.4 × 4–4.8 μm) while ca. 30–45 % of the basidiospores are almost smooth and distinctly larger (7.2–8.8 × 4.8–5.6 μm). *C. lateritius* also shows this odd dimorphism in spore form.

Section ***Decurrentes*** (Konrad & Maubl.) Contu, in Bollettino AMER 77–78: 29. 2009.

Rhodocybe sect. *Decurrentes* (Konrad & Maubl.) Singer in Lilloa 22: 609. 1951.

Basidiocarps centrally stipitate; lamellae decurrent; pileus grey to grey-brown; pseudocystidia absent.

Type species: *Clitopilus mundulus* (Lasch) P. Kumm.

8. **Clitopilus pallidogriseus** (T.J. Baroni & G.M. Gates) Noordel. & Co-David, Persoonia 23: 162. 2009.

Rhodocybe pallidogrisea T.J. Baroni & G.M. Gates, Austr. Syst. Bot. 19: 348. 2006.

MycoBank #509941.

Original diagnosis: Pileus uniformiter griseus vel griseo-brunneus nitore micaceo ornatus, 16–35(−75) mm latus, convexus vel et convexus et non profunde depressus, deinde plano-convexus vel planus, ut videtur glaber sed sub lente × 10 leniter asper vel pannosus, siccus, opacus. Lamellae cremeo- bubalinae vel cinerascenter bubalinae vel griseae, aetate propter sporas subroseolae, adnatae vel saepius subdecurrentes vel decurrentes, arctae vel coarctatae, lamellulis in 2–3 gradibus dispositis. Stipes griseolo-bubalinus vel aetate griseolo-brunneus, omnino pruina alba obtectus tomentoque albo basali, ad apicem 4–5(−10) mm latus, ad basem 1.5–5(−9) mm latus, 14–46 mm longus, aequus vel deorsum vel seorsum contractus, teres vel lateraliter compressus, glaber, siccus, in contextu griseolo-bubalinus vel intus albescens farctus. Odor plerumque valde farinaceus. Sapor plerumque valde farinaceus. Superficies pilei sicci in 3 % KOH non reagens. Basidiosporae 4.7–7.5 × 3.8–5.6 μm, globosae vel subglobosae vel late ellipticae, ad polum visae angulares vel obscure angulares (faciebus minutis 8–10), omnino obscure vel distincte undulato-pustulatae, parietibus aeque cyanophilicis. Basidia 4-sterigmata. Cystidia hymenialia nulla. Pileipellis 2-stratosa, suprapelle ex hyphis hyalinis laxe implicatis, 3–5 μm diametro, subpelle ex hyphis atro-brunneis, repentibus, cylindricis, modice vel valde vel interdum leniter crustulis brunneis ornatis, 2.4–6.4(−14) μm diametro. Fibulae nullae.

Holotype: Chauncy Vale, 42° 37′ S, 147° 16′ E, 20 Nov. 2001, G. Gates E 1321 (HO 536111; isotype, CORT) (Plate 4.5).

Description:

Pileus 16–35(−75) mm diam., convex, sometimes slightly depressed at centre, finally plano-convex, with inrolled then decurved or straight margin, not hygrophanous, not translucently striate, greyish brown or uniformly grey, glabrous but appearing felted under a hand lens, dry, opaque or with micaceous sheen. Lamellae adnate to subdecurrent or decurrent, segmentiform or arcuate, close to crowded, 2–6 mm broad, thin to moderately thick, creamy buff at first then becoming greyish buff or grey with pink tinge, lamellulae in 2–3 tiers. Stipe 15–50 × 1.5–6(−9) × 4–10 mm, sometimes broadened at base, cylindrical or laterally compressed, cartilaginous,

Plate 4.5 *Clitopilus pallidogriseus* (Photo Michael Pilkington)

pale yellowish brown or greyish brown, glabrous, dry. Context white. Odour usually strongly farinaceous. Taste usually strongly farinaceous.

Spores 4.7–7.5 × 4.0–5.5 μm, Q = 1.0–1.5, Qav = 1.20, globose or subglobose (or broadly elliptical in profile and face views), angular or obscurely angular by minute facets (8–10) in polar view, obscurely or clearly undulate-pustulate in all views, walls evenly cyanophilic, inamyloid. Basidia 22–40 × 6.5–9 μm, clavate, 4-spored, clampless. Lamella edge fertile, cystidia absent, but occasionally with long projecting contorted undulate hyphal ends scattered along lamella edges. Lamella trama hyaline or pale brown, parallel, cylindrical or slightly inflated hyphae 2.4–14(–24) μm in diameter. Pileipellis bilayered, suprapellis a hyaline layer up to 40 μm thick of loosely interwoven cylindrical hyphae 3–5 μm in diameter; subpellis a 140–160 μm layer of repent, cylindrical, hyphae 2.4–6.4(–14) μm in diameter, with moderate to heavy brown pigment encrustations or occasionally encrustations fine and obscure. Pileitrama interwoven, made up of cylindrical or inflated hyphae 3.0–14 μm in diameter. Stipitipellis a cutis of 2.5–5 μm wide hyphae, with scattered clusters of distinct, cylindrical or cylindrical-contorted, entangled terminal elements, up to 11 μm wide. Clamp-connections absent (Fig. 4.8).

Habitat & distribution: in wet sclerophyll gullies, and in wet sclerophyll and dry sclerophyll forests.

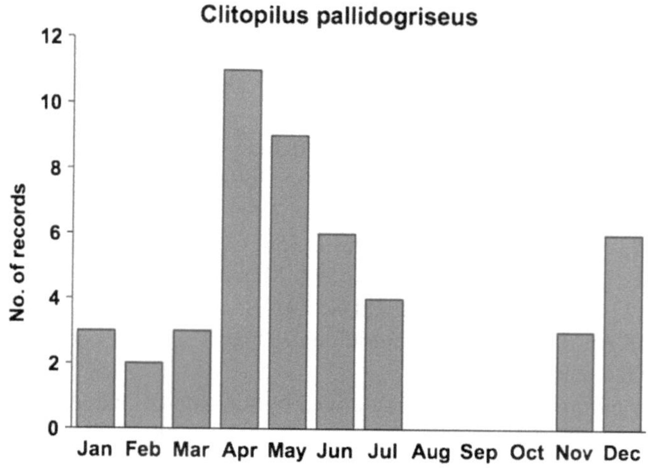

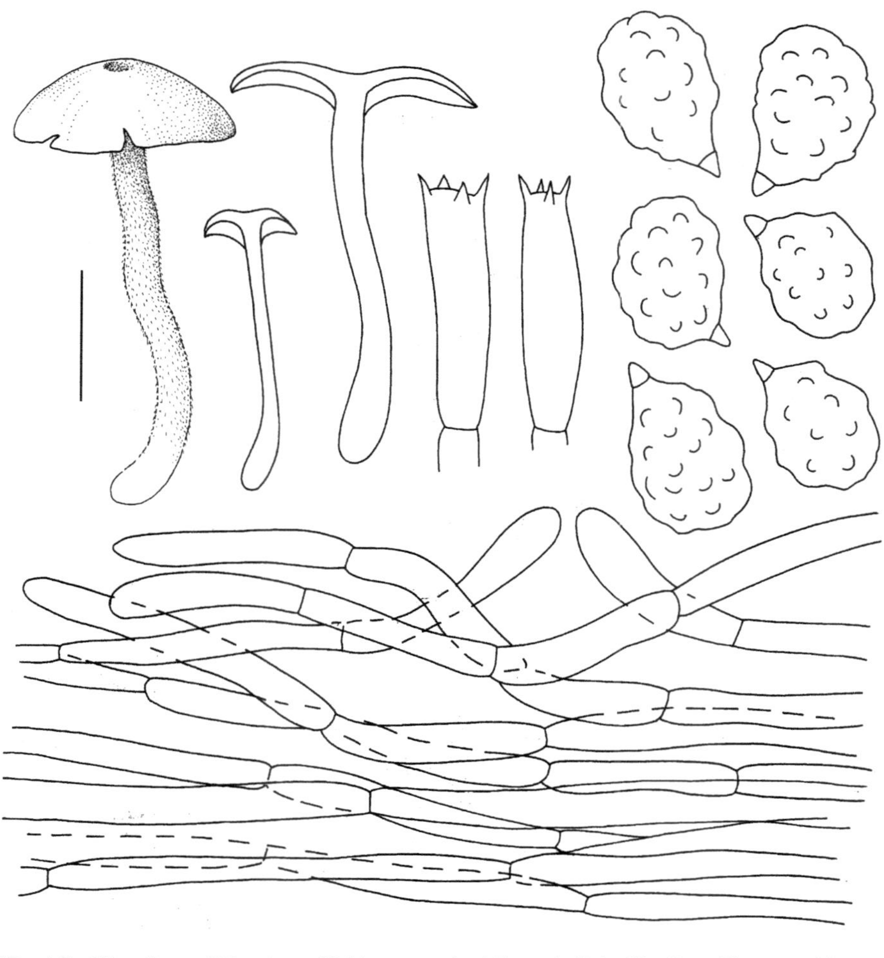

Fig. 4.8 *Clitopilus pallidogriseus*. Habit, spores, basidia, and pileipellis. Bar = 10 mm or 10 μm

Collections examined. Australia, Tasmania, Adamsons Peak Track, 43° 20′ S, 146° 54′ E, 1 May 1999, G. Gates E 463; Bluff River State Forest, 42° 31′ S, 147° 40′ E, 14 June 2003, G. Gates E 1867; Chauncy Vale, 42° 37′ S, 147° 16′ E, 20 Nov. 2001, G. Gates E 1321 (holotype, HO 536111); —, 22 Nov. 2001, G. Gates E 1340; —, 29 Nov. 2001, G. Gates E 1351; —, 10 June 2003, G. Gates E 1861; Clarks Cliffs, Koonya, 43° 06′ S, 147° 47′ E, 29 Dec. 2001, G. Gates E 1381; Collinsvale, Myrtle Forest, 42° 52′ S, 147° 09′ E, 28 May 1999, G. Gates E 538; Ellendale, 42° 37′ S, 146° 43′ E, 5 April 1999, G. Gates E 358; Geeveston, Donnellys Rd, 43° 07′ S, 146° 54′ E, 12 June 1999, G. Gates E 594; Marriotts Falls, 42° 43′ S, 146° 39′ E, 2 June 2001, G. Gates E 1213; Mt Field, 42° 41′ S, 146° 42′ E, 6 July 1999, G. Gates E 652; —, 28 April 2011, M.E. Noordeloos 2011018 (L); —, Lady Barron Falls, 4 May 2000, G. Gates E 895.

Clitopilus pallidogriseus is characterized by the pale colours of the basidiocarp, isodiametric to subisodiametric spores, and bilayered pileipellis. It was placed in section *Decurrentes* (Baroni and Gates 2006). However, molecular-phylogenetic studies place it outside the /decurrentes clade, and closer to section *Rufobrunnei* (Co-David et al. 2009).

9. **Clitopilus reticulatus** (Cleland) Noordel. & Co-David, Persoonia 23: 163. 2009.
Rhodocybe reticulata (Cleland) E. Horak, Sydowia 31:58. 1979 (1978).
MycoBank #509944.

Original diagnosis: Pileus 2.5–5 cm., convexus, subumbilicatus, lineis concentricatus et interdum rimosis reticulatus vel adversum marginem portionibus irregularibus elevatis et fissuris pallidioribus, cinereus. Lamellae dentibus subdecurrentibus subsinuatae, confertae, angustae, cinereo-carneo-luteae. Stipes 1.8 cm, subtenuis, infra striatus, infarctus vel interdum cavus, particularis pallido-fumoso-cinereis, subter, subniger. Caro tenuis, albida. Sporae subsphaerico-pyriformes, subangulatae, subcoloratae, 4.5–5.5 μm. Ad terram.

Holotype: South Australia, Mt Lofty, 25 April 1924, Herb. AD 3950, J. B. Cleland No. 13938 (WAITE) (Plate 4.6).

Description:
Pileus 30–47 mm diam., planoconvex or applanate, with undulate surface and incurved margin, hygrophanous, grey-brown (5D3–4) but with a darker watery brown reticulate pattern of fine lines concentrically distributed over the surface, becoming greyish on drying, dry, opaque, soft and suede-like, felted under a hand lens. Lamellae crowded, with 2 tiers of lamellulae, decurrent or arcuate-decurrent, moderately thick, up to 4 mm broad, pale yellowish grey with pinkish hues. Stipe 35 × 8 (apex) × 6 mm (base), centrally attached, tapering towards base, slightly compressed, greyish brown, entirely whitely pruinose, with white basal mycelium, and white rhizomorphs, dry, cartilaginous, stuffed yellowish grey inside. Context 6 mm thick. Odour and taste cucumber or farinaceous-cucumber.

Spores 4.5–7.5 × 4–6.5 μm, Q = 1.0–1.3, Qav = 1.15, broadly elliptical or subglobose-ovate and conspicuously apiculate (to 1.6 μm long) in profile and face

Plate 4.6 *Clitopilus reticulatus* (Photos Genevieve Gates)

views, angular in polar view (8–10 facets), undulate-pustulate, walls evenly cyanophilic, inamyloid. Basidia 22–31 × 7–10 µm, clavate, 4-spored, mostly filled with small lipoidal bodies, also occasional sclerobasidia present, walls 1.6–2.4 µm thick. Hymenial cystidia absent. Hymenophoral trama hyaline, made up of parallel, cylindrical hyphae 4–13 µm wide. Pileipellis a cutis of 4–10 µm wide, dark brown hyphae, heavily dark brown encrusted, and with dark brown intracellular pigment. Pileitrama hyaline, made up of loosely entangled, cylindrical or slightly inflated hyphae 4–14 µm in diameter, occasionally with a distinctively inflated, clavate cell embedded in the context, to 22 µm in diameter. Clamp-connections absent (Fig. 4.9).

Habitat & distribution: solitary on soil and humus in wet sclerophyll forest. Rare.

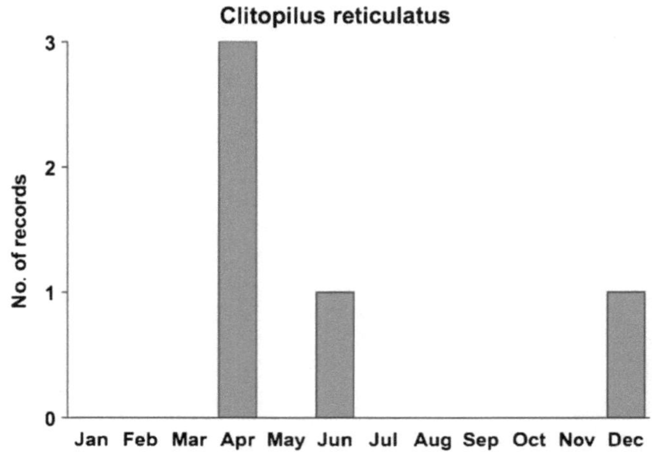

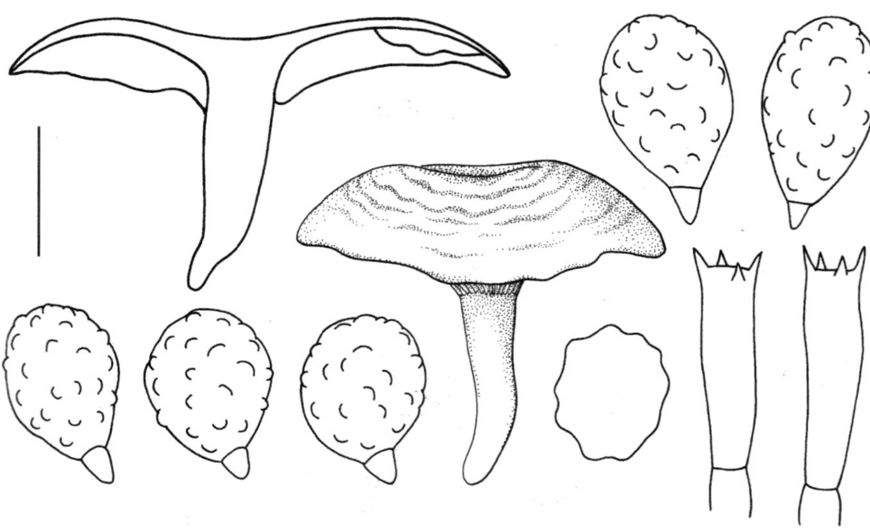

Fig. 4.9 *Clitopilus reticulatus*. Habit, spores, and basidia. Bar = 10 mm or 10 μm

Collections examined. Australia, Tasmania, North West Bay River, 42° 57′ S, 147° 12′ E, 14 June 2005, G. Gates E 2183; —, 29 Dec. 2005, G. Gates s.n.; —, 2 April 2011, G. Gates s.n.; —, 20 April 2011 G. Gates & M. Pilkington s.n. — South Australia, Mt Lofty, 25 April 1924, Herb. AD 3950, J. B. Cleland No. 13938 (WAITE).

Clitopilus reticulatus belongs to section *Decurrentes* because of the greyish colours, decurrent lamellae, absence of clamps, and lack of hymenial cystidia. It is a very distinctive species due to the dark watery brown, concentrically distributed lines and cracks on the pileus, and the subglobose basidiospores. *Clitopilus mundulus* (Lasch) P. Kumm. from the Northern Hemisphere is similar, but differs by turning

Plate 4.7 *Clitopilus conchatus* (Photos Genevieve Gates)

black when bruised and staining red when a drop of 3 % KOH is applied on the fresh and dried fruit bodies.

Section ***Crepidotoides*** (Singer) Contu, in Bollettino AMER 77–78: 29. 2009.
Rhodocybe sect. *Crepidotoides* Singer, Beih. Sydowia 7: 96. 1973.
Basidiocarps crepidotoid; hymenial pseudocystidia present.

Type species: *Clitopilus crepidotoides* (Singer) Noordel. & Co-David.

10. **Clitopilus conchatus** (E. Horak) Noordel. & Co-David, in Co-David, Langeveld & Noordeloos, Persoonia 23: 16. 2009.
Rhodocybe conchata E. Horak in New Zealand J. Bot. 17: 275. 1979.

Original diagnosis: Pileo –15 mm dimidiato vel conchiformi, convexo vel subdepresso, albidulo dein griseo vel pallide griseo-brunneo, tomentoso. Lamellis adnato-adnexis. griseo-argillaceis roseo tinctis. Stipite –8 × –1 mm, cylindrico, laterali, curvato, pileo concolori. Odore ingrato. Sporis 5–7 × 4–5 µm ovatis, rugulosis. Pseudocystidiis –70 × –7 µm, lanceolato-fusoideis. Hyphis fibulatis. Ad truncos putridos, Novazelandia.

Holotype: New Zealand, Coromandel, Kauaeranga Valley, 8 July 1968, E. Horak (PDD 27143) (Plate 4.7).

Description:
Habit crepidotoid. Pileus 5–6 mm diam., applanate with straight, entire margin, white, tomentose with a soft jelly-like greyish context revealed upon the collapse of the aeriferous tomentum with handling. Lamellae adnexed, up to 0.5 mm broad, crowded, brownish grey, soft, jelly-like, sometimes bifurcate, whitely grandinoid on faces and margins, lamellulae in a single tier. Stipe rudimentary attached to substrate with white, cottony rhizomorphs. Taste and odour not recorded.

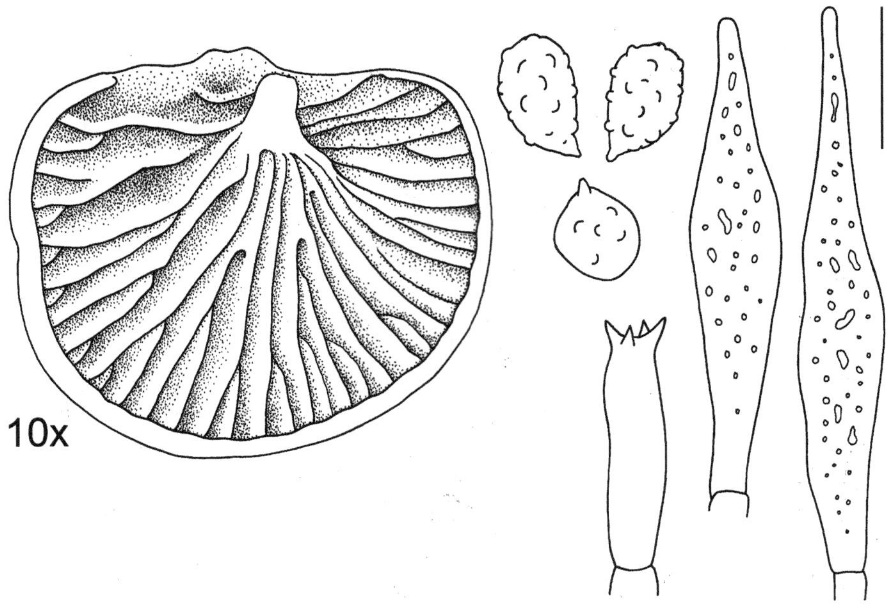

Fig. 4.10 *Clitopilus conchatus*. Habit, spores, basidia, and pseudocystidia. Bar = 10 mm or 10 µm

Spores 5–7 × 4–5 µm, Q = 1.2–1.5, undulate-pustulate, thin-walled. Basidia 20–30 × 5–7 µm, 4-spored. Pseudocystidia 40–75 × 4–8 µm, fusoid with yellow-brown refractive contents. Pileipellis an entangled cutis of narrow hyphae without visible pigment. Clamp-connections absent (Fig. 4.10).

Habitat & distribution: on the inside of a very rotten log in a wet sclerophyll forest beside a river. Found only once in Tasmania. Known from New South Wales and New Zealand.

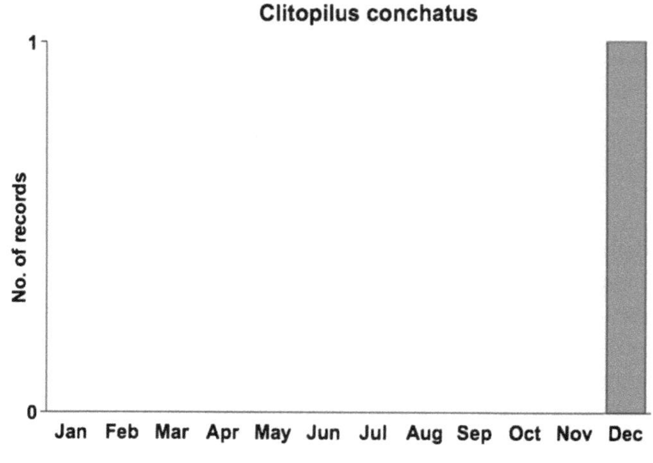

Collection examined. Tasmania, North West Bay River, 42° 57′ S, 147° 12′ E, 29 Dec. 2011, G. Gates E 2308.

Clitopilus conchatus is the only pleurotoid species with rhodocyboid spores found in Tasmania. Originally described from New Zealand, it has also been recorded from New South Wales (Horak 2008).

References

Baroni, T. J. (1981). *A revision of the genus Rhodocybe Maire (Agaricales)* (Beihefte zur Nova Hedwigia, Vol. 67). Vaduz: Cramer.

Baroni, T. J., & Gates, G. M. (2006). New species and records of *Rhodocybe* (Entolomataceae, Agaricales) from Tasmania. *Australian Systematic Botany, 19*, 343–358.

Baroni, T. J., Desjardin, D. E., & Hywel-Jones, N. (2001). *Clitopilus chalybescens,* a new species from Thailand. *Fungal Diversity, 6*, 13–17.

Co-David, D., Langeveld, D., & Noordeloos, M. E. (2009). Molecular phylogeny and spore evolution of *Entolomataceae*. *Persoonia, 23*, 147–176.

de Haan, A. (1998). *Clitopilus amarus* nov. spec., een bittere Molenaar. *Sterbeeckia, 18*, 32–35.

Hartley, A. J., de Mattos-Shipley, K., Collins, C. M., Kilaru, S., Foster, G. D., & Bailey, A. M. (2009). Investigating pleuromutilin-producing *Clitopilus* species and related basidiomycetes. *FEMS Microbiology Letters, 297*(1), 24–30.

Hausknecht, A., & Noordeloos, M. E. (1999). Neue oder seltene Arten der *Entolomataceae* (*Agaricales*) aus Mittel- und Südeuropa. *Österreichische Zeitschrift für Pilzkunde, 8*, 199–221.

Horak, E. (2008). Agaricales of New Zealand 1: *Pluteaceae – Entolomataceae*. In *The fungi of New Zealand* (Fungal Diversity Research Series, no. 19, Vol. 5). Hong Kong: Fungal Diversity Press.

Kühner, R. (1980). Les grandes lignes de la classification des Pluteales et leur bases. *Bulletin Mensuel de la Sociéte Linnéenne de Lyon, 49*, 357–430.

Matheny, P. B., Curtis, J. M., Hofstetter, V., Aime, M. C., Moncalvo, J.-M., Ge, Z.-W., Yang, Z.-L., Slot, J. C., Ammirati, J. F., Baroni, T. J., Bougher, N. L., Hughes, K. W., Lodge, D. J., Kerrigan, R. W., Seidl, M. T., Aanen, D. K., Denitis, M., Daniele, G., Desjardin, D. E., Kropp, B. R., Norvell, L. L., Parker, A., Vellinga, E. C., Vilgalys, R., & Hibbett, D. S. (2006). Major clades of Agaricales: A multilocus phylogenetic overview. *Mycologia, 98*, 982–995.

Moncalvo, J.-M., Vilgalys, R., Redhead, S. A., Johnson, J. E., James, T. Y., Aime, M. C., Hofstetter, V., Verduin, S. J. W., Larsson, E., Baroni, T. J., Thorn, R. G., Jacobsson, S., Clémençon, H., & Miller, O. K., Jr. (2002). One hundred and seventeen clades of euagarics. *Molecular Phylogenetics and Evolution, 23*, 357–400.

Noordeloos, M. E. (1988). Entolomataceae. In C. Bas, T. W. Kuyper, M. E. Noordeloos, & E. C. Vellinga (Eds.), *Flora agaricina neerlandica* (Vol. 1, pp. 77–182). Rotterdam: Balkema.

Noordeloos, M. E. (1993). Studies in *Clitopilus* (*Basidiomycetes, Agaricales*) in Europe. *Persoonia, 15*, 241–248.

Noordeloos, M. E. (2008). Clitopilus. In H. Knudsen & J. Vesterholt (Eds.), *Funga Nordica: Agaricoid, boletoid and cyphelloid genera* (p. 965). Copenhagen: Nordsvamp.

Vizzini, A., Musumeci, E., Ercole, E., & Contu, M. (2011). *Clitopilus chrischonensis* sp. nov. (Agaricales, Entolomataceae), a striking new fungal species from Switzerland. *Nova Hedwigia, 92*, 425–434.

Yang, Z. L. (2007). *Clitopilus amygdaliformis*, a new species from Tropical China. *Mycotaxon, 100*, 241–246.

Index

M.E. Noordeloos and G.M. Gates, *The Entolomataceae of Tasmania*, Fungal Diversity Research Series 22, DOI 10.1007/978-94-007-4679-4,

MIX
Papier aus verantwortungsvollen Quellen
Paper from responsible sources
FSC® C105338

If you have any concerns about our products,
you can contact us on
ProductSafety@springernature.com

In case Publisher is established outside the EU,
the EU authorized representative is:
Springer Nature Customer Service Center GmbH
Europaplatz 3, 69115 Heidelberg, Germany

Printed by Libri Plureos GmbH
in Hamburg, Germany